2nd International Conference on Biomass (ICB 2017)

Toward Sustainable Biomass Utilization for Industrial and Energy Application

IOP Conference Series: Earth and Environmental Science
Volume 141

Bogor, Indonesia
24 – 25 July 2017

ISBN: 978-1-5108-6108-4
ISSN: 1755-1307

Printed from e-media with permission by:

Curran Associates, Inc.
57 Morehouse Lane
Red Hook, NY 12571

Some format issues inherent in the e-media version may also appear in this print version.

Printed by Curran Associates, Inc. (2018)

Institute of Physics
Dirac House, Temple Back
Bristol BS1 6BE UK

Phone: 44 1 17 929 7481
Fax: 44 1 17 920 0979

techtracking@iop.org

Additional copies of this publication are available from:

Curran Associates, Inc.
57 Morehouse Lane
Red Hook, NY 12571 USA
Phone: 845-758-0400
Fax: 845-758-2633
Email: curran@proceedings.com
Web: www.proceedings.com

Table of contents

Volume 141

2nd International Conference on Biomass: Toward Sustainable Biomass Utilization for Industrial and Energy Applications

24–25 July 2017, Bogor, Indonesia

Accepted papers received: 19 March 2018
Published online: 6 April 2018

Preface

Papers

IOP Conf. Series: Earth and Environmental Science **141** (2018) 011001 doi:10.1088/1755-1315/141/1/011001

Preface

The 2nd International Conference on Biomass (ICB 2) 2017 was successfully conducted by Surfactant and Bioenergy Research Center (SBRC) LPPM Bogor Agricultural University (IPB) in cooperation with Indonesian Bioenergy Expert Association (IKABI), Indonesian Biogas Association (AbgI), Bandung Institute of Technology, University of Lampung, University of Indonesia, Sriwidjaya University, Universiti Sains Malaysia, University of Villanova, University of Hiroshima and was supported by The Ministry of Research, Technology, and Higher Education Republic of Indonesia and Indonesia Estate-Crop Fund for Palm Oil on 24-25 July 2017 in Bogor, Indonesia. This conference with the theme of **Toward Sustainable Biomass Utilization for Industrial and Energy Application** was the second scientific conference on biomass in Indonesia. This biomass-themed conference will be conducted annually to raise current global issues in biomass.

The committee accepted 39 of over 60 papers which were presented in the conference and the selected papers are published in this ICB 2 proceeding. These papers were classified into the following groups:

1. Feedstock and Bioenergy,
2. Biosurfactant and Biomaterial, and
3. Environment and Sustainability, Policy, Management, and Business.

Many appreciations were given to Rector and Vice Rector of Bogor Agricultural University for their support to the conference, to the reviewers, committee members, and supporters who had worked dedicatedly to make the conference and the publication of this proceeding successful. Supports from Bogor Agricultural University, Indonesia Estate Crop Fund-Palm Oil (BPDPKS), and The Ministry of Research, Technology, and Higher Education Republic of Indonesia were acknowledged. We tried our best to accomplish manuscript editing by following the writing format used in IOP Conference Series: Earth and Environmental Science. However, there might be some shortcoming editing found in this proceeding. Therefore, inputs from readers and writers are greatly appreciated so that editing can be performed better in the future.

We hope this ICB 2 2017 proceeding brings knowledge and benefits to academics, scientists, industrial stakeholders, especially in field of biomass. Thank you for your attention and cooperation.

Bogor, February 2018
The 2nd International Conference on Biomass 2017
Organizing Committee

Reviewers of International Conference on Biomass 2017

1. Prof. Erliza Hambali
2. Prof. Yandra Arkeman
3. Dr. Dwi Setyaningsih
4. Dr. Mujizat Kawaroe
5. Dr. Yohanes Aris Purwanto
6. Dr. Titi Candra Sunarti
7. Dr. Uju
8. Dr. Obie Farobie

ICB
2017

IOP Conf. Series: Earth and Environmental Science **141** (2018) 011002 doi:10.1088/1755-1315/141/1/011002

List of committees ICB2 2017

Steering Committee

Vice Rector for Research and Collaboration IPB	: Prof. Dr. Ir. Anas M Fauzi, M. Eng
Head of SBRC-IPB	: Prof. Dr. Erliza Hambali
Biomass Resources and Conversion Technology - Villanova University, USA	: Dr. Justinus Satrio
Chairman of the Indonesian Association of Bioenergy Experts	: Dr. Tatang Hernas
Chairman of Indonesian Biogas Association	: Prof. Dr. Udin Hasanudin
Ministry of Research, Technology, and Higher Education Republic of Indonesia	: Dr. Dadan Kusdiana
Chairman of MAKSI	: Dr. Darmono Taniwiryono

Organizing Committee

Chairman	: Dr. Y. Aris Purwanto
Secretary	: Dr. Mira Rivai
Treasurer	: Athin Nuryanti, STP
Programs	: Dr. Endang Warsiki; Dr. Mujizat Kawaroe; Agatha Maria Gadi, ST
Papers	: Dr. Ika Amalia Kartika; Dr. DwiSetyaningsih; Dr. Siti Nikmatin; Prof. Dr. Ani Suryani; Dr. Uju; Dr. Hariyadi; Dr. Herdhata Agusta; Dr. Irzaman; Prof. Dr. Dadang;
Publication and Documentation	: Dr. Obie Farobie; Dr. YandraArkeman; Hendri Wijaya, M.Si; Dr. Aiyen Tjoa; Prof. Dr. Basuki Wirjosentono; Prof. Dr. Misri Gozan; Prof. Dr. Ahmad Zuhairi Abdullah;
Logistics	: Ratno, Dudung Sulaeman, Heri, Ferry Eka Pawitra; Nurdin
Secretariat	: Fifin N.Nisya, MSi; Ristia Fitria, STP; Ari Imam Sutanto, MSi; Ganjar Saefurahman, M.Phil

Peer review statement

All papers published in this volume of *IOP Conference Series: Earth and Environmental Science* have been peer reviewed through processes administered by the proceedings Editors. Reviews were conducted by expert referees to the professional and scientific standards expected of a proceedings journal published by IOP Publishing.

IOP Conf. Series: Earth and Environmental Science **141** (2018) 012001 doi:10.1088/1755-1315/141/1/012001

Glycerol as an additional carbon source for bacterial cellulose synthesis

Y E Agustin, K S Padmawijaya, H F Rixwari and V A S Yuniharto

Chemical Engineering Department, University of Surabaya, Surabaya, Indonesia
E-mail : yuana@staff.ubaya.ac.id

Abstract. Bacterial cellulose, the fermentation result of *Acetobacter xylinus* can be produced when glycerol was used as an additional carbon source. In this research, bacterial cellulose produced in two different fermentation medium, Hestrin and Scharmm (HS) medium and HS medium with additional $MgSO_4$. Concentration of glycerol that used in this research were 0%; 5%; 10%; and 15% (v/v). The optimum conditions of bacterial cellulose production on each experiment variations determined by characterization of the mechanical properties, including thickness, tensile strength and elongation. Fourier Transform Infra Red Spectroscopy (FTIR) revealed the characterization of bacterial cellulose. Results showed that the growth rate of bacterial cellulose in HS-$MgSO_4$-glycerol medium was faster than in HS-glycerol medium. Increasing concentrations of glycerol will lower the value of tensile strength and elongation. Elongation test showed that the elongation bacterial cellulose (BC) with the addition of 4.95% (v/v) glycerol in the HS-$MgSO_4$ medium is the highest elongation value. The optimum bacterial cellulose production was achieved when 4.95% (v/v) of glycerol added into HS-MgSO4 medium with stress at break of 116.885 MPa and 4.214% elongation.

1. Introduction

Bacterial cellulose (BC) is a strong and ultrapure form of cellulose produced naturally by several species of *Acetobacteraceae*. Its high strength, purity, and biocompatibility make it of great interest to material science. It is currently used to create materials for tissue engineering, medicine, electronics, paper manufacture, food industry and fabrics. The enhanced mechanical properties of BC occur due to the uniform, continuous and nano-scalar network of cellulosic fibers. These properties are affected by various factors, such as culture conditions, the microorganism and the fermentation media employed. [1][2][3]

In microbial fermentations, the cost of substrate normally accounts up to 50-65% of the total cost of production. Despite its enormous potential in various applications, the high cost of BC production is the main drawback that hinders industrial implementation [4]. The utilization of industrial wastes and by-product streams as fermentation media could improve the cost-competitiveness of BC production. In recent years, many studies have focused on developing cost-effective fermentation media for BC production, such as pineapple peel juice, sugarcane juice, coffee cherry husk extract (a by product from the coffee processing) and corn steep liquor (a by product from the starch processing industry) as less expensive sources of carbon and nitrogen [5][6][7].

In the current work, production of bacterial cellulose by *Acetobacter xylinus* used glycerol (to represent by-product from biodiesel production) as carbon sources. Aim of this research is formulating a general, simple, and inexpensive medium to produce BC. Furthermore, the physical and mechanical properties of the bacterial celluloses produced from different fermentation media were determined.

IOP Conf. Series: Earth and Environmental Science **141** (2018) 012001 doi:10.1088/1755-1315/141/1/012001

2. Materials and Methods

Glycerol p.a was provided by PT. Brataco Chemica with purity 87% (in weight). While strain of *Acetobacter xylinus* was provided by Agrotekno Sarana Industri. Two different fermentation media were used in this research, Hestrin and Scharmm (HS) media [2%w/v glucose, 0.5%w/v peptone, 0.5%w/v yeast extract, 0.5%w/v disodium phosphate, 0.115w/v citric acid] and HS-$MgSO_4$ [0.115%w/v magnesium sulphate]. Culture media used for BC production were HS modified by repleacing D-glucose with other carbon source, such as glycerol. Concentration of glycerol that used in this research was 0%; 5%; 10%; 15% and 20% (v/v).

2.1. Cultivation media and conditions

Inoculum was cultured for 24 hours in Mc. Cartney bottle containing fermentation media [HS-glycerol or HS-Glycerol-$MgSO_4$] at 26°C and 150 rpm orbital speed. For BC production, static incubations were performed in Erlenmeyer's flasks for 14 days. To purified BC production, pellicles were rinsed with water to remove the culture medium, and then boiled in 1N NaOH solution at 90°C for 20 minutes in order to eliminate the bacteria cells from the cellulose matrix. Then, pellicles were washed with distilled water till neutralization. Dry weight was measured after drying the films at 50°C till constant weight.

2.2. Analytical method

Bacterial cellulose films were characterized for Fourier transform infrared spectroscopy (FTIR), thickeness and mechanical properties as described below.

2.2.1 Fourier Transform Infrared Spectroscopy

The chemical structure of bacterial cellulose films were analyzed by FTIR (Bruker). The bacterial cellulose produced by *Acetobacter xylinus* from our research was mixed well with potassium bromide (KBr) powder and press into small tablet.

2.2.2 Thickness measurement

Thickness of each bacterial cellulose film was measured at eight different positions by a thickness gauge, and the values were averaged.

2.2.3 Tensile Strength and Elongation at break

Mechanical properties of the films were investigated using autograph in University of Airlangga, Surabaya. The dimensions of the test specimen were 7 cm x 2 cm. each test was performed in duplicate.

3. Results and Discussions

3.1. Bacterial cellulose production

Acetobacter xylinus formed a cellulose layer on the fermentation media containing HS medium with glycerol as an additional carbon source at the end of 14 days incubation at 26°C. NaOH and water were used to purified the pellicles of bacterial cellulose. After purification bacterial cellulose was dried at 50°C. Two different media were used for bacterial cellulose production in this research, such as HS with various concentration of glycerol (Fig 1) and HS-$MgSO_4$ with various concentration of glycerol (Fig 2).

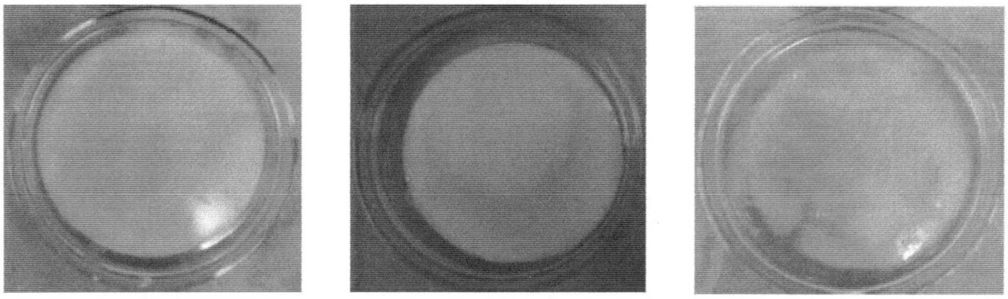

(a) (b) (c)

Fig 1. Bacterial cellulose production from HS media with addition of (a) 5%v/v(b) 10%v/v(c) 15% v/v glycerol

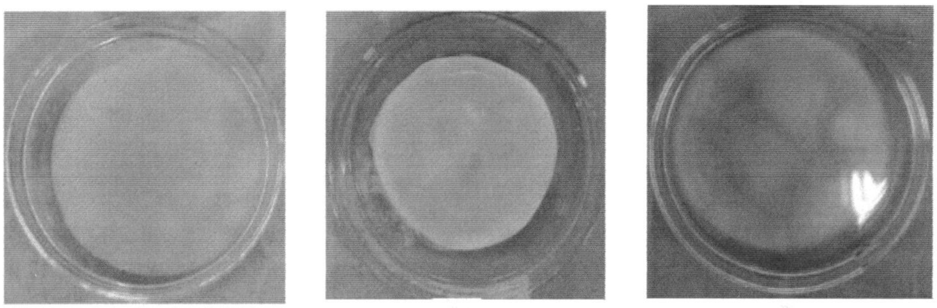

(a) (b) (c)

Fig 2. Bacterial cellulose production from HS-$MgSO_4$ media with addition of (a) 5%v/v (b) 10% v/v (c) 15% v/v glycerol

3.2. Characterization of bacterial cellulose (BC)

FTIR spectra obtained from bacterial cellulose films are shown in Fig 3. There were 3 main groups in bacterial cellulose, such as O-H group, pyranose cyclic group and C-O (β-1,4 glycosidic bond) group. The result showed that bacterial cellulose from this research have O-H bond at 3450-3400 cm^{-1}. The peaks at 1640-1504 cm^{-1} were attributed to pyranose cyclic. Another intense peak located at 1000 cm^{-1} was attribute to C-O (β-glycosidic bond) group. Moreover, some additional peaks appeared in the spectrum. The peaks include 3000-2900 cm^{-1} (C-H group) and 1480-1400 cm^{-1} (bending O-H) group.

IOP Conf. Series: Earth and Environmental Science **141** (2018) 012001 doi:10.1088/1755-1315/141/1/012001

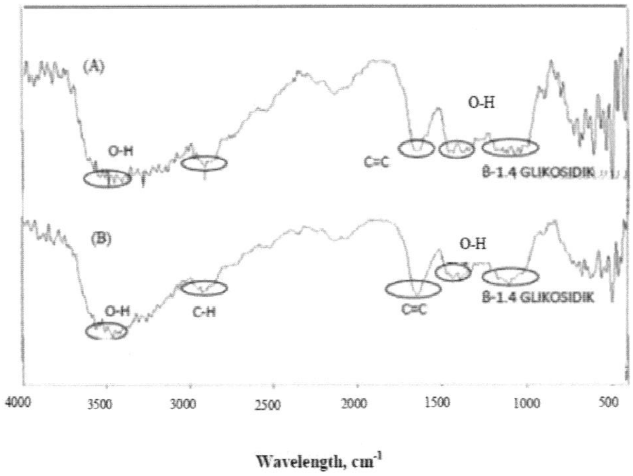

Fig 3. Fourier transform infrared (FTIR) spectra of bacterial cellulose on (a) HS (b) HS-$MgSO_4$ media with addition of 5% v/v glycerol

3.3. Thickness measurement

Fig 4 showed the thickness of bacterial cellulose produced on this research using two different media at various glycerol concentrations. The maximum thickness of bacterial cellulose was produce on HS-$MgSO_4$ media with 5% v/v glycerol concentration. Hence, HS-$MgSO_4$ media with additional of glycerol as a carbon source gave maximum thickness when compared with HS media with addition of glycerol. MgSO4 act as a nutrient that can increased growth rate of *Acetobacter xylinus*.

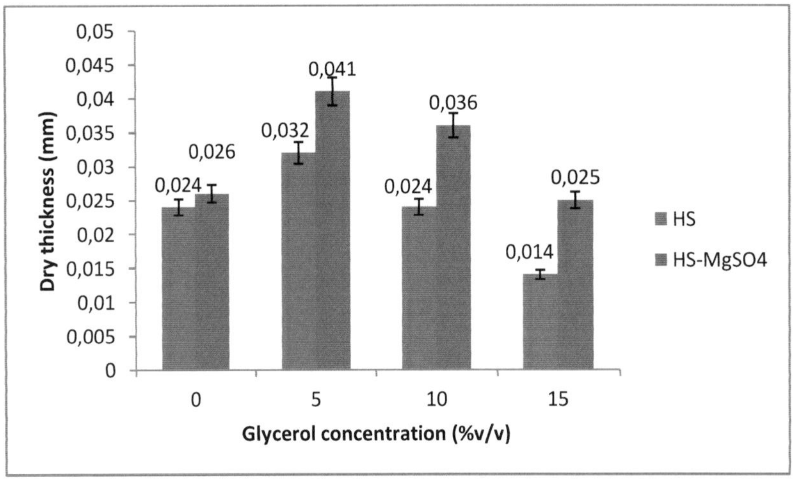

Fig 4. Bacterial cellulose thickness on HS and HS-$MgSO_4$ media at various glycerol concentrations

3.4. Mechanical test

The mechanical properties of bacterial cellulose at various media were presented in Table 1. For pure BC, the tensile strength was about 99 MPa and the elongation at break was about 3.5%. The value of bacterial cellulose vary depend on situations such as culture time, medium supplement.

Table 1. Mechanical properties of bacterial cellulose at various media

Material	Glycerol concentration (%v/v)	Young's modulus (N)	Tensile strength (MPa)	Elongation at break (100%)
HS	0	48.045	99.06	3.514
	5	87.755	86.458	2.157
	10	23.042	85.3398	2.14
	15	4.420	10.43	0.567
HS-$MgSO_4$	0	50.50	97.107	2.429
	5	103.443	116.885	4.214
	10	54.418	105.665	3.876
	15	8.534	21.086	1.934

4. Conclusions

Bacterial cellulose composite was successfully prepared in this research. HS-MgSO4 media with addition of glycerol produced thicker films compared to HS media. Bacterial cellulose from HS-MgSO4 with addition of glycerol also produced films with better mechanical properties (higher tensile strength and higher elongation of break number). These results showed that glycerol can be used as a potential substrate for bacterial cellulose production.

5. References

[1] Lin W C, Chun C L, Hsiu J Y, Chao M Y, Shan H H 2013 *Carbohydrate Polymers* **94** 603-611
[2] Czaja W K, Young D J, Kawecki M, Brown R M 2007 *Biomacromolecules* **8** 1-12
[3] Ostadhossein F, Nafiseh M, Gabriel M C, Elnaz T, Fransisco J N M, Belen S C, Jose M L P, Abdolreza S 2015 *Materials* **8** 6401-6418
[4] Vazques A, Maria L F, Patricia C, Miguel G 2013 *J. Polym. Environ.* **21** 545-554
[5] Rani M U and K A Anu A 2013 J. Food. Sci. Technol **50(4)** 755-762
[6] Mohammadkazemi F, Azin M, Ashori A 2015 *Carbohydrate Polymers* **117** 518-523
[7] Esa, Tasairin S M, Rahman N A 2014 *Agriculture and Agricultural Science Procedia* **2** 113-119

IOP Conf. Series: Earth and Environmental Science **141** (2018) 012002 doi:10.1088/1755-1315/141/1/012002

The comparison of antimicrobial packaging properties with different applications incorporation method of active material

R W Anwar[1], Sugiarto[1] and E Warsiki[1*]

[1]Department of Agroindustrial Technology, Faculty of Agricultural Engineering and Technology, Bogor Agricultural University, Bogor, Indonesia

*Email : endang.warsiki@apps.ipb.ac.id

Abstract. Contamination after the processing of products during storage, distribution and marketing is one of the main causes of food safety issues. Handling of food products after processing can be done during the packaging process. Antimicrobial (AM) active packaging is one of the concept of packaging product development by utilize the interaction between the product and the packaging environment that can delay the bacterial damage by killing or reducing bacterial growth. The active system is formed by incorporating an antimicrobial agent against a packaging matrix that will function as a carrier. Many incorporation methods have been developed in this packaging-making concept which were direct mixing, polishing, and encapsulation. The aims of this research were to examine the different of the AM packaging performances including its stability and effectiveness of its function that would be produced by three different methods. The stability of the packaging function was analyzed by looking at the diffusivity of the active ingredient to the matrix using SEM. The effectiveness was analyzed by the ability of the packaging to prevent the growing of the microbial. The results showed that different incorporation methods resulted on different characteristics of the AM packaging.

1. Introduction

Active packaging sometimes referred to a smart packaging, are designed to safeguard food quality. The concept of active packaging is intended to sense internal environmental change and to respond by changing its own properties of package. Active packaging systems are developed with the goal of extending shelf life for foods and increasing the period of time that the food is still fresh. Active packaging technologies include some physical, chemical, or biological action which changes interactions between a package, product, and/or headspace of the package in order to get a desired outcome [1]. An exciting innovation in active packaging is antimicrobials packaging. Antimicrobials packaging systems based on the application of packaging materials with incorporated antimicrobial agents in the packaging matrix and/or using antimicrobial polymers. When the packaging system acquires antimicrobial activity, the packaging system (or material) limits or prevents microbial growth by extending the lag period and reducing the growth rate or decreases live counts of microorganisms [2]. The active agents could usually derive from chemical and nature compound, such as chili pepper.

Chili pepper contains many biological active compounds, including capsaicinoids [3]. The famous compounds enlisted as capsaicinoids are capsaicin, dihydrocapsaicin, nordihydrocapsaicin, dihydrocapsaicin and capsaicin. Major part (greater than 90%) of chili pepper capsaicinoids consists of two most potent compounds, capsaicin and dihydrocapsaicin [4]. Natural antimicrobials, such as chili peppers, are receiving a good deal of attention for a number of microorganism-control issues. Recent reports stated that the Capsicum genus, among other plant genera, was good source of antimicrobial and antifungal compounds [5]. Antimicrobial agents can be incorporated into a packaging system through simple blending with packaging materials, immobilization or coating differently depending on the characteristics of packaging system, antimicrobial agent and food [6].

These techniques are performed in an effort to maximize the potential for active ingredients used when it has been incorporated into the packaging matrix. The use of antimicrobial packing will be achieved if during the packaging process the microorganism's inhibitory activity can lasted until the product is consumed. The effectiveness and stability of antimicrobial agents should further be known by the analytical approach to the things indicating that the incorporated active ingredients are maximized. This is the basis of the development of this research, which is expected to be a consideration for further antimicrobial packaging innovations. The aim of the study was to know the different of the AM packaging performances that would be produced by three different methods of incorporation of active agent into the matrix in order to know its stability and effectiveness of AM packaging function.

2. Materials and Methods

2.1. Material preparation

The material preparation included active compound and matrix. The active compound used was oleoresin from chili pepper containing capcaisin as an anti-microbial active compound. The use of this compound referred to the research of Panji [7] with the best result of using oleoresin of Capsicum™. This compound would also be used in the step of anti-microbial film incorporation. The matrix was prepared based on the research of Lestari [8], which consisted of powder agar 2.0% (m/v), glycerol 1.0% (v/v), and tapioca powder 0.5% (b/v).

2.2. The incorporation of active compound

The production of anti-microbial film was carried out with three principles of the incorporation of active compound to film matrix i.e. direct mixing, spreading and encapsulating. The method of direct mixing was referred from Maharani [9], in which the active compound was incorporated into the film solution made in the previous step. At the end of film solution production (50°C), as much as 0.6% (v/v) chili pepper oleoresin was added to the solution with a 5-minute stirring. The film solution was then poured into a sterilized glass plate to form it. Then, it was left to dry at room temperature (ambient temperature) for 10 minutes prior to oven-drying at 50°C from 24 hours.

The spreading method followed Nofrida [10] with a modification. The same film solution was printed in the similar sterilized glass plate and was left to dry for 10 minutes in ambient temperature prior to the spread on the film with oleoresin to obtain an evenly spread anti-microbial film by visual mean. Next, it was left to dry at ambient temperature for 10 minutes before undergoing oven-drying at 50°C for 24 hours. The encapsulating method was done by encapsulate the active compound before incorporating it into the matrix. The encapsulation followed the coacervation method by using 3.0% alginate (m/v), 1.0% casein (m/v) and 1.0% active compound (v/v).

2.3. The measurement of incorporation respond

The measurements of film characteristics after three distinct incorporation methods were carried out, which were the measurement of film thickness, pH, water content and color as well as anti-microbial property test and microstructure using Scanning Electron Microscopy (SEM).

IOP Conf. Series: Earth and Environmental Science **141** (2018) 012002 doi:10.1088/1755-1315/141/1/012002

3. Results

3.1. The thickness of anti-microbial film

This parameter was measured by using digital thickness gauge to understand the influence of incorporation method to the thickness. The sample (A1) was made from direct mixing, (A2) was from spreading method and (A3) was the sample from encapsulating method. The amount of solution used in each sample was the same.

Table 1. The result of film thickness

Sample	Thickness (mm)
A1	0.100 ± 0
A2	0.086 ± 0
A3	0.237 ± 0.01

Analysis of variance revealed that the incorporation method significantly influenced the thickness (α=5%). [9] investigated the influence of chili pepper variance to the thickness of the film, which resulted in a non-significant influence of it. The thickness of anti-microbial film influenced the effectiveness of the film to maintain the active compound incorporated in it. Owing to the active compound that can be released to function as an anti-microbial property, the thickness is then related to the theory of mass transfer which observes the kinetics of the absorption and release of a compound to a matrix. The thickness measurement in this research indicated that A3 resulted from encapsulation method had the highest thickness level. As encapsulation aimed to prevent a material/compound loss from its core, it might be assumed that encapsulation method prevent water loss during the drying process, hence it obtained a thicker film.

3.2. Anti-microbial film color by quantitative measurement

The quantitative method using colorimeter was employed to minimize the human error in organoleptic test (qualitative method).

Table 2. Result of quantitative color test

Sample	°Hue value	Picture
A1	58.32 ± 0.15	
A2	59.87 ± 1.72	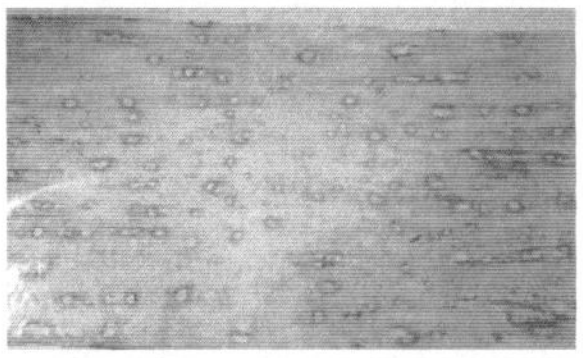
A3	55.70 ± 0.18	

Based on the result and the chromatic value [11], the three of the samples had colors ranging from yellow to red with °hue value of 54-90. This was significantly affected by the addition of capsaicin which in fact had a natural color of red. Though they had a relatively similar color range, A2 had the highest range of °hue value amongst the others, followed by A1 and finally A3. The value of °hue below 54 means that the AM film has a color range of red, while the value of °hue above 90 describes a yellowish color. Encapsulated film had a more reddish color. This color is more identical to capsaicin natural color as the active compound. This could prove that the encapsulated process was able to maintain the active compound incorporated in it. Thus, our research corresponded to the research of [12], explained that the encapsulation was a technique to protect a core from its surrounding.

3.3. SEM Analysis of anti-microbial film

SEM analysis aimed to observe the surface or the microscopic structure of the film in detail. Figure 1 displayed the comparison of each sample as seen by SEM.

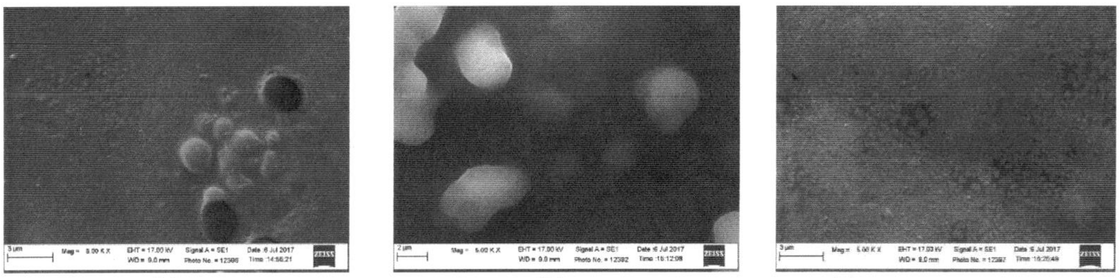

(a) (b) (c)

Figure 1. The surface of the anti-microbial film made from the method of (a) direct mixing; (b) spreading; and (c) encapsulating

It can be seen in Figure 1(a), the surface of the sample consisted of white globes which spread unevenly but still physically bound to the matrix. This was probably due to the active compound – capsaicin, a volatile compound, had been suspended in the matrix could still evaporate. Thus the structure might be disturbed during the drying process. Meanwhile, Figure 1(b) showed the white globes appear on the matrix surface, clearly due to the process of incorporation which caused incomplete absorption of the active compound to the matrix. Moreover, this method might result a more evaporated active compound than other two methods which allowed the active compound to react with the solution directly. Figure 1(c) exhibited a more homogenous result compared to 1(a) and 1(b). Owing to the encapsulation process prior to the incorporation into the solution – which it would protect the active compound from the surrounding, thus the drying process might have little or no effect to the structure by means of visual microscopy.

3.4. The pH of anti-microbial film

The pH of the film was measured to investigate the influence of pH of the anti-microbial property, which was conducted with three repetition for each film.

Table 3. The result of pH test to the solution

Sample	pH
Control	6.2 ± 0.11
A1	4.8 ± 0.11
A2	5.9 ± 0.10
A3	5.1 ± 0.06

All of the sample including the control had pH values below 7.0 as showed in Table 3. It is known, the favorable pH for E. coli growth is in the range of 7.0 – 7.5 [13]. Our results showed that the film can provide an anti-microbial properties based on the pH test result. The control showed a higher pH compared to the capsaicin-incorporated sample, meaning the addition of the active compound resulted in pH decrease. Thus we can conclude that the lower the pH, the higher the anti-microbial activity. The anti-microbial compound with acidic condition could penetrate the cell of the microbe and destroy the cell [14]. The mechanism of the anti-microbial property in preventing the growth of E. coli was described as follows. It is related to the mechanism of *E. coli* to maintain its internal pH to be neutral. If the pH is low, the high amount of proton will penetrate into the cytoplasm of the cell. This causes the cytoplasm pH decrease. The pH decrease induces the enzymes to return the pH into normal [15]. These protons should be removed out of the cell to prevent the denaturation of cell. The activity to return pH to normal requires such high energy that it can interfere the metabolism of the cell, which in longer period will kill the cell [13]. Describes that the extract of betel leaf at pH 4 was the most effective to prevent *E. coli* [16].

3.5. Anti-microbial property test of the film

The test was undergone using *E. coli* as subject culture and the testing method was distinguished for the active compound itself and for the produced anti-microbial film. The active compound anti-microbial test was conducted using well diffusion method, while the test to the produced film employed the agar diffusion method. The principle has a quite similar method in which an agar media was used where in well method, the agar media was holed, whereas in diffusion method the agar was not holed.

The activity of anti-microbial was measured by observing the surface area of clear zone, which indicated no bacterial growth in the zone, around the "well" (hole) and the anti-microbial film. If there is no clear zone, it is assumed that there was no anti-microbial activity. The results of both test (to the solution and to the film) showed that there was a clear zone on the sample. Furthermore, the test was conducted to the three sample of different incorporation method. Based on the result, there was significance (α=5%) in between the different method, as show in Table 4. This indicated that there was significant influence of the incorporation method to the anti-microbial activity.

Table 4. The result of anti-microbial test

Sample	Clear zone (mm^2)
Control	0
A1	0.61 ± 0.03
A2	0.68 ± 0.02
A3	0.89 ± 0.03

The mechanism of the inhibition of bacterial growth is either (i) by interacting with cell membrane; (ii) by inactivating the essential enzymes; and (iii) by destructing or inactivating the genetic function and material [17]. Capsaicin is an anti-microbial compound associates with the film matrix. When the anti-microbial film has a contact with the *E. coli*-inoculated media, capsaicin will diffuse from the film to the cell wall of *E. coli*, as the cell wall is consisted of 80% fat content while capsaicin is a fat-soluble compound [18]. The dissolution of capsaicin to the cell wall of *E. coli* is called penetration. This penetration will inhibit the synthesis of protein and destroy the DNA band of the cell, causing the cell to stop from growing. Thus, the area around the film and on the film itself, *E. coli* cannot grow. That area is observable as a clear zone or no metallic green color because there is no E. coli growth [19].

4. Conclusions

In this research, it is proved that the incorporation method significantly influenced the thickness of the produced antimicrobial film. Each of the film had a chromatic color range of yellow-red as shown by the quantitative test of the film color. Particularly, the product from encapsulation method had the highest °hue value which indicated its more reddish color than the other products, SEM analysis on the sample exhibited the difference of each's surface structure in which the encapsulation method resulted a more homogenous structure. Meanwhile, the pH test revealed that all of the samples were acidic, indicating they has the property of antimicrobial activity. Moreover, the appearance of a clear zone on each sample in the antimicrobial test substantiated the presence of antimicrobial activity in any incorporation method.

IOP Conf. Series: Earth and Environmental Science **141** (2018) 012002 doi:10.1088/1755-1315/141/1/012002

5. References

[1] Yam K L, Takhistov P T and Miltz J 2005 *Intelligent packaging: concepts and applications* Journal of Food Science 70: R1-R10

[2] Han J H 2000 *Antimicrobial food packaging* Food Technol 54(3): 56-65

[3] Whiting S, Derbyshire E and Tiwari B K 2012 *Appetite* 59, 341

[4] Bernal M A, Calderon A A, Pedreno M A, MuÒoz R and Ros BarcelÛ A 1993 *J. Agric Food Chem.* 41: 1041

[5] Tajkarimi M M, Ibrahim S A and Cliver D O 2010 *Antimicrobial herb and spice compounds in food* J Food Control 21:1199-1218

[6] Ahvenainen R E, Hurme K, Randall and M Eilame 2000 *The Effect of Leakage on the Quality of Gas-Packed Foodstuffs with the Leak Detection* VTT Reasearch Note. Finland

[7] Panji C, Warsiki E, Rini P and Laras W S 2013 *Oleoresin extraction techniques from various chili* Proceeding of Seminar on Research Result LPPM IPB. Vol 1. Field of Food, Energy, Technology and Engineering: 177 – 187

[8] Lestari I A 2013 *Making a Smart Label As Detector Escherichia coli* (Bogor.ID: IPB)

[9] Maharani U 2015 *Utilization of Chilli Oleoresin for Growth Inhibitor of Escherichia coli Antimicrobial Film* (Bogor.ID: IPB)

[10] Nofrida R, E Warsiki and I Yuliasih 2013 *The effect of storage temperature on the changing color of Erpa leave indicator smart label* Agroindustrial Journal. Vol 23(3): 232 – 241

[11] Hutching J B 1999 *Food Color and Appearance* Second Edition (Maruland: US) Chapman Hall Food scl.

[12] Krasaekoopt W B, Bhandari H and Deeth H 2003 *Evaluation of encapsulation techniques of probiotics for yoghurt* Int. Dairy J. 13:3-13

[13] Fardiaz S 1992 *Mikrobiologi Pangan 1* (Jakarta: Gramedia Pustaka Utama)

[14] Jay J M 1997 *Modern Food Microbiology* 5th*ed* (New York: Chapman and hall)

[15] Booth I R 1985 *Regulation of cytoplasmic pH in bacteria* Microbial Rev. 49:359-378.

[16] Parhusip A J N, Anugrahati N A and Nathalia T 2005 *Aktivitas Antimikroba Ekstrak Sereh (Cymbopogon citrarus DC Stapf) terhadap Bakteri Patogen* Jurnal Ilmu dan Teknologi Pangan. 3(2):27-28

[17] Davidson P M and Branen A L 1993 *Antimicrobial in Food* (New York: Marcel Dekker)

[18] Jawetz E, Melnick J L and Aldenberg E A 2005 *Mikrobiologi Kedokteran* (Jakarta: Salemba Medika)

[19] Krishna De and Amit 2003 *Capsicum, the genus capsicum* (New York: Tailor & Francis)

IOP Conf. Series: Earth and Environmental Science **141** (2018) 012003 doi:10.1088/1755-1315/141/1/012003

Kinetic studies of potassium permanganate adsorption by activated carbon and its ability as ethylene oxidation material

F Aprilliani[1], E Warsiki[2] and A Iskandar[2]

[1]Post Graduate Program of Agroindustrial Technology, Bogor Agricultural University, Bogor, Indonesia
[2]Department of Agroindustrial Technology, Faculty of Agricultural Engineering and Technology, Bogor Agricultural University, Bogor, Indonesia

Email: fennyaprilliani20@gmail.com

Abstract. Generally, ethylene production in many horticultural products has been seen to be detrimental to the quality during storage and distribution process. For this reason, removing ethylene from storage or distribution atmosphere is needed to maintain the quality. One of the technologies that can be applied is the use of potassium permanganate ($KMnO_4$). $KMnO_4$ is an active compound that can be used as an oxidizing agent on ethylene removal process. $KMnO_4$ is not recommended for direct used application. As the result, additional material is required to impregnate the potassium permanganate. The inert materials used are commercial activated carbon. Activated carbon is chosen because it has high surface area. The purpose of this research is to determine kinetics adsorption and oxidation model of ethylene removal material. The kinetics adsorption was determined using the pseudo-first and second-order kinetic models. The data on adsorption process show that the second-order equation is more suitable to express the adsorption process on this research. The analyzing of the ethylene oxidation capacity increased with time until it reaches an optimal value. The ethylene oxidation rate is able to be estimated by the formula $r = 0.1967 [C_2H_4]^{0.99} [KM_nO_4]^{0.01}$; MSE = 0.44 %. The actual and estimation data of ethylene oxidation show that the model is fitted to describe the actual ethylene oxidation under same experimental conditions.

1. Introduction

Ethylene is a chemical compound that plays role in many physiological processes including undesirable ripening of fruits. Generally, ethylene production in many horticultural products has been seen to be detrimental to the quality during storage and distribution process. Horticultural products have different sensitivity to the presence of ethylene. A concentration of ethylene in the air of 0.1 $\mu L\ L^{-1}$ is often quoted as the threshold level for physiological activity [1]. For this reason, good postharvest handling and removing ethylene from storage or distribution atmosphere are needed to maintain the quality. One of the technologies that can be applied is removing or reducing the level of ethylene in the atmosphere around horticultural products during storage and distribution process. The use of ethylene adsorber materials is one of the postharvest technology that can be used to maintain the horticultural product's quality by removing or reducing the level of ethylene in the atmosphere. The materials that can remove ethylene have been identified and some have been succeeded in the commercial application, such as activated carbon [2], nano zeolite-$KMnO_4$ [3] and the used of potassium permanganate and alumina beads [4].

Potassium permanganate ($KMnO_4$) is an active compound that can be used as an oxidizing agent on ethylene removal process. However, potassium permanganate is not recommended for direct used application. As the result, additional material is required to impregnated the potassium permanganate. The inert materials that can be used are activated carbon, alumina, zeolite, and clay. Activated carbon from coconut shells is used in this research as inert material to adsorb potassium permanganate. Activated carbon has been chosen because it has high surface area and significant pore volumes which relatively large [5]. Nowadays, most published studies have been evaluated the effect of potassium permanganate on products, reporting the effect of potassium permanganate concentration, determining the effect of ethylene concentration but not determining the oxidation rate model of the absorber material itself.

This paper will evaluate the effect of corrugated cardboard ethylene absorber. The purpose of this research was to determine the ability of activated carbon to adsorb potassium permanganate (used kinetics adsorption models), determine the effect of impregnated activated carbon on the absorption of ethylene gas under some environmental conditions, and also to develop the ethylene oxidation rate model. The absorber materials will be applied on corrugated cardboard. This method will provide the undirected contact between the absorber and horticultural products. The corrugated cardboard ethylene absorber can be used as an inner on packaging box with high stability against mechanical effects and it may be easily recycled.

2. Materials and Methods

2.1. Kinetic Studies

The experiment was carried out using an untreated commercial activated carbon (coconut shells). First, activated carbon was washed with distilled water and dried overnight at 105°C. The kinetic studies were carried out by the modified method. The effect of contact time was determined by dipping 25 g of activated carbon in 100 mL of potassium permanganate solution (6.4 g of $KMnO_4$ on 100 mL water) under stirring [6]. The temperature of the solution was held constant at 20°C. After different time intervals (5-60 minutes), the volume of 1 mL supernatant was taken for spectrophotometrically measurements to determine the residual concentration of potassium permanganate. The solution was placed into a cuvette and determined with a spectrophotometer. The adsorption rate was analyzed using the pseudo-first-order [7]:

$$\ln(q_e - q_t) = \ln q_e - k_1 t \qquad (1)$$

q_e and q_t are the amount of potassium permanganate adsorbed (mg/g) at equilibrium and at time (min). The k_1 is the constant rate of adsorption. The values of k_1 were calculated between the graph of ln ($q_e - q_t$) vs time. If the result showed the $q_{e\text{-exp}}$ values have similarity with $q_{e\text{-cal}}$, it means the adsorption of potassium permanganate is the first-order reaction. However, if the value of $q_{e\text{-exp}}$ does not match with the calculated one, the adsorption model is the second-order reaction. The second-order kinetic model is expressed as:

$$\frac{t}{qt} = \frac{1}{k_2 q_e^2} + \frac{1}{qe} t \qquad (2)$$

k_2 is the constant rate of the second-order adsorption. If the second-order kinetic is acceptable, the graph of t/q_t vs t should show a linear relationship.

2.2. Corrugated Cardboard Ethylene Absorber-Making Procedures and It's Ability as Ethylene Absorber

The corrugated cardboard ethylene absorber is made by glued 5 g of adsorbent on the surface of the corrugated paper (10 × 5 cm). The corrugated cardboard is glued to single paper ply. The fixing will holds the corrugation in shape and ethylene absorber materials are not visible to the end user. The glue is made from cassava starch and gelatin. Then, the corrugated cardboard ethylene absorber is applied as an inner box on the horticultural packaging. The evaluation of ethylene removal rate was determined by the modified method. The absorber (5 plies of corrugated cardboard ethylene absorber) was placed in 2.2 L of a glass jar and then sealed with a rubber stopper [8]. Gas containing 100 ppm (1 mL) of ethylene was injected into the jar. The ethylene concentration in the jar was determined by this formula:

$$C_1.V_1 = C_2.V_2 \quad (3)$$

V_1 and V_2 are the volume of ethylene injected and ethylene on the glass jar (mL), C_1 and C_2 are the concentration of ethylene injection and concentration of ethylene on a glass jar (ppm). During incubation at 18°C, after different time intervals (15-75 minutes), 1 mL of sample gas from headspace was injected into a gas chromatograph (GC) to measure the residual ethylene concentration inside the glass jar. Jars with corrugated cardboard activated carbon were used as controls. The GC was equipped with FID (flame ionization detector). Ethylene concentration was measured by the following formula:

$$\text{Ethylene concentration (ppm)} = \left(\frac{\text{peak sample}}{\text{peak standard}}\right) \text{x concentration of ethylene standard} \quad (4)$$

The GC was equipped with FID, 8 ft unibead column with the permissible temperature at 85°C. The inlet temperature is set at 150°C and the outlet temperature is set at 200°C. Helium gas carrier resulting a total gas flow of 30 (mL/min) and the pressure maintained at 100 psi. The ethylene removal rate (mL/g.h) in the jar was calculated to determine the ethylene absorption rate [9].

2.3. The oxidation rate of Corrugated Cardboard Ethylene Absorber

The oxidation rate of corrugated cardboard ethylene absorber was processed into a model to estimate the rate of ethylene absorption. The actual data of absorption was analyzed to determine the oxidation rate model of corrugated cardboard ethylene absorber activated cardboard-$KMnO_4$. The oxidation process can be described by the following formula [10, 11]:

$$A + B \rightarrow C \quad (5)$$

A and B are the two reactants that will react to produce C products. From the oxidation process, the A described the concentration of ethylene, B is the concentration of potassium permanganate and the products from oxidation process described with C.

3. Results and Discussions

3.1. Kinetic Study

The effects of dipping times of activated carbon on potassium permanganate are shown in Figure 1. Based on Figure 1, the residual concentration of potassium permanganate solution was decreased. This condition is supported by the increasing number of potassium permanganate accumulation on the surface and pore area of activated carbon (q_t). This process continues until reaches certain point which is no more potassium permanganate that is lost from the solution [5,12].

IOP Conf. Series: Earth and Environmental Science **141** (2018) 012003 doi:10.1088/1755-1315/141/1/012003

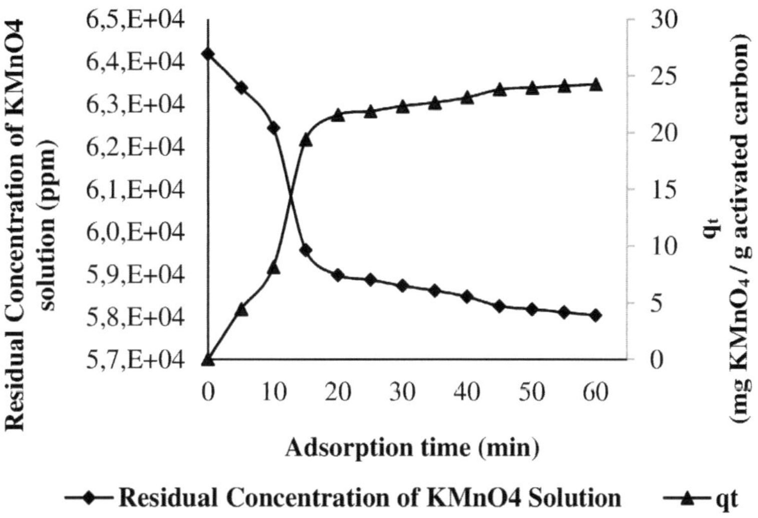

Figure 1. The effect of dipping time on residual concentration and the accumulation of $KMnO_4$ on the surface and pore area of activated carbon

The amount of $KMnO_4$ on activated carbon (mg/g) has increased along with the addition of dipping time until it reaches certain point which is no more $KMnO_4$ that loss from the solution (10-15 minutes). The amount of $KMnO_4$ on activated carbon (mg/g) on 15 minutes of dipping time indicated the optimum capacity of $KMnO_4$ adsorption under this operating conditions. The results show that the amount of $KMnO_4$ adsorbed increases from 8.1555 mg (10 minutes of dipping time) to 19.3735 (15 minutes of dipping time) for each gram of activated carbon used on adsorption process.

The kinetic adsorption of $KMnO_4$ on activated carbon was analyzed with pseudo-first-order and pseudo-second-order. The value of k_1 on the pseudo-first-order kinetic model was obtained from the slopes of the linear graph between ln $(q_e - q_t)$ vs times. According to pseudo-first-order kinetic models, the q_e values did not match with the calculated values that were obtained from the linear graph (Table 1). This results indicated that the kinetic adsorption of $KMnO_4$ onto activated carbon did not follow the pseudo-first-order equation. If the pseudo-second-order kinetic model is acceptable, the graph of t/q_t versus times should show a linear relationship. The relationship between ln $(q_e - q_t)$ vs times and the linear graph of t/q_t vs times are shown in Figure 2.

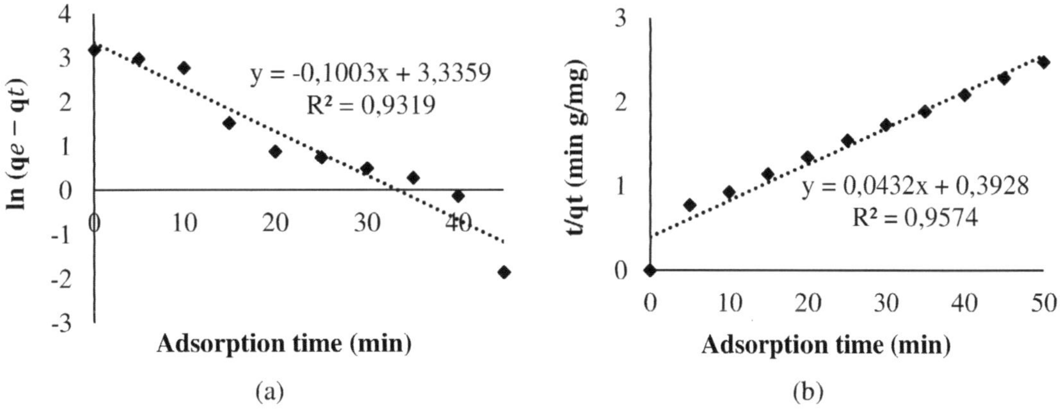

Figure 2. The graph model of first (a) and second (b) kinetic order of $KMnO_4$ adsorption

Table 1. Comparison of the experimental and calculated data on kinetic models

$q_{e\text{-}ob}$ (mg g^{-1})	*Pseudo-First-Order*			*Pseudo-Second-Order*		
	$q_{e\text{-}cal}$ (mg g^{-1})	k_1	R^2	$q_{e\text{-}cal}$ (mg g^{-1})	k_2	R^2
23.9469	28.1037	0.1003	0.9319	23.1481	0.0048	0.9574

According to Table 1, the calculation data show the pseudo-second-order model is more acceptable to describe the actual condition of $KMnO_4$ absorption on this experimental. This condition is supported by the higher coefficient correlation of calculated data on pseudo-second-order if it's compared to coefficient correlation data on pseudo-first-order.

3.2. Efficiency of Corrugated Cardboard Ethylene Absorber

The coconut shell activated carbon was impregnated by the saturated solution of potassium permanganate for 15 minutes under stirring. Activated carbon than allowed to dry at 60 °C for 16 h. The corrugated cardboard ethylene absorber is made by glued 5 g of adsorbent on the surface of corrugated ply (10×5 cm) and then the surface was glued with the liner to holds the corrugated cardboard shape. The absorber (5 plies of corrugated cardboard) was placed in a glass jar and the ethylene absorption rate was determined. The ethylene absorption capacities of corrugated cardboard are shown in Figure 3.

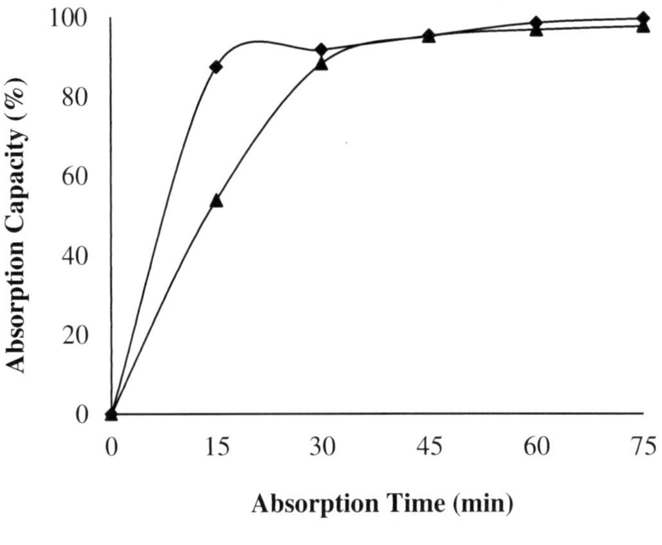

Figure 3. Ethylene absorption capacity of corrugated cardboard; —▲— Activated carbon (AC),—◆— Activated carbon-$KMnO_4$ (AC-$KMnO_4$)

Figure 3 shows the increasing number of ethylene absorption capacity from corrugated cardboard activated carbon and activated carbon-$KMnO_4$. This condition indicated the ability of corrugated cardboard ethylene absorber to reduce and even eliminated ethylene gas on the atmosphere packaging. The corrugated cardboard activated carbon-$KMnO_4$ has higher absorption capacity if compared with corrugated cardboard activated carbon. The absorption capacity on 15 minutes of absorption time was 87.41 % for corrugated cardboard activated carbon-$KMnO_4$ and 53.87 % for corrugated cardboard with activated carbon. The experimental data show that during an hour of absorption times, corrugated cardboard activated carbon-$KMnO_4$ is able to absorb 0.0745 ppm gas ethylene for each gram material ethylene absorber, while the ability of corrugated cardboard activated carbon is 0.0008 (ppm ethylene/g activated carbon).

IOP Conf. Series: Earth and Environmental Science **141** (2018) 012003 doi:10.1088/1755-1315/141/1/012003

3.3. The Oxidation Rate Model of Corrugated Cardboard Ethylene Absorber

The ethylene oxidation rate generally describes as a process of two reactants (ethylene and potassium permanganate) that produce some products (ethylene glycol, mangan dioxide and calcium hydroxide) which is influenced by a coefficient (k). The ethylene oxidation rate is calculated by the formula:

$$r = -\frac{d\,[C_2H_4]}{dt} = -\frac{d\,[KM_nO_4]}{dt} = k\,[C_2H_4]^x\,[KM_nO_4]^y \qquad (6)$$

where r is the oxidation rate (ppm/g.h), $[C_2H_4]$ and $[KM_nO_4]$ are the concentration of ethylene and potassium permanganate (mg/L), x and also y are the kinetic order with respect to ethylene and potassium permanganate. Based on the experimental data, the ethylene oxidation rate model of corrugated cardboard activated carbon-$KMnO_4$ is given by the equation:

$$r = 0.1967\,[C_2H_4]^{0.99}\,[KM_nO_4]^{0.01} \qquad (7)$$

where r is the equation model to estimate the ethylene oxidation rate (ppm/g.h), $[C_2H_4]$ and $[KM_nO_4]$ is the concentration of ethylene and potassium permanganate (mg/L) with the kinetic order for each component is 0.99 for ethylene and 0.01 for potassium permanganate.

The validation model was performed with mean square error (MSE) calculation from actual and estimation data. The advantage of validation model using MSE is to avoid the use of undesirable absolute data on mathematical calculations [14]. The validation models are shown in Figure 4.

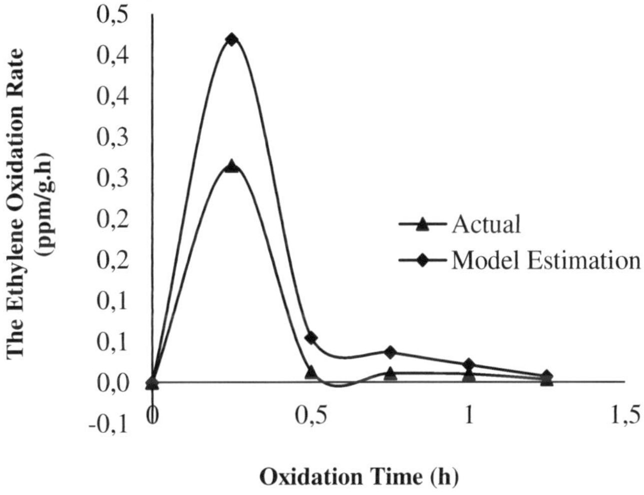

Figure 4. The ethylene oxidation rate; —▲— actual, —◆— and model estimation

The validation data on Figure 4 show the similarities between actual and model estimation of the ethylene oxidation rate. According to Suryani [13], the model is valid to estimate the actual ethylene oxidation process with the value of MSE is under and equal to 5 %. From the calculation data on this experiment, the model is able to determine the ethylene oxidation rate with 0.44 % of MSE value. This condition indicated that the ethylene oxidation rate model is able to describe the actual ethylene oxidation process under same experimental conditions.

IOP Conf. Series: Earth and Environmental Science **141** (2018) 012003 doi:10.1088/1755-1315/141/1/012003

4. Conclusions

The results of this research reveal that the effect of dipping time on 10-15 minutes increased the amount of adsorbed from 8.1555 to 19.3735 (mg/g) which is fitted well to pseudo-second-order model. The analysis efficiency of corrugated cardboard ethylene absorber proves that the effect of corrugated cardboard ethylene absorber from treated activated carbon with $KMnO_4$ (AC-$KMnO_4$) has higher and faster ability to oxidizing the ethylene from the atmosphere of storage when it is compared with untreated activated carbon corrugated cardboard ethylene absorber (AC). The results show that the ethylene oxidation rate model can be estimated by formula $r = 0.1967\ [C_2H_4]^{0.99}\ [KM_nO_4]^{0.01}$; MSE = 0.44 % under same experimental conditions.

References

[1] Kader A A 1985 Ethylene-induced senescence and physiological disorders in harvested horticultural crops HORTSCI **20** 54-57

[2] Cao J, Xiaohong L, Kunsheng W, Weibo J and Guiqin Q 2015 Preparation of novel $PdCl_2$-$CuSO_4$-based ethylene scavenger supported by acidified activated carbon powder and its effect on quality and ethylene metabolism of broccoli during shelf-life *J. Postharvest Biol. Tec.* **99** 50-57

[3] Syamsu K, Endang W, Sri Y and Siti M W 2016 Nano zeolite-$KMnO_4$ as ethylene adsorber in active packaging of horticulture products (*Musa Paradisiaca*). IJSBAR **30(1)** 93-103

[4] Wills R B H and M A Warton 2004 Efficacy of potassium permanganate impregnated into alumina beads to reduce atmospheric ethylene ASHS **129(3)** 433-438

[5] Tan I A W, A L Ahmad and B H Hameed 2008 Adsorption of basic dye on high-surface-area activated carbon prepared from coconut husk: equilibrium, kinetic and thermodynamic studies *J. Hazard. Mater.* **154(1-3)** 337-346

[6] Shi Q, Jian Z, Changlu Z, Wei N, Bo Z and Hwayoung Z 2010 Adsorption on basic violet 14 in aqueous solution using $KMnO_4$-modified activated carbon *J. Colloid Interface Sci.* **343(1)** 188-193

[7] Langergren S and B K Svenska 1898 Zur theorie der sogenannten adsorption geloester stoffe *Veternskapsakad Handlingar* **24(4)** 1-39

[8] Rahman A N 2007 *Kajian penggunaan sistem kemasan aktif penyerap etilen untuk memperpanjang masa simpan buah alpukat* (*Persea americana* Mill) [*Study of using active packaging system for ethylene adsorber to prolog the storage life of Avocado fruits* (*Persea americana* Mill)], *Thesis* (Bogor: Bogor Agricultural University)

[9] Ayu A P K 2010 *Kajian pola penyerapan etilen dan oksigen untuk penyimpanan buah segar* [*The study of ethylene and oxygen adsorption for fresh fruits storage*], *Thesis* (Bogor: Bogor Agricultural University)

[10] Damm J H, Christopher H, Robert M K and Kayleen P W 2002 Kinetics of the oxidation of methyl tert-butyl ether (MTBE) by potassium permanganate *J. Water Res.* **36** 3638-3646

[11] Hernandez A B, Edralina P S and Ernesto J R 2007 Kinetic studies of ethylene oxidation by potassium permanganate adsorbed on rice hull ash, lahar ash or coconut coir dust *J. Philipp. Agric. Sci.* **90(1)** 28-39

[12] Namasivayam C and D Kavitha 2002 Removal of congo red from water by adsorption onto activated carbon prepared from coir pith, an agricultural solid waste *J. Dyes Pig.* **54(1)** 47-58

[13] Suryani E 2006 *Konsep dasar simulasi* [*Basic concepts of simulations*] (Yogyakarta: Graha Ilmu Press)

IOP Conf. Series: Earth and Environmental Science **141** (2018) 012004 doi:10.1088/1755-1315/141/1/012004

Biomaterial compounds and bioactivity of horseshoe crab *Carsinoscorpius rotundicauda* biomass harvested from the Madura Strait

Eka Nurrahema Ning Asih[1], Mujizat Kawaroe[2,3] and Dietriech G Bengen[2]

[1]Post Graduate Program of Marine Science and Technology, Faculty of Fisheries and Marine Science, Bogor Agricultural University, Bogor 16680, Indonesia
[2]Departement of Marine Science and Technology, Faculty of Fisheries and Marine Science, Bogor Agricultural University, Darmaga Campus, Bogor 16680, Indonesia
[3]Surfactant and Bioenergy Research Centre (SBRC), Bogor Agricultural University, Bogor 16680, Indonesia

Email: ekanurrahemaningasih@gmail.com

Abstract. *Carsinoscorpius rotundicauda* or horseshoe crab biomass has great potential in pharmaceutical aspects, one of them as an antibacterial substance. Information related to the benefits of *Carsinoscorpius rotundicauda* biomass such as meat and blood is essential because in fact, this species is considered a pest by fishermen, a low market value and has no legal protection in Indonesia. The purpose of this study was to determine the content of biomaterial compounds of meat and bioactivity of *Carsinoscorpius rotundicauda* plasma on bacterial inhibition from three different stations harvested from the waters in Madura Strait. The observation of the utilization of the potential from horseshoe crab biomass ie meat and plasma was performed by measuring the content of biomaterial compound in horseshoe crab meat by HPLC method and zone of inhibition test for gram-positive and gram-negative bacteria in horseshoe crab plasma. Analysis of the relationship between the two parameters used the Principal Component Analysis. The highest content of biomaterial compounds of monoterpenoid and zoosterol is found in horseshoe crab from Bangkalan waters, namely monoterpenoid (18.33 ppm) and zoosterol (22.67 ppm), while the smallest compound content obtained in horseshoe crab from Probolinggo waters, namely monoterpenoid (13.67) ppm and zoosterol (17.33 ppm). The bioactivity of Dark Blue Plasma (BDP) and Light Blue Plasma (LBP) samples of horseshoe crab obtained around the Madura Strait has the ability to inhibit gram-positive bacteria higher than gram-negative bacteria. The total average of DBP plasma inhibitory power on *Staphylococcus aureus* was 10.00 mm and 10.07 mm on *Bacillus*, while that in LBP sample, *Staphylococcus aureus* was 9.11 mm and *Bacillus* was 9.67 mm. The high biomaterial compound content of horseshoe crab is in line with the ability of horseshoe crab plasma to inhibit *Bacillus* and *Staphylococcus aureus*.

1. Introduction

Horseshoe crab is a unique animal as it morphologically resembles a spider, has a shell, and live in the sea. There are 4 species remaining that identified belongs to the classification system of horseshoe crab in the world's, namely *Limulus polyphemus* (Linnaeus, 1758), *Tachypleus gigas* (Muller, 1785), *Tachypleus tridentatus* (Leach, 1819), and *Carcinoscorpius rotundicauda* (Latreille, 1802). *Carcinoscorpius rotundicauda* or mangrove horseshoe crab belongs to the class *Merostomata*, an ancient class of animals, and ordo *Xiphopsurida* which have smaller body size than other types of horseshoe crab [1]. Horseshoe crab is spread in some intertidal waters of Asia and America. These fossil organisms are distributed in Asian waters including Hong Kong to China waters [2], Singapore waters [3], and Malaysian waters [4]. This species is also found around the Madura Strait (Indonesia), namely Bangkalan waters [5], Sumenep waters [6] and Gresik waters up to Surabaya waters [7].

Horseshoe crab plays important roles in the intertidal ecosystem and community life such as the fulfillment of food and medicine. Eggs and meat horseshoe crab are used as expensive proteins in Malaysia and Thailand [8]. The nature of horseshoe crab which selective in choosing food has the potential to maintain the configuration of organism in the mangrove ecosystem in order to avoid blooming [9]. Horseshoe crab is also widely used as a traditional and modern pharmaceutical ingredient. Utilization of biomass and biomaterials from horseshoe crab in the form of meat (flesh) and blood as a modern pharmaceutical ingredient has been developed in America since the 20th century. The need for LAL extract (Limulus Amebocyte Lysate) from horseshoe crab plasma biomass *Limulus polyphemus* (endemic species of America) as a bacterial endotoxin test for medical devices and drug testing of about 500.000 liters annually gives a profit of about 50 million USD/year [10]. The same potential is also seen in *Carcinoscorpius rotundicauda*. Plasma from *Carcinoscorpius rotundicauda* biomass is used by Indian society to cure joint pain [11]. Plasma biomass extracts of these organisms are able to neutralize tetrodotoxin in neuroblastoma cells tested in mice [12]. Several studies have shown that plasma horseshoe crab type *Carcinoscorpius rotundicauda* contains haemagglutinin compounds [13], tachyplesia [14], C-reactive protein 2 [15], which acts as an active ingredient for inhibiting antibacterial activity and agglutination. The carapace extract of this species is also potentially used as antifouling and antibacterial [16].

The protection status of horseshoe crab in Indonesia is listed in Peraturan Pemerintah No. 7 tahun 1999 which mentions that *Tachypleus gigas* belongs to the category of endangered marine animals [17]. *Carsinoscorpius rotundicauda* is a species that is not yet protected. The absence of laws to protect this species may lead to the potential for population decline due to unintentional exploitation by the community because it is considered pests and has no economic value, so it is generally killed by most fishermen when it caught by nets. The habit of killing horseshoe crab mangrove when caught by the net makes this species untapped and wasted. This phenomenon can indirectly disrupt the population of horseshoe crab in the location, so it is necessary to study the utilization of this organism biomass so that people know the importance of horseshoe crab one of them in the pharmaceutical field such as the potential of biomaterial compounds and bioactivity. The purpose of this study was to analyse the content of bioactive compounds of horseshoe crab from three different locations harvested from the Madura Strait and to know the plasma bioactivity of horseshoe crab against gram-negative and gram-positive bacteria.

2. Materials and Methods

2.1. Materials and Tools

Materials used in this research is nets, calipers, label, autoclaves, petri disk, tweezers, micro pipettes, bunsen, incubators, paper discs, gloves, 1 set of syringes, falcon tubes, oven, pipette drops, measuring tube, erlenmeyer, centrifuge, HPLC (High Performance Liquid Chromatography), refrigerators, knives, aquades, aluminum foil, plastic bag, tissue, methanol, biomass of horseshoe crab ie meat and plasma (blood or hemolymph) *Carsinoscorpius rotundicauda*, spirits, alcohol, ampicillin, tetracycline, *Escherichia coli*, *Staphylococcus aureus, Vibrio harvey, Bacillus*, TSA (Tryptone Soya Agar), NB (Nutrient Broth).

2.2. Determination of station and Sampling

This research was conducted from October 2016 until January 2017. Sampling was conducted at 3 stations in Madura Strait namely Kertasada-Sumenep waters, Dakiring Socah-Bangkalan waters, and RanduPutih-Probolinggo waters (Figure 1). Samples were obtained by spreading 3 belt transects in the form of nets with the size of 200 meter x 1 meter with a mesh size of 3 inches at 3 stations. The transect is extended parallel to the coastline and the distance of each sub-station is 150 meter. The sample caught was taken to the laboratory with alive conditions for further analysis.

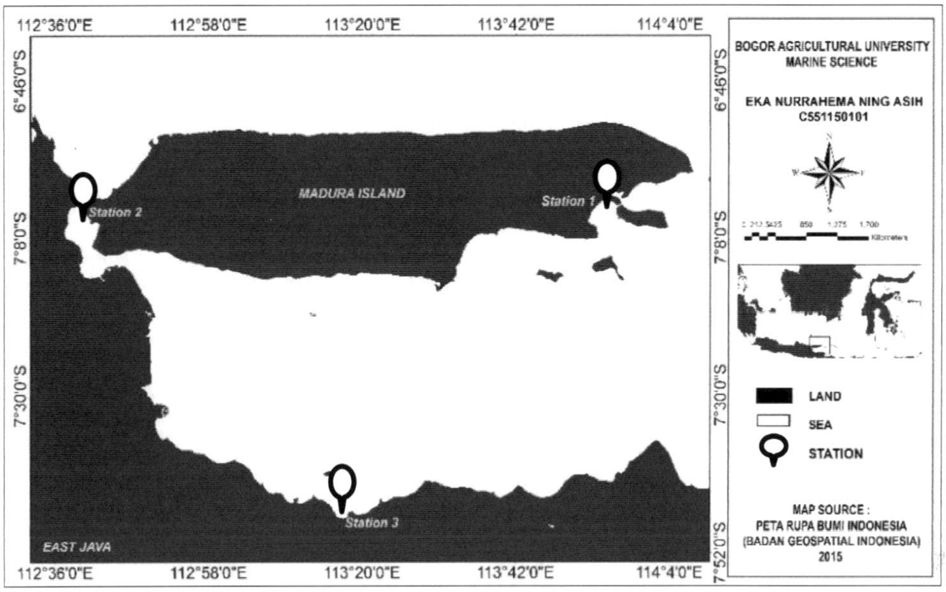

Figure 1. Research Sites in Madura Strait Waters.

2.3. Estimation of Biomaterial Compounds

2.3.1. Criteria of test samples

Morphometric measurements in horseshoe crab were performed by measuring the width of the horseshoe crab's prosoma using the sliding term (calipers). Horseshoe crab sample criteria used for testing of biomaterial compounds was female horseshoe crab with a maximum width of prosoma of 12 cm (Figure 2), whichis numerous width captured by the net in three research sites. On this size, horseshoe crab considered as the mature at the waters of the Madura Strait.

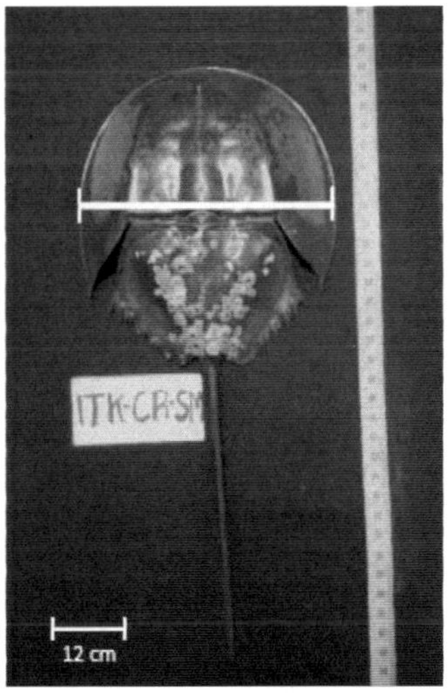

Figure 2. The size of *Carsinoscorpius rotundicauda* prosoma

2.3.2. Biomaterial compounds test
The biomaterial compound test was performed by measuring the levels of secondary metabolite compounds or bioactive compounds including the terpenoids and steroids according to the Harborne method (1996) [18]. The sample preparation was done by weighing 10 grams of horseshoe crab meat biomass *Carsinoscorpius rotundicauda*, then put into erlenmeyer with 100 ml of methanol. Horseshoe crab meat biomass that has been mixed with methanol is smoothed using a mini blender. The fine sample was then eluted using a methanol solution in the preparative column to bind the fats and proteins contained in the solution. The eluted sample was then refluxed using methanol at 50°C for 72 hours. The refluxed sample was filtered using a 0.45 micron filter to separate residues and filtrate. The filtrate obtained was centrifuged at a rate of 5000 rpm for 5 min to separate the supernatant and natan contained in the filtrate. The supernatant was kept at 4°C for further analysis. The readings of terpenoids and steroids are performed using HPLC. The reading of terpenoid content is conducted by taking a sample solution of 0.5 ml plus 10 ml of methanol using an injector made of stainless copper. The solution is injected directly into HPLC in the mobile phase with λ maximum of 230 nm dynamic column C18 UV VIS detector flowing under pressure to the column and mobile phase λ maximum 254 nm dynamic column C18 UV VIS detector .

2.4. Bioactivity Test
2.4.1. Plasma of horseshoe crab
The sampling biomass of *Carsinoscorpius rotundicauda* blood plasma as the sample of the antibacterial test was performed according to Derek method (1993). The samples of horseshoe crab from each research site were sterilized with sterilized water and alcohol. Biomass of plasma was collected by injecting the cardiac tissue around the prosoma and opisthosoma muscles. The biomass plasma collection was taken from several horseshoe crab samples which then stored in a sterile polypropylene tube. The plasma was then centrifuged at 4×6000 rpm for 1 hour at 4°C to separate the supernatant (liquid plasma of horseshoe crab) and pellet (the settling cell). Plasma was stored at -20°C, then tested on agar media to observe the inhibition zones against gram-negative and gram-positive bacteria.

2.4.2. Inhibition zones of the bacteria
Gram-negative bacteria used for the inhibition zone test were *Eschericia coli* and *Vibrio harveyi*, whereas the gram-positive bacteria used were *Staphylococcus aureus*, and *Bacillus*. The antibacterial activity test was done by using 3 stages that are media preparation to make Tryptic Soy Agar media, bacterial rejuvenation and bacterial inhibition test using Nutrient Broth buffer media. The bacterial inhibition test was performed by inoculating the rejuvenated bacteria on the Nutrient Broth medium (the buffer medium used to reactivate the bacteria) and then homogenized them using vortex and incubated for 18 hours. Nutrient Broth which has contained bacteria on medium then placed for 1 hour at room temperature (27-28°C). The next stage was to put the paper disk then 20 micron of *Carsinoscorpius rotundicauda* plasma from each station was dropped in the center of the media. The last stage was done by reincubating the media for 24 hours then measured the inhibition zone. The bacterial inhibitory test was performed 3 replicates each (tryplo) on each test bacteria to see the resulting inhibitory zones

2.5. Data Analysis
Analysis of bioactive compounds and bioactivity of horseshoe crab from Kertasada-Sumenep waters, Socah-Bangkalan and Randuputih-Probolinggo waters were conducted using descriptive analysis. Data obtained during the study are presented in the form of tables and histogram graphs. The correlation of bioactive compounds and bioactivity of *Carsinoscorpius rotundicauda* from the three location sites was analyzed using Pricipal Component Analysis (PCA) with XLSTAT software 2016.2.01 [19].

3. Results and Discussions

3.1. Biomaterial Compounds from Carsinoscorpius rotundicauda Biomass

The identification of biomaterial compounds from the secondary metabolites of *Carsinoscorpius rotundicauda* meat biomass is the levels of terpenoids and steroids. Terpenoids are the largest secondary metabolite based on the amount of compounds and variations of their basic structural framework. These compounds are also produced by insects and marine organisms [20]. Terpenoids are produced by organisms as the synthesis result of acetic acid through a biochemical process with an enzymatic system or called metabolism [21]. Other biomaterial compounds also found in marine organisms are steroids. Steroids in marine animals are commonly found in many-legged animals (insect-shaped) having molting hormones [22] such as horseshoe crab. This biomaterial compound can be used as a drug substance [23].

The results of terpenoid and steroids tests obtained monoterpenoid and zoosterol compounds. The content of monoterpenoid and zoosterol compounds in horseshoe crab varies based on the study site (Figure 3). The highest content of *Carsinoscorpius rotundicauda* monoterpenoid and zoosterol is found in horseshoe crab existing in Bangkalan waters. Horseshoe crab in Bangkalan waters has monoterpenoid levels of 18.33 ppm and zoosterol of 22.67 ppm. The second highest biomaterial compound found in horseshoe crab living in Sumenep waters with monoterpenoid levels of 14.67 ppm and zoosterol levels of 20.33 ppm, while the smallest compounds of the three study sites found in horseshoe crab of Probolinggo waters with monoterpenoid and zoosterol levels of 13.67 ppm and 17.33 ppm respectively.

Differences in monoterpenoid and zoosterol levels of female horseshoe crab at three sites are might be caused by environmental conditions of horseshoe crab. Increased production of secondary metabolites by marine organisms is an attempt to protect themselves against environmental conditions, interactions with biotic and abiotic environments and chemical weapons against other organism disorders [24]. The high levels of monoterpenoid and zoosterol in *Carsinoscorpius rotundicauda* biomass from Socah-Bangkalan waters are suspected because this location has a higher environmental stress than the other two locations. Socah-Bangkalan waters nearby to the community settlement, Kamal waters and in direct contact with Gresik waters which is an area with dense industrial activity and ship repair area potentially effect the high concentration of aquatic pollutant such as harmful bacteria and heavy metal content.

Some research states that Socah-Bangkalan and surrounding waters condition as gresik waters have high content of heavy metals in sediments and heavy metal content in water exceeds the standard quality threshold for continuity of marine organism by KMNKLH NO. 51 year 2004. The average weight of heavy metals on sediments in high Socah-Bangkalan waters ie Cd 0.36-0.38 mg/L, Cu 0.2 mg/L, Pb 47.31-50.49 mg/L, and Hg 9.01-9.07 mg/L [25]. The content of metal Pb in water at Manyar Gresik waters ranges from 0.31-0.57 mg/L and the metal content of Pb in the base of sediment ranges from 2.78-3.37 mg/kg [26]. The range can indirectly disrupt the physiological system of marine organism that performs an activity in sediments around Socah waters, one of which is *Carsinoscorpius rotundicauda*. Another indicator suggesting the high environmental stresses of horseshoe crab around this station is the abundance of coliform bacteria in the waters. The abundance of bacteria living on the profile of sea level around Kamal waters is 26-73 x 10 cfu/ml [27].

The average content of monoterpenoid and low horseshoe crab zoosterol was found in the Probolinggo sample. The low content of bioactive horseshoe crab from the sound of RanduPutih-Probolinggo because the condition at this location there is no pressure of heavy pollution and still feasible to survive horseshoe crab. The location of the Randu Putih-Probolinggo adjacent to Bentar beach in Gending sub-district is the gathering area of whale sharks [28]. The concentrations of metal Pb and Cd in Gending range between <0.001-0.002 mg/l, while the concentrations of Pb and Cd metal in the sediments are 0.09-0.53 mg/kg [29]. The concentration of the heavy metal is still below the limit by KMNKLH stipulated No. 51 of 2004 [30].

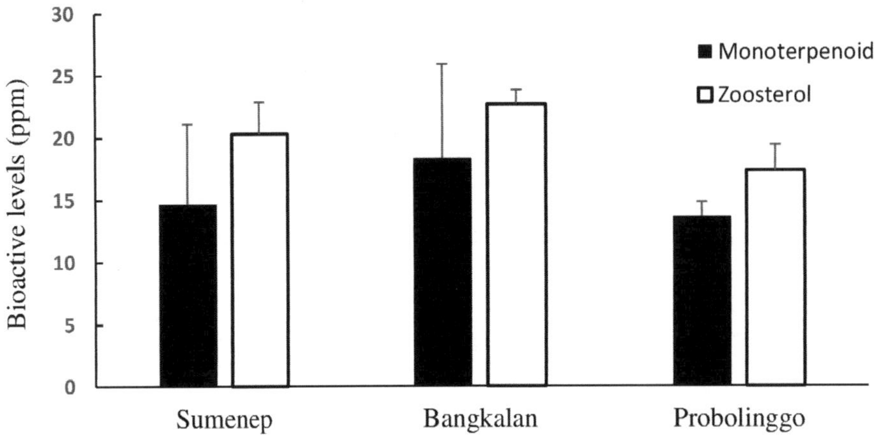

Figure 3. The average and standard deviation of biomaterial compound content of horseshoe crab based on study location.

3.2. Bioactivity Biomass from Plasma of Carsinoscorpius rotundicauda Against Bacteria

3.2.1. Extracted plasma of horseshoe crab

Liquid plasma containing hemocytes and protein [31]. Hemolymph or blood of horseshoe crab contains amebocyte, protein, sugar, salt, enzyme, creatinine, cholesterol and triglyceride cells [32]. Extracted plasma of *Carsinoscorpius rotundicauda* that centrifugation at 24.000 rpm is divided into 2 fractions namely light blue plasma and dark blue plasma. The light blue plasma is the top layer of plasma after dark blue plasma. Total plasma samples obtained from each study site were 35 ml, while the volume of light blue plasma sample is 25 ml (less than 10 ml compared to dark blue plasma) (Figure 4). The percentage value of both samples was 71% for dark blue plasma and 29% for light blue plasma samples. The protein content of two centrifugated plasma of horseshoe crab suggested vary, of which clear plasma of *Carsinoscorpius rotundicauda* (top layer) contains 0.6% protein of total crude plasma protein, while that bright blue plasma is 20% [33].

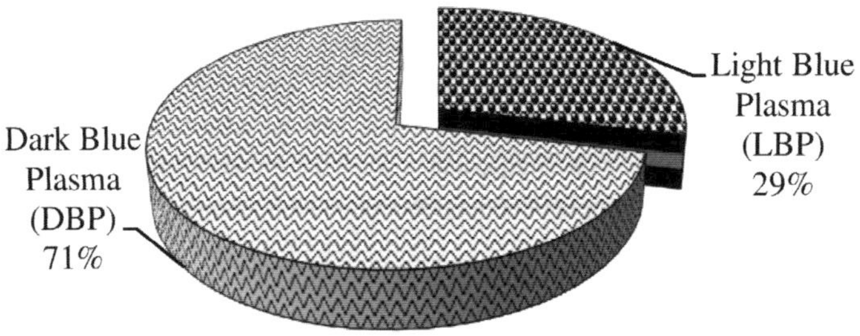

Figure 4. Plasma component proportion of horseshoe crab (%) from the study site

3.2.2. Bacterial inhibition activity of extracted plasma biomass of Carsinoscorpius rotundicauda

The sample fraction of Dark Blue Plasma (BDP) and Light Blue Plasma (LBP) from *Carsinoscorpius rotundicauda* around the Madura Strait have a higher inhibition zone against gram positive bacteria than gram negative bacteria. The total average of plasma inhibitory activity of this organism in gram-positive bacteria; *Staphylococcus aureus* 10.00 mm and *Bacillus* 10.07 mm for DBP while that *Staphylococcus aureus* 9.11 mm and *Bacillus* 9.67 mm for LBP samples (Table 1). The same potential is also present in extracted carapace of *Carsinoscorpius rotundicauda* using acetone palate which is capable of inhibiting *Bacillus* bacteria by 6 mm/30 µl and *Staphylococcus aureus* bacteria 7 mm/30µl [16].

One of the factors affecting the high activity of inhibition zone of two plasma fraction of horseshoe crab against gram-positive bacteria is wall structure of the bacterial cell and the existence of bacteria on the body organ of horseshoe crab. *Staphylococcus aureus* bacteria have a single-cell-layer with relatively simple structure thereby allowing the entry of substances potentially damage bacterial cells [34]. *Bacillus* is commonly found in the dorsal and ventral portions of horseshoe crab larvae [35], so plasma horseshoe crab is used as a material to protect their body from bacterial infection. *Bacillus* infections may cause changes in egg color (red, gray, black), interfere the egg's cells growth, disrupt the movement of larvae and cause high mortality after molting [35]. This bacterial infection can also cause stress on horseshoe crab by interrupting mobility and slowing down blood clotting reaction of horseshoe crab [36].

Inhibitory activity of extracted horseshoe crab plasma biomass on bacteria in each sample also showed variation (Table 1). Extracts of DBP and LBP of *Carsinoscorpius rotundicauda* formed the highest inhibition zone on gram-positive bacteria are samples from Bangkalan waters. The lowest inhibition zone of the DBP sample is found in horseshoe crab from Sumenep waters, while the lowest inhibition zone of LBP is present in horseshoe crab from Probolinggo waters. The inhibition zone of DBP from extracted horseshoe crab samples of Bangkalan waters against *Staphylococcus aureus* bacteria was 10.67 mm while that *Bacillus* was 10.89 mm. The LBP samples resulted in inhibition zone of 9.33 mm for *Staphylococcus aureus* and 10.00 mm for *Bacillus*.

Table 1. Average inhibition of biomass from *Carsinoscorpius rotundicauda* plasma extract on test bacteria

Station	**Average Zone of Inhibition (mm)**			
	Escherichia coli	*Staphylococcus aureus*	*Vibrio harveyi*	*Bacillus*
		Dark blue plasma* (DBP) *sample		
Sumenep	8.00	10.00	9.67	10.00
Bangkalan	-	10.67	-	10.89
Probolinggo	9.00	9.33	9.00	9.33
Total	**8.50**	**10.00**	**9.33**	**10.07**
		Light blue plasma* (LBP) *sample		
Sumenep	7.33	9.67	9.00	9.67
Bangkalan	-	9.33	-	10.00
Probolinggo	7.33	8.33	7.67	9.33
Total	**7.33**	**9.11**	**8.33**	**9.67**
Control	**15.00**	**20.00**	**14.67**	**27.67**

The high capability of DBP and LBP horseshoe crab from Bangkalan waters to inhibit bacteria is suspected due to the horseshoe crab living in this location has high environmental stress compared to the horseshoe crab from Probolinggo and Sumenep waters. The high capacity of the samples from this location to inhibit bacterial growth is an indication of self-defense efforts of horseshoe crab against gram-positive bacteria in the environment. Horseshoe crab will produce anti endotoxin and anti-bacterial compounds within their body's immune system by conducting an intracellular freezing system to defend against pathogenic organisms [37]. Plasma *Carsinoscorpius rotundicauda* produces Carsinoscorpius Amoebocyte Lysat (CAL) or called carsinoscorpin, a compound in horseshoe crab blood that acts as a growth inhibitor or bacterial killer and detects endotoxins in horseshoe crab [38]. The pure extract of this compound is capable of tolerating and maintaining the life of *Carsinoscorpius rotundicauda* from high concentrations of *Escherichia coli* bacterial infections of 10^6 CFU/ml in laboratory scale [39].This shows that the production of immunity horseshoe crab will increase along with the environmental stress conditions, one of which is to increase the inhibition of bacteria in their body.

Plasma of *Carsinoscorpius rotundicauda* has a moderate inhibitory ability against *Escherichia coli* and *Vibrio harveyi*. This is evidenced by the small diameter of inhibition zones of these two bacterial in BDP and LBP of horseshoe crab samples from Sumenep and Probolinggo waters. Based on the clear zone diameter, the capability of bacterial inhibition is classified into 3 groups, that is 10-20 mm as strong inhibitory power, 5-10 mm as moderate inhibition power and <5 mm as low inhibitory power [40]. The low inhibition zone of horseshoe crab from from the waters of Madura on *Escherichia coli* and *Vibrio harveyi* is thought to be caused by layers of bacterial cell wall structures. *Escherichia coli* has a three-layered cell wall structure composed of lipopolysaccharides, peptidoglycan, and proteins that cause difficult entry of cell destructive substances [40]. In contrast to *Escherichia coli*, *Vibrio* sp is a bacterium associated with horseshoe crab within a metabolic system and has a mutualistic symbiotic relationship in producing bioactive compounds [41].

3.3. Correlation between Biomaterial Compound and Bioactivity of Plasma

Production of the biomaterial compound of horseshoe crab is closely related to the potency of horseshoe crab bioactivity to inhibit bacteria. Several factors affect the size of the inhibition zone are biomaterial compounds, the types of biomaterial compounds, the nature of bacterial resistance to the biomaterial substance levels, and bacterial density [42]. The result of analysis of the principal components on correlation between the biomaterial compound and inhibitory activity of light blue plasma of horseshoe crab was formed by two main axes, F1 with the variance of 50.85% and F2 of 20.05% (Figure 5).

The same results were also obtained on dark blue plasma horseshoe crab samples. Analysis of the principal components of the biomaterial compound and the inhibitory activity of the dark blue plasma from horseshoe crab sample were formed by two major axes namely F1 and F2 with a total range of 86.75% (Figure 6). Based on the results obtained it is known that the high content of zoosterol of horseshoe crab from Bangkalan waters (ST2U1, ST2U2, ST2U3) is parallel to the ability of DBP sample to inhibit *Bacillus* and *Staphylococcus aureus*. This corresponds to the close correlation value of the zoosterol compound to *Bacillus* (0.768) and *Staphylococcus aureus* (0.734) (Table 2). The high level of zoosterol compounds in horseshoe crab meat can inhibit gram-positive bacteria higher than the gram-negative bacteria. The inhibitory power of DBP against gram-negative bacteria is formed on the negative axis F1 on ST1 and ST3, indicates that the inhibitory of the sample on *Escherichia coli* and *Vibrio harveyi* is highest in Sumenep and Probolinggo waters. The width of the inhibitory zone diameter indicates the stronger inhibition of biomaterial compounds on bacterial growth [43]. Steroid compounds play a role in inhibition of bacterial activity by destroying bacterial cell membranes [44]. The ability of these compounds to penetrate cell membrane layers can lead to leakage and damage of bacterial cells [45]. The three most important types of zoosterol are cholesterol, steroid hormone, and coprostanol [46]. Cholesterol in the blood circulates in lipoprotein particles which are complex compounds between fat and protein [47]. Horseshoe crab from *Tachypleus tridentatus* plasma contains tachycitin [48], a type of protein that has a molecular weight of 380-440 kDa that is potential as an antibacterial. Some amino acid derivatives sterols (zoosterol) in cholesterol synthesized from animal bile acids have the ability to inhibit the strong growth of *Staphylococcus aureus* bacteria with a minimum concentration of 6.25 MIC inhibition, but require considerable concentrations to inhibit *Escherichia coli* >50 MIC [49].

Table 2. The correlation of samples and test bacteria on the principal components of F1 and F2

Variable Bacteria	***Dark blue plasma sample***		***Light blue plasma sample***	
	Monoterpenoid	Zoosterol	Monoterpenoid	Zoosterol
Escherichia coli	-0.3736	-0.6656	-0.3270	-0.6549
Staphylococcus aureus	0.4554	**0.7336**	0.1845	0.0880
Vibrio harveyi	-0.3541	-0.6111	-0.4160	-0.6369
Bacillus	0.3368	**0.7680**	-0.1534	0.3747

The high inhibition zone of LBP and DBP of *Carsinoscorpius rotundicauda* on bacteria from Bangkalan waters is supported by the high content of biomaterial compounds produced by horseshoe crab. The conditions of Bangkalan waters is suspected to be the cause of horseshoe crab living in this location produce higher secondary metabolite compounds in the form of monoterpenoid and zoosterol to adapt and defend themselves from environmental stresses, one of which is antibacterial.

Environmental conditions of the organism are able to stimulate secondary metabolic processes as a response to increase self-defense activities [50]. The response form is carried out chemically by producing secondary metabolite compounds as an adaptation material to its environmental conditions [51], defense materials from other organism attacks [20] eg predators, parasites and microorganisms such as bacteria forming biofilms, fungi , epiphytic algae, and invertebrates [52]. Bioactivity of extract from organism againts bacteria is influenced by environment condition of the waters. Organisms living in polluted aquatic environments tend to have higher bioactivity than that organisms in relatively clean waters. The high bioactivity of organism in polluted waters since these organisms produce a lot of secondary metabolite as a survival substance for infection prevention of pathogenic microorganism [53].

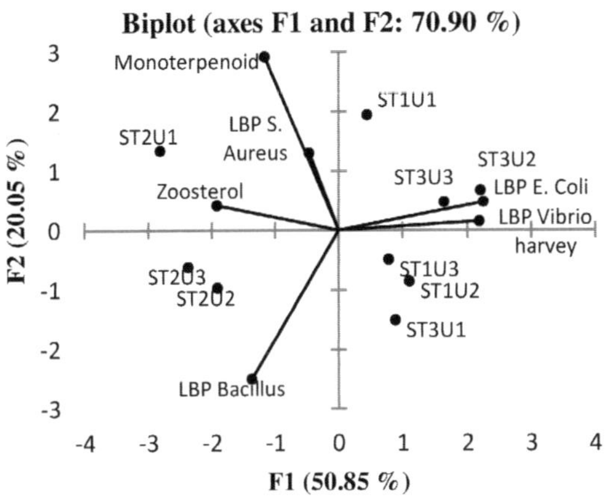

Figure 5. Principal Component Analysis of biomaterial content and bacterial inhibition of LBP sample of horseshoe crab.

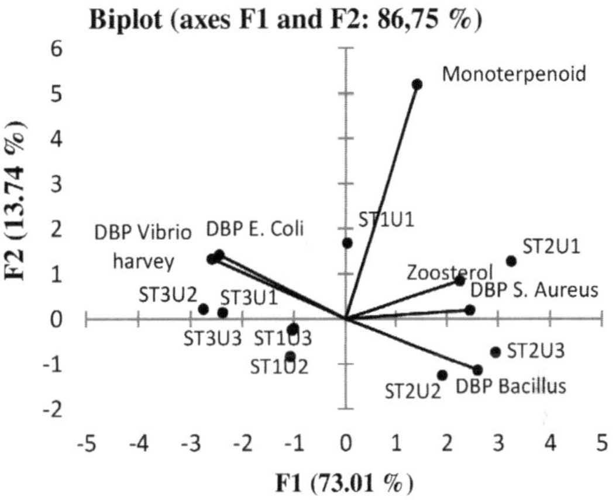

Figure 6. Principal Component Analysis of biomaterial content and bacterial inhibition of DBP sample of horseshoe crab.

4. Conclusions

The highest content of monoterpenoid and zoosterol in *Carsinoscorpius rotundicauda* biomass is obtained by horseshoe crab living in Bangkalan waters, while the smallest of three locations found in horseshoe crab living in Probolinggo waters. Bioactivity of sample from both extracted plasma produced the highest bacterial inhibition zone in *Staphylococcus aureus* and *Bacillus*, which is samples from Bangkalan waters, while that the highest *Escherichia coli* and *Vibrio harveyi* inhibition zones was found from Sumenep waters. The lowest gram positive and gram negative bacterial inhibitor zone is found in Probolinggo waters.

Acknowledgments

The Author acknowledges to the LPDP (Lembaga Pengelola Dana Pendidikan Republik Indonesia) as a scholarship institution that sponsors and fund this research up to completion. The author also expressed his gratitude to the Lembaga Ilmu Pengetahuan Indonesia (LIPI) who has supported and provided recommendations for sampling permits and the smooth running of the research.

References

[1] Lembaga Ilmu Pengetahuan Indonesia. 2013. *Biota perairan terancam punah di Indonesia prioritas perlindungan* (Jakarta: Lembaga Ilmu Pengetahuan Indonesia).

[2] Helen M, C Chiu and B Morton. 2003. The morphological differentiation of two horseshoe crab species, *Tachypleus tridentatus* and *Carcinoscorpius rotundicauda* (Xiphosura), In Hong Kong with a regional asian comparison. *J.Nat.Histo.* **37** 2369–82.

[3] Taylor LC, J Lee and C Hsu. 2009. Population structure and breeding pattern of the mangrove horseshoe crab *Carcinoscorpius rotundicauda* in Singapore. *J. Aqua. Bio.* **8** 61-69.
[4] Robert R, Muhammad A and Amelia. 2014. Demographics of horseshoe crab populations in Kota Kinabalu, Sabah, Malaysia with emphasis on *Carcinoscorpius rotundicauda* and some aspects of its mating behaviour. *J.Trop.Agric.Sci.* **37** 375 -88.
[5] Yamasaki T. 1988. *Morfology in biology of horseshoe crab by K Sekiguchi.* (Tokyo: Science House Ltd)
[6] Mashar, NA Butet, B Juliandi, Y Qonita, AA Hakim and Y Wardiatno. 2017. *IOP Conf. Series: Earth and Environmental Science.* vol 54 (Indonesia: LISAT) pp 1-8
[7] Meilana L. 2015. Kajian morfologi dan genetik horseshoe crab (Xiphosura, Limulidae) sebagai dasar konservasi dan pengelolaan di pulau Jawa. (Bogor: Institut Pertanian Bogor).
[8] Christianus A, Saad CR. 2007. Horseshoe crabs in Malaysia and the world. *Fishery Mail.* **16** 8-9.
[9] John BA, Kamaruzzaman BY, Jalal KC and Zaleha K. 2012. Feeding ecology and food preferences of *Carcinoscorpius rotundicauda* collected from the pahang nesting grounds. *Sains.Malay.* **41** 855–61.
[10] Doljs P, Vanousova K. 2015. A collection of horseshoe crabs (Chelicerata: Xiphosura) in the National Museum, Prague (Czech Republic) and a review of their immunological importance. *Arachno.Mitteil.* **49** 1-9.
[11] Mishra J.K. 2009. Biologi and conservation of horseshoe crab. *Horseshoe crabs, their eco-biological status along the northeast coast of India and the Necessity for ecological* conservation. vol 1, ed Tanacredi *et al* (New York: Springer Science+Business Media) chapter 5 pp 89-96.
[12] Derek SAY, Jeak LD and Ho B. 1996. Neuroblastoma cell culture assay shows that Carcinoscorpius rotundicauda haemolymph neutrales tetrodotoxin. *J.toxicon.* **34** 1054-57.
[13] Subita S, Toshiyuki M, Shun-Ichiro K. 1985. The complete amino acid sequence of coagulogen isolated from Sotheast Asia horseshoe crab, *Carcinoscorpius rotundicauda. J. Biocem.* **98** 305-08.
[14] Muta T, Fujimoto T, Nakajima H. 1990. Tachyplesins isolated from hemocytes of Southeast Asian horseshoe crabs (*Carcinoscorpius rotundicauda* and *Tachypleus gigas*): Identification of a new tachyplesin, tachyplesin III, and a processing intermediate of its precursor. *J Biochem.* **108** 261-66.
[15] Tincu JA, Taylor SW. 2004. Antimicrobial peptides from marine invertebrates. *J. Antimicrob Agents Chemother.* **48** 3645-54.
[16] Ismail, Nadzari, Izzatul H, Nazeerah, Taib, Chatterji and Shamsuddin. 2011. Potential of antibacterial and antifouling from horseshoe crabs, *Tachypleus gigas* and *Carcinoscorpius rotundicauda* against gram-positive and negative bacteria. *Emp.Sci.Tech.Innov.Tow.Better Tom.* **111** 717-20.
[17] Peraturan Pemerintah Republik Indonesia Nomor 7 Tahun 1999. 1999. *Pengawetan jenis tumbuhan dan satwa.* (Jakarta: Presiden Republik Indonesia).
[18] Harborne JB. 1996. Metode Fitokimia Edisi 2. (Indonesia:ITB Press).
[19] Bengen DG. 2000. *Sinopsis teknik pengambilan contoh dan analisa data biofisik sumberdaya pesisir.* (Bogor: Pusat Kajian Sumberdaya Pesisir dan Lautan, Fakultas Perikanan dan Imu Kelautan IPB).
[20] Kristanti AN, Aminah NS and Kurniadi B. 2008. *Fitokimia.* (Surabaya: Airlangga Press).
[21] Muniarsih T. 2005. Substansi kimia untuk pertahanan diri hewan laut tak bertulang belakang. *Oseana.* **30** 19-27.
[22] Bhakuni DS, Rawat DS. 2005. *Bioaktive marine natural products.* (India: Anamaya Publishers).
[23] Harborne JB. 1987. *Metode Fitokimia, Penuntun Cara Modern Menganalisa Tumbuhan.* K Padmawinata, I Sudiro, penerjemah. (Bandung: Institut Teknologi Bandung).

[24] Harper MK, Bugni TS, Copp BR, James JD, Lindsay BS, Ricardson AD, Schnabel PC, Tasdemir D, Van Wagoner FM, Verbitski SM and Ireland CM. 2001. *Introduction to the Chemical Ecology*. in ed James B *et al.* (USA: CRC Press).
[25] Nugraha WA. 2009. Kandungan logam berat pada air dan sedimen di perairan socah dan kwanyar kabupaten Bangkalan. *J Kelautan*. **2** 158-64.
[26] Putri ZL, Wulandari SY, Maslukah L. 2014. Studi sebaran kandungan logam berat timbal (Pb) dalam air dan sedimen dasar di perairan muara sungai Manyar Kabupaten Gresik, Jawa Timur. *J. Oce*. **3** 589-95.
[27] Wahyuni EA. 2015. Bakteri indikator pencemaran di perairan kabupaten Bangkalan. *J Kelautan*. **8** 33-36.
[28] Ruwaidah A. 2017. Kesuburan perairan dan potensi makanan pada habitat hiu paus (*Rhincodon typus* Smith, 1828) di pesisir Kabupaten Probolinggo, Jawa Timur (Bogor: Institut Pertanian Bogor).
[29] Fahmi, Setyono DED. 2015. Kondisi lingkungan pesisir dan perairan Probolinggo, Jawa Timur. (Indonesia: LIPI Press).
[30] Kantor Menteri Negara Lingkungan Hidup. 2004. Pedoman penetapan baku mutu lingkungan, keputusan menteri negara lingkungan hidup: Kep-51/MENEG LH/2004. (Jakarta: Seketariat Negara).
[31] Chapman RF, Simpson SJ and Douglas AE. 2013. *The Insects: Structure and Function*. (Jakarta: Cambridge Univ Press).
[32] Smith SA, Berkson JM and Barratt RA. 2002. *Proc. Int. Conf. 33rd on Association for Aquatic Animal Medicine*. vol 35 (Portugal: ZooMarine).
[33] Derek SAY, Ding JL and Ho B. 1993. An antimicrobial factor from the plasma of the horseshoe crab, *Carsinoscorpius rotundicauda*. *J.microbios*. **73** 45-58.
[34] Pelczar MJ, Chan ESC. 1993. *Microbiology:Consept and Aplication*. (New York: Graw-Hill Inc).
[35] Faizul MI, Eng HT, Christianus A and Abdel H. 2015. Changing global perspektives on horseshoe crab biology, concervation and management. *Bacteria and fungi identified on horseshoe crabs, Tachypleus gigas and Carcinoscorpius rotundicauda in the laboratory*. vol 2, ed Carmichael RH *et al* (New York: Springer International Publishing) chapter 17 pp 303-311.
[36] Bang FB. 1956. A bacterial disease of *Limulus polyphemus*. *Bull.Johns.Hopkins.Hosp*. **98** 17-21.
[37] John BA, Jalal KC, Kamaruzzaman BY and Zaleha K. 2010. Mechanism in the clot formation of horseshoe crab blood during bacterial endotoxin invasion. *J.App.Scie*. **10** 1930-36
[38] Basu S, Ranjan B. 1990. Immunomodulatory stimulation of an invertebrate circulatory lectin by its haptenic molecules of pathogenic origin. *J. Dev Comp Immunol*. **14** 31-38.
[39] Basu S, D Datta, R Bhadra. 1999. Role of Carcinoscorpin, a haemolymph lectin of horseshoe crab *Carcinoscorpius rotundacauda* as humoral factor. *J Biosci*. **24** 471-76.
[40] Davis WW, TR Stout. 1971. Disc plate methods of microbiological antibiotic assay. *Microbiology* **22** 659-65.
[41] Noraznawati I, Mohamad F, Amirrudin A, Amirul AA, Sheng K, Muhd FA, Mohd S and Azwarfarid M. 2015. Changing global perspektives on horseshoe crab biology, concervation and management. *Marine bacteria associated with horseshoe crabs, Tachypleus gigas and Carcinoscorpius rotundicauda*. Vol 2, ed Carmichael RH *et al* (New York: Springer International Publishing) chapter 18 pp 313-320.
[42] Susanna. 2006. Kajian kualitas perairan terhadap kelimpahan dan senyawa bioaktif antibakteri spons demospongiae di kepulauan Seribu DKI Jakarta. (Bogor: Institut Pertanian Bogor).
[43] De Voo gd NJ. 2005. Indonesian Sponges: Biodiversity and Mariculture Potential. (Amsterdam: Febodruk B.V, Ensched).
[44] Cowan MM. 1999. Plant products as antimicrobial agents. *Clin Microbio Reviews*. **12** 564-82.
[45] Hardiningtyas SD. 2009. Aktivitas antibakteri ekstrak karang lunak *Sarcophyton* sp yang difragmentasi tidak difragmentasi di perairan Pulau Pramuka, Kepulauan Seribu. (Bogor: Institut Pertanian Bogor).

[46] Bergmann W, Harold PS, Eva M. 1945. Contributions to the study of marine products XVIII microcionasterol and other sterols of sponge. *J. T. Baker Chemical Co.* 580-586
[47] Delvin, M Thomas. 1992. Textbook of Biochemistry, with Clinical Correlation 3rd Edition edition. (New York: John Wiley and Sons Press).
[48] Kawabata SI, Iwanaga S. 1999. Role of lectins in the innate immunity of horseshoe crab. *Dev Comp Immunol.* **23** 391-400
[49] Nilkanth GA, Vandana SP, Nripendra NM, Praveen KS, Rajesh GG. 2009. Design and synthesis of bile acid-based amino sterols as antimicrobial agents. *Bioorganic & Medicinal Chemistry Letters.* **19** 5411-14
[50] Triyulianti I. 2009. Bioaktivitas ekstrak karang lunak *Sinularia* sp dan *Lobophytum* sp hasil fregmentasi di perairan pulau Pramuka, Kepulauan Seribu, Jakarta (Bogor: Institut Pertanian Bogor).
[51] Harper MK, Bugni TS, Copp BR, James JD, Lindsay BS, Ricardson AD, Schnabel PC, Tasdemir D, Van Wagoner FM, Verbitski SM, Ireland CM. 2001. Introduction to the Chemical Ecology. (USA: CRC Press).
[52] Tringali C. 2012. Bioaktive compounds from natural sources. (Singapura: CRC Press).
[53] Soedharma D, Kawaroe M and Haris A. 2005. Kajian potensi bioaktif karang lunak (*Octorallia*: *Alcyonacea*) di perairan kepulauan Seribu, DKI Jakarta. *JIPPI.* **12** 121-28.

IOP Conf. Series: Earth and Environmental Science **141** (2018) 012005 doi:10.1088/1755-1315/141/1/012005

The effect of amphipilic lignin derivatives addition on enzymatic hydrolysis performance of kraft pulp from sorghum bagasse

Widya Fatriasari[1,3], D T Nugroho Adi[2], R P B Laksana[1], T Fajriutami[1], R Raniya[1], M Ghozali[2], E Hermiati[1]

[1]Research Center for Biomaterials LIPI, Jl Raya Bogor KM 46 Cibinong Bogor 16911, Indonesia

[2]Research Center for Chemistry, Puspiptek Serpong, Indonesia

[3]Widya.fatriasari@biomaterial.lipi.go.id

Abstract. Previously, the chemical characteristics of isolated lignin from *Acacia mangium* black liquor of kraft pulping was characterized. This lignin was blended with natural rubber latex (NR-L) as adhesive in laminated wood. In addition, lignin has potency for biosurfactant materials by modification of the hydrophobic into hydrophilic properties. Therefore, this study was intended to develop lignin as material for amphipilic lignin derivatives (A-LD) biosurfactant by reacting lignin with epoxilated polyethylene glicol (PEG). A-LD addition in slurries was used to improve the enzymatic hydrolysis (EH) of kraft pulp sweet bagasse sorghum (SSB). The main observation in EH performance was to investigate the effect of lignin isolation method (one and two step) in A-LD and A-LD loading addition on reducing sugar yield (RSY) of SSB kraft pulp. The pulp was hydrolyzed at 50°C and 150 rpm for 72 h with 10 FPU cellulase loading in the shaking incubator. A-LD from one (L_1S) and two step (L_2S) lignin was added with A-LD loading of 0, 1, 2, 5, and 10% (b/v). The RSY of hydrolyzate has been observed after EH. A-LDs addition in EH of SSB kraft pulp enhanced RSY. L_1S worked better in reaction performance with PEDGE compared to L_2S and LS. A better performance was showed by PEDGE 500 than that of PEDGE 6000. Generally, the higher A-LDs loading resulted higher RSY. The highest RSY (81.33%) was resulted in addition of 10% A-LD L_1S using PEDGE 500. A 5% A-LD loading was more considered to be added in EH because the RSY was comparable with 10% A-LD loading.

1. Introduction

Previously, pulp kraft of Sweet sorghum bagasse (*Sorghum bicolor L* Moench) has been converted succesfully to into fermentable sugar via enzymatic hydrolysis [1]. Pulping condition with 17% active alkali and 20% sulfidity resulted the highest reducing sugar yield (RSY) of 45.57%. Considering this result, this study was only used kraft pulp of sweet sorghum bagasse (SSB) as substrate in enzymatic hydrolysis (EH) by biosurfactant addition.

Acacia mangium is known as main raw materials in pulp and paper mills in Indonesia with kraft processas pulping method. Lignin is abundant materials that can be recovered from black liquor during pulping process. The main utilization of black liquor isto support energy generation in the mill with contribution of 66.2% from total energy needed at 2012 PT. Tanjung Enim Lestari pulp and paper (TELPP) mill. TELPP produces 1760 m^3 per day black liquor with total solid content of 70% [2]. It is indicated that there is high potency of lignin to be converted into value added lignin-based materials. This isolated lignin has been also utilized as filler in Aqueous Polymer Isocyanate (API) adhesive with natural rubber latex (NRL) and polyvinyl alcohol (PVA) as based polymer (registered patent

P00201507466) [3]. Besides bioadhesives, isolated lignin can also utilize as biosurfactant materials to improve the enzymatic hydrolysis (EH) performace of biomass. Up to now, the various method has been employed to increase hydrolysis effectivity by removing the inhibitor during pretreatment stage. To improve the hydrolysis performance of pretreated biomass, surfactant can be added. Some of commercial surfactant such as Trixton X, tween, PEG *(poly ethylene glycol)* resulted from polymerization reaction [4, 5, 6] often become addition materials in EH. Surfactant can reduce surface tension thus enzyme activity will improve and increase reducing sugar yield (RSY) [7].

Biosurfactant which can be sulfonated directly from lignin such as sodium lignosulphonate (Na-LS). However, the major application of Na-LS in industry is for emulsifier, corrosion inhibitor, foaming prevention, or detergent. Lignin based biosurfactant concern in EH application is still limited. Therefore, the conversion lignin for Amphipilic Lignin Derivatives (A-LD) surfactant to improve theEH performance is prospective to be developed. A-LDs are a water-soluble polymer that serves to prevent non-productive absorption of enzymes in EH process, and maintain high enzyme activity during the process. Recovery of enzyme after EH has occurred [8, 9, 10]. The improvement of EH by addition A-LDs made polyethylene glicol (PEG) and lignin in EH has reported previously [10]. This approach resulted increase hydrolysis efficiency as much as 80% and cellulase enzyme can be reused [8]. In addition, A-LDs can also accelerate the bioethanol production on large scale fed-batch SSF with substrate loading of 30% (w/v) producing high ethanol concentration of 87.9 g/L [11].

In Indonesia, A-LD synthesized from lignin as a byproduct of the kraft pulping mill has been conducted before. In the future, integration between pulp and paper mill with bioethanol pilot plant will support development of biorefinery concept. Patent No US 8911976-B2 [12] reported that lignin was reacted with hydrophilic compound i.e lauryl alcohol polyethylene oxide-glycidyl ether to result lignin derivative surfactant. This invention covered several lignin source including kraft lignin, acetic lignin, organosolve, steam explosion lignin, sulphate lignin and alkali lignin from softwood (cedar, cypress and pine), hardwood (beech and oak), and graminae (ricestraw, fir, bagasse) to be used in this reaction. The addition of0.5-20% A-LDs can improve enzymatic digestibilty of biomass. Up to now, there is no report in utilization of lignin from *Acacia mangium* black liquor for A-LD surfactant to improve hydrolysis performance of kraft pulp SSB.Besides that this research emphasis was to evaluate the effect of lignin isolation method in A-LD performance during EH. The effect of molecular weight of PEDGE in A-LD was also has been reported before. Therefore, this research was required to observe the effect of A-LD addition from one and two step isolated lignin with different molecular weight of PEDGE in enzymatic hydrolysis of kraft pulp SSB.

2. Materials and Methods

The kraft pulp of SSB with active alkali (AA) of 17% and sulphidity of 20% was prepared as described in Fatriasari *et al.* [13]. The cellulose, hemicellulose, acid insoluble lignin (AIL), acid soluble lignin (ASL) of SSB kraft pulp *Acacia mangium* was 64.65%, 29.18%, 0.81%, 0.68%, respectively [1] . *Acacia mangium* black liquor with 70% solid content was obtained from PT TELPP, South Sumatera. Epoxylated polyethylene glicol (PEDGE) with Mn 500 and 6000 and celluclast enzyme was purchased from Sigma Aldrich. The chemicals for lignin isolation, polyethylene glicol (PEG) content analysis and reducing sugar determination were in analytical grade purchased from Sigma Adrich and Merck.

2.1 Lignin Isolation

One and two step acid isolation was used to isolate lignin from the *Acacia mangium* black liquor following Hermiati *et al.* [2] with modification in frequency of solution washing. Principally, the difference of one and two step isolation of lignin was in the precipitation of sugar embadded in solution by addition of ethanol when pH of solution reached 7. Lignin isolates, namely L_1S (one step) and L_2S (two step) were then grounded and shieved through 40 mesh size shieve. The moisture content of them was also determined before used in A-LD synthesis.

2.2 A-LD Synthesis

Synthesis A-LD was performed based on Cheng *et al.* [14] with technical modification. Firstly, the lignin stock wasprepared by dissolving 5 g of L_1S and L_2S in 50 mL of 1 M aqueous NaOH in room temperature. Before addition of 3 g PEDGE with Mn 500 and 6000 in 10 mL in lignin stock, the aquades placed in beaker glass was heated until 70^0C. After that, the solution in reaction tubes was clamped and placed in heated water. The mixture was heated at 70^0C for 2 h with mechanical stirring until homogenous. To stop the reaction, acetic acid was dropped wise about 2-4 mL in the solution to pH 4 and formed an un-settled emulsion. The A-LD mixtures were kept in refrigerator before using in EH. Besides that, A-LD from commercial lignin was also prepared following the mention syntesis method. It was used as control to compare with the EH performance of SSB kraft pulp by addition of A-LDs from L_1S and L_2S.

2.3 Enzymatic Hydrolysis

As substrate of EH, a 150 mg oven dry (OD) of SSB kraft pulp was placed into vial bottle with final volume of 15 mL. Each substrate was added 300 µl sodium azide (0.02 g/mL), 0.05 M citrate buffer (pH 4.8), 10 FPU/g substrate of cellulase from *Trichoderma reesei* ATCC 26921 with enzyme activity 42.3 FPU/mL (NREL/TP-510-42628) and A-LD dissolved in citratebuffer with concentration of 1-10% v/w. The samples were put in the shaking incubator (WiseCube WIS-30R) set at 50 ºC with 150 rpm of shaking rate for 72 h.After hydrolysis finished, the samples were then centrifugated at 10.000 rpm for 5 min to separate hydrolyzate and solid fraction and then decantated the supernatant. Afterward the residue was added 10 mL aquades followed by centrifugation and decantation. The final residues were then oven dried at 60^0C for 3 days. Weight loss of samples was calculated by substracting the initial weight of samples with the final dried weight of residue. While, an aliqouts were taken from each hydrolyzates in microtubes and stored at 5-10 ºC in freezer before sugar analysis (DNS method, [15]).

2.4 Reducing Sugar Analysis

Reducing sugar concentration of hydrolyzate was determined with DNS method [15]. Previously, standard stock glucose of 10 mg/mL was prepared to make glucose standard curve. And then glucose stock was dilluted to obtain the series of glucose solution with a concentration of 2, 3.3, 5, and 6.7 mg/mL. The linier regresion with its equation can be determined from the absorbance of glucose series. About 0.5 mL aliquouts were placed in reaction tube and then added 1.0 mL destillated water and 3 mL DNS reagent. The mixture was vortexed to mix completely and then boiled for 5 min at 100^0C. Subsequently, the mixture was transferred into ice-water bath for 20 min.A 0.2 mL sample was taken to be reacted in reaction tube then added 2.5 mL distilled water and then vortexed again. The absorbance was measured by UV VIS Hitachi U-2001 spectrophotometer at 540 nm[15] in absorbance range of 0.1-1 [15].Each concentration of glucose standard solution was taken 0.5 mL and added with 1 mL aquades and 3 mL DNS reagent. The reducing sugar concentration of samples was calculated by using the regression equation which was built in series of standard curve. The RSY per biomass (g/100 g) wascalculated with considering the weight loss of SSB during kraft pulping. The theoretical RSYwas calculated by conversion factor of holocellulose into reducing sugar of 1.111 [16].

3. Results and Discussions

3.1 The effects of A-LD L_1S Addition in Enzymatic Hydrolysis

As seen in Fig. 1, the addition of A-LD L_1S caused the positive effect on RSY per biomass. The highest RSY was resulted after addition of 10% A-LD L_1S using PEDGE 500 which having similar RSY with addition of 5% A-LD L_1Sloading. Theoretically, conversion of reducing sugar of SSB with hydrolysis rate of 100% can produce 88.09g reducing sugar/100g initial SSB. It wasconsidered the holocellulose content of SSB as much as 79.29% as reported by Solihat *et al.*[1] and conversion factor of holocellulose into reducing sugar [16]. After hydrolysis of SSB kraft pulp with addition of A-LDs, the highest RSY can obtain 92.32% of theoritical RSY of initial SSB. This result is in agreement with an improvement of EHefficiency of unbleached cedar pulp with addition of A-LDs. This A-LD was synthesized from acetic acid lignin (AL) and PEG-epoxide [8]. The improvement of enzymatic efficiency from 16% to 70.1% has been reported on application of water-soluble lignin based polyoxyethylene ether (EHL-PEG) of corn stover. In this present study, the sugar yield was relatively better than previous reports. Almost of potential cellulose can be converted into sugar monomer. The difference of lignin content in cedar pulp and SSB kraft pulp might affected it. The proposed mechanism was interaction EHL-PEG with cellulase and dispersion cellulase aggregates to form smaller aggregates thatcan reduce nonreproductive adsorption of cellulase on lignin and then EH of lignocellulose[17]. During EH, there was decrease in enzyme activities because of non-productive binding between enzyme and cellulose, and non-spesific hydrofobic interaction between enzyme and lignin contained biomass [18]. Therefore, the presence A-LDs in the coctails plays important roles to improve RSY by that proposed mechanism.

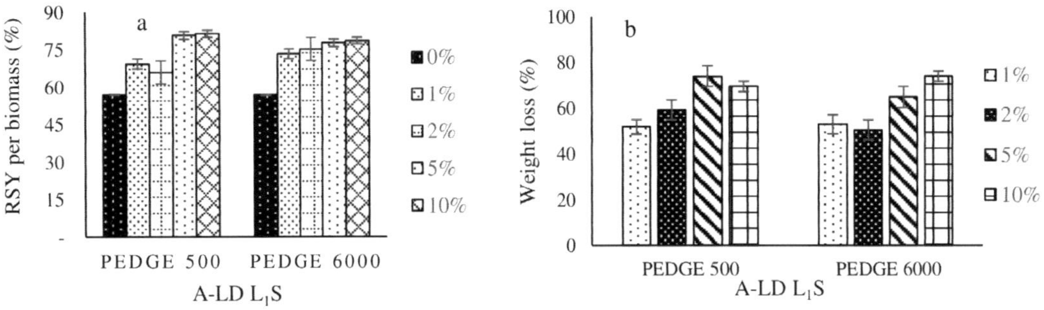

Figure 1. The effects of A-LD L_1S loading and molecular weight of PEDGE on reducing sugar yield (RSY) after enzymatic hydrolysis (EH) (a) and weight loss (b)

The higher A-LD L_1S loading, the higher of RSY obtained. It can be understood caused by more presence of A-LD to improve digestibility in enzymatic hydrolysis of SSB kraft pulp. A-LD with PEDGE 500 showed better performance of RSY per biomass compared to PEDGE 6000. A-LD was resulted from reaction of PEDGE and lignin solution with heating process. In same reaction condition, L_1S particlewith bigger size (visual observation) might more difficult to react completely. PEDGE 500 itself has lower molecular weight than that of PEDGE 6000 (unpublished data), therefore it might easier to react with bigger particle size of L_1S. The subsequent isolation of lignin resulted light colour of lignin as shown in Fig. 2.

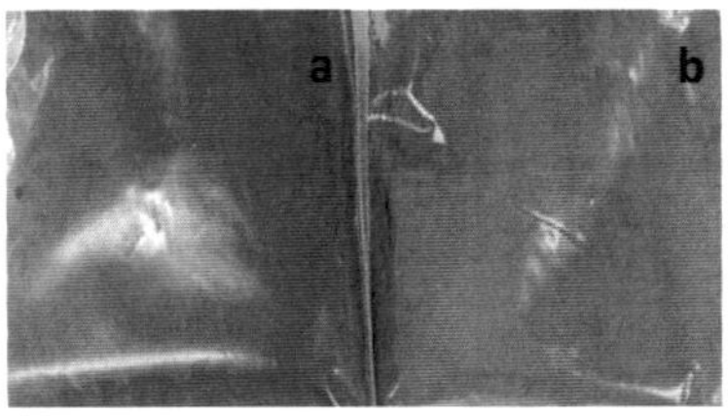

Figure 2. Physical performance of L_1S (a) and L_2S (b) isolated from black liquor of *Acacia mangium* wood

Six A-LD kinds (PEDGE 500-L_1S, PEDGE 6000-L_1S, PEDGE 500-L_2S, PEDGE 6000-L_2S, PEDGE 500-LS, PEDGE 6000-LS) were added in enzymatic hydrolysis of kraft pulp SSB. Two glycidyl groups of PEDGE interacted with lignin polymer. Winarni *et al.* [8] also reported the hydrophobic properties of PEDGE-acetic lignin (AL) has becomed as inhibitor of cellulase covering the active sites of enzymes and then prevented the substrate to insert into these sites. Amongst lignin derivatives including EPEG-, DAEO-, PEDGE-ALs, the most recovery of cellulose activity wasresulted by utilization of PEDGE-AL [8].DOPEG-SL and EPEG-SL derived from Japanese cedar has been reported capable to reduce enzyme adsorption and end-product inhibition, and improve bioethanol production [11].

The weight loss of SSB kraft pulp after enzymatic hydrolysis is presented in Fig. 1b. This loss indicated that there was breaking down process of cellulose polymer in SSB kraft pulp into its glucose monomer. The presence of glucose in the hydrolyzate was determined in reducing sugar concentration. Generally, the higher A-LD loading results the higher weight loss both PEDGE 500 and 6000 (Fig. 1b). These results are inline with higher RSY obtained by increasing of A-LD loading. At same A-LD loading except 10%, A-LD with PEDGE 500 demonstrates higher weight loss than that of A-LD PEDGE 6000. It means that PEDGE 500 worked more effective with L_1S compared to PEDGE 6000.

3.2 The effects of A-LD L_2S Addition in Enzymatic Hydrolysis

Fig. 3a shows the effects of the addition of A-LD L_2S in the EH of SSB kraft pulp on RSY per biomass. The RSY improvement after addition of A-LD L_2S was occurred that has similar tendency with effect of A-LD L_1S addition. Preventation of non-productive enzymesabsorption during EH facilitated high enzyme activity during the process [8, 9, 10] thus theRSY improvement occured.

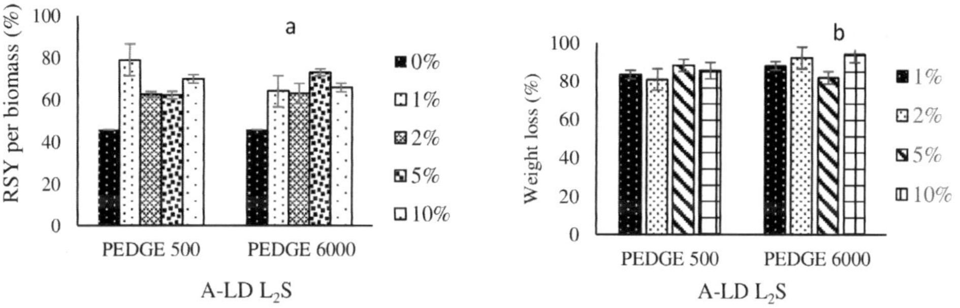

Figure 3. The effects of A-LD L_2S loading and molecular weight of PEDGE on reducing sugar yield (RSY) after enzymatic hydrolysis (EH) (a) and weight loss (b)

The highest RSY was obtained in 1% A-LD loading using PEDGE 500 as much as 78.89%. This yield is also higher than that of 5% A-LD loading using PEDGE 6000. It was proposed that PEDGE 500 resulted better efficiency compared to PEDGE 6000. And these results arehigher than that of Winarni *et al.* [8] and Lin *et al.* [17] by using AL-PEG epoxide and EHL-PEG. In the same reaction condition, the RSY of A-LD L_2S produces lower RSY slightly compared to A-LD L_1S. It indicates that the further step in the lignin isolation is not required. Furthermore, PEDGE 500 is better to beused in the reaction with L_1S and L_2S as well. With hydrolysis rate of 100%, the highest RSY can produce 89.56% of theoretical RSY of initial SSB. Compared to RSY of A-LD L_1S, this yield has lower value. Previously, Cheng *et al.*[11] reported that the utilization of GC 220 enzyme in EH demonstrated higher sugar yield (73.7%) than that of meicellase enzyme in EH (63.9%) of softwood unbleached kraft pulp (NUKP) by addition DOPEG-SL (soda lignin). It was indicated that GC 220enzyme relatively prefer to be proposed in improving performance reaction with DOPEG-SL, one of A-LD types. In this present research, we used celluclast with PEDGE-KL (kraft lignin) in which the RSY was higher than that of DOPEG-SL with GC 220 enzymeand meicellase.

Weight loss has also occurred during EH of SSB kraft pulp with addition of A-LD L_2S (Fig. 3b). As mention before, cellulose in the pulp has been hydrolyzed into sugar monomer. A-LD L_2S using

PEDGE 6000 tended to yield higher weight loss than that of A-LD L_2S using PEDGE 5000. The highest weight loss (92.14%) was found in A-LD loading of 2% with PEDGE 6000. Eventhough, weight loss after EH with A-LD L_2S is higher than that of A-LD L_1S, the RSY isnot higher anymore. Thus, high weight loss isnot always related with high RSY after EH. The other factors might affected it which need to be further studied.

3.3 The effects of A-LD LS Addition in Enzymatic Hydrolysis

In the same method, commercial lignin (LS) was also reacted with PEDGE used as comparison to observe the effect of isolation method in the performance of A-LDs. A-LD LS can alsoincrease RSY compared to without A-LD addition. The higher A-LD LS loading in EH, the higher RSY obtained (Fig 4a). The highest RSY (76.02%) is still lower than that of the highest RSY of A-LD L_1S and L_2S. These results has been supported by previous report of Winarni *et al.*[8]; Cheng *et al.* [11] and Lin *et al.* [17]. This highest RSY can produce about 86.29% of the theoretical RSY of initial SSB with hydrolysis rate of 100%. It is indicated that lignin which was isolated from black liquor of *Acacia mangium*kraft pulping for A-LDs can become new prospective utilization compared to commercial lignin. Eventhough commercial lignin and lignin two step physically have finer particles, however prior to the synthesis process, they were dillutedin alkali solution. Thus, they can react with PEDGE easily to yield A-LDs.

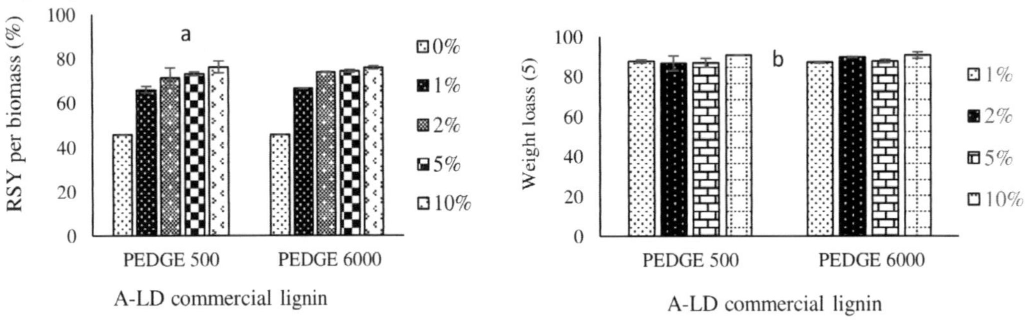

Figure 4. The effects of A-LD LS loading and molecular weight of PEDGE on reducing sugar yield (RSY) after enzymatic hydrolysis (EH) (a) and weight loss (b)

As mention before, A-LDs was effective to allow enzyme in working effectively by preventing non-productive enzymeabsorption [8, 9, 10]. PEDGE 500 worked better reaction with LS than that of PEDGE 6000. Therefore, based on the results, PEDGE 500 is prefer to be used than that of PEDGE 6000. EH of SSB kraft pulp with addition of A-LD LS caused weight loss (Fig. 3b). The formation of sugar monomer from cellulose contained in the pulp contributed on the loss. There is no significant difference in weight loss between A-LD LS using PEDGE 500 and 6000. The higher A-LD LS loading, the higher weight loss. This weight losstendency is in line with RSY. During EH, the weight lossoccurred due to conversion cellulose in the SSB pulp in the sugar monomer.

4. Conclusions

In this present study, A-LD L_1S, L_2S and LS was successfully made by reacting PEDGE and lignin. The addition of A-LDs in EH of SSB kraft pulp improved the RSY compared to control. Further step lignin isolation was not required to yield higher RSY. Isolated lignin performed more effective with PEDGE than that of LS.PEDGE 500 showed better reaction performance with lignin compared to PEDGE 6000. The higher A-LD loading tends to result the higher RSY. Utilization of 5% A-LD loading was more considered to achieve high RSY yield.

Acknowledgements
The authors would like to thank the Indonesian Institute of Sciences (LIPI) forfunding this research through Competitive Program on Material, Energy, and Manufacturing Engineering Development Program 2017.

5. References

[1] Solihat N N, Fajriutami T, Adi D T N, Fatriasari W and Hermiati E 2017 Reducing Sugar Production of sweet sorghum bagasse kraft pulp *AIP Conf. Proc.* **1803** p. 020012-1–020012-8
[2] Hermiati E, Risanto L, Lubis M A R and Dewi A R 2017 Chemical characterization of lignin from kraft pulping black liquor of *Acacia mangiumAIP Conf. Proc.* **1803**020005-1 - 020005-7
[3] Hermiati E, Fatriasari W, Laksana R P B, Puspita Sari F, Jayadi, Darmawan T, Sudarmanto, Risanto L, Akbar F, and Lubis M A R 2015 Formulasi dan proses pembuatan perekat kayu *aqueous polymer isocyanate* berbasis lateks karet alam Patent Reg. P 00201507466 19 November 2015
[4] Börjesson J, Engqvist M, Sipos B and Tjerneld F 2007 Effect of poly (ethylene glycol) on enzymatic hydrolysis and adsorption of cellulase enzymes to pretreated lignocellulose *Enzyme and Microbial Technology* **41** 186–195
[5] Kristensen J B, Börjesson J, Bruun, M H, Tjerneld F, Jørgensen H 2006 Use of surface active additives in enzymatic hydrolysis of wheat straw lignocellulose *Enzyme and Microbial Technology* **40**888–895
[6] Zhou Y, Chen H, Qi F, Zhao X and Liu D 2015 Non-ionic surfactants do not consistently improve the enzymatic hydrolysis of pure cellulose *Bioresource Technology* **18** 136–14
[7] Li J, Li S, Fan C, and Yan Z 2012 The mechanism of poly (ethylene glycol) 4000 effect on enzymatic hydrolysis of lignocellulose *Coloids and surfaces B: Biointerfaces* **89**203-210
[8] Winarni I, Oikawa C, Yamada T, Igarashi K, Koda K and Uraki Y 2013 Improvement of enzymatic saccharification of unbleached cedar pulp with amphipathic lignin derivatives *Bioresources* **8** (2) 2195-2208
[9] Winarni I, Koda K, Waluyo T K, Pari G and Uraki Y. 2014 Enzymatic saccharification of soda pulp from sago starch waste using sago lignin-based amphipathic derivatives *J. Wood Chem.Technol.* **34** (3) 157-168
[10] Uraki Y, Ishikawa N, Nishida M, and Sano Y 2001 Preparation of amphiphilic lignin derivative as a cellulase stabilizer *J. Wood Sci.* **47** (4) 301-307
[14] Cheng N N, Yamamoto Y, Koda K, Tamai Y and Uraki Y 2014 Amphipathic lignin derivatives to accelerate simultaneous saccharification and fermentation of unbleached softwood pulp for bioethanol production *Bioresour. Technol.* **173** 104-109
[11] Cheng N, Koda K, Tamai Y, Yamamoto Y, Takasuka T E, and Uraki Y 2017 Optimization of simultaneous saccharification and fermentation conditions with amphipathic lignin derivatives for concentrated bioethanol production *Bioresour. Technol.* **232** 126-132
[12] Uraki Y, Honma H, Yamada T, Kubo S and Nojiri M 2014 Lignin-based enzyme stabilizer US 8911976 B2 Granted 16 December 2014
[13] Fatriasari W, Supriyanto, Iswanto A H 2015 The kraft pulp and paper properties of sweet sorghum bagasse (*Sorghum bicolor* L Moench) *J. of Eng. and Technol. Sci.* **47** (2) 149-159
[15] Adney B and Baker J 2008 Measurement of cellulase activities laboratory analytical procedure (LAP) NREL/TP-510-42628. NREL/TP-510-42628
[16] Fatriasari W, Syafii W, Wistara N J, Syamsu K, and Prasetya B 2014 Digestibility of betung bamboo fiber following fungal pretreatment *Makara Journal Technol.* **18**(2) 51-58
[17] Lin X, Qiu X, Yuan L, Li Z, Lou H, Zhou M, and Yang D 2015 Lignin-based polyoxyethylene ether enhanced enzymatic hydrolysis of lignocelluloses by dispersing cellulase aggregates *Bioresour.Technol.* **185** 165–170
[18] Eriksson T, Borjesson J and Tjerneld F 2002 Mechanism of surfactant effect in enzymatic hydrolysis of lignocellulose *Enzyme Microb. Technol.* **31** (3) 353-364

Coffee agroforestry for sustainability of Upper Sekampung Watershed management

Fitriani[1)], Bustanul Arifin[2)], Wan Abbas Zakaria[2)], R Hanung Ismono[2)]

[1)] State Polytechnicof Lampung, Jl. Soekarno Hata No. 10 Rajabasa ,Bandar Lampung; Lampung, Indonesia
[2)]Agribusiness Department, Agriculture Faculty of Lampung University, Jl. Soemantri Brojonegoro No. 1 Rajabasa, Bandar Lampung;Lampung, Indonesia
[1)]fitriani@polinela.ac.id;[2)]barifin@uwalumni.com; [2)]wan_abbas@yahoo.com; [2)]hismono@yahoo.com

Abstract

The main objective of watershed management is to ensure the optimal hydrological and natural resource use for ecological, social and economic importance. One important adaptive management step in dealing with the risk of damage to forest ecosystems is the practice of agroforestry coffee. This study aimed to (1) assess the farmer's response to ecological service responsibility and (2) analyze the Sekampung watersheds management by providing environmental services. The research location was Air Naningan sub-district, Tanggamus, Lampung Province, Indonesia. The research was conducted from July until November 2016. Stratification random sampling based on the pattern of ownership of land rights is used to determine the respondents. Data were analyzed using descriptive statistics and logistic regression analysis. Based on the analysis, it was concluded that coffee farmers' participation in the practice of coffee agroforestry in the form of 38% shade plants and multiple cropping (62%). The logistic regression analysis indicated that the variables of experience and status of land ownership, and incentive-size plans were able to explain variations in the willingness of coffee growers to follow the scheme of providing environmental services. The existence of farmer with partnership and CBFM scheme on different land tenure on upper Sekampung has a strategic position to minimize the deforestation and recovery watersheds destruction.

Keywords: coffee agroforestry, willingness, Sekampung watershed, Indonesia

1. Introduction

Way Sekampung is well known as the main watershed, troughs to seven districts in Lampung Province. Batu Tegi Dam at Upper Sekampung provides water irrigation for 66.573 ha wet paddy system (Public Work and Settlement Ministry, 2015). Most of Upper Sekampung watershed areas include protection forest Register 39 under Batutegi Forest Management Office (FMO) jurisdiction. Another part is private land. Several factors were being caused the damage to an ecosystem in WS

IOP Conf. Series: Earth and Environmental Science **141** (2018) 012006 doi:10.1088/1755-1315/141/1/012006

Seputih Sekampung. The carrying capacity of the environment has been declining due to the pressure on land use and lack of conservation efforts by the community. Significant changes in land use utilization, from the forest into the plantation, cultivated/farmland also turned into land residential, industrial, etc. Approximately 60% of forest lands converted into gardens and settlements in Batutegi watershed. The main water source of Batu Tegi Dam was supplied by Upper Sekampung watershed. The land degradation caused by erosion and deforestation in Seputih-Sekampung reach to138.026,38 ha (Mesuji-Sekampung River Basin Agency, 2010).

Erosion and sedimentation threaten the sustainability of the function of water resources as well. Erosion problems not only decreased land productivity in the on-site area but also increased sedimentation in the off-site area. The more severe watersheds need to be carried out in an effort to manage the watersheds more integrated. Upstream agricultural activities that do not heed the rules conservation, including illegal forest clearing for land agriculture have triggered the process of erosion and sedimentation (1). The quantification of land use affected dominant erosive processes (2). Ecological services will be threatening worst by those factors as the negative externalities. Watershed management is important to minimize negative externalities.

Watersheds management means to ensure the optimal hydrological and natural resource use for ecological, social and economic. One important adaptive watersheds management in dealing with damage ecosystems is the practice of agroforestry. Agroforestry is recognized as an integrated land-management to solve environmental problems in both developing and industrialized nations today. Agroforestry systems can be found specific differentially based on climatic conditions and socio-economic factors such as human-population pressure (3); (4); (5). Agroforestry systems retain much higher quantities of carbon in above and belowground biomass in comparison to crop and grazing land use systems (6). Agroforestry systems also have been shown in reducing erosion through their canopy cover and their contribution to the litter layer (7).

Most of the stakeholders at upper stream watersheds involved agricultural sector production, with the main cash crop such as coffee, cacao, banana, and other wood plantation. Upper watersheds mostly have specific topography elevation. For more than 5o elevation, the agricultural practise must consider conservation principals (1). Coffee and others cash crop could cultivate as well with the Good Agricultural Practices (GAP) with conservation.

Internalizing the negative externalities of ecological service reduction has been discussed broadly on Payments for Ecosystem Services (PES) domain (8); (9). It is commonly defined as a market-based environmental policy instrument to efficiently achieve ecosystem services provision. The conceptualization of PES cannot be easily generalized and implemented in practice. The gap between

IOP Conf. Series: Earth and Environmental Science **141** (2018) 012006 doi:10.1088/1755-1315/141/1/012006

PES theory and practice to increase sustainable ES provision and improve livelihoods may bridge the neither fairness nor efficiency (10).

The contribution of coffee farmers in the provision of ecosystem services is important to gain recognition, rewards and incentive rewards from stakeholders who enjoy the positive externalities of this agroforestry coffee. However, information on the willingness of coffee farmers and user communities to meet in the mechanism of rewards for environmental services is still very limited. The possibility of coffee farmer participation in environmental services needs to dig deeper. The certainty of agroforestry coffee practice in upstream of Sekampung River Basin will be providing water availability and other environmental services. Agroforestry coffee is a part of internalizing externalities of local land-use in upper Sekampung. The paper was aims to (1) assess the farmer's response to ecological service responsibility and (2) analyze the Sekampung watersheds management by providing environmental services.

2. Materials and Methods

2.1 Survey design

The research location was in Air Naningan sub-district, Tanggamus. Two villages were chosen as a representation of upper Sekampung territory in Air Naningan sub-district. There were Datar Lebuay and Sinar Jawa village. Both of the villages lay down between Sangarus and Sekampung river. Sangarus and Sekampung rivers was the main water sources for Batu Tegi DAM. Respondent represents the land tenure of coffee farming. There were private, CBFM, and non-CBFM land tenure. CBFM is standing for Community Based Forest Management. Upper Sekampung watershed was under the Batu Tegi Forest Management Office (FMO) territory.

Most of the upper Sekampung territory is protected forest under the jurisdiction of Batutegi FMO included all territory in Register 39 with area reached 58,162 ha. Air Naningan sub-district has population 30,185 people. There was 32 farmers group had been licensed as CBFM in Batutegi FMO jurisdiction. CBFM territory reached 14,609.15 ha and involved 16,169 farmers. Respondents of CBFM were HKm Mandiri Lestari and Wana Tani Lestari Utama. The management area of CBFM was located on upstream of Sekampung and Sangarus rivers. The total number of respondents is 400 coffee farmers. The field survey conducted from July until November 2016. The site location relevant to describe the upper Sekampung land-use related to Sekampung watersheds management. Site location can see at figure1 as follows:

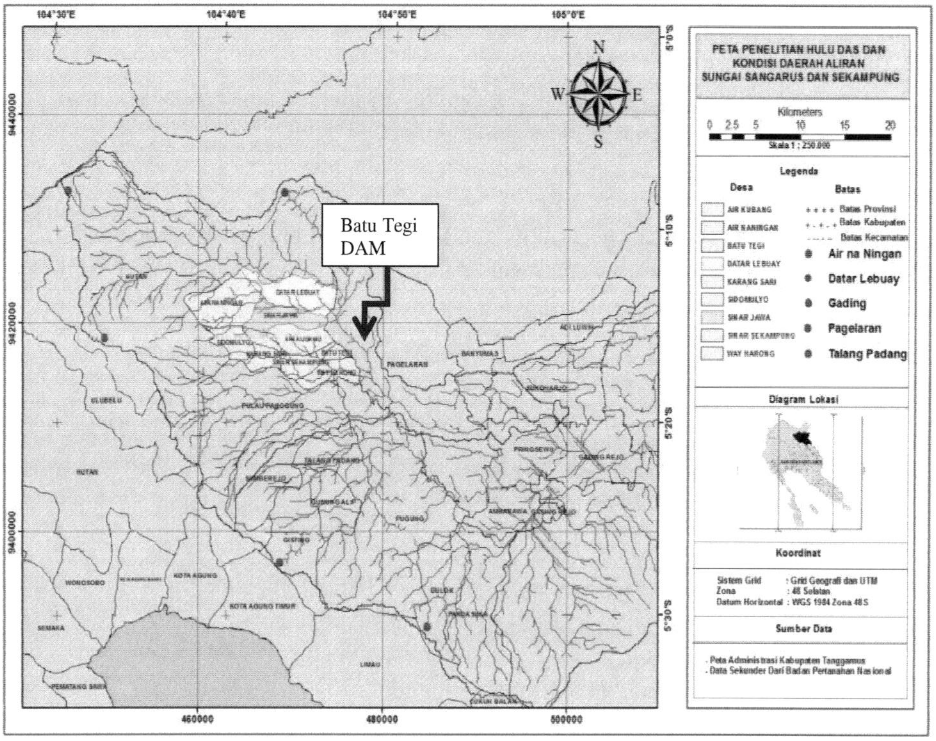

Figure 1. Site location of Datar Lebuay and Sinar Jawa village at Air Naningan Sub-district

2.2 Data Analysis

Data were analyzed by descriptive statistics and logistic regression analysis. Investigation of the logistic regression equation model is used to test the model of community willingness to participate in Sekampung ecological services (WTP). The model was developed for logistic regression function. Logistic regression had well known as a good predictor to analyze the probability of willingness to participation (11)(12) (13). Describing the farmer's response to ecological service responsibility built as a logistic regression model, binner logistic regression. Formulation model predicted WTP as Y, dependent variable as binner, (1 = Yes, 0 = No) to participate in ecological services scheme. The logistic regression equation model is formulated as follows:

$$Pi = E\left(Y = \frac{1}{Xi}\right) = \frac{e\ ln(\frac{Px}{1-Px})}{1+e\ ln(\frac{Px}{1-Px})} \quad \text{...(1)}$$

Furthermore logistic regression equations are modified in multiple regression equations (14)to be:

$$ln \frac{Px}{Pi/Px} = Y \text{ ....} (2)$$

$$Y = \beta 0 + \beta 1\, X1 + \beta 2 X2 + \beta 3\, X3 + \varepsilon \text{ ...}(3)$$

Information:
PX / P1-PX = Odd ratio, is a comparison of the opportunity of people who are willing to accept participate (WTP=Y) with the community (respondents) who are not willing to participate.
P (xi) = Community Opportunity in its willingness to participate or Y(1 = Yes, 0 = No)
P (xi) = Yp
β0 = Constant
X1 = Experiences
X2 = Land ownership
X3 = Compensation
β1..β3 = Regression coefficient,
E = Error in equation or interruption

3. Results and Discussions

3.1 Farmer's response on ecological service responsibility

Air Naningan sub-district has territory 116.5 km2 residence by 30,185,000 people. There were 10 villages under the Air Naningan sub-district jurisdiction. Datar Lebuay and Sinar Jawa were represented as a village which borders with protection forest Register 39 Batu Tegi. Most of the population cultivated annual plantation such as coffee, pepper, and cocoa as multiple cropping systems. There were also appeared the shade trees combination for the multipurpose benefit. Based on Table1 recognized that there no monoculture coffee cultivation. All of the respondents applied combined coffee plantation with another crop as shade trees and also multiple cropping for optimum benefit economically and ecologically.

Table 1. Coffee agroforestry form

Agroforestry form	Frequency	Valid Percent
Shade trees	168	43.6
Multiple cropping	217	56.4
Total	385	100.0

Coffee farmers have participated in agroforestry coffee practice with 43.6% shade trees form and multiple cropping (56.4%). Biologically coffee needs shade trees for their optimum productivity (15).

The combination of coffee and shade trees is a must to fulfill high vegetation requirement for the sustainability of protected forest and watersheds area, especially the land with elevation more than 5% (1). Coffee agroforestry system was well known as a management strategy to minimize ecological negative impact in Central America (16). Some of the major agroforestry systems in the tropics are grouped according to such a framework. The scheme appears a logical, simple, pragmatic and purpose-oriented approach to classification of agroforestry systems (5). Modifying agro forests according to their knowledge and tree preferences, and that the resulting agro-forest was very important. The higher proportion of pioneer trees relative to the forest is mostly explained by farmers' tree selection decisions (63%) rather than as a byproduct of management practices (37%) that disturb the soil and open the canopy, altering light penetration and microclimate conditions. Farmers gradually replace canopy trees of neutral and disliked species by preferred species, in particular, Inga spp. This indicates that farmers are receptive to incorporate outside knowledge into their knowledge systems and adapt their resource management practices accordingly (29), (30), (31).

The duration of agroforestry coffee practice showed in figure 2. Most of the respondents have been experienced in practicing agroforestry coffee for more than 5 years. The experience in agroforestry coffee is important to keep good agriculture practice on coffee cultivation.

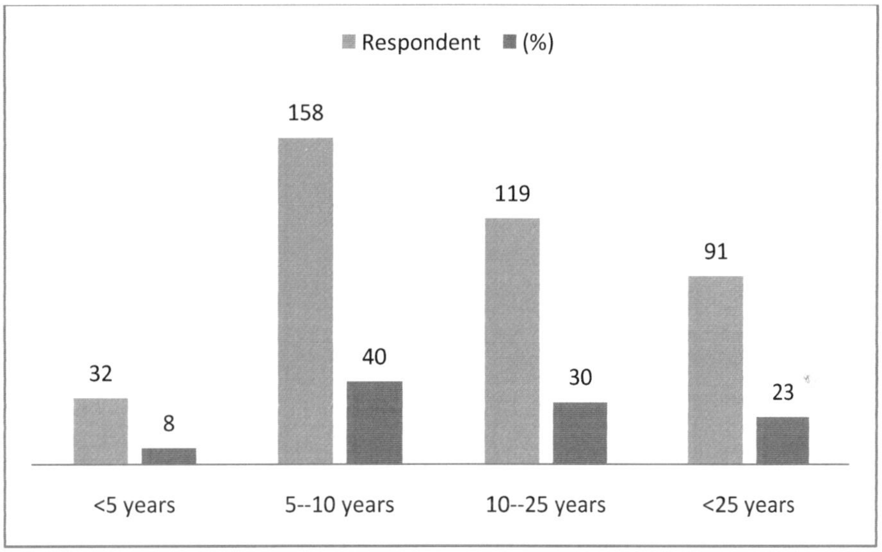

Figure 2. Duration of farmer's experiences in agroforestry coffee practiced

Based on the land ownership status, private/clan was the smallest. Private/clan lay on village jurisdiction land territory. The ownership of certification of private land still limited. Both of two villages follow the national program on land certification. Almost 500 land certificate has been released for the community in 2016, the others still on issued process. There was CBFM dominantly (figure 3).

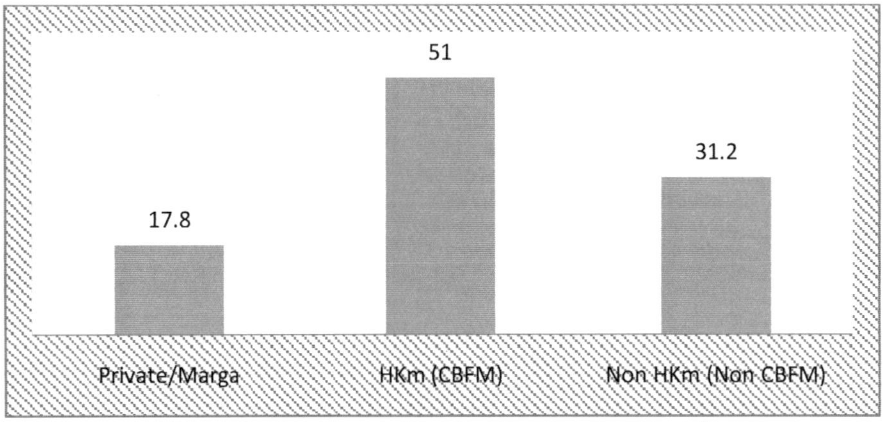

Figure 3. Respondent based on land ownership (%)

Table 2. Duration agro forestry adoption and land status cross tabulation

Duration years		Land status			Total
		Private/clan	CBFM	Non CBFM	
	<5	9	16	10	35
	5-10	35	76	47	158
	10-25	22	65	32	119
	<25	5	45	36	86
Total		71	202	127	400

The cultivation system of community on upper Sekampung watershed will be determining the sustainability Sekampung ecological services. Coffee farmers have been realized that agroforestry coffee gives them tobenefit more than monoculture. Coffee plantation must plant with other trees for biological needs.Agroforestry systems have been shown to reduce erosion through their canopy cover and their contribution to the litter layer. The agroforestry system examined, for the cultivation of coffee and mixed shade trees (Musa *spp* and Inga *spp*), an average of 10.4% of the area is affected by erosion. The soil was relatively little affected by erosion, around 13% of the cultivated area (7).

Describing the farmer's response to ecological service responsibility built as a binner logistic regression model. Based on Table 3 had revealed that most of the farmers (74.75%) prepared to participate inSekampung ecological services scheme, while the other was still considered about the

IOP Conf. Series: Earth and Environmental Science **141** (2018) 012006 doi:10.1088/1755-1315/141/1/012006

scheme. The analysis of coffee farmer willingness to participate onSekampung ecological services scheme presented at Table 4.

Table 3. Coffee farmerwillingness to participate inSekampung eco services scheme

	Alternative answer	N	(%)
	Yes	299	74.75
Coffee farmer willingness to participate Sekampung eco services scheme	No	101	24.25
	Total	400	100

Table 4. Logistic regression of coffee farmerswillingnessfor Sekampung Ecoservices scheme

	Estimate	Std. Error	Wald	Odd
[EcoServices = 1.00][Y]	16.46	0.48	1,174.10	
Experiences [X1]	0.05	0.01	19.28***	1.04
Land ownership [X2]	0.40	0.18	5.20***	1.5
[Compensation] [X3]	13.55	0.29	2,132.37***	
Link function: Logit.				
Chi-Square 34.09 Sig 6.46E-06				
Nagelkerke value 0.12068				

At the Table 4 was showed that Chi-Square value 34.09 with Sig. value 6.46E-06 (p-value <0.05), significant with α 5%. It means that the variation of the coffee farmer to participate in the ecological services scheme significant influenced by variable experiences, land ownership, and compensation.

$$Y = a + b_1 X_1 + b_2 X_2 + b_3 X_3 + e$$

$$Y = 16.46 + 0.05X_1^{***} + 0.40X_2^{***} + 13.55X_3^{***}$$

From the logistic regression analysis, indicates that the variables of experience (X1), the status of land ownership (X2), and compensation (incentive-size plans) (X3) were significantly able to explain variations in the willingness of coffee growers to follow the scheme of providing environmental services. Odd ratio of experience variable (X1) was 1.04. It means that long duration of practicing coffee agroforestry will influence the probability to participate in ecological services scheme. The odd value of land status variables was 1.5. It means that if every each changing in land tenure status, odds of private/clan ownership status 1.5 times greater to participate on environmental services scheme than CBFM and non CBFM land tenure, assuming all variables in the model were constant. The variable amount of the compensation plan was significant in determining the willingness of the coffee farmers participates in the scheme.

3.2 Sekampung watersheds management

Batutegi Forest Management Office (FMO) was established in 2011 with an area of 58,162 hectares (based on Forest Ministry Policy No: SK.650/ Menhut-II/2010). Based on Landsat imagery in 2010, the vegetation cover is dominated by non-forest vegetation (76%) and dominated by farmers' fields with the staple crop of coffee. The existence of these tenant farmers is one of the important considerations for Batutegi FMO in formulating its management plan. Thus, the vision of Batutegi's FMO is "Achieving the Function of Protected Forests Providing Community Welfare".

BFM group in Tanggamus regency in 2015 was 32 groups. According to the Head of Batutegi FMO, currently, there is many farmer groups and farmer groups association that are in the process of applying for partnership permit to manage the land in Reg. 39. Strong public support is needed to create private incentives for exploring economic and environmental win-win innovations (17).

Mr Yayan Ruchyansah as Batutegi FMO head officer mentioned that they have been developed an excellent partnership farmers program. The program has been pioneered since 2012. It is directed to embrace the illegal farmers to become farmers established by providing the legality of cultivation in the form of partnership cooperation. Until the year 2016 has been successfully built 11 farmer groups with an area of 2,582 hectares and the number of members of more than 1,546 peasants. They have maintained seedlings planted more than 500,000 stems with various types of plants such as nutmeg, cloves, areca nut, aren (Arenga pinnata), jengkol (Archidendron pauciflorum) and gaharu (Aquilaria malaccensis). Eleven groups have planted, only 3 groups have been verified and agreed on the number of crops to be imaged, the rest is the stage of verification. This is done to clarify the responsibility of each farmer in the calculation of profit sharing. Figure 4 described one sample design of agroforestry plot by Batu Tegi FMO-farmers partnership coverage 1,200 ha.

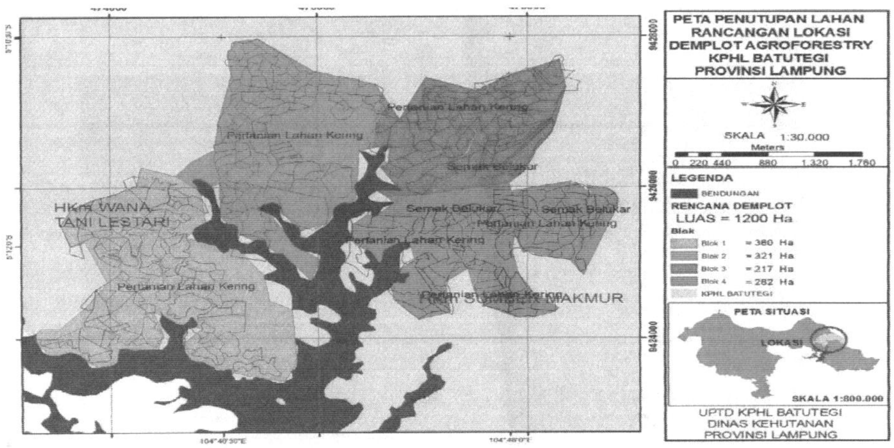

Figure 4. Design of agro forestry plot by Batu Tegi FMO
Sources: Batu Tegi FMO, 2015

The program emphasized not just on planting, but the program for harvest. It means that the plant must be maintained in order to grow properly and harvest optimally. Cooperation arrangement was carried out by involving farmer actively through intensive discussion in the field. It set agreement on yield distribution between farmers and FMO was 20-25% for FMO and 75-80% for farmers. The difference is based on the amount of cost sharing in each location. The first goal is sustainability on financial benefit for the farmer and Batutegi FMO. The last but the most important is environment benefit for ecological services sustainability..

The existence of farmer with partnership and CBFM scheme on different land tenure on upper Sekampung has a strategic position to minimize the deforestation and recovery watersheds destruction. Previous research also found the fact that the scenario of land management of CBFM with the pattern of agroforestry cultivation is the best scenario on rehabilitating the forest area and changing the management of monoculture coffee land into a mixed coffee planting pattern (18). In the watershed management circuit, there is a management activity involving mutual relationships between natural resources and the community. In the execution of the impact of reciprocal relationships between humans and the environment specifically identified with externalities (19).

These results indicate that it is important to improve forest area monitoring during the early establishment of a participatory forest management association to maximize the effect of CBFM formation (20). Awareness among group members of CBFM is higher than those who have not received permission. CBFM member and household groups believe that participation in the forestry program will substantially increase their security of land ownership, land value, land investment, and income. CBFM existence has a positive impact on non-coffee tree planting and contributes to increasing their income (21).

Well, watersheds management is primary action due to protecting the hydrological cycles. The worst management will effects on dominant erosive processes and contribute to more targeted efforts and relevant incentives to reduce (or live with) sediment load of the rivers (2). Damage to watersheds in Indonesia increasing every year so had caused very harmful excesses. Every year there is always a flood, landslides, droughts and disasters. Watershed management activities have been a long time practised, but the results achieved not optimal yet. One source of caused failure is emphasizing patterns command and control with approach top-down policy, operational level, and implementation. Watershed management in new paradigm is the deep community empowerment watershed management efforts at the level operational and implementation with using a bottom-up approach(22).

Based on information on Table3 revealed that land ownership between CBFM and non CBFM status was significant in determining the willingness of the coffee farmers participate in the scheme. This fact had relief that coffee farmer responsible with their obligation to take good care the Upper Sekampung watersheds environment. Based on Indonesian Law No. 7, 2004, the water resources management pattern is the basis for planning, implementing, monitoring and evaluation of conservation activities, utilization and control of natural resources damages. The management pattern becomes the foundation of coordination among related parties based on the principle of sustainability, the balance of social-economic-environmental function and the general benefit principle and involves the role of society at large.

4. Conclusions

Based on the analysis, it was concluded that coffee farmers' participation in the practice of coffee agroforestry in the form of 38% shade plants and multiple cropping (62%). The logistic regression indicated that the variables of experience, the status of land ownership, and incentive-size plans were able to explain variations in the willingness of participation coffee growers to follow the scheme of providing environmental services. The odd value of land status variables means that if every each changed in land tenure status, odds of private/property ownership status 1.5 times greater following the environmental service reward scheme than others land tenure, assuming all variables in the model were constant. The variable amount of the compensation plan was also significant in determining the willingness of the coffee farmer's participation. The existence of farmer with partnership and CBFM scheme on different land tenure on upper Sekampung has a strategic position to minimize the deforestation and recovery watersheds destruction.

References

1. Banuwa IS, Sinukaban N, Tarigan SD, Darusman D. Evaluation of Upper Seaport Capacity of Sekampung. 2008;13(1):145–53.
2. Verbist B, Poesen J, van Noordwijk M, Suprayogo D, Agus F, Deckers J. Author's personal copy Factors affecting soil loss at plot scale and sediment yield at catchment scale in a tropical volcanic agroforestry landscape. [cited 2017 May 23];
3. Ramachandran Nair PK, Kumar BM, Nair VD. Agroforestry as a strategy for carbon sequestration. J Plant Nutr Soil Sci [Internet]. 2009 [cited 2017 May 23];172:10–23.
4. Hardcastle PD. P. K. R. Nair (ed.). 1989. Agroforestry systems in the tropics. Kluwer Academic Publishers, in cooperation with ICRAF. 664 pages. ISBN 90-247-3709-7. 1991 Feb 10 [cited 2017 May 23];7(1):84
5. Nair PKR. Classification of agroforestry systems. Agrofor Syst [Internet]. 1985 [cited 2017 May 23];3(2):97–128.
6. Abbas1 F, Hafiz Mohkum Hammad, Shah Fahad, Artemi Cerdà, Muhammad Rizwan, Wajid Farhad, et al. Agroforestry: a sustainable environmental practice for carbon sequestration under the climate change scenarios—a review. Env Sci Pollut Res [Internet]. 2016 [cited 2017 May 23];24(March):11177–91.
7. Blanco Sepúlveda R, Aguilar Carrillo A. Soil erosion and erosion thresholds in an agroforestry system of coffee (Coffea arabica) and mixed shade trees (Inga spp and Musa spp) in Northern Nicaragua. Agric Ecosyst Environ [Internet]. 2015 Dec [cited 2017 May 23];210:25–35.
8. Bernard F, Noordwijk M Van, Luedeling E, Villamor GB, Sileshi GW, Namirembe S. ScienceDirect Social actors and unsustainability of agriculture. Curr Opin Environ Sustain

[Internet]. 2014;6:155–61.
9. Namirembe S, Beria Leimona, Meine van Noordwijk, Florence Bernard, Kukunda E Bacwayo. Co-investment paradigms as alternatives to payments for tree-based ecosystem services in Africa. Curr Opin Environ Sustain 2014, [Internet]. 2016 [cited 2017 May 23];6(December):89–97.
10. Leimona B, Lusiana B, Van Noordwijk M, Mulyoutami E, Ekadinata A, Amaruzaman S. Boundary work: Knowledge co-production for negotiating payment for watershed services in Indonesia. Ecosytem Serv. 2015;15:45–62.
11. Kotimah MK, Wulandari SP. Binary Logistic Regression Model Stratification on Women's Economic Participation in East Java Province. J Sains dan Seni POMITS. 2014;3(1):D1–6.
12. Beharry-borg N, Smart JCR, Termansen M, Hubacek K. Evaluating farmers ' likely participation in a payment programme for water quality protection in the UK uplands. Reg Environ Chang. 2013;13(February):633–47.
13. Kong F, Xiong K, Zhang N. Determinants of Farmers' Willingness to Pay and Its Level for Ecological Compensation of Poyang Lake Wetland, China: A Household-Level Survey. Sustainability. 2014;6(September):6714–28.
14. Hosmer DW, Lemeshow S. Applied Logistic Regression [Internet]. Wiley Series in Probability and Sattistics. 2000. 373 p.
15. Hairiah K. Sistem agroforestri di indonesia. 1995;1–19.
16. Gomez-Delgado F, Roupsard O, Le Maire G, Taugourdeau S, Pérez A, Oijen M Van, et al. Modelling the hydrological behaviour of a coffee agroforestry basin in Costa Rica. Hydrol Earth Syst Sci Eur Geosci Union [Internet]. 2011 [cited 2017 May 23];15(1):369–92. Available from: https://hal.archives-ouvertes.fr/hal-01189535
17. Cerin PC. Bringing economic opportunity into line with environmental influence : A discussion on the Coase theorem and the Porter and van der Linde hypothesis. Ecol Econ. 2006;56(23 May):209–25.
18. Maryanto A, Murtilaksano K, Rachman M. Land Use Planning and the Impact to Water Resource at Way Besai Watershed - Lampung . J Kehutan Wallacea. 2014;3. No.2(Juni):85–96.
19. Kerr J. Watershed Management : Lessons from Common Property Theory. 2007;1(1):89–109.
20. Takahashi R, Todo Y. Impact of a Shade Coffee Certification Program on Forest Conservation : A Case Study from a Wild Coffee Forest in Ethiopia Ryo Takahashi and and Yasuyuki Todo. Vol. 55. 2013.
21. Pender, John, Suyanto, John Kerr EK. Impacts of the Hutan Kamasyarakatan Social Forestry Program in the Sumberjaya Watershed , West Lampung District of Sumatra , Indonesia. 2008.
22. Syam T, Nishide H, Salam AK, Utomo M, Mahi AK, Lumbanraja J, et al. Soil Science and Plant Nutrition Land use and cover changes in a Hilly Area of South Sumatra , Indonesia (from 1970 to 1990) Land Use and Cover Changes in a Hilly Area of. Soil Sci Plant Nutr. 1997;43 (3):587–99.
23. Napitupulu DF, Asdak C, Budiono. The mechanism for rewarding environmental services in the Cikapundung Sub-watershed (Case Study in Cikole Village and Suntenjaya Village, West Bandung Regency). J Ilmu Lingkung. 2013;11(2):73–83.

24. Bellver-domingo A, Hernández-sancho F, Molinos-senante M. Geoforum A review of Payment for Ecosystem Services for the economic internalization of environmental externalities : A water perspective. GEOFORUM [Internet]. 2016;70:115–8.
25. Leimona B, Lusiana B, Van Noordwijk M, Mulyoutami E, Ekadinata A, Amaruzaman S. Boundary work: Knowledge co-production for negotiating payment for watershed services in Indonesia. Ecosyst Serv [Internet]. 2015 [cited 2017 May 23];15:45–62.
26. Milder JC, Scherr SJ, Bracer C. Trends and Future Potential of Payment for Ecosystem Alleviate Rural Poverty in Developing Countries. Ecol Soc. 2010;15(2):4.
27. Hergoualc 'h K, Blanchart E, Skiba U, Hénault C, Harmand J-M. Changes in carbon stock and greenhouse gas balance in a coffee (Coffea arabica) monoculture versus an agroforestry system with Inga densiflora, in Costa Rica. Agric Ecosyst Environ [Internet]. 2012 [cited 2017 May 23];148:102–10.
28. Albrecht A, Kandji ST. Carbon sequestration in tropical agroforestry systems. Agric Ecosyst Environ [Internet]. 2003 [cited 2017 May 23];99:15–27.
29. Valencia V, West P, Steling EJ, Garcia-Barrios L, Naem S. The use of farmers ' knowledge in coffee agroforestry management : implications for the conservation of tree biodiversity. Ecosphere. 2015;6(July):1–17.
30. Ajayi OC, Centre ET. Role of externality in the adoption of smallholder agroforestry : Case studies from Southern Africa and Southeast Asia . In : Sunderasan S (Ed .) Externality : Economics , Management and ...Role of Externality in the adoption of smallholder agrofor. In: Sunderasan S, editor. Externality. Nova Science Publisher; 2016. p. 167–88.
31. Corradini M, Costantini V. Interacting innovation investments and environmental performances : a dynamic impure public good model. 2015;109–29.

IOP Conf. Series: Earth and Environmental Science **141** (2018) 012007 doi:10.1088/1755-1315/141/1/012007

Physical and mechanical properties of parallel strand lumber made from hot pre-pressed long strand oil palm trunk waste

Inayah Fridiyanti[1] and M Y Massijaya[1]

[1]Biocomposite Laboratory, Forest Products Department, Faculty of Forestry, Bogor Agricultural University, Bogor, West Java, Indonesia.

Email: fridiyanti.inay@gmail.com; mymassijaya@yahoo.co.id

Abstract. This research was focused on the utilization of oil palm trunk waste as a Parallel Strand Lumber (PSL) raw material. This research aimed to analyze the effect of adhesive types and glue spreads to the physical and mechanical properties of PSL. The adhesive types used were isocyanate and urea formaldehyde adhesives. The glue spreads used were 150 g/m^2 and 300 g/m^2. The research results showed that the moisture content of PSL ranged from 9.30% to11.80%. The PSL density ranged from 0.64 to 0.78 g/cm^3. The volume shrinkage ranged from 5.69 to 7.17%. Modulus of Elasticity (MOE) parallel to the grain and edge side ranged from 51.6 x 10^3 to 98.3 x 10^3 kg/cm^2, and 62.1 x 10^3 to 99.9 x 10^3 kg/cm^2, respectively. The Modulus of Rupture (MOR) parallel to the grain and edge side ranged from 269 to 724 kg/cm^2 and 342 to 728 kg/cm^2, respectively. The PSL hardness perpendicular to the grain, parallel to the grain and the edge side ranged from 135 to 300 kg/cm^2, 87 to 321 kg/cm^2, and 128 to 251 kg/cm^2, respectively. The compressive strength ranged from 181 to 231 kg/cm^2. The best adhesive and glue spreads of PSL was isocyanate with glue spread 300 g/m^2. PSL made from hot pre-pressed long strand of oil palm trunk waste bonded by isocyanate fulfill JAS 1152: 2007. However, those of bonded by urea formaldehyde failed to fulfill the standard. The physical and mechanical properties of PSL made from oil palm trunk were better compared to those of solid oil palm trunk.

Keywords: glue spreads, isocyanate, oil palm trunk, PSL, urea formaldehyde

1. Introduction

Oil palm has a productive lifespan is up to 30 years. Several researches proved that after 30 years the productivity of oil palm classified as unproductive. The unproductive oil palm will be felled and most of the oil palm biomass including trunk are left to rot on plantation areas and considered to be oil palm agricultural waste [1]. Up till now, oil palm trunk waste has not been utilized optimally in Indonesia. On the other hand, utilization of wood as biocomposite products raw material increased but not supported by sufficient wood supply from our forests. Utilization of oil palm trunk waste as raw material of composite products such as Parallel Strand Lumber (PSL) is one of an effective and smart solution. In addition, PSL made from oil palm trunk waste is classified as a green product. As a substitute for sawn timber, PSL has the advantage not only of high utilization of wood/raw material but also in strength, size availability, and dimensional stability [2]. PSL can be used for structural applications such as poles, beams, and columns [3].

Previous research stated that the portion of oil palm trunk that is suitable for use as a lightweight construction material is 1/3 of the outer part while the remaining 2/3 is not feasible because it is too soft [4, 5]. In addition, oil palm trunk has low properties such as specific gravity, low durability, poor dimensional stability, low strength and machining properties [6]. The weaknesses of oil palm trunk in physical and mechanical (basic) properties led the limitation of waste oil palm trunk utilization for furniture and buildings applications. Our previous researches shown that the basic properties of oil palm wood improved significantly after hot pressed. Wood compaction is one of the efforts to increase the strength and durability of low-density timber by pressing wooden boards to become more solid [7].

Adhesive used in the PSL production is one of the important components which influence significantly the properties of PSL [1]. The adhesive types and glue spreads used should be adjusted to the PSL manufacture requirements. The selection of adhesives in PSL manufacturing depends on end use, technology, dimensions, treatment, and design requirements [8]. For exterior applications preferable use weather and water-resistant adhesives such as isocyanates, whereas for interior applications can use urea formaldehyde adhesives [9]. Isocyanate adhesive is one of the best adhesives suitable for the manufacture of PSL, especially from oil palm trunk material [1]. According to research before the isocyanate adhesive has the advantage of being more tolerant of high moisture content, more stable in terms of board dimensions stability and the absence of formaldehyde emissions [10]. While the urea formaldehyde adhesive has the advantage of water soluble, good adhesion, and the price is cheap [11]. This allows the urea formaldehyde adhesive to be widely used in the composite products production.

Based on the problems described above, the objective of this research was to analyze the effect of isocyanate and urea formaldehyde adhesives and the influence of glue spreads on the physical and mechanical properties of Parallel Strand Lumber made from oil palm trunk waste.

2. Materials and Methods

2.1 Materials

The oil palm trunk used in this research were harvested from a plantation in Bunar area, Jasinga, Bogor, West Java, Indonesia. All together, fifteen (15) oil palm trunks of 25 to 30 years old (unproductive) were harvested. The adhesives used were isocyanate and urea formaldehyde adhesives. The equipment used was chain saw, band saw, circular saw, planner, moisture meter, hot press, tape measure, plastic, scales, oven, caliper, Universal Testing Machine (UTM), and camera.

2.2 Methods

2.2.1 Production of Parallel Strand Lumber

1. Harvesting
   Oil palm trunk that have reached 25 to 30 years old were harvested using chain saw.
2. Strand making
   Then the palm oil trunks were cut with the size of 120 cm long, 6 cm wide and 1 cm thick.
3. Drying and compacting
   The produced strand then dried and compacted using hot pressing technique for 60 minutes with a temperature of 150°C and pressure of 9.45 kg/cm^2. Drying and compaction process done until strand reach 0.6 cm thickness.
4. Glue spreading
   The strand that has been dried and then spreaded with isocyanate and urea formaldehyde adhesives. The glue spreads used were 150 g/m^2 and 300 g/m^2.
5. Mat forming
   The strand that has been spreaded by adhesive then arranged into sheets manually to be 4 cm in thickness, 6 cm in width, and 120 cm in length.
6. Hot pressing
   The obtained PSL mat then hot pressed according to the condition of each adhesive type. PSL bonded by isocyanate and urea formaldehyde adhesives were hot pressed in temperature of 150°C and 130° C, respectively. The pressure of hot press was 9.45kg/cm^2 for 40 minutes.

7. Conditioning

   After hot pressing, the produced PSL to be conditioning for at least 2 weeks to remove the residual stresses during the PSL production.

8. Trimming

   PSL that has been conditioned and then trimmed according to JAS 1152 : 2007 standard.

2.2.2 Testing Physical and Mechanical properties of PSL

The PSL was tested according to JAS 1152: 2007 standard. The properties tested include physical properties (moisture content, density, and volume shrinkage) as well as mechanical properties (MOE, MOR, hardness, and compressive strength).

1. Moisture content

   The sample test dimensions were 5 cm in width, 5 cm in length, and 4 cm in thickness. The water content was calculated from the initial weight (A) and the final weight (B) after oven dried for 24 hour at 103 ± 2° C to constant weight. The value of moisture content calculated using the formula:

$$MC = \frac{A-B}{B} \times 100\% \quad (1)$$

2. Density

   The sample test dimensions were 5 cm in width, 5 cm in length, and 4 cm in thickness. The density was calculated by weighting the sample (m), and calculated the sample volume (v) by multiplying the dimensions of length, width and thickness. Density calculated using the formula:

$$\rho = \frac{m}{v} \quad (2)$$

3. Volume Shrinkage

   The sample test dimensions were 5 cm in width, 5 cm in length, and 4 cm thickness. The samples were measured as initial volume at dry air condition (v1), and then the samples were oven dried for 24 hours with temperature of 103 ± 2° C until the constant weight and then measured its dimension again by using caliper (v2). Volume shrinkage calculated using the formula:

$$\text{Volume shrinkage} = \frac{(v1-v2)}{v1} \times 100\% \quad (3)$$

4. MOE (Modulus of Elasticity) and MOR (Modulus of Rupture)

   The samples size were 61 cm in length, 5 cm in width and 4 cm in thickness. The samples were tested with centralized load at the center of the panel, with a span length of 56 cm. Testing has been done in parallel to the grain and side edge. MOE and MOR calculated using the formula:

$$MOE = \frac{\Delta PL^3}{4\Delta Ybh^3} \qquad MOR = \frac{3Pmax\,L}{2bh^2} \quad (4)$$

Where :

MOE = Modulus of Elasticity (kg/cm^2)
MOR = Modulus of Rupture (kg/cm^2)
ΔP = Load under Proportion Limit (kg)
ΔY = Deflection (cm)
Pmax = Maximum load (kg)
L = Span length (cm)
B = Sample width (cm)
h = Sample thickness (cm)

5. Hardness

The sample test dimensions were 4 cm in thickness, 5 cm in width, and 10 cm in length. Testing has done by load half a steel ball into the wood. Hardness calculated using the formula:

$$H = \frac{P\ max}{Total\ area} \quad (5)$$

6. Compression strength Parallel to the grain

The sample test dimensions were 4 cm in thickness, 5 cm in width, and 20 cm in length was loaded in the parallel direction of the grain at the position of the vertical and load slowly until the test sample damaged. The load is the maximum allowable strength by the test sample (Pmax). The value of compression strength parallel to the grain calculated using the formula:

$$CS = \frac{P\ max}{Total\ area} \quad (6)$$

2.2.3 Data analysis

The experimental design used in this research was a factorial completely randomized design with two factors. Factor A was adhesive types with two levels (isocyanate and urea formaldehyde) and factor B that was glue spreads with two levels (150 g/m^2 and 300 g/m^2). All the physical and mechanical properties were conducted in 5 replications and processed by statistical programs.

3. Results and Discussions

3.1 Density

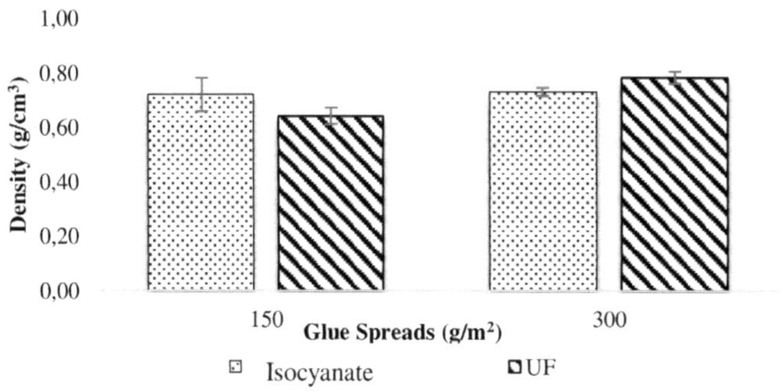

Figure 1. Histogram of PSL density.

Density is defined as the mass unity of volume in kilograms per cubic meter [12]. Figure 1 shows that the density values ranged from 0.64 to 0.78 g/cm^3. Variations of density due to the difference in mass used and PSL volume variations. Strand which contains many vascular bundles who have thick cell wall and perfectly compacted can increase the density and produce a higher mass. PSL density bonded by isocyanate adhesive was higher (0.73 g/cm^3) compared to those of bonded by UF adhesive (0.71 g/cm^3). This is due to the chemical bonds that occur in the isocyanate adhesive. The isocyanate is capable of producing products that have better physical and mechanical properties, because isocyanate form the chemical bonds [13]. Statistical analysis with 95% confidence interval showed that the adhesive types and the glue spreads didn't have significant effect on density. However, the interaction of the both factors produced significant effect. Solid oil palm trunk density ranged from 0.23 to 0.32 g/cm^3 [1], while previous research stated that density values ranged from 0.31 to 0.52 g/cm^3 [14]. PSL density in this research was higher compared than the solid oil palm trunk and hot pressing treatment increased remarkably the density of PSL.

3.2 Moisture content

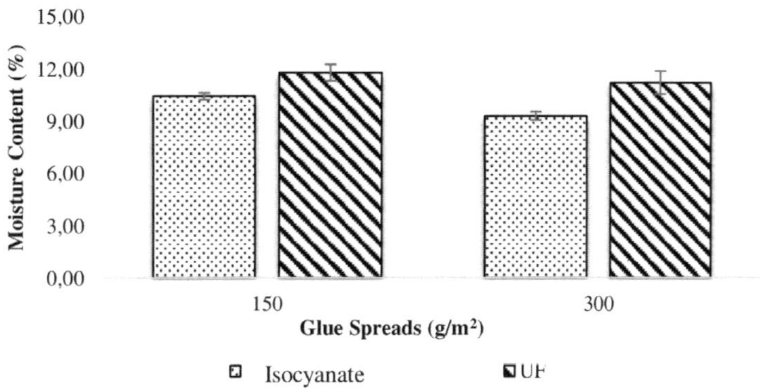

Figure 2. Histogram of PSL moisture content.

Moisture content is defined as the percentage of water contained in wood. Wood strength will increase with the decrease in the water content of the wood below the saturation point of fiber [12]. Figure 2 shows that the moisture content of PSL ranged from 9.30% to 11.80%. Statistical analysis with 95% confidence interval, adhesive types and glue spreads influence significantly the PSL moisture content. However, the interaction of both factors did not significant. Utilization of higher glue spreads decreased the PSL moisture content. This was because the adhesive was able to fill the empty space in the strand so that the water was difficult to get in or get out from the PSL. According to [15], utilization of higher adhesives decreased the value of water content due to many adhesives will close space on the strand perfectly and not easily hydrolyzed. The moisture content of PSL bonded by UF adhesive was higher compared to those of bonded by isocyanate adhesive. This was due to isocyanate adhesive more resistant to water compared to those of UF adhesive which was not resistant to water because the water were able to hydrolyzed the UF adhesive and damaged the bond between the strand and UF adhesive. The average value of the moisture content was 10.69%, this value was lower than the previous study of 16.76% [14]. According to research before the value of moisture content of solid oil palm trunk was about 311.0% [16], this shows that the moisture content of the PSL was much better. The moisture content of PSL fulfilled JAS 1152: 2007 standard, JAS requires maximum moisture content by 15% [17]. PSL produced with hot pressing pre treatment succesfully reduced the moisture content of oil palm as well as prevented fungal attack.

3.3 Volume shrinkage

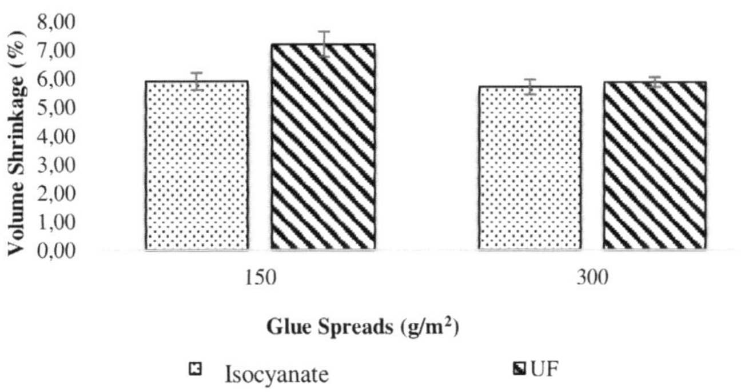

Figure 3. Histogram of PSL volume shrinkage.

Figure 3 shows that the volume shrinkage values ranged from 5.69 to 7.17% with an average of 6.15%. The previous research [16] stated that the volume shrinkage of solid oil palm trunk ranged from 17.5 to 39.1%, so the volume shrinkage value of this research was better compared to those of solid oil palm trunk. The volume shrinkage of PSL bonded by isocyanate adhesive smaller than those of bonded by UF adhesive. The previous research states that the isocyanate adhesive has the advantage of being more tolerant of high moisture content, more stable stability of the board dimensions and the absence of formaldehyde emissions [10]. However, the volume shrinkage did not differ significantly between PSL bonded by isocyanate and those of bonded by UF in this research.

3.4 Modulus of Elasticity

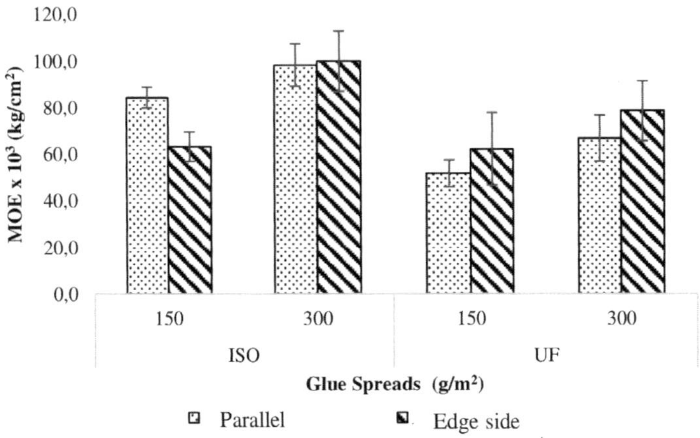

Figure 4. Histogram of PSL MOE parallel to the grain and edge side.

Figure 4 shows that the MOE values parallel to the grain ranged from 51.6×10^3 to 98.3×10^3 kg/cm^2, while the MOE edge side values ranged from 62.1×10^3 to 99.9×10^3 kg/cm^2. The average MOE values parallel to the grain and the edge side were 75.3×10^3 kg/cm^2 and 75.9×10^3 kg/cm^2, respectively. The variations of MOE values caused by uneven distribution of density in the parallel to the grain. The MOE values of the edge side were higher compared to those of the parallel to the grain. This was due to the effect of thicker PSL thickness when it was tested on edge side compared to those of parallel side. [1] reported that the MOE value of laminated wood of oil palm trunk was determined by the density and MOE of the constituent laminae. The PSL from oil palm trunk was higher compared to those of solid oil palm trunk. Statistical analysis with 95% confidence interval showed that the adhesive types produced a significant effect on parallel to the grain MOE values and glue spreads had a very significant effect on the edge side MOE values. PSL bonded by isocyanate adhesive at 300 g/m^2 glue spreads produced highest MOE. The isocyanate was capable of producing products that have relatively better physical and mechanical properties, because isocyanate form chemical bonds [13]. Isocyanate form a chemical bond with a strong polarity, not only having good adhesion potential but also the potential to form covalent bonds with substrates having reactive hydrogen groups. Chemically the isocyanate reacts with water contained in a strand to form a mechanically bonded polyurea glue [18]. Referring to the JAS 1152: 2007 standard that requires a minimum MOE value of 75×10^3 kg/cm^2, PSL bonded by isocyanate adhesive at 300 g/m^2 glue spread fulfilled the JAS standard [17].

3.5 Modulus of Rupture

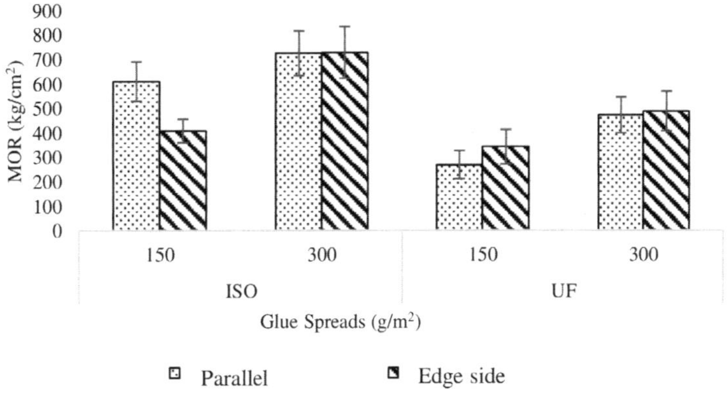

Figure 5. Histogram of PSL MOR parallel to the grain and edge side.

Figure 5 shows that MOR parallel to the grain and edge side ranged from 269 to 724 kg/cm^2 and 342 to 728 kg/cm^2, respectively. Variations in MOR values can be affected by the density and the number of adhesives. The greater number of adhesives used will increase the laminate's hardness properties as the adhesive plays an important role in strengthening the resulting laminated wood [19]. The strength of wood can also be affected by the pattern of lamina preparation. Preparation of palm wood lamina which put the stronger part in the outer position improved the mechanical properties significantly [1]. According to previous research that in the anatomical structure of oil palm trunk, the center portion of the trunk was dominated by the parenchymal base network while in the middle and the edge of the trunk were composed by thick-walled vascular bundles [20]. The average MOR values of parallel to the grain and the edge side were 519 kg/cm^2 and 492 kg/cm^2. The MOR values of PSL bonded by isocyanate adhesive at glue spreads of 150 g/m^2 and 300 g/m^2 fulfilled the JAS 1152: 2007 standard which requires a minimum MOR value is 300 kg/cm^2 [17]. Statistical analysis with 95% confidence interval showed that the adhesive types and the glue spreads had a very significant effect on the parallel to the grain MOR values as well as the edge side. The PSL MOR values was higher compared to those of solid oil palm trunk which has MOR ranged from 97 to 151 kg/cm^2 [1].

3.6 Hardness

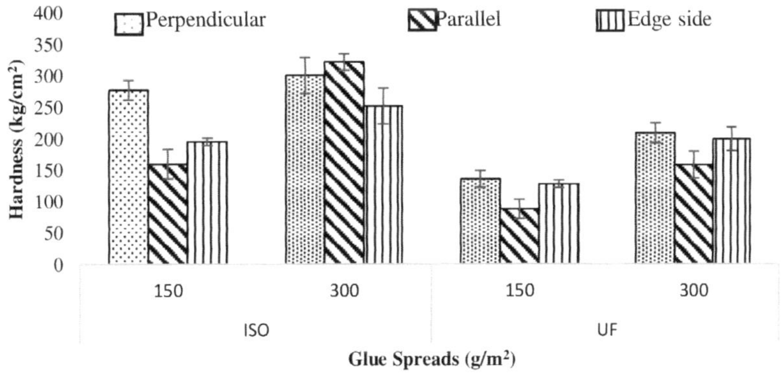

Figure 6. Histogram of PSL hardness perpendicular, parallel and edge side.

Hardness is defined as the ability of boards to withstand scraping and destruction on its surface [21]. Figure 6 shows that the PSL hardness perpendicular to the grain, parallel to the grain and the edge side ranged from 135 to 300 kg/cm^2, 87 to 321 kg/cm^2, and 128 to 251 kg/cm^2, respectively. The mean values of perpendicular, parallel and edge side are 230 kg/cm^2, 182 kg/cm^2 and 194 kg/cm^2, respectively. The density of the strand affected the PSL hardness. According to [21] that the hardness is affected by the density, the size of the fiber, the interconnection between fibers and the arrangement of its wood fibers. From the three directions of testing, the highest value to the smallest PSL was the direction of perpendicular, edge side and parallel to the grain, respectively. This was due to the density in the strand on the core was lower compared to those of the outer strand of the PSL. The coating system affects the value of wood strength, such as the preparation of lamina wood palm placing the stronger part in the outer position [1]. The values of oil palm trunk hardness ranged from 64.3 to 137.3 kg/cm^2 [16]. PSL hardness was higher compared to those of solid oil palm trunk. Statistical analysis with 95% confidence interval showed that the adhesive types and the glue spreads have a very significant effect on hardness values.

3.7 Compression strength

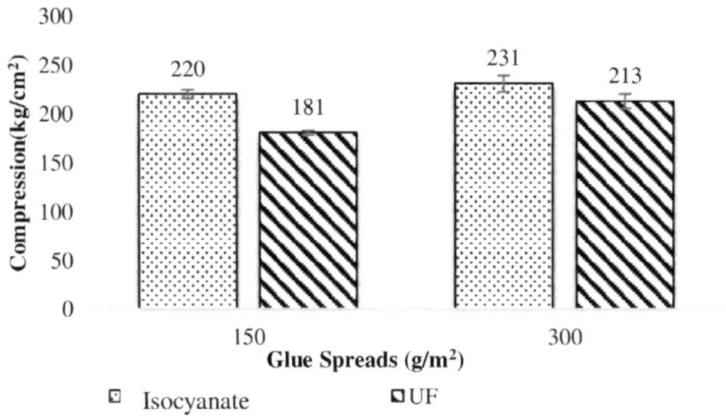

Figure 7. Histogram of PSL compression strength.

Figure 7 shows that the compression strength ranged from 181 to 231 kg/cm^2 with an average of 211 kg/cm^2. The compressive strength of the fiber is required to determine the load that a pole can carry. Parallel compressive strength of the fiber is the ability of the board to withstand loads or pressures that try to reduce their size until the board is damaged [22]. The highest compressive strength resulted from PSL bonded by isocyanate adhesive at 300 g/m^2 glue spread. PSL which has more dense vascular bundles and higher fiber cells will has higher compressive strength. Statistical analysis with 95% confidence interval showed that the adhesive types, glue spreads, and interaction of the research factors influence compressive strength significantly. Previous research reported that the compressive strength solid oil palm trunk ranged from 38.8 to 141.8 kg/cm^2 [16]. The PSL compressive strength was better than those of solid oil palm trunk. This was proved that hot pressing pretreatment to the long strand increased the value of compression strength.

4. Conclusions

The adhesive types and glue spreads has no significant effect on the PSL physical properties except the values of moisture content. In general, the adhesive types and the glue spreads influenced significantly the PSL mechanical properties. The best adhesive types and glue spreads for PSL was isocyanate adhesive at a glue spread of 300 g/m^2. PSL made from hot pre-pressed long strand oil palm trunk waste bonded by isocyanate adhesive fulfilled JAS 1152: 2007, while those of bonded by UF adhesive failed to fulfill the standard. PSL quality was better than those of solid oil palm trunk. Producing PSL using hot pressing pretreatment was successfully improve the mechanical properties of the produced PSL.

Acknowledgment
The authors gratefully acknowledge financial support from the Directorate of Higher Education, Ministry of Research, Technology and Higher Education of the Republic of Indonesia. The authors also thank for Bogor Agricultural University (IPB) for supporting research facilities.

References

[1] Darwis A, Massijaya MY, Nugroho N and Alamsyah EM 2014 *J. Ilmu Teknol. Kayu Lapis.* **12** (2): 157–168
[2] Nelson S 1997 *Structural composite lumber.* Engineered Wood Products: A Guide for Specifiers, Designers, and Users, S. Smulski (ed.), (Madison, Wisconsin (AS): PFS Research Foundation)
[3] Canadian Wood Council 1997 *Wood Reference Handbook* (Ottawa (CA): Canadian Wood Council)
[4] Bakar ES, Rachman O, Hermawan D, Karlinasari L and Rosdiana N 1998 *J. Teknologi Hasil Hutan* **11** (1): 1-12
[5] Bakar ES, Massijaya MY, Rahman O and Bahruni 2000 *Laporan Penelitian Hibah Bersaing VI Dirjen Dikti.* Bogor (ID): Jurusan Teknologi Hasil Hutan. Fakultas Kehutanan IPB
[6] Dungani R, Islam MN, Abdul Khalil HPS, Davoudpour Y, and A Rumidatul 2014 *Bioresources.* **9** (1): 455–471
[7] Sulistyono, Nugroho N and Surjokusumo S 2003 *Buletin Keteknikan Pertanian* **17**(1):32-45
[8] Kurt R 2010 *BioResources* **5**(3), 1868-1878
[9] Kurt R and Çavuş V 2011 *Wood Res* **56**(1), 137-144
[10] Marra AA 1992 *Technology of Wood Bonding : Principles in Practice.* (New York (US): Van Nostrand Reinhold)
[11] Dungani R, Islam MN, Abdul Khalil HPS, Davoudpour Y, and A Rumidatul 2014 *Bioresources.* **9** (1): 455–471
[12] Iswanto AH, Sucipto T, Azhar I, Coto Z and Febrianto F 2010 *J. Teknologi Hasil Hutan* **3**(1):1-7 (Medan (ID): Universitas Sumatera Utara)
[13] Sucipto T, AH Iswanto and I Azhar 2010 *J. Ilmu dan Teknologi Hasil Hutan* Vol **3**(2) (Medan (ID): Universitas Sumatera Utara)
[14] Balfas J 2008 *Teknologi Pengolahan Batang Kelapa Sawit menjadi Produk Kayu Utuh (solid wood dan kayu lapis (plywood).* (Bogor (ID): Badan Penelitian dan Pengembangan Kehutanan)
[15] Mawardi I 2009 *Jurnal Teknik Mesin* Vol **11**(2). (Banda Aceh (ID): Politeknik Negeri Lokseumawe)
[16] Iswanto AH, Sucipto T, Azhar I, Coto Z and Febrianto F 2010 *J. Teknologi Hasil Hutan* **3**(1):1-7 (Medan (ID): Universitas Sumatera Utara)
[17] [JAS] Japanese Agricultural Standard 2007 *Japanese Agricultural Standard for Glued Laminated Timber Notification No 1152* (Tokyo (JP): JPIC)
[18] Nuryawan A, Massijaya MY and Hadi YS 2008 *J. Ilmu dan Teknologi Hasil Hutan* **1**(2): 60-66
[19] Persson M and Wogelberg S 2011 *Analytical models of pre-stressed and reinforced glulam beams: A competitive analysis of strengthened glulam beams* [Tesis] (Göteborg (SE): Chalmers University of Technology)
[20] Bakar ES 2003 *Forum Komunikasi Teknologi dan Industri Kayu* Vol **2**. (Bogor (ID): Jurusan Teknologi Hasil Hutan Fakultas Kehutanan IPB)
[21] Mardikanto TR, Karlinasari L and Bahtiar ET 2011 *Sifat Mekanis Kayu* (Bogor (ID): IPB Pr)
[22] Tsoumis G 1991 *Science and Technology of Wood. Structure, Properties, Utilization* (New York (US): van Nostrand Reinhold)

Comparison of isolate dadih with yeast dadih in improving nutrition quality of Cassava Waste (CW)

N Ginting

Faculty of Agriculture, University of Sumatera Utara, Indonesia Kampus USU Medan 20155

Email : nurzainahginting@gmail.com

Abstract. The cassava industry in North Sumatra Province was one of the most significant agricultural industries. Waste from the cassava industry which was called cassava waste/CW/Onggok was used as feed for ruminants such as cattle, sheep and monogastric such as pigs. The low nutrients in CW caused the need to find a way for improving the nutrients quality. This research was conducted with the aim to help livestockers to ferment their livestock feed. This study compared the ability of fermentation between dadih isolate with dadih yeast. Dadih is traditional food in Indonesia where milk is fermented in bamboo tube. Dadih yeast was made by mixing dadih and whey with flour, made in around shape and sun dried. The results showed that pH of CW by dadih isolate was the lowest while crude protein, crude fiber and fat in CW treated with dadih isolate were improved significantly compared either to control or to dadih starter while fermented CW was better than non-fermented CW. It was recommended livestockers to ferment CW by using either by dadih isolate or dadih starter.

Keywords: dadih isolate, dadih starter, CW, nutritions quality

1. Introduction

In 2015, the province of North Sumatera was rated as the second biggest cassava producer in Sumatera together with Lampung [1]. Cassava production was 1,619,405 ton which mostly processed into tapioca starch by many big tapioca factories in North Sumatera Province and the raw materials are supplied by perkebunan ubi masyarakat dari kabupaten yang berdekatan.

There were several regencies that significantly produced cassava, namely Serdang Bedagai, Simalungun, Deli Serdang,Tapanuli Utara and Tapanuli Tengah. Cassava production were 447,990 tons, 380,701 tons, 178,790 tons, 64,853 tons, 33,648 tons respectively in 2014 [2]. There were orther regencies that also produced cassava however as they produced only small number, usually tapioca produced locally in small factories.

From the cassava industry were produced abundant cassava waste (CW) because for every ton of cassava were produced 114 kg waste. In North Sumatera, CW was used for animal feed, such as beef cattle, sheep, pigs and poultry. Livestock industry in North Sumatera generally utilizes CW in dry conditions that were ordered directly from tapioca factories. For small-scale livestockers generally they buy CW in wet conditions. On the edge of the road in North Sumatera, it could be found pile of CW sell. Small livestockers who used CW for feed often complained of slow livestock production. This was due to the low nutrient content of the CW which was estimated to be only able to meet the basic needs of livestock life, so for growth, pregnancy and lactation required additional feed to meet protein and energy needs of livestock [3].

In North Sumatera, livestockers community have been familiar with the application of fermentation feeding. Fermentation was a very old way/thousand years ago which supported higher final concentration of products with environmental advantages [4]. Until now the simple fermentation technique in North Sumatera that was socialized to livestockers was to utilize the yeast of traditional food which called tape which contained *Saccharomyces cerevisiae* on CW. But some breeders have the assumption that fermentation with yeast tape could caused reproductive disorders. CW could be hosted by microbial diversity and the true microbial would support effective utilization of CW [5]. Therefore, it is necessary to find more kind microorganisms especially those which could applied by simple fermentation and especially those microorganism that could easily to have.

This study aims to investigate alternative microorganism for CW fermentation. The result would help the livestock industry,i.e (1). producer which were CW factories (2). user which were livestockers. Microorganims were isolated from dadih which is a traditional food in Indonesia. Livestock milk either buffalo or goat was fermented in bamboo tube. After 2 or 3 days dadih was emerged which contained of 36 strain of lactic bacteria [6] and was dominated by *Lactobacillus plantarum* [7] while [5] found *Bacillus licheniformis* and *Bacillus substilis* were the dominant bacterial species in cassava peel waste, *Lactobacillus fermentum* and *Lactobacillus plantarum* were the dominant bacterial species in liquid CW. *L plantarum* was also found in liquid CW [8]. It was assumed that by using BAL from dadih, either CW producer or CW user will be greatly get benefit because bamboo could be found in the whole area of Indonesia up to a very remote area while milk also was everywhere either fresh milk especially commercial milk/dry milk. The output of this study was also that livestockers in rural area could do fermentation without depend on commercial fermentor.

2. Materials and Methods

This research started with making dadih which was a process of fermentation by entering goat milk into bamboo tube. The development of microorganisms spontaneously was occured that was dominated by lactic acid bacteria (BAL) although there were also mold and yeast. The fermentation was occured within 2 days. Furthermore, in this research, the isolation of microorganisms from fermented whey was conducted. Next, the preparation of yeast dadih which was made by mixing sticky rice flour, dadih and whey then left alone for 1 night. The next day the dough was formed into a sphere weighing 10 grams and dried in the sun to dry.

The identification of microorganisms only focused on lactic acid bacteria through the slide culture method on microorganisms growing on PDA medium which only selected 3 directly solitary. The next three selected BALs were developed and processed into bioactivators which be used to ferment CW.

Fermentation was done by steaming CW. CW that was cold then put in tupperwares each of 5 replication. The research method was used Completely Randomized Design (RAL). Fermentation was carried out for 10 days. Treatments were as follows: T1 (without bioactivator), T2 (bioactivator BAL), T3 (curd yeast bioactivator). Parameters of research were pH, crude protein, crude fiber and fat.

3. Results and Discussions

In this study fermented CW was dominated by lactic acid bacteria. This was consistent with Sunarlim et.al [7] that on dadih was found BAL which was dominated by *Lactobacillus plantarum*. In the fermentation of materials that were inoculated by microbes would undergo changes in their chemical composition. Fermentation of carbohydrate-containing substances would produce alcohol. In this study also found the smell of alcohol. By Sunarlim [7] it was stated that any particular fermentation process was caused by certain microbial activity, which was specific to the fermentation process. An example lactic acid fermentation by *Lactobacillus* bacteria. The working bacteria had their own features that were related to the benefits it provides to the environment. In this study observations were on pH, crude protein, fiber and fat.

IOP Conf. Series: Earth and Environmental Science **141** (2018) 012008 doi:10.1088/1755-1315/141/1/012008

Table 1. pH and nutrient content of CW and fermented CW by isolat dadih and yeast dadih

	pH	Crude Protein	Crude Fiber	Fat
CW	4.8	2,28 C	20,23 B	0,68 C
CW + Isolate	3.4	7,48 B	17,86 A	2,64 A
CW + Yeast	4.1	4,94 A	17,59 A	1,26 B

3.1. pH

From Table 1 it was known that there was a decreased in pH, initial pH 4.8 to 3.4 and 4.1 in line with increasing fermentation time. Araujo [9] fermented with BAL from Youghurt found a decrease in pH between 3.84 and 3.98. According to Sunarlim [7] dadih were dominated by BAL especially *Lactobacillus plantarum. Lactobacillus* was a class of lactic acid-producing bacteria, including gram-positive bacteria, anaerobic faculty that had two mechanisms to obtain energy. When there was oxygen, energy was obtained by aerobic respiration, when no energy oxygen was obtained by anaerobic fermentation. Lactobacillus in animal feed materials was an indication of a healthy environment because these bacteria were normal microflora in the environment and in the digestive tract of living things. The resulting lactic acid will inhibit various pathogenic microorganisms. In the feed industry *lactobacillus* bacteria was widely used as a probiotic directly applied as a mixture of animal feed. For the feed industry, the decreasing pH indicates that the feed material was in good condition because the acidic pH causes the decay bacteria to be inhibited. Some other benefits of *Lactobacillus plantarum* were inhibiting pathogenic bacteria, improving immunity, can produce vitamin B and bacteriocin, antimutagenic, anticarcinogenic [6].

For ruminants, feed that has an acid pH is not a problem because when the cattle chew the feed, saliva of cattle that has an alkaline pH causes the process of neutralizing the feed ingredients. Observations made by Ginting on fermentation CW application with BAL-containing bioactivators on Barbados Blackbelly sheep, fermented CW for 2 months remained palatable.

3.2. Protein

On livestock feed, conventional sources of protein were very expensive, for example fish meal and also inadequate.This was why, efforts to find nonconventional sources were of much concern, for example through the state of fermentation. In this research, protein was increased significantly from 2.28 (control) into 7.48 (isolate dadih) and 4.94 (yeast dadih). It was assumed that isolate dadih contains lots more microorganisms than in yeast dadih thus was followed by enzymaztic-pools. This was in concordance with Soccol[4] which mentioned that fermentation was how to produce enzymatic-pools which mainly proteases, amylases and lipases to hydrolized sub-strate macromolecules like starch, proteins, fatty acids. Aruna et.al [10] found an increased on protein in yam after 96 h, using Saccharomyces cerevisiae (BY4743). Samples from fermented biomass were collected at 24 h interval for 96 h, oven dried at 60 C, cooled, milled and assayed for chemical and amino acid composition.The crude protein, true protein, fat and ash contents increased significantly from 6.60%, 4.38%, 1.12%and 4.45% to 15.54%, 13.37%, 2.09% and 8.02% respectively when fermentation reached 96 h in the presence of ammonium sulphate. Any differencies in nutrient quality after fermentation were depended on most suitable microrganisms with substrated, nutrient content of subtrates and treatment addition [11]. However, any kind of microorganisms had their own specific good, for example *L. plantarum* which could be acted as probiotic. Also, the most important thing was the technology could be easily be implemented in field, easily to be found and cheap.

3.3. Fiber

Crude fiber from CW control was about 20.23% and there was a decreased about 3% due to fermentation. The initial crude fiber depends on the milling process taking place at tapioca factory. In North Sumatera, generally the new generation machines cause crude fiber smaller in CW. Fiber in this study became better about 3%. This was probably because BAL did not produce cellulase enzymes to degrade fibers. Actually the fiber in this study was about 17% which good enough for ruminants [3]. That was why feedlot industry of beef cattle found in North Sumatera generally use CW as animal feed material. The need for feedlot CW in North Sumatra was very large. In addition to feedlot, pig farmers for the Simalungun and Toba areas also use CW. For example in North Sumatra there were 6 feedlot/cow feed industries, among others, Lembu Andalan Langkat Enterprises that maintained 3500 cows. Daily dry daily needs of about 10,500 kg because livestock feed everyday were 3,500 x 10 kg = 35,000 kg and CW was 30% or about 300 ton dry CW every month. In North Sumatera, based on BPS [2], there were 184,612 ton of CW and which were predicted not enough to supply feedlot industries. That was why, part of CW supply came from out of North Sumatra which was Lampung. Only small livestockers in North Sumatra were used mostly local CW.

3.4. Fat

The fat in this study differed significantly between controls by treatment of isolate dadih and with whey yeast dadih. This was occured because BAL microorganisms that work to remodel starch in CW result in increased fat from the activation of the lipase enzyme [4]. According to Ghaffar et al [12] BAL made eco-friendly processing and BAL also had a fermentable capability on many of the agricultural and agro-industrial by-products. BAL became attractive candidates in fermentation biotechnology to produce a value added product with multiple applications.

3.5. Utilization of BAL Dadih, strategic effort to improve the quality of animal feed ingredients

As mentioned before, the result of this research will help livestocker industry. In Figure 1 it can be seen the process in tapioca industry. CW was generated after the pressing process. The water content of the pressing depends on the machine type in the cassava factory. However for fermenting process the water content of CW should be about 60%. In the interest of raising the quality of CW then some strategies can be made by utilizing the results of this study. For feedlot industries that require dry CW, CW fermentation became difficult due to space problems. CW should be fermented at least 4 days [4] while the volume of CW were so enormous that the factory must invest to build a fermentation room. Therefore the most feasible CW was implemented for non dried CW. In Indonesia, the factory directly packs the CW into a plastic sack.

In cassava factory, when CW coming out of the exhaust pressing pipe, the application of microorganisms can be done by spraying bioactivator on CW. Automatically microorganisms will develop inside the plastic sack because BAL was anaerobic facultative bacteria. Usually about 2 months time to spend stock of CW sell. Ginting discovers from unpublished studies that up to 2 months CW fermentation still in good condition and favored by livestock. CW purchased by small livestockers can also be fermented alone by livestockers by utilizing yeast dadih. With yeast dadih that can be made by livestockers themselves, they can save time, and bamboo plants become sustainable because livestockers do not need to cut bamboo everytime thay need dadih. Livestockers just need to open the CW plastic sack a little for dadih yeast entry, flatten yeast and close the plastic sack for a fermentation process.

The advantages of dadih microorganisms because these microorganisms can be developed very easily through milk and bamboo tube. People even who live in very remote area which usually cannot have biofermentor can buy available commercial milk powder, pasteurized it, process it in bamboo tube and after 2 or 3 days become dadih.

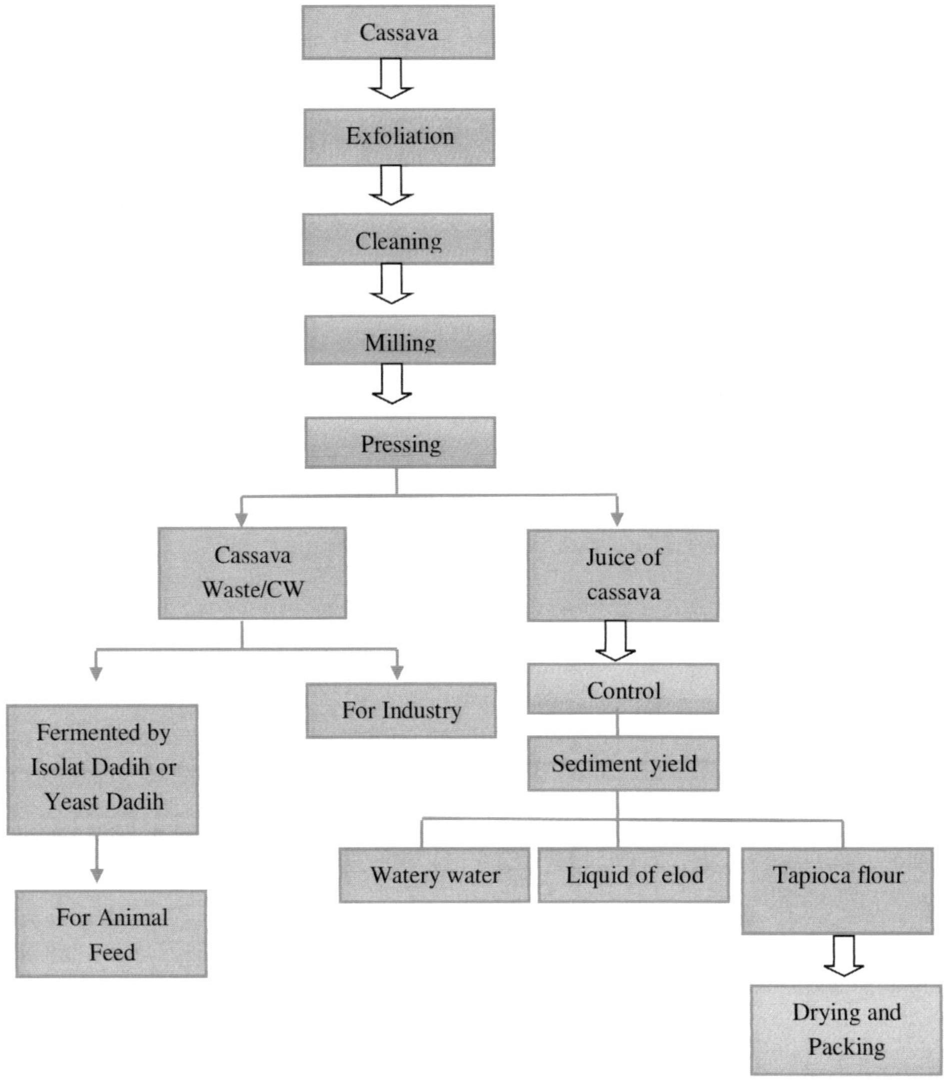

Figure 1. Flow Chart CW and Tapioca Processing

Furthermore, the dadih was processed into dadih yeast by mixing the dadih with glutinous flour (250 g) mixed with cornstarch (50 g), dadih 30 g, 230 cc whey left to stand for 1 night, rounded and dried to dry. Preliminary research conducted by Ginting by using rice flour, cornstarch and glutinous flour, then the best result was to use glutinous starch mixed with cornstarch. If people already have whey yeast, people can keep it in the refrigerator to keep it fresh. Thus the community does not need to sacrifice many bamboo trees every willing to prepare the microorganisms of dadih.

4. Conclusions

Due to fermentation, there was a decreased in pH, initial pH 4.8 to 3.4 and 4.1. Protein was increased significantly from 2.28 (control) into 7.48 (isolate dadih) and 4.94 (yeast dadih). Fat improved from 0.68 (control) to 2.64 (isolate dadih) and 1.26 (yeast dadih). CW could be fermented by isolat dadih or yeast/yeast dadih while dominant microorganism in dadih was *Lactobacillus plantarum.* Conventional sources of protein were very expensive, this was why, efforts to find nonconventional sources were of much concern. By using microorganism from dadih, livestocker could fermented CW even if they live in the very remote area without need to buy commercial bioactivator.

Recommendation

It is necessary to test the fermented CW in the sack by using a dadih starter with an incubation period of 1 month and 2 months with the condition of sack lies at the location of the sale. In vivo research is worth trying for cow, sheep and pig.

References

[1] BPS-Statistic Indonesia. 2016. Production of cassava by Province (ton), 1993-2015.
[2] BPS-Statistic Indonesia. 2015. Production of cassava by Regencies (ton), 1993-2015.
[3] Campbell, J.R., M.D. Kenealy, & K.L. Campbell. 2006. Animal Science. 4th Edition. McGraw-Hill, New York.
[4] Soccol, Carlos Ricardo, Eduardo Scopel Ferreira da Costa,Luiz Alberto Junior Letti, Susan Grace Karp, AdeniseLorenciWoiciechowski, Luciana Porto de Souza Vandenberghe.2017. Recent developments and innovations in solid state fermentation. http://www.journals.elsevier.com/biotechnology-research-and-innovation/
[5] Elijah, A.I.1, Atanda, O.O.2, Popoola, A.R.3 and Uzochukwu, S.V.A. 2014. Molecular Characterization and Potential of Bacterial Species Associated with Cassava Waste. J. Nigerian Food Vol. 32 No. 2, Pages 56-65.
[6] Pato, U. 2003. Potensi bakteri asam laktat yang diisolasi dari dadih untuk menurunkan resiko penyakit kanker. (Potential lactic acid bacteria isolated from the curd to reduce the risk of cancer). J Natur Indonesia 5 (2) 162 -166.
[7] Sunarlim, R., Triyantini, Abubakar, M. Poeloengan, dan H. Setiyanto. 1999. Peningkatan teknologi pembuatan inokulum mikroba pengolahan dadih untuk menunjang agroindustri pedesaan. (Improved technology for making curd microbial inoculum to support rural agroindustry). Laporan Penelitian Balai Penelitian Ternak Bogor TA 1998/1999. (Research Report of Bogor Livestock Research Institute FY 1998/1999).
[8] Tosungnoen, Sirirat, KannikaChookietwattanaa, SomchaiDararatc.2013. Lactic Acid Production from Repeated-Batch and Simultaneous Saccharification and Fermentation of Cassava Starch Wastewater by Amylolytic *Lactobacillus Plantarum*MSUL 702. APCBEE Procedia 8 (2014) 204–209.
[9] Araújo, Diego D. Alessandro B. Amorim, Mayra A.D. Saleh, Felipe Curcelli,Pedro L. Perdig_on, Silvio J. Bicudo, Dirlei A. Berto. 2016. J of Animal Nutrition 2 (2016) 149-153.
[10] Aruna, Tawakalitu E.,Ogugua C. Aworh b, Akeem O. Raji a,⇑, Aderonke I. Olagunju. 2017. Protein enrichment of yam peels by fermentation with Saccharomyces cerevisiae (BY4743). Annals of Agricultural Science 62 (2017) 33-37.
[11] Soccol, C. R., &Vandenberghe, L. P. S. (2003). Overview of solid-state fermentation in Brazil. J. Biochemical Engineering13 (2003), 205-218.
[12] Ghaffar, Thayyba, Muhammad Irshad, Zahid Anwar, Tahir Aqil, ZubiaZulifqar, Asma Tariq, Muhammad Kamran, Nudrat Ehsan, SajidMehmood .2014. Recent trends in lactic acid biotechnology: A briefreview on production to purification.Journal of radiation research and applied Sciences 7 (2014) 222-229.

IOP Conf. Series: Earth and Environmental Science **141** (2018) 012009 doi:10.1088/1755-1315/141/1/012009

Biomass by-product from crystal sugar production: A comparative study between Ngadirejo and Mauritius sugar mill

Gunawan[1], T Bantacut[2], M Romli[2] and E Noor[2]

[1]Post Graduate Program of Agroindustrial Technology, Bogor Agricultural University, Bogor
[2]Department of Agroindustrial Technology, Bogor Agricultural University, Bogor, Indonesia
*gunawan.samin@gmail.com

Abstract. Sugarcane has been used as raw material in crystal sugar industry. Sugar cane that contains high sugar will be utilized into crystals sugar. In addition, the productivity of sugarcane is large enough in 2016 approximately 360 713 tons/year. Crystal sugar itself is a daily necessity for its use in the food and beverage industry. Problem that occurs in Indonesia is the energy consumption. The sugar mills supposed to be an independent energy source which means it can produce its own energy by utilization the material that is available in the sugar mills such as by-product (bagasse, molasses, filter cake, etc.), the by product in every production stage are quite a lot in sugar industry especially in Indonesia. In this paper, a comparison between two sugar mills was examined between Ngedirejo sugar mill and Mauritius sugar mill which has the same geological state as Indonesia. The results of comparison between the two sugar factories demonstrated the difference in terms of productivity of the sugar that has been produced and the effectiveness of the production process in a sugar mill seen from the amount of waste and the by product.

1. Introduction

Sugar cane is one of the large available commodity in Indonesia. The area of sugar cane in Indonesia reached 445.520 ha in 2016 with a productivity of 2 222 971 ton [1]. Such area makes sugarcane as potential raw material in the process of making white crystal sugar.

The production process of white crystal sugar is through several stages such as grinding, refining, evaporation, cooking, and crystallization. Each stage produces output and by-product in the form of biomass. Process of white crystals sugar production was known as close loop system, indicating that there is energy and material rotation in production process system characterized by the use of by-product from the production process for the sake of the sugar mills.

The use of biomass in sugar mills has long been known. Sugar mill produce biomass in the form of bagasse, filter cake and molasses which can be reused as fuel to generate energy for the production process of white crystal sugar. Ngadirejo and Mauritius sugar mill have conducted biogas cogeneration process to utilize by-product of production process into electrical energy that can be used to support the production process of white crystal sugar. Both sugar mills produce considerable amount of biomass from sugar production process. Sugar cane is associated with sugar production, such as in Mauritius, where most of the country's power plant comes from efficient system installed in sugar mills. This paper examines the difference of white crystal sugar production process at Ngadirejo and Mauritius sugar mill concerning how much output and by product produced in sugar cane extraction process at each sugar mill and how much energy can be generated from the biomass conversion process in the form of bagasse produced at the mills.

2. Materials and Methods

The research method is analyzing mass and energy balance of Ngadirejo and Mauritius sugar mill as benchmark. Benchmarking is a very important parameter of performance comparison measurement showing the level of performance of the mills to the industry standard. The research stages conducted through literature study, field observation, interview and data processing.

3. Results and Discussions

3.1 Ngadirejo and Mauritius Sugar Mill

The case study is conducted by comparing the production process of white crystal sugar at the Ngadirejo and Mauritius sugar mill. Ngadirejo sugar mill is one of the sugar factories located in Kediri, East Java, Indonesia. Mauritius sugar mill is located on the African Continent which has similar geological condition to Indonesia. Therefore, the production process of white crystal sugar has the same characteristics due to similarity of geological structures of both regions. Generally, the production process of white crystal sugar is almost the same in every sugar mill, as well as between Ngadirejo and Mauritius sugar mill. The difference is in the raw materials capacity, production and by-product produced. Stages of white crystal sugar production process in Ngadirejo and Mauritius sugar mill is similar that is through several stages including the process of milling, refining, evaporation, cooking and crystallization.

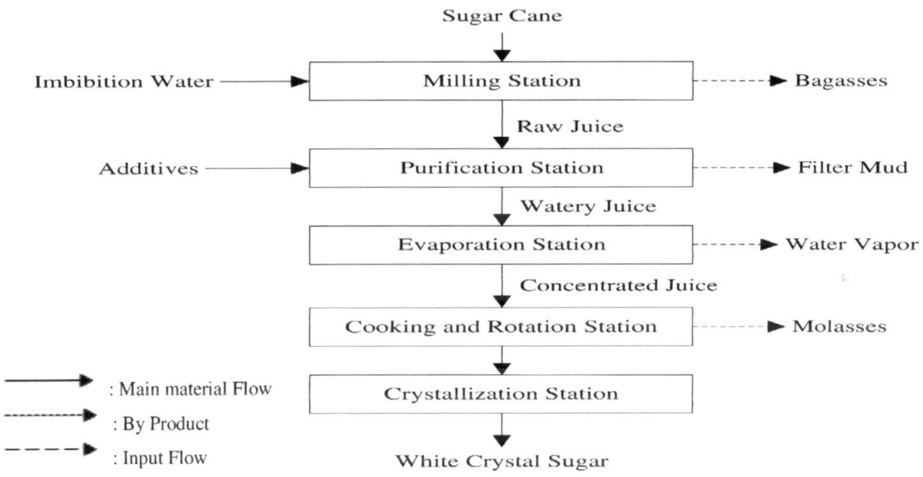

Figure 1. Flowchart of white crystal sugar production

It appears that most of sugar factories produce white crystalline sugars through the same stages (Figure 1). It is assumed that both sugar mills have an input of 1 ton, and after calculating the mass balance (Figure 2) it is known that Ngadirejo sugar mill produces white crystal output of 8.22% and Mauritius sugar mill of 10.96%. Differences in output show that Mauritius sugar mill is more effective than Ngadirejo sugar mill in the process of sugar production. The difference of output at the sugar mills shows that the less white crystal sugar produced the more losses it generates. Ngadirejo sugar mill with total production of 8.22% resulted in losses of the whole process of 55.4% compared to Mauritius sugar mill by 26.3%, a significant difference from the two factories. According to Subyanto [2], decrease in crystal sugar productivity and increased losses in sugar mills occurs due to inefficiency of machine or tool. It is confirmed by Wauters and Mathot [3] which states a decrease in the productivity of sugar crystals can be caused due to many losses produced. Decreased productivity of crystalline sugars can occur due to malfunctioning machines, loss in process, and external factors.

3.2 Baggase Availability in White Crystal Sugar Production

Sugar crystals production is essentially done through several stages namely milling, purification, evaporation, cooking, and crystallization [4]. Extraction is the initial stage of sugar crystals production process that occur at the mills. The main raw materials of sugar cane will be processed first at a milling station which produces a watery syrup (*nira*) output and by-product bagasse. Nira will be processed at the next station while bagasse as by-product will be reused by sugar mills. Bagasse produced from milling stations is potential to be used in some industries such as whiteboard, paper, fuel and especially for combusting which will generate energy. The last role of bagasse is combusting which is usually used by sugar mill to generate additional energy for white crystal sugar production process. Bagasse used as fuel in boiler station as energy source of sugar mills.

Historically, sugar mills have been designed to meet their energy needs by burning bagasse. Bagasse is a potential biomass as a renewable energy source and available substantially in sugar mill. In 100% of sugarcane, the resulting bagasse varies from 23% - 40% [5,6]. This depends on sugarcane fiber, which usually ranges from 12 % to 19%. The average composition of sugar industry in Indonesia is 52.9% of waste water, filter mud 3.5%, bagasse 32.0%, molasses 4.5%, sugar 7% and ash 0,1%. Water content of Bagasse from sugar milling process is 49% -52%, sucrose is 1.5% -3.0%, brix is 4% -7%, not sugar is 2.5% -4 %, and coir by 10% -16%. The value of bagasse as a fuel depends heavily on the gross calorific value (GCV), which is depend on the composition, particularly water content and calorific value of sugar cane which is primarily depend on its sucrose. GCV bagasse ranges from 2275 kcal/kg for bagasse moisture 50% to around 3426 kcal/kg for moisture content 25%. The average ratio of steam to bagasse is usually 2. At a density of 130 kg/m^3, bagasse storage requires a lot of space, therefore need a boilers to burn as much as possible. According to Deepchand [7], Mauritius sugar mill has produced 105 GWh of electricity from 210 tons of bagasse (0.5 GWh/ton).

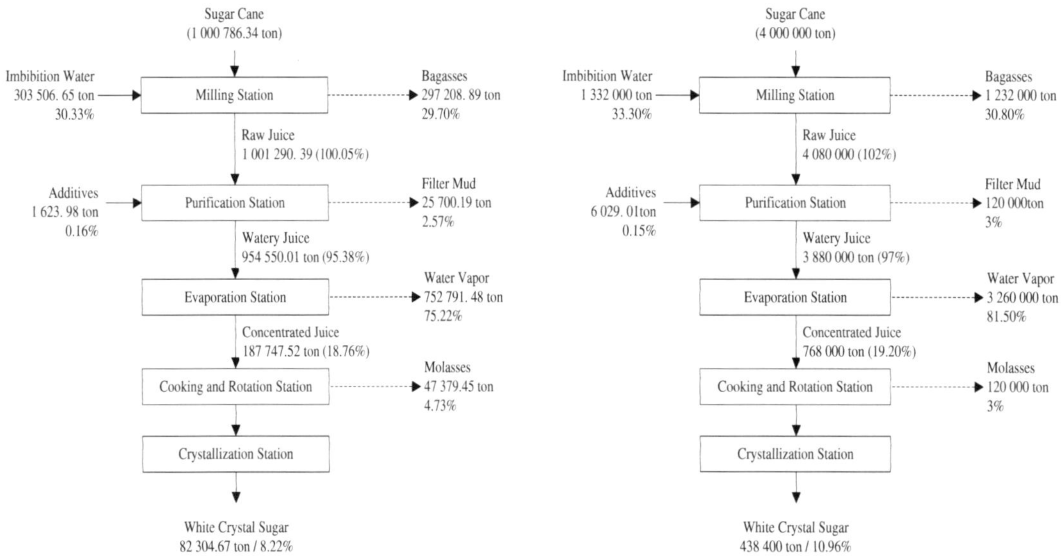

(a) (b)

Figure 2. Mass balance between Sugar mill Ngadirejo (a) and PG.Mauritius (b)

IOP Conf. Series: Earth and Environmental Science **141** (2018) 012009 doi:10.1088/1755-1315/141/1/012009

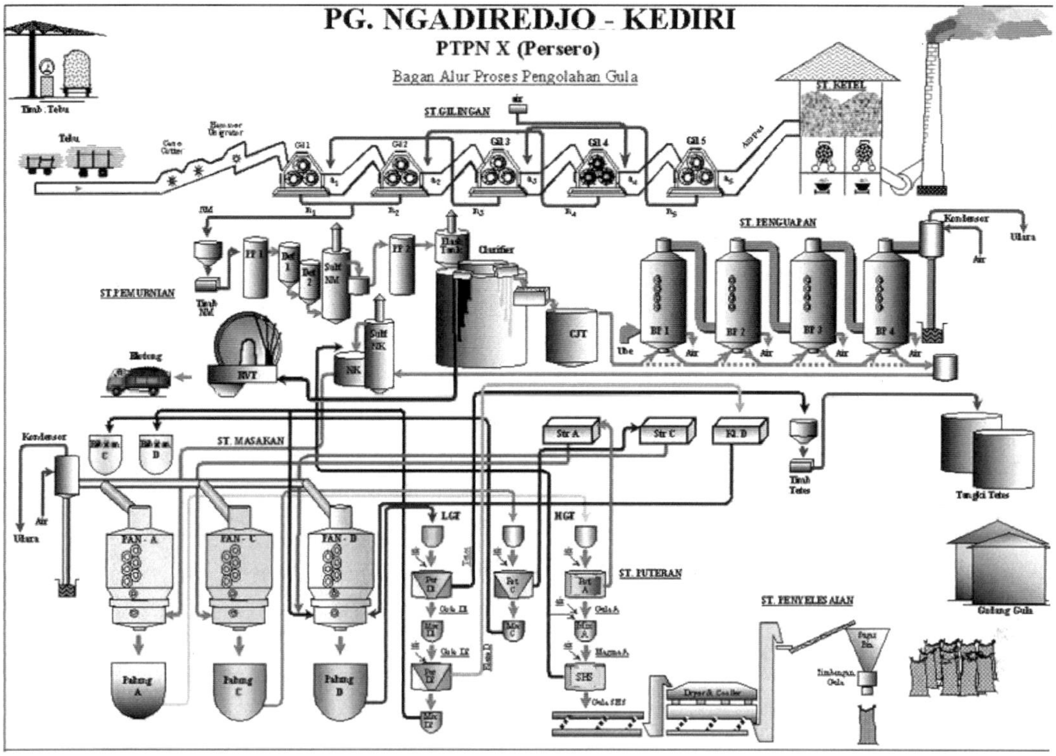

Figure 3. Flowchart of sugar production processing at Ngadirejo sugar mill

The process of white crystal sugar production with sugarcane raw materials at Ngadirejo and Mauritius sugar mill is using 1 ton of sugar cane as the initial input (Figure 3). During the extraction process, dilute juice, bagasse and waste/losses will be generated. There is a difference of output, by product and losses generated between two sugar mill though having the same input bases. Table 1 shows a clear distinction between Mauritius and Ngadirejo sugar mill. It appears that Mauritius sugar mill produces more output and bagasse than Ngadirejo sugar mill. In addition, due to the losses, Mauritius sugar mill produces fewer losses. This shows that the production process of white crystal sugar at Mautitius sugar mill is more effective and efficient compared to Ngadirejo sugar mill.

Table 1. Comparison of mass balance at Milling station

Input	**Unit**	**Ngadirejo Sugar Mill**	**Mauritius Sugar Mill**
Sugar cane	Ton	1,00	1,00
Imbibition Water	%	30.33	33,3
Additives	%	0,0016	0,0015
Loss Milling Station	%	0,444	0,375
Output			
Raw Juice	%	100.05	102.00
By Product			
Bagasse	%	29.70	30.80

IOP Conf. Series: Earth and Environmental Science **141** (2018) 012009 doi:10.1088/1755-1315/141/1/012009

Bagasse difference between the two sugar mill is 1.1%. This suggests that more energy can be generated at Mauritius sugar mill than Ngadirejo sugar mill. By-product or losses produced at a production stage can be affected by the effectiveness of a machine used [2]. In addition to the effectiveness of the machine, it can also be affected by the amount of imbibition water added during grinding. Bagasse as by-product at milling station can be converted into energy for sugar mill during the production process. Utilization of bagasse as additional energy source at sugar mill has been done because provide sufficient high energy where the electricity generated increases with the high percentage of fiber [8]. Bagasse generated from the biomass produced by the sugar mill itself, renewable in nature and will benefit economically [9]. The use of bagasse as energy source will directly contribute to clean energy use compared to fossil-based energy that can damage the environment.

3.3 Convertion Process

Sugar mill is also known as self-sufficiency energy which means supply its own energy[10]. Most of sugar milling the world already supply the energy needed to process sugar from sugar cane biomass as raw material of white crystal sugar. The process of utilizing bagasse as energy source for Sugar Mill is also known as cogeneration. Cogeneration is activity to produce electrical and thermal energy simultaneously from a fuel combustion process. The process of bagasse cogeneration is shown in Figure 4.

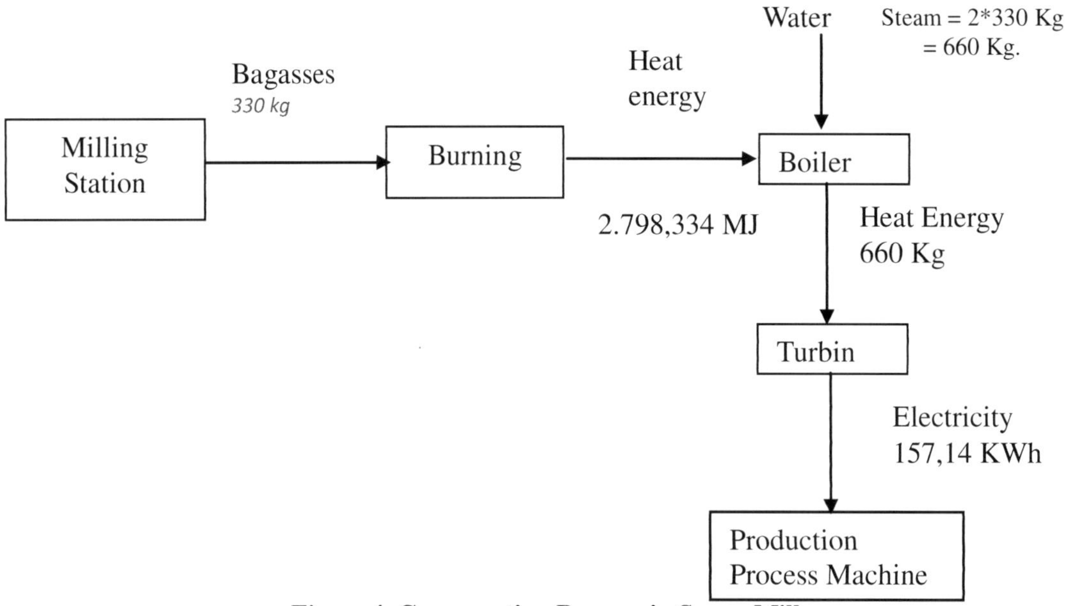

Figure 4. Cogeneration Process in Sugar Mills

The process of bagasse conversion into energy at sugar mills generally uses two important components namely boilers and turbines. Conversion begins in combustion process at the boiler, using bagasse as fuel generated from the milling station. The boiler will produce steam which the vapor will directly enter the turbine to convert into energy. The resulting electrical energy will then be used for the production process of white crystal sugar.

How much electrical energy generated from Bagasse in sugar mills can be calculated. Saechu [11] explains 1 kg of bagasse can be converted in to 2 kg of steam. Meanwhile 4.2 kg of bagasse can generate 1 KWh of electricity. Calculations of bagasse conversion to energies as as follow [11]:

1 kg bagasses → 2 kg vapor - - - - -→ 4.2 kg vapor ———→ 1 KWh

If we assumed that in milling station produces 330 kg bagasse, it will show that from 330 kg bagasse can produce 2772 KWh (330 * 2 * 4.2). Bagasse conversion process in Ngadirejo and Mauritius sugar mill to electrical energy is 141.43 KWh and 146.7 KWh respectively. Mauritius sugar mill produces more additional energy compared to Ngadirejo sugar mill due to more by-product generated. In addition, the overall calculation shows that Mauritius sugar mill produces more by-products and generates more electrical energy. The results of by-product calculations are presented in Table 02.

Table 2. Calculation of by-product potential

Sugar mill	Electricity (KWh)		
	Generated	Required	Surpluss
Ngadirejo	944,93	25	919,93
Mauritius	971,63	20	951,63

Source : [10,12, 13, 14]

Table 02 shows calculations of potential by-product during the production process of white crystal sugar. In addition to bagasse, other by products of sugar mill can also be used as energy source such as molasses, can be converted into energy for sugar mill production process. Both sugar mill produces enough electrical energy to be used during the production process. It appears that the sugar mill is capable to supply its own electrical energy and can be used for other purposes. This confirms that sugar mills are generally able to supply their own energy and even have considerable energy reserves.

4. Conclusions

Mass balance between two sugar mill showed that both of sugar mill produce by product that potential to become energy raw material especially in Mauritius sugar mill that produce 30.80% by product (bagasse) higher than Ngadirejo sugar mill. The higher by product production by the sugar mill, the higher energy that can be produce and convert to electricity. It can be concluding that Mauritius sugar mill is more efficient and effective in produce as we can see in the by product and losses during the white sugar production process. The low productivity of a sugar mill can be caused due to several things such as ineffective machine condition.

References

[1] Directorate General of Estate Crops *Tree crop estate statistic of Indonesia 2015-2017* (Indonesia : Jakarta)

[2] Subiyanto S 2014 Analisis efektifitas mesin/alat pabrik gula menggunakan metode overall equirement effectiveness *Jurnal Teknik Industri* **16** 41-50.

[3] Wauters F and Mathot J 2002 *Overall Equipment Effectiveness,* A Whitepaper http:// www05.abb.com/global.

[4] Ensinas AV Nebra SA Lozano MA Serra L 2006 *Proc. of ECOS 2006* July 12-14 (Greece: Aghia Pelagia, Crete)

[5] Suparjo 2008 Degradasi Komponen Lignoselulosa oleh Kapang Pelapuk Putih *Artikel. Laboratorium Makanan Ternak* Fakultas Peternakan Universitas Jambi (Jambi).
[6] Mbohwa C and Fukuda S 2003 *Biomass and Bioenergy* **25** 197-207
[7] Deepchand K 2005 Sugar cane bagasse energy cogeneration - Lessons from Mauritius Parliamentarian Forum on Energy Legislation and Sustainable Development, Cape Town, South Africa. 10/2005: Mauritius Sugar Authority
[8] Mbohwa C 2013 *Proc. of the World Congress on Engineering* July 3 - 5 (United Kingdom : London)
[9] Damodaran C 2008 Energy audit in Sugar Industry. A Passport to Prosperity. Avante –Garde, Energy consultants. www.avantgarde-india.com/techpapers/Energy Last accessed 13 March 2013
[10] Bantacut T dan Novitasari D 2016 Energy and water self-sufficiency assessment of the white sugar production process in Indonesia using a complex mass balance model *J. Cleaner Production* **126** 478 492
[11] Saechu M 2009 Bagasse energy optimation at sugar cane plant *Jurnal Teknik Kimia* **4** 274-280
[12] Pippo WA and Luengo CA 2013 Sugarcane Energy Use: Accounting of Feedstock Energy Considering Current Agro-Industrial Trends and Their Feasibility. *International Journal of Energy and Environmental Engineering* **4**.
[13] Hariyanto 2011 Case Study Indonesian Sugar Industry: Productivity Gains From Efficient Waste to Energy Schemes: Energy Technology Center, Agency forThe Assessment and Application of Technology. Badan Pengkajian dan Penerapan Teknologi Presented in German Indonesian Renewable Energy Days Oct 24th – 26th 2011 Jakarta
[14] Bhatt M and Rajkumar N 2001 Mapping of Combined Heat and Power Systems in Cane Sugar Industry *Applied Thermal Engineering* **21** 1707-1719

IOP Conf. Series: Earth and Environmental Science **141** (2018) 012010 doi:10.1088/1755-1315/141/1/012010

Calcium soap from palm fatty acid distillate for ruminant feed: The influence of water temperature

Lienda A. Handojo, Antonius Indarto, Dian Shofinita, Anggina Meitha, Rakhmawati Nabila, Harry Triharyogi, Leonardus Kevin

Department of Chemical Engineering, Faculty of Industrial Technology, Institut Teknologi Bandung, Bandung, Indonesia
lienda@che.itb.ac.id

Abstract. As the largest palm oil producing country in the world, Indonesia also produces abundant amount of Palm Fatty Acid Distillate (PFAD), a by-product of Crude Palm Oil (CPO) refining process. PFAD can be utilized as the raw material for calcium soap, ruminant feed that is widely used to increase milk yield, as well as to increase the ruminant's fertility. However, the practice of feeding ruminants with calcium soap has not been practiced in Indonesia, which makes it imperative to develop calcium soap production process from PFAD within the country. This research aimed to study the effect of operating conditions of the saponification reaction using PFAD and CaO as reactants on the quality of the calcium soap obtained. The saponification reaction was carried out by modified fusion method. A range of stoichiometric mole ratios of CaO to PFAD (1.0 to 1.6) and the temperature of water (60-90°C) were studied in this research. An increase in the stoichiometric mole ratio of CaO/PFAD was observed to cause a decrease in the acid value, which indicates an increase in the reaction conversion. In contrast, the temperature of water was found to have little impact on the acid value of the product.

1. Introduction

Indonesia is the largest producer of palm oil in the world. In 2016, it contributed to approximately 54% of the world total palm oil production, with annual palm oil production of 32.5 million metric ton per year [1,2]. The crude palm oil (CPO) is refined to obtain Refined, Bleached, and Deodorized (RBD) palm oil that takes 95% of the initial mass of the CPO. The 5% remaining fraction is the by-product, which is known as palm fatty acid distillate (PFAD) [3]. Most of the produced PFAD in Indonesia is exported directly without further processing. According to Central Bureau of Statistics, approximately 94-99 thousand metric tons PFAD was exported each month in the second quarter of 2017 [4]. In fact, fat content in PFAD can reach 81.7% and has a lot of potentials to be used as raw materials for oleochemical, biofuel, and ruminant feed production [4]. Among these three alternatives, the utilisation of PFAD for ruminant (animals with four stomachs) feed has not been explored in Indonesia, which makes it necessary as well as promising to develop the production process of this product.

Calcium soap is widely used in countries other than Indonesia as supplement for ruminants, as feeding ruminants with calcium soap can increase the yield of milk up to 5-8% and fat content in the milk up to 0.2-0.3%. Ruminants need high energy diets for producing milk, which is ideally supplied by fat. However, their rumen cannot tolerate the presence of free fatty acids as they disturb fiber digestive process in the rumen. Unlike free fatty acids, calcium soap is inert in the rumen of ruminants due to its 'protected fat' form, enabling it to be transported to the later part of the digestive system. Later in aabomasum, PFAD will be broken into fatty acids that can be absorbed and used for milk production or as an energy source. Hence, feeding ruminants with calcium soap is expected to increase the absorption of fat and calcium in ruminant's digestion system. [6]. Moreover, it has been proven to increase the ruminant's fertility by 20% [7].

From commercial point of view, the benefits of calcium soap for ruminants will be the answer for the challenges faced by Indonesian dairy farmers. In 2014, milk productivity in Indonesia was only 1.6 tonnes of milk per cow per year; making Indonesia is left behind China, Thailand, and Australia (2.9, 4.6 and 5.7 tonnes per cow per year, respectively) in terms of its local dairy production [8]. Since this low productivity can be attributed to the absence of calcium soap in the diets of Indonesian milk cows, the introduction of calcium soap for ruminant feed in Indonesia is potential to increase the milk productivity, as well as give added values to non-exported PFAD.

Calcium soap is a product of saponification reaction that can be produced through several saponification method alternatives. Modified fusion reaction is a popular method for saponification, which is carried out by reacting fatty acids with alkaline or alkaline earth ions source, which can be in the form of hydroxide, oxide, or salt [9]. Particularly for calcium soap production, $Ca(OH)_2$ and CaO are the most common calcium source. With $Ca(OH)_2$, the highest conversion obtained is 98% at stoichiometric mole ratio $Ca(OH)_2$/PFAD of 3 [10]. Meanwhile, CaO was shown to give better result than $Ca(OH)_2$ in terms of reaction time, yield, and energy consumption. In PFAD saponification reaction, water is also added as the catalyst to initiate the reaction [11].

This study aims to evaluate the production of calcium soap from PFAD and CaO for ruminant feed. Specifically, this research looks into the influence of water temperature on the saponification reaction of PFAD to produce calcium soap. Reaction conversion is selected as the quantitative parameter to be observed, which is expressed by the acid value of the calcium soap. Additionally, the effect of water temperature on the saponification reaction time required was also studied.

2. Materials and Methods

2.1 Materials

Materials used in this research were technical grade CaO and PFAD as the source of fatty acid. The PFAD was obtained without any further treatment from PT Tunas Baru Lampung Tbk – Sidoarjo, a CPO refining plant in Indonesia.

2.2 Calcium soap preparation

The calcium soap was prepared using modified fusion method. 50 g PFAD was preheated to 60°C until completely melted, then CaO powder (d ≤100 mesh) with a mole ratio of CaO/PFAD that ranged from 1 to 1.6 was added. Immediately after homogeneous mixing was achieved, water at various temperature within the range of 60-90°C was added to the mixture.

2.3 Acid value determination

For all experiments, the analysis of the extent of reaction was focused on acid value determination of the calcium soap product. The analysis was done according to the standard method described in ISO 660:1990. One gram of calcium soap was dissolved in ethanol at 60°C for 10 min. The resulting mixture was titrated with 0.01 N potassium hydroxide using phenolphthalein indicator. The acid value in the product was calculated with the following equation:

$$Acid\ value = \frac{(56.1 \times c\, V)}{m} \tag{1}$$

where c is the exact concentration of the potassium hydroxide solution used (mol/L), V is the volume of the potassium hydroxide solution (mL), and m is the mass of the test portion (grams).

3. Results and Discussions

3.1 The effect of water temperature on reaction time

Water was added to the PFAD/CaO mixture at four different initial water temperatures: 60, 70, 80, and 90°C. Stoichiometric mole ratio of CaO/PFAD was fixed at 1.4, and water:PFAD (w/w) ratio was fixed at 1:5. The change of reaction temperature with respect to reaction time was plotted for each water temperature, as depicted in figure 1. The plot was terminated as the temperature reached its peak.

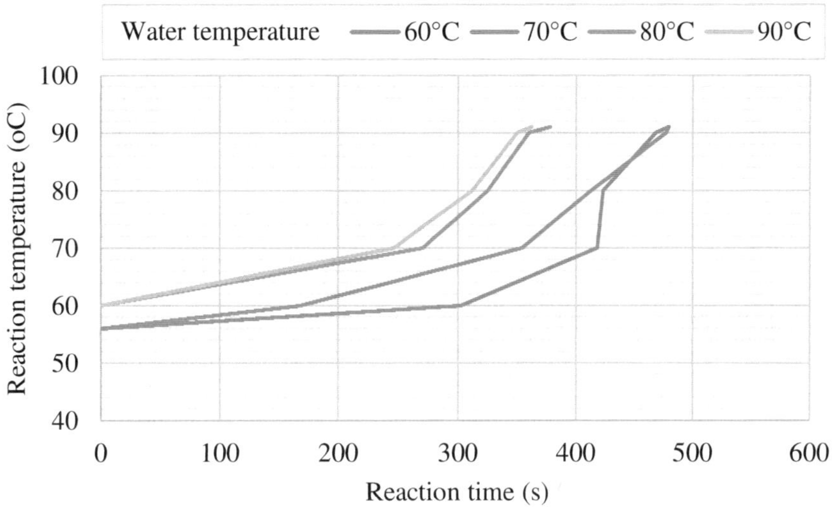

Figure 1. Reaction temperature vs reaction time at various initial water temperature

Generally, saponification reaction requires high temperature of about 150-200°C. However, in modified fusion reaction this high temperature can be reduced up to 90-100°C by the addition of small amount of water. The exothermic reaction of CaO and water produces heat which helps the reaction reach the saponification temperature. As a result, the modified fusion reaction that initiates at 90-100°C and reaches the peak temperature of 100-200°C. Hence, the energy to heat up the reaction mixture can be lowered [12].

From figure 1, the effect of various initial water temperature on the time needed for the system to reach the peak temperature can be observed. The peak temperatures of reaction mixture made with water temperature of 60, 70, 80, and 90°C are relatively similar (approximately 91°C), which are different from the statement made by Scott [12]. Additionally, there was no significant effect of initial water temperature on the peak temperature values reached by the reaction mixture. This phenomenon is mainly caused by the different initial temperature condition. In this research, the initial temperature is 60°C for the molten PFAD and 60-90°C for the water, while the initial temperature described by Scott is in the range of 90-100°C [12].

Figure 1 shows that higher water temperature will boost the reaction time needed to reach the peak temperature significantly. However, the use of water at higher temperature will consume more energy, as longer time is required to heat the water. Therefore, economic analysis and the heat balance of the reaction need to be researched further for making the most economical process.

IOP Conf. Series: Earth and Environmental Science **141** (2018) 012010 doi:10.1088/1755-1315/141/1/012010

3.2 *The effect of water temperature on acid value*

Water temperatures that are used on experiments are 60, 70, 80, and 90°C. The acid value of the product with various water temperature and stoichiometric mole of CaO/PFAD are shown in figure 2.

At same CaO/PFAD stoichiometric mole ratio, it was observed that an increase in water temperature from 60-90°C decreased the acid value of calcium soap produced. The acid value of the product decreased as the stoichiometric mole ratio of CaO/PFAD increased from 1.2 to 1.6. Various stoichiometric mole ratio have greater effect than water temperature on the acid value of the product as water temperature was found to have negligible effect on the acid value of the product.

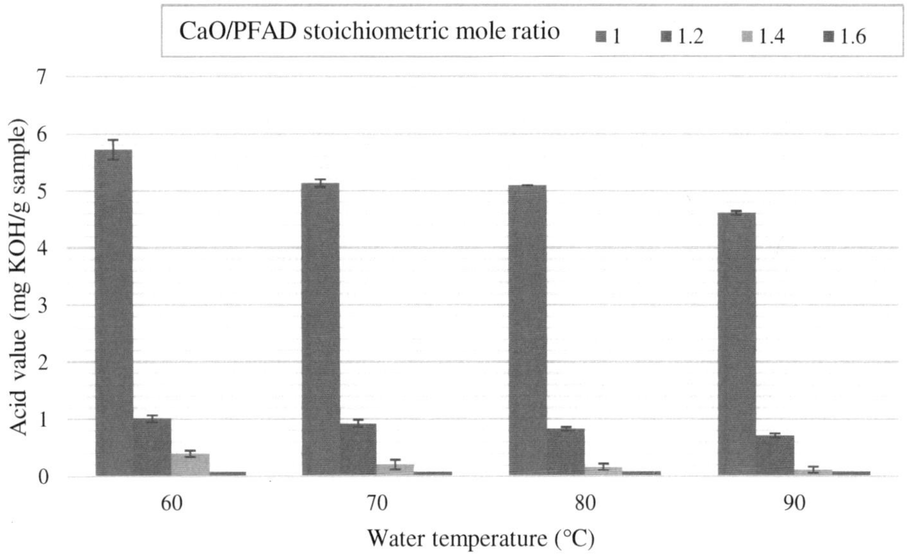

Figure 2. The effect of water temperature on acid the acid value of the product at various stoichiometric mole ratio

4. Conclusions

Increasing water temperature shortens the time needed to reach peak temperature significantly, but has little influence on the acid value of the product. Meanwhile, the acid value changes significantly with the variation of CaO/PFAD stoichiometric mole ratio. Taking into account the energy requirement for the production of calcium soap from PFAD, it is then better to go for the combination of water temperature and reaction time that gives the most optimum energy requirement, which can be done through economic analysis based on energy consumption.

Acknowledgements

The authors would like to express their sincere gratitude to BPDPKS Indonesia 2016 for the research grant that enable the authors to conduct this research and PT Tunas Baru Lampung Tbk for providing PFAD for this research.

References

[1] Indonesian Palm Oil Producers Association 2016

[2] US Department of Agriculture 2016. *Oilseeds: world markets and trade* (United Stated: Foreign Agricultural Service)

[3] Zero and Rainforest Foundation Norway 2016 *Palm Fatty Acid Distillate (PFAD) in Biofuels*

[4] Export Statistic Subdirectory 2017 *Export Trading Bulletin According to Harmonized System March-May 2017* (Jakarta: Indonesia Central Bureau of Statistics)
[5] Top A G 2010 *Production and Utilization of Palm Fatty Acid Distillate* (*Lipid Technology)* vol 22 (Wenheim: Wiley) p 11
[6] Volac 2016 *Megalac Product Sheets*, Retrieved 30 September 2016, from http://www.volac.com/megalac
[7] Suksombat W 2009 *Improving the Productivity of Lactating Dairy Cows through Supplementation* (*International Dairy Topics*) vol 8 p 7
[8] Food and Agriculture Organizations 2014, from http://www.fao.org/faostat/en/#data
[9] Perez, E P 2007 *Method of producing calcium, sodium, or magnesium soaps from fatty acids or oleins from animal or vegetable fats and use thereof as nutrients in monogastric animal feed.* EP Patent 800 546
[10] Handojo L A, Indarto A, Shofinita D, Meitha A, Nabila R, Triharyogi H, Hakim M G A and Saadi M R 2017 *Calcium soap from palm fatty acid distillate for ruminant feed: $Ca(OH)_2$ as calcium source* (to be published)
[11] Handojo L A, Indarto A, Shofinita D, Meitha A, Nabila R and Triharyogi H 2017 *Calcium soap from palm fatty acid distillate for ruminant feed: reaction method* (to be published)
[12] Scott, L. 1974. *United States of America Patent No. US3803188 A.*

IOP Conf. Series: Earth and Environmental Science **141** (2018) 012011 doi:10.1088/1755-1315/141/1/012011

Biogas production from anaerobic codigestion of cowdung and elephant grass (*Pennisetum Purpureum*) using batch digester

Agus Haryanto[1,*], Udin Hasanudin[2], Chandra Afrian[1], Iskandar Zulkarnaen[1]

[1] Agricultural Engineering Department, Faculty of Agriculture, The University of Lampung. Soemantri Brojonegoro Road #1, Bandar Lampung, Indonesia.
[2] Agricultural Product Technology Department, Faculty of Agriculture, The University of Lampung. Soemantri Brojonegoro Road #1, Bandar Lampung, Indonesia.

[*] E-mail: agus.haryanto@fp.unila.ac.id

Abstract. This study aimed at determining biogas production from codigestion of Elephant grass and cowdung using batch digester. Fresh grass was manually chopped with a maximum length of 3 cm. Chopped grass (25 kg) was perfectly mixed with fresh cowdung (25 kg). The mixture was introduced into a 220-liter batch drum digester. The substrate was diluted with water at different rates (P1 = 50 L, P2 = 75 L, and P3 = 100 L) and was stirred thoroughly. Six digesters were prepared as duplicate for each treatment. Two other digesters containing only 25 kg cowdung diluted with 25 L water were also provided as control treatment (P0). The digesters were air tightly sealed for 70 days. Observation was conducted on daily temperature, substrate pH (initial and final), TS and VS content, biogas yield and biogas composition. Results showed that final pH of grass containing substrate was in the acidic range, namely 4.50, 4.62, 6.82, whereas that of control (P0) was normal with pH of 7.30. Digester with substrate composition 25:25:100 (cowdung:grass:water) produced the highest biogas total (524.3 L). Biogas yield of codigestion, however, was much lower as compared to that of control, namely 7.35, 16.75, and 111.72 L/kg VS_r respectively for treatment P1, P2, P3. with dilution rate of 50, 75, and 100 L. Biogas produced from control digester had methane content of 53.88%. In contrast, biogas resulted from all treatments contained low methane (the highest was 31.37%). Methane yield of 39.3 L/kg TS removal was achieved from digester with dilution 100 L (P3). Mechanical pretreatment is suggested to break Elephant grass down into smaller particles prior to introducing it into the digestion process.

1. Introduction

Biogas is mixture of flammable gas produced through anaerobic digestion process of organic materials. During this process, complex organic matter such as protein, fats, and carbohydrate polymers (cellulose and starch) are first hydrolysed into monomer like amino acids, long-chain fatty acids, and sugars. These monomers are then fermented to form volatile fatty acids (lactic, propionic, and butyric acids) during acidogenesis (second step). In the third step (acetogenesis), bacteria consume these volatile fatty acids and generate acetic acid, carbon dioxide (CO_2), and hydrogen (H_2). Finally, methanogenic organisms consume the acetate, H_2, and some of the CO_2 to produce methane (CH_4) [1].

The biogas is mainly composed of CH_4 (45-70%) and CO_2 (30-45%) and traces of H_2, water vapour (H_2O), ammonia (NH_3), and hydrogen sulphide (H_2S) [1, 2]. Biogas can be used to fuel several applications, from cooking stove to generating electricity. Biogas technology is now an ecologically sound option to reduce environmental burden by decomposing organic material and producing not only energy but also good quality organic fertilizer [3, 4].

Grasses are efficient at cellulosic biomass production. Grasses give more benefit because of sequestering more carbon, requiring less tillage and consuming less fertilizers and pesticides [5], consuming less water and can be cultivated on non-arable soil [6]. In recent years, perennial grasses have been identified as leading candidates for bioenergy production including switchgrass (*Panicum virgatum* L.) in North America [7], giant miscanthus (*Miscanthus giganteus*) in Europe [8], and Elephant grass or Napier grass (*Pennisetum purpureum* Schumach) in Thailand [9–11].

Elephant grass, originating from Africa, has been successfully cultivated in Indonesia for cattle feed purposes. Thc plant has a high posture, wide (3 cm) and length (30-90 cm) leaves, and it is one of the highest-yielding tropical forage grasses [12]. Munasik *et al.* [13] reported that this grass can achieve a height of 259.6–289.2 cm and yield of 70–100 ton fresh biomass or 1.11–1.57 ton dry matter per hectare at 60 days old after planting. Biomass yield is influenced by age, the older the higher [11]. Field observation at PT. Great Giant Livestock, Central Lampung showed that the grass is able to be ratooned 6–7 times before replanting [14]. In addition to high yield, Elephant grass grows fast and is highly adaptive, and is easy in propagation and management [11]. The grass is also tolerant in acidic soil [13], copper [15], and drought condition [12].

Studies on biogas production by co-digestion of animal wastes with grass have attracted special interest. The grass has such high in organic substrate that it is good potential as co-substrate for biogas production. [11] Jewell *et al.* [16] examined the production of biogas from energy crops including Elephant grass with *TS* of 25-30% and obtained maximal methane production levels. Ekpenyong *et al.* [17] reported biogas production from anaerobic fermentation of ground dry elephant grass stems about 450 ml for 5 days with 4 g of substrate. An increase in biogas production by 40% was obtained by the addition of 0.01 g Urea. Ahn *et al.* [18] examined the anaerobic fermentation performance of dry grasses of switchgrass mixed with different livestock manure (pigs, poultry and dairy cattle) at 15% *TS* and thermophilic temperature (55°C). Mixture with pig manure resulted in a decomposition of *VS* (volatile solids) of 52.9%, higher than poultry manure mixture (9.3%) and dung cow dung (20.2%). For 62 days of residence time, the mixture of grass with pig manure produced the highest methane (0.337 L/g VS), whereas the mixture with dairy cow manure and poultry droppings respectively yielded 0.028 and 0.002 L CH_4/g VS.

Recently, Sinbuathong *et al.* [11] stated that cutting age affected methane yield with grass cut at 30 days after planting produced the highest, achieving 197 (L/kgTVS added). Total methane yield per hectare (6500 m^3/ha), however, was achieved at cutting age of 60 days. Sawanon *et al.* also reported that amount of the composition of substrate also influenced the biogas yield with the highest rate of 169 L/kg TVS added was achieved at a composition of 10:10:20 (grass:cowdung:water) [10]. Olugbemide *et al.* [19] noted that co-digestion of maize leaves and Elephant grass at a ratio of 60%:40% produced 67.3% higher biogas as compared to digestion of maize leaves alone. Evaluation of large scale biogas production using anaerobic co-digestion of pig manure and elephant grass silage for car transportation has been conducted in Thailand with a conclusion that co-digestion of pig manure and grass silage is a promising approach for improving biogas production [9].

Utilization of Elephant grass for biogas substrate will actually compete with animal feed, especially for grass fed cattle and buffalo. With high productivity and digestibility, however, this grass can be dedicated solely for energy crop in remote areas. Tufvesson *et al.* (2013) concluded that if there is no demand for them as animal feed, all residues studied are well suited for biogas production. Otherwise, growing dedicated biogas crops directly will be more efficient [20].

The objective of our research is to determine the maximum biogas yield from anaerobic co-digestion of cowdung and fresh elephant grass by means of batch experiments.

2. Materials and Methods

2.1. Digester preparation

Digester was prepared using plastic drum with a volume of 220 L as presented in Figure 1. The lid was wrapped with rubber band as a seal to make the digester airtight. A hole on the lid of the drum was provided to deliver biogas into a storage balloon trough a plastic tube. The tube was facilitated with a stop valve to close the piping for biogas volume measurement and biogas sampling. Another hole was also made on the drum to insert a thermometer through a rubber cock.

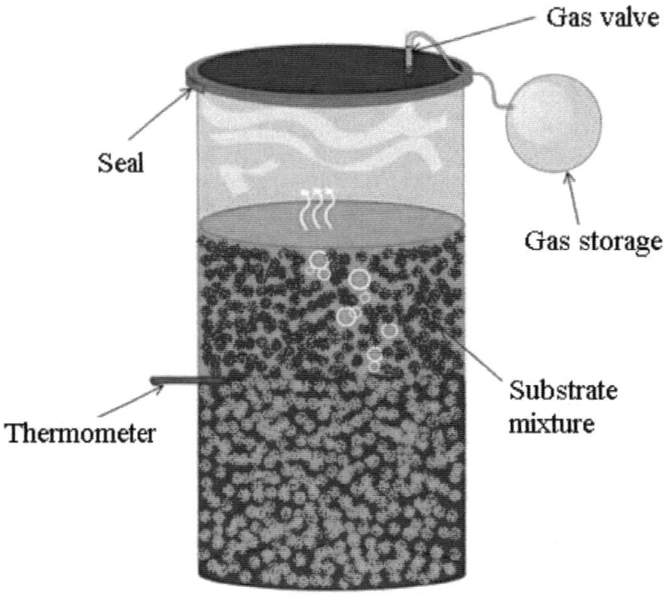

Fig. 1 A 220-L drum digester prepared for the experiment

2.2. Substrate preparation

Fresh cowdung was taken from the Department of Animal Husbandry, the University of Lampung. Elephant grass was cut at around 2 months of age from a farmer field in Pesawaran District, Lampung. Before starting the anaerobic digestion experiments, samples of grass and cowdung were analysed for total solids (*TS*), volatile solids (*VS*), and carbon (*C*) and nitrogen (*N*) contents. Table 1 shows characteristic of separate substrate.

Table 1. Fresh substrate characteristic

Characteristic	Cowdung	Elephant grass
Water content (%, wet basis)	71,32	87,03
Total solid (TS) (%, wet basis)	28,68	12,97
Ash (% TS)	25,04	14,41
Volatile solid (VS) (% TS)	74,96	85,59
C (%)	39,87	55,51
N (%)	1,42	1,81
C/N Ratio	28,08	30,62

2.3. Treatments

The fresh grass was chopped into maximum length of 3 cm. Twenty five kg of chopped grass were mixed thoroughly with cowdung at a mass ratio of 1:1. The mixture was introduced into 220-L drum digester and then diluted with water at three different levels, namely 50, 75, and 100 L. The mixture was manually stirred in order to homogenize the substrate. The lid of the drum was then closed with a rubber seal. Two drum digesters were provided for each treatment as duplicate. Two other digesters were also prepared for control treatment containing only cowdung and water with mass ratio of 1:1 and thoroughly mixed. Table 2 shows substrate composition of all treatments.

Table 2. Treatment and substrate composition

Treatment	Cowdung (kg)	Elephant grass (kg)	Water (L)	TS (%, wb)	TS (Kg)	VS (%, wb)	VS (Kg)
P0 (Control)	25	0	25	14.34	7.17	10.75	5.38
P1	25	25	50	10.41	10.41	8.15	8.15
P2	25	25	75	8.33	10.41	6.52	8.15
P3	25	25	100	6.94	10.41	5.43	8.15

2.4. Analysis and calculations

For determining the total solids (*TS*), samples with certain weight were placed in ceramic vessels and dried in a drying oven (Memmert, type UM 500, Germany) at 105 °C for 24 hours until constant weight. After cooling in the desiccators, the samples were weighed for *TS* measurement. The samples was then burnt in a furnace (Barnstead Thermolyne FB1300, USA) at 550 °C for 2 hours for volatile solids (*VS*) determination. The *VS* is determined by subtraction of the minerals content of the sample (residual ash after oxidation) from the total solids content. Total solid, *VS*, and removed VS_r are calculated by using Eq. (1) to (3), respectively:

$$TS\ (\%,\ \text{wb}) = \frac{W_2}{W_1} \times 100 \quad (1)$$

$$VS\ (\%\ \text{TS}) = \frac{W_3 - \text{Ash}}{W_3} \times 100 \quad (2)$$

$$VS_r\ (\%) = \frac{VS_{in} - VS_{out}}{VS_{in}} \times 100 \quad (3)$$

Carbon and nitrogen contents of each substrate were measured using element analyzer (Elementar Vario EL Cube, Germany). Carbon to nitrogen ratio (*C:N*) of the mixture is calculated using Eq. (4):

$$C:N = \frac{(C_c \times m_c) + (C_g \times m_g)}{(N_c \times m_c) + (N_g \times m_g)} \quad (4)$$

where *m* is dry mass and subscripts *c* and *g* denote cowdung and Elephant grass, respectively.

In order to evaluate process condition, temperature and pH of the substrate during experiment were also checked. Daily temperature was monitored from a thermometer inserted in the digester. The pH values were determined using pH meter (PHMETER, PH_009(I), China) for both fresh and spent substrates.

Biogas production was determined daily using simple water displacement method. Biogas composition was measured using gas chromatograph (Shimadzu GC 2014, Japan) with thermal conductivity detector (TCD) and 4-m length of shin-carbon column. Helium gas was used as carrier gas with flow rate 40 ml/min. Biogas yield (*BY*) was calculated from biogas productivity (*BP*) and VS_r by using Eq. (5):

$$BP\,(\text{L/kg}\ VS_r) = \frac{BY}{VS_r} \quad (5)$$

3. Results and Discussions

3.1. Operation condition

Biogas is produced during biological process involving a group of bacteria working in an anaerobic condition. The interaction of several factors affects the performance of biogas process. Temperature and pH are among the important factors. Figure 2 shows daily temperature of the digesters. All digesters operated in the mesophilic temperature region with a range of 30.7 – 35.5 °C for P1, 28.6 – 30.5 °C for P2, 28.6 – 29.8 °C for P3, and 28.8 – 31.2 °C for P0 (control). These digesters worked at temperature higher than ambient temperature (26.4 – 28.6 °C) mainly due to exposure to sun rays. Different shading conditions also caused temperature differences among the treatments. Even though all digesters placed under the same roof overhang of a building, there were different shading conditions due to the existence of surrounding trees and buildings. This is confirmed by digester temperature in each treatment, which differs from morning to noon to afternoon. Digester temperatures at noon and afternoon are higher than that at morning. Digesters for treatment P1 show the highest temperature than the other treatments. This is caused by digesters position that are more exposed to sun radiation than the others especially during day and afternoon.

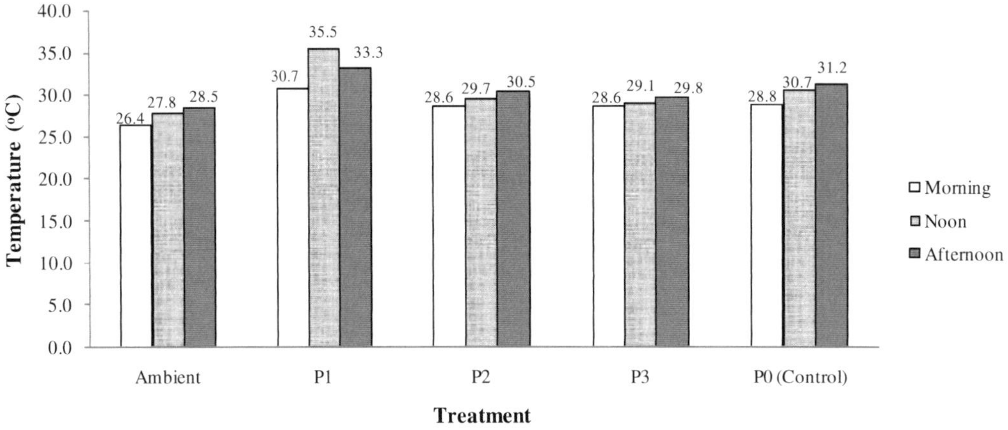

Fig. 2 Average working temperature of batch drum digesters

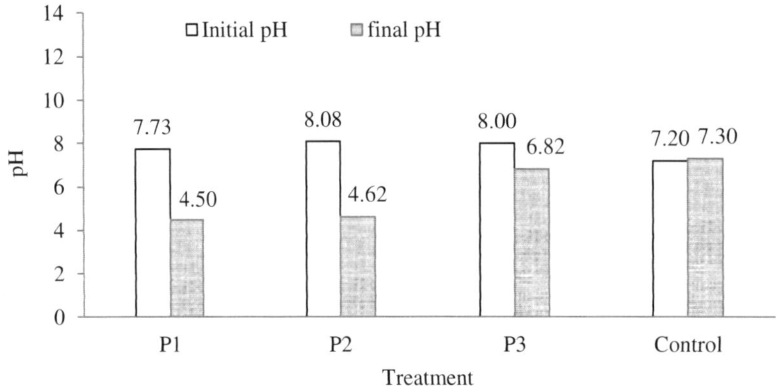

Fig. 3 Acidity of fresh and final substrate for different treatments

Figure 3 presents pH values of fresh and final substrates used in this experiment. Control treatment using only dilute cowdung shows ideal condition for anaerobic digestion process with pH values of 7.20 to 7.30, respectively from the beginning to the end. Stafford [21] reported the effects of pH upon methane production from anaerobic digestion of dairy cattle manure maintained at pH levels of 5.0 to 7.6 and found that biogas and methane production was highest at pH of 7.0. In our experiment, digesters for codigestion of cowdung and Elephant grass have initially basic pH (closes to 8) that actually in the range for good anaerobic digestion process. Abbasi *et al.* [1] noted that anaerobic degradation processes meet the requirement for both activities and cell growth of anaerobic microorganisms at pH of 5.5–8.5.

During the experiment lasted for 70 days, however, the pH decreases to acidic condition, especially for digesters with dilution rate of 50 L and 75 L. This indicates that dilution rate influences acidity of the substrate. More dilution resulted in better pH value in term of anaerobic process. At the end process, acidity value of digester P1 and P2 revealed acidic condition. Digester P3 showed final pH of 6.82 that closes to normal and meet the condition for anaerobic decomposition process. This asserted that dilution rate influences anaerobic decomposition process, the higher the better.

During anaerobic digestion process, substrate decomposition occurs. Table 3 presents characteristic of spent substrates for each treatment. Control digester has the highest decomposition of organic matter, achieving 50% (represented by VS_r). This value closes to the performance of family size digesters having average organic material removal of 51.32% [22]. The addition of fresh chopped Elephant grass resulted in the decrease in decomposition rate. The table showed that organic material degradation of substrate containing grass was in the range of 27.40% to 37.13%. Increase in dilution rate has improved organic material decomposition.

Table 3. Total solid (*TS*) and volatile solid (*VS*) of spent substrates

Treatment	*TS* (%, wb)	*TS* (Kg)	*VS* (%, wb)	*VS* (Kg)	VS_r (%)
P0 (Control)	13.51	6.76	5.31	2.66	49.91
P1	9.78	9.78	7.10	7.10	27.40
P2	7.50	9.38	5.27	6.59	29.74
P3	6.14	9.21	3.86	5.79	37.13

*) wet basis

3.2. Biogas production

Figure 4 presents daily biogas production resulted from different treatments. During first week the digesters show decreasing trend of biogas production. Initially, gas is produced from respiration of the substrates due to the existence of air filling void space in the digester. The figure reveals that digesters with grass co-substrate have more extensive respiration process. Oxygen in the air has resulted oxidation reaction that produce CO_2 and H_2O. The produced gas is exerted out from the digester. The amount of oxygen in the digester continuously decreases and finally vanishes at which time the anaerobic process starts. The figure implies that anaerobic process starts at the beginning of week two.

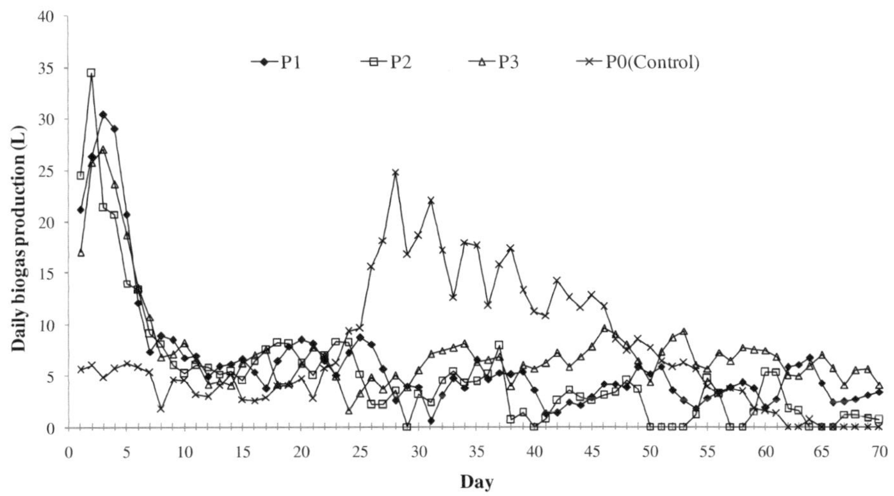

Figure 4. Daily biogas production from different treatments

Figure 4 also reveals that during anaerobic process, digesters containing grass show fluctuated biogas production at an average of 4.67 L/d for P1, 3.47 L/d for P1, and 6.16 L/d for P3. During week three, control digester (P0) demonstrates a significant increase of biogas production. From the beginning of week four, however, biogas production of P0 decreases until practically zero at day 60. On the other hand, digesters containing grass still produce biogas at significant amount, especially digester P1 and P3.

Figure 5 presents cumulative biogas production. During the first month, control digester shows the lowest cumulative biogas yield. This may relate to low respiration during first week due to the absence of fresh grass in the substrate. Starting week fifth, biogas production from control digester lead but stuck at a total of 519 L after day 61. On the other hand, biogas productions with grass co-digestion accumulatively increase for longer time. Digester P3 shows higher cumulative biogas yield than those of P1 and P2. Even by day 68, cumulative biogas yield of P3 (535 L) has surpassed that of P0. This indicates that addition of Elephant grass as a co-substrate is promising to improve total biogas yield. Digester P3 with dilution rate 100 L (twice of the substrate mixture) also produces higher total biogas than those of P1 (439 L) and P2 (353 L) with dilution rate 50 L and 75 L, respectively.

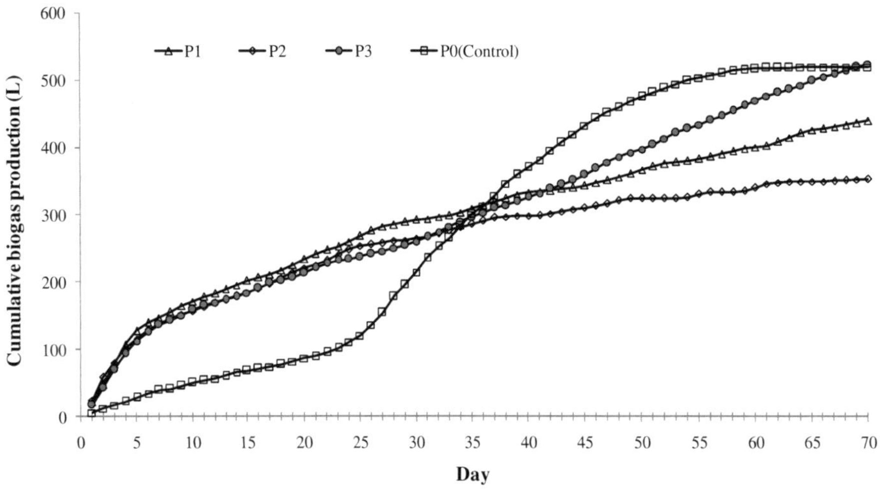

Figure 5. Cumulative biogas production from different treatment.

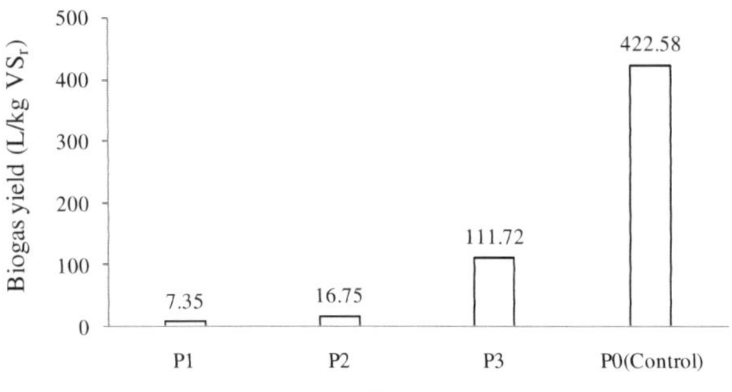

Figure 6. Biogas yield codigestion of cowdung and Elephant grass.

Figure 6 shows biogas yield on the basis of removed *VS*. Digester P0 demonstrates the highest biogas yield (422.58 L/kg VS_r). Biogas yield from grass containing digesters is substantially lower than that of P0. The figure also reveals that the degree of dilution affects the biogas yield, the higher the better. Increasing dilution (meaning reducing TS content) is possibly increase biogas yield.

3.3. Methane yield

The methane content of control digester (53.9%) is much higher as compared to those of grass containing digesters. Biogas produced from digesters containing grass has very low methane content. Methane content is measured twice, at day 30 and day 60. As depicted by Figure 7, at day 30, all digesters with grass have methane content lower than 4%. At day 60, a significant increase in methane content is presented by digester P3 with 33%. No significant changes of methane content occur at digesters P1 and P3 at day 60. This implies that acidification steps still dominate the decomposition process of organic materials. Methanogenesis step has not optimally gone on. This suggests that the presence of fresh grass has slowed down the methane formation and it requires longer time to favour decomposition.

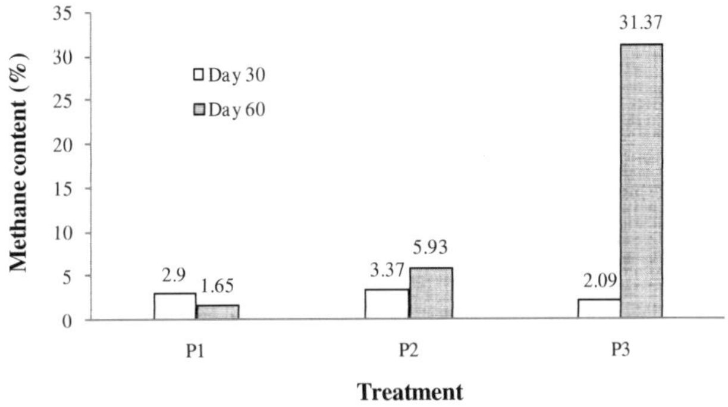

Figure 7. Methane content in the biogas produced from codigestion of cowdung and Elephant grass.

Total solid content of grass containing digesters can be a factor that causes low methane content. Figure 8 shows that in general increasing *TS* content has resulted in the reduction of methane yield. Even though the graph contains limited data, it provides finding complying with other studies. Liotta *et al.* (2014), for example, reported that anaerobic digestion of food waste produces the highest methane yield at *TS* content of 4.52% and a reduction of the specific methane yield of 4.3% and 40.8% occurs at *TS* content of 12.87% and 19.20%, respectively [23]. In our experiment, the highest methane yield of 39.3 L/kg *TS* removal is achieved at *TS* content of 6.94% or 69.4 g/L. Sawannon *et al.* [10] reported high biogas yield of 116 LCH_4/kg TS_r was achieved from co-digestion of cowdung and Elephant grass at *TS* content of 22.72 g/L and 142 LCH_4/kg TS_r at *TS* content 33.08 g/L. This means that there is a room to increase methane yield by decreasing *TS* content or increasing dilution rate.

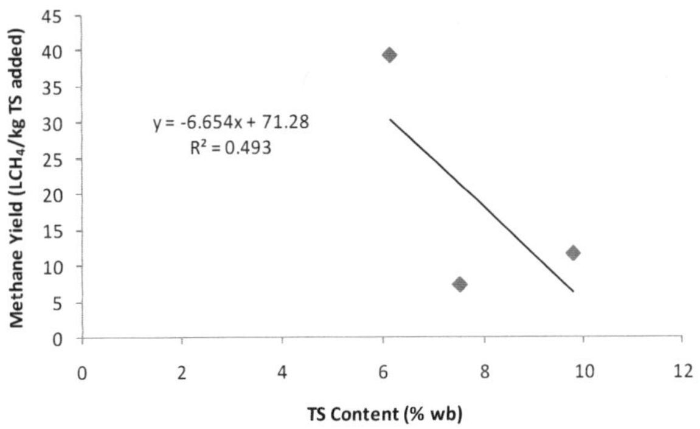

Figure 8. Effect of *TS* on the methane yield from codigestion of cowdung and Elephant grass.

The size of chopped grass can be an important key to be taken into consideration. The accessibility of microorganisms to the fermentable sugars is restricted because of complex structure of lignocellulosic materials [24]. Reduction of particle size favours the contact between the enzymes with the substrate and reveals new zones which initially inaccessible [6]. In this case, mechanical pretreatments breakdown particle size of complex cellulose and increase the substrate specific surface area available to the microorganisms, and therefore improve the biogas yield [24]. It was reported that the rate of biogas production is highly correlated to particle size and the highest biogas production from agricultural residues was obtained from raw materials of the smallest particles [25-27]. Low biogas yield in our experiment could be resulted from particle size of grass that was barely chopped into a maximum length of 3 cm, longer than those reported sizes.

4. Conclusions

The addition of Elephant grass as co-substrate potentially improve total biogas yield. Dilution rate greatly influences biogas yield and its methane content. Digester with substrate composition 25:25:100 (cowdung:grass:water) produced the highest biogas (524.3 L) with methane content (31.37%). Methane content of biogas from co-digestion elephant grass and cowdung, however, substantially lower than that of only diluted cowdung which achieve 53.9%. Biogas yield of codigestion was 7.35, 16.75, and 111.72 L/kg VS_r respectively for treatment with dilution rate of 50, 75, and 100 L. Biogas yield from control digester 422.58 L/kg VS_r, much higher than those of grass containing digesters. Methane yield of codigestion process decreases with TS content. Mechanical pretreatment is suggested to break Elephant grass down into smaller particles prior to introducing it into the digestion process.

Acknowledgement

This work was financially supported by DGHE (Directorate General of Higher Education), Ministry of National Education though Strategis Nasional (STRANAS) research scheme. Contract number: 419/UN26/8/LPPM/2016 (June 6, 2016).

References

[1] Abbasi T, Tauseef S M and Abbasi S A 2012 *Biogas Energy* (New York: Springer) pp 1–10.
[2] Gomez C D C 2013 Biogas as an energy option: an overview. *The Biogas Handbook: Science, Production and Applications*, eds A Wellinger, J Murphy and D Baxter. (Cambridge: Woodhead Publishing Limited) pp 1–16.
[3] Higgins K T 2008 Anaerobic gas and cash *Food Engineering* **80(4)** 119–20.
[4] Petis M 2006 Anaerobic digestion plant yields nutrients and power *BioCycle* **47(12)** 61–2.
[5] Moore K J, Fales S L and Heaton E A 2008 Biorenewable energy: new opportunities for grassland agriculture *Multifunctional Grasslands in a Changing World Proc. XXI Int. Grassland Conf. vol. II* eds X Haining and H Jiehua (Guangzhou: Guangdong People's Publication House) pp 1023–30.
[6] Rodriguez C, Alaswad A, Benyounis K Y and Olabi A G 2017 Pretreatment techniques used in biogas production from grass *Renew Sustain Energy Rev* **68(2)** 1193-204.
[7] McLaughlin S B and Kszos L A 2005 Development of switchgrass (*Panicum virgatum*) as a bioenergy feedstock in the United States *Biomass Bioenergy* **28** 515–35.
[8] Lewandowski I, Clifton-Brown J C, Scurlock J M O and Huisman W 2000 Miscanthus: European experience with a novel energy crop *Biomass Bioenergy* **19** 209–27.
[9] Dussadee N, Reansuwan K and Ramaraj R 2014 Potential development of compressed bio-methane gas production from pig farms and elephant grass silage for transportation in Thailand *Bioresour Technol.* **155** 438-41.
[10] Sawanon S, Sangsri P, Leungprasert S and Sinbuathong N 2017 Methane production from Napier grass by co-digestion with cow dung *Energy Solutions to Combat Global Warming. Lecture Notes in Energy 33* ed X-R Zhang and I Dincer (Switzerland: Springer International Publishing) chapter 7 pp 169–80.

[11] Sinbuathong N, Sangsil Y and Sawanon S 2016 Biogas production from Napier prass at parious cutting intervals *Energy, Transportation and Global Warming, Green Energy and Technology* ed P Grammelis (Switzerland: Springer International Publishing) chapter 29 pp 375–85.
[12] El-Bassam N 2010 Handbook of Bioenergy Crops: A Complete Reference to Species, Development and Applications (London: Earthscan) pp 378–9.
[13] Munasik, Sutrisno C I, Anwar S and Prayitno C H 2012 The growth, yield and quality of elephant grass (*Pennisetum purpureum*) specific tolerant of acid soils by mutagenesis with ethyl methane sulfonate *Animal Production* **14(2)** 87–91.
[14] GGL (Great Giant Livestock) 2014 *Personal communication* (January 8, 2014).
[15] Liu X, Shen Y, Lou L, Ding C and Cai Q 2009 Copper tolerance of the biomass crops Elephant grass (*Pennisetum purpureum* Schumach), Vetiver grass (*Vetiveria zizanioides*) and the upland reed (*Phragmites australis*) in soil culture *Biotechnol Adv.* **27** 633–40.
[16] Jewell W J, Cummings R O and Richards B K 1993 Methane fermentation of energy crops: maximum conversion kinetics and *in situ* biogas purification *Biomass Bioenergy* **5(3-4)** 261–78.
[17] Ekpenyong K I, Arawo J D E, Melaiye A, Ekwenchi M M and Abdullahi H A 1995 Biogas production potential of unextracted, nutrient-rich Elephant-grass lignocelluloses *Fuel* **74(7)** 1080–82.
[18] Ahn H K, Smith M C, Kondrad S L and White J W 2010 Evaluation of biogas production potential by dry anaerobic digestion of switchgrass–animal manure mixtures *Appl. Biochem. Biotechnol.* **160**: 965–75.
[19] Olugbemide A D, Imasuen A O, Oleghe P O and Efosa J O 2012 Anaerobic co-digestion of fresh maize leaves with Elephant grass *J Appl Sci Environ Manage.* **16(1)** 133–35.
[20] Tufvesson L M, Lantz M, and Börjesson P 2013 Environmental performance of biogas produced from industrial residues including competition with animal feed – life-cycle calculations according to different methodologies and standards *J Clean Prod.* **53** 214–23.
[21] Stafford M G 1982 *The Effect of pH on Methane Production from Dairy Cattle Manure*, Masters Thesis (USA: University of Central Florida).
[22] Haryanto A, Cahyani D, Triyono S, Murdapa F and Haryono D 2017 Economic benefit and greenhouse gas emission reduction potential of a family-scale cowdung anaerobic biogas digester *Int. Journal of Renewable Energy Development* **6(1)** 29–36.
[23] Liotta F, d'Antonio G, Esposito G, Fabbricino M, van Hullebusch E D, Lens P N L, Pirozzi F, and Pontoni L 2014 Effect of total solids content on methane and volatile fatty acid production in anaerobic digestion of food waste *Waste Manag Res.* **32(10)** 947-53.
[24] Rodriguez C, Alaswad A, El-Hassan Z and Olabi A G 2017 Mechanical pretreatment of waste paper for biogas production. *Waste Manag.* **68** 157–64.
[25] Sharma S K, Mishra I M, Sharma M P, and Saini J S 1988 Effect of particle size on biogas generation from biomass residues *Biomass* **17** 251–63.
[26] Hills D J and Nakano K 1984 Effects of particle size on anaerobic digestion of tomato solid wastes. *Agric Wastes* **10** 285–95.
[27] Xiao X, Zhang R, He Y, Li Y, Feng L, Chen C, and Liu G 2012 Influence of particle size and alkaline pretreatment on the anaerobic digestion of corn stover *BioResources* **8(4)** 5850–60.

Performance of composite boards from long strand oil palm trunk bonded by isocyanate and urea formaldehyde adhesives

Indra Hermanto[1] and M Y Massijaya[1]

Biocomposite Laboratory, Forest Products Department, Faculty of Forestry, Bogor Agricultural University, Bogor, West Java, Indonesia
Email : her2indra@gmail.com ; mymassijaya@yahoo.co.id

Abstract. In this research, the obtained long strand were produced from the outer part of oil palm trunk and then hot-prepressed. The three-ply composite boards were made from hot-prepressed long strand and use bonded by isocyanate and urea formaldehyde adhesives with a glue spread variation of 150 g/m², 225 g/m², and 300 g/m². The board target density was 0.65 g/cm^3, face and back layers orientation is the same and the core layer was perpendicular to the face and back layers. The research results showed that : (1) composite boards bonded by isocyanate performed better physical and mechanical properties compared to those of bonded by urea formaldehyde, (2) utilization of higher glue spread level would improve the physical and mechanical properties of the composite board. (3) composite boards bonded by isocyanate and urea formaldehyde adhesives at glue spread of 225 g/m²,300 g/m², respectively were enough to fulfill the JIS A 5908 (2003) standard.

Keywords : composite board, long strand, isocyanate, urea formaldehyde, oil palm trunk.

1. Introduction

Indonesia has a vast oil palm plantation. Based on data from Statistic Indonesia the area of oil palm plantations in Indonesia reached 11.3 million ha, spread over 24 provinces with an average growth rate of 2.78% per year [1]. Oil palm plantations have grown rapidly since the 1980s. One of the reason was the increasing demand for crude oil palm to meet the needs of local and global markets. Increasing of oil palm plantation area was directly proportional to the waste production. Oil palm trees at the age of 25 to 30 years old was no longer productive so needs to be re-planting. The oil palm at that time normally reach the height of 7 to 13 meters and possess diameter from 45 to 65 cm. Oil palm trunk has some problems if used as furniture or light construction, especially in terms of dimensional stability, strength, durability, and machining properties [2]. The utilization of oil palm trunk waste has several constraints, such as poor dimensional stability, strength and resistance to weak organism and poor machining properties [3]. The stability of the oil palm trunk dimension was very low with the variation of shrinkage of 9.2-74%, its strength was in strong class III-IV, durability of durable class V and the properties of class V machining [4,5,6].

The trunk used was 1/3 of the outer part of the trunk which has better anatomical structure. Only 1/3 of the outer part of the oil palm trunks have good strength properties that can be used as building materials and furniture, while 2/3 of the inside was not good [7]. The saw yield with modified sawmill was 30% in average [8]. One of techniques to improve the quality of oil palm trunk as materials was by hot pre-pressed technique which was successfully applied for solid wood. Wood compaction was a process for increasing wood density by altering cell structure using physical or mechanical means

[9]. The compacting that has been done on some types of solid wood before has been able to improve its quality. Based on the results of previous studies using solid wood, solidification was able to improve physical and mechanical properties significantly for example Agathis [10]. Therefore, in this research, the compaction treatment on the oil palm trunk was expected to improve its quality as a lightweight construction material.

Utilization of oil palm trunk waste for composite boards was a smart effort to handle the Indonesian wood shortage problems and environmental problems caused by tremendous oil palm trunk waste in the plantation area. The composite board was ideally developed as a substitute for the main wood product because it has advantages such as raw materials can be from various non-timber waste (agricultural waste, plantation waste and household waste) [11]. The manufacture of composite boards can also improve the quality of waste oil palm trunk so that it will increase economic value. In addition, the use of composite products will improve the properties of oil palm trunk as raw material board. The advantages of composite boards are: it can be produced in large dimensions based on the requirement, high dimensional stability, and can be produced from low quality wood or non-wood raw materials [12].

2. Materials and Methods

2.1 Materials

The materials used in this research were oil palm trunks (*Elaeis guineensis* Jacq.) with age 25 to 30 years. Trunk waste harvested from oil palm plantations in Jasinga, Bogor West Java. The adhesives used were isocyanate and urea formaldehyde adhesives.

2.2 Methods

The composite boards sample dimensions were 35,0 cm in length, 35,0 cm in width, and 1.5 cm in thickness, and the target density was 0.65 g/cm^3.The boards production was started from the stage of oil palm trunk harvesting using chain saw. The trunk used was 1/3 of the outer part of the trunk. Then the sortiment was made long strand pieces with dimension 120 cm in length, 6 cm in width, and 1 cm thickness. Long strands were hot pressed for 60 minutes with a temperature of 150 °C and a pressure of 20 kg/cm^2 to getting a-0.6 cm strand thickness. The long strand was then spreaded by adhesive of 150 g/m^2, 225 g/m^2, and 300 g/m^2 in each adhesive type and arranged perpendicular between layers of three layers. The next stage was the process of hot pressing according with the type of adhesive. Composite boards bonded by isocyanate and urea formaldehyde were hot pressed at 150°C and 130°C, respectively. All composite boards were hot pressed for 15 minutes, at 20 kg/cm^2 pressure. After hot pressing, the board was conditioning for two weeks, and subsequently the physical and mechanical properties were tested according to JIS A 5908 (2003) standard [13].

2.2.1 Testing of Properties of Composite Board

Testing of Physical Properties

The physical properties of the board tested include: density, moisture content, water absorption, and thickness swelling. Density testing was performed during dry air board conditions JIS A 5908 (2003). The samples dimensions were 10 cm in length, 10 cm in width, and 1.5 cm in thickness.

The density test can be calculated by the equation:

$$\rho\ (g/cm^3) = \frac{\text{Mass}}{Volume}$$

Testing of moisture content was calculated based on baseline weight (A) and oven dry weight (B) for 24 hours and temperature 103 ± 2 ° C. The value of moisture content was calculated by the equation:

$$\text{MC (\%)} = \frac{A-B}{B} \times 100\ \%$$

The samples dimensions of water absorption were 5 cm in length, 5 cm in width, and 1.5 cm in thickness. The testing was conducted according to JIS A 5908 (2003) standard. Initial weight as baseline of the samples determine by weighed it (m1) then immersed in water at room temperature for 2 and 24 hours. The value of water absorption (WA) was calculated by the equation:

$$\text{WA (\%)} = \frac{m2-m1}{m1} \times 100\ \%$$

Thickness swelling was calculated based on the thickness before immersion (t1) measured on all four sides and averaged in dry air and thickness after immersion (t2) in water at room temperature for 2 hours and 24 hours. The value of thickness swelling (TS) was calculated by the equation:

$$\text{TS (\%)} = \frac{t2-t1}{t1} \times 100\ \%$$

Testing of Mechanical Properties

The mechanical properties of the board were tested according to JIS A 5908 standard. The mechanical properties tested were Modulus of Elasticity (MOE) and Modulus of Rupture (MOR).

a. MOE test with destructive testing method used Universal Testing Machine (UTM) instrument. The MOE test used span length of 22.5 cm. The test sample dimensions were 5 cm in width, 27.5 cm in length, and 1.5 cm in thickness.

b. MOR test was performed together with MOE testing using the same sample. In MOR testing, when loading on the MOE test was continued until the test sample was damaged (broken). The MOE and MOR values were obtained from the equation:

$$\text{MOR (kg/cm}^3) = \frac{3PL}{2bh^2} \qquad \text{MOE (kg/cm}^3) = \frac{\Delta PL^3}{4\Delta Ybh^3}$$

Explanation:
MOE = Modulus of Elasticity (kg/cm^2)
MOR = Modulus of Rupture (kg/cm^2)
ΔP = Loading under Proportion Limit (kg)
ΔY = Deflection load (cm)
P = Maximum load (kg)
L = Span length (cm)
b = Sample width (cm)
h = Sample thickness (cm)

2.2.2 Data Analyze Procedure

This research used a complete factorial randomized design model with two factors and confidence level of 5%. If the results of statistical analyzes above 0.05 then the results have not significant effect, if the results of analyze below 0.05 then have significant effect, and if below 0.01 then the results have a very significant effect. All data measured in 5 replications and processed using statistical program.

$$\mathbf{Yijk = \mu + Ai + Bj + (AB)ij + \epsilon ijk}$$

Yijk: The observed value of type of adhesive factor of the stage i and the glue spread factor of the stage j with the repetition stage k.
M: The average value of expectations.
Ai: The main effect of the type of adhesive on the board.
Bj : The main effect of the glue spread on the board.
(AB)ij: Influence of interaction from combination of adhesive type and glue spread on the board.
Eijk: Experiment error.

3. Results and Discussions

3.1 Density

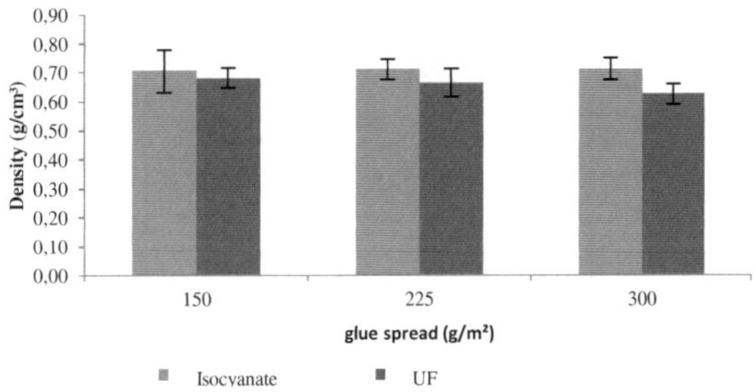

Figure 1. Histogram of composite boards density.

The density of wood is the mass contained in the unit of wood volume [14]. Density is one of the board properties used to predict board strength. Figure 1 shows the board density varies between adhesive types and glue spread treatment. Boards bonded by isocyanate adhesive ranged from 0.70 to 0.71 g/cm³ while those of bonded by urea formaldehyde adhesive ranged from 0.62 to 0.68 g/cm³. The board density increased compared to those of solid oil palm trunks. The solid oil palm trunk density ranged from 0. 26 to 0.45 g/cm³ [15]. Higher of composite boards density caused by the application of hot pressed treatment during composite board production. The composite board density increases with increasing of pressure on hot pressing process [9]. The result of statistical analysis showed that the glue spread and adhesive type had no significant effect on the composite board density.

3.2 Moisture content

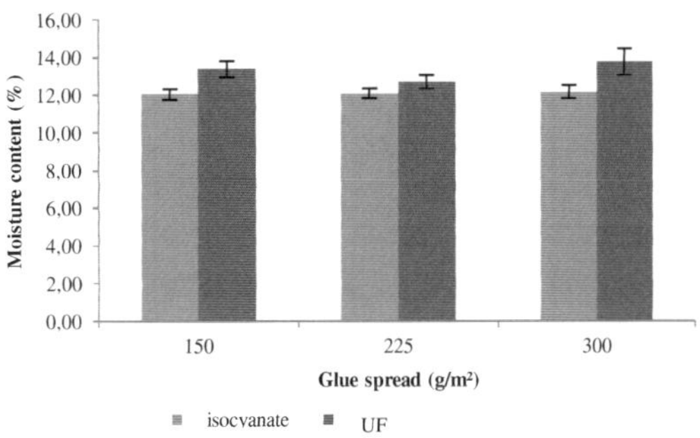

Figure 2. Histogram of composite boards moisture content.

Moisture content is a value used to express the amount of water present in a material or a lignocellulose product [16]. Figure 2 shows the moisture content of the composite boards bonded by isocyanate adhesive ranged from 12.04 to 12.16% with an average of 12.09%. The moisture content of composite boards bonded by urea formaldehyde adhesive ranged from 12.70 to 13.75% with an average of 13.28%. The boards bonded by urea formaldehyde adhesive has an average moisture content higher than JIS A 5908 (2003) standard which requires moisture content below 13%. This was to be influenced by the environmental conditions and urea formaldehyde adhesive failed to prevent the boards absorbing excessive water vapor from the environment. The value of moisture content of oil palm trunks ranged from 219.9 to 379.4% [15]. In addition, according to Putra [17] said that the moisture content of composite board was very dependent on the surrounding air conditions, because the raw material was ligno-cellulosic which had hygroscopic characteristic. The strength of the board will increase when the moisture content decreases below the fiber saturation point. That happened because the condition of the cell wall will be more compact so that the cellulose bond was stronger. The structural unit (micro-fibril) gets close and the tensile force between cellulose molecules gets stronger [14]. Composite board bonded by urea formaldehyde with glue spread 150 g/m² and 300 g/m² failed to meet JIS A 5908 (2003) standard which set up moisture content ranged from 5 to 13%, while those of bonded by isocyanate fulfilled the standard. Based on the statistical analysis, the adhesive type had significant effect on the moisture content of the composite board, while glue spread did not.

3.3 Thickness swelling

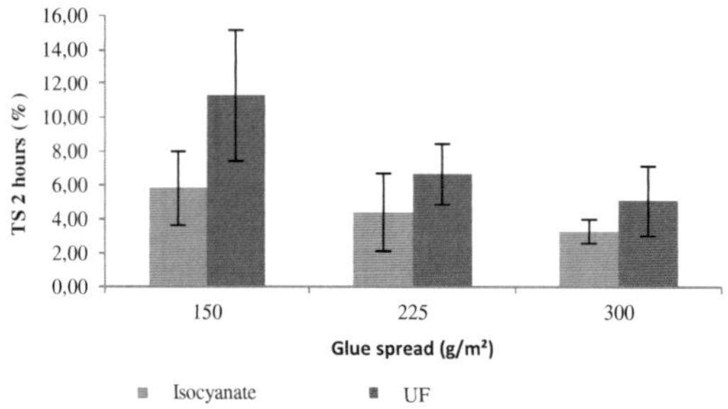

Figure 3. Histogram for thickness swelling after 2 hours of water immersion.

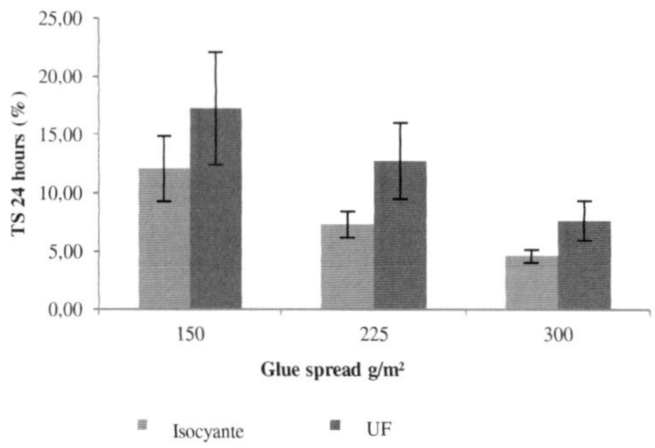

Figure 4. Histogram for thickness swelling after 24 hours of water immersion.

Thickness swelling is one of the parameters that determine the quality of a composite board. Thickness swelling is a quantity that expresses the thickness of the test sample in percent to the initial thickness after the test sample was immersed in water at room temperature for 24 hours [18]. Figures 3 and 4 show the composite board bonded by isocyanate and urea formaldehyde adhesives at two hours of water immersion has a thickness swelling ranged from 3.30 to 5.80% and 5.09 to11.28%, respectively. While at the water immersion of 24 hours ranged from 4.62 to 12.06% and 7.65 to 17.24%, respectively. The composite board bonded by isocyanate adhesive has a thickness swelling less than compared to those of bonded by urea formaldehyde adhesive. This was because, the isocyanate adhesive has waterproof characteristics, quickly hardened and rapidly matures at lower temperatures and was more tolerant of particulate matter or high-water veneer [19]. The JIS A 5908 (2003) standard set up thickness swelling not more than 25% for 24 hours of water immersion. Statistical analysis showed that the glue spreads and adhesive types had significantly affected the thickness swelling of the composite boards.

3.4 Water Absorption

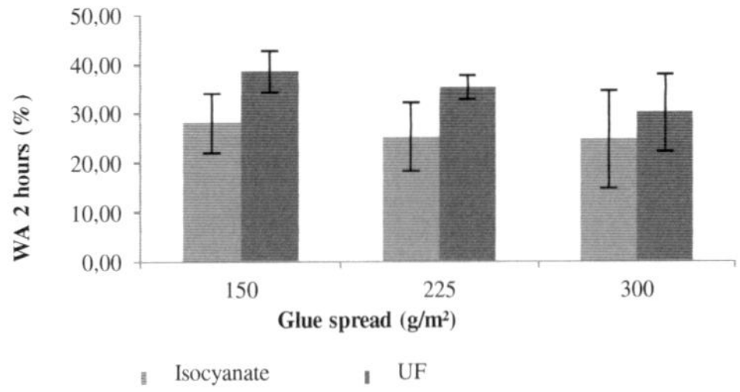

Figure 5. Histogram for water absorption after 2 hours water immersion.

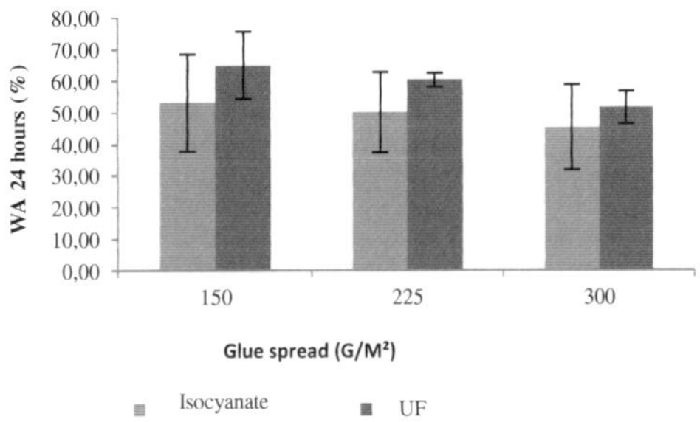

Figure 6. Histogram for water absorption after 24 hours of water immersion.

The water absorption represents the amount of water absorbed by the test sample in percent to its initial weight after the test sample was immersed in water at room temperature for 24 hours [4]. Water absorption was not required in JIS A 5908 (2003). Figures 5 and 6 show that the water absorption of 2 hours and 24 hours of composite boards bonded by isocyanate adhesive ranged from 24.72 to 27.98% and 45.28 to 53.18%, respectively. Whereas those of bonded by urea formaldehyde adhesive ranged from 30.11 to 38.48% and 51.53 to 64.99%, respectively. That water entering the board was divided into two kinds, namely water entered the empty hollow inside the board and into the material of the board [11]. The research results showed that compaction in both adhesive types made the boards more compact, preventing water from entering the board [20]. Although glue spread of 300 g/m² has a smaller water absorption, the statistical analysis indicated that the glue spread levels and adhesive types had no significant effect on composite board water absorption.

3.5 Modulus of Elasticity (MOE)

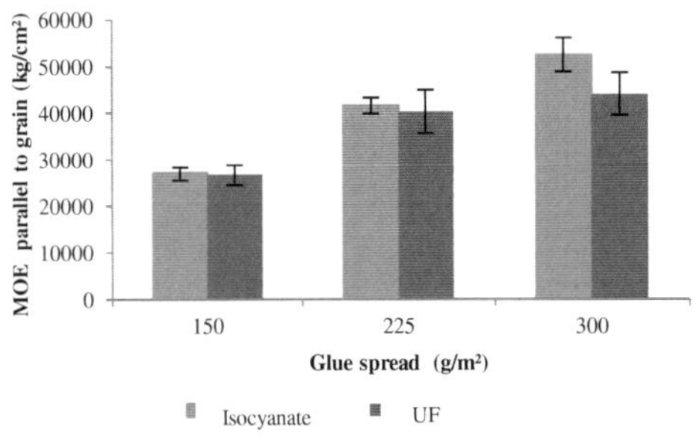

Figure 7. Histogram for MOE parallel to the grain.

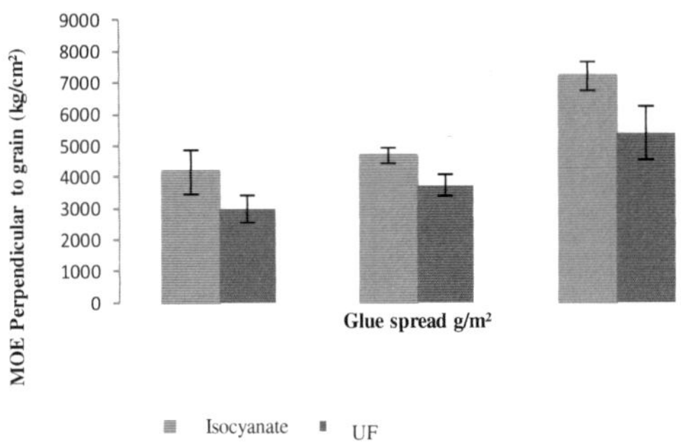

Figure 8. Histogram for MOE perpendicular to the grain.

Modulus of Elasticity (MOE) is a measure of material stiffness or material in resisting form changes or flexures that occur due to loading and only valid in the extent of proportion [17]. Figures 7 and 8 show that the composite boards bonded by isocyanate adhesive had MOE values which were parallel to the grain and perpendicular to the grain ranged from 28.9x10³ to 56.2x10³ kg/cm² and 4.2x10³ to 7,2x10³ kg/cm², respectively. Whereas those of bonded by urea formaldehyde adhesive ranged from 28.5x10³ to 47.3x10³ kg/cm² and 2,9x10³ to 5,4x10³ kg/cm², respectively. In both types of adhesive, glue spread level at 150 g/m² resulted the smallest MOE values while the largest values resulted from those of glue spread at 300 g/m². Composite boards of oil palm trunk waste bonded by isocyanate adhesives had an average MOE better than those of bonded by urea formaldehyde adhesives. The isocyanate adhesive was one of the water-based adhesives that so well bonds with porous materials such as wood and other ligno-cellulosic materials such as oil palm trunk[9,21]. The MOE values of boards increased linearly following the increasing of glue spread used in board production. Based on JIS A 5908 (2003) standard, the composite board with glue spread at 150 g/cm² failed to fulfill the standard, while the glue spread of 225 g/m² and 300 g/m² fulfilled the standard. The MOE parallel to the grain of the composite boards were higher compared to those of solid oil palm trunk. The average MOE of solid oil palm trunks was 15,7x10³ kg / cm² [21]. Statistical analysis indicated that glue spreads and adhesive types had significantly affected the MOE values.

3.6 Modulus of Rupture (MOR)

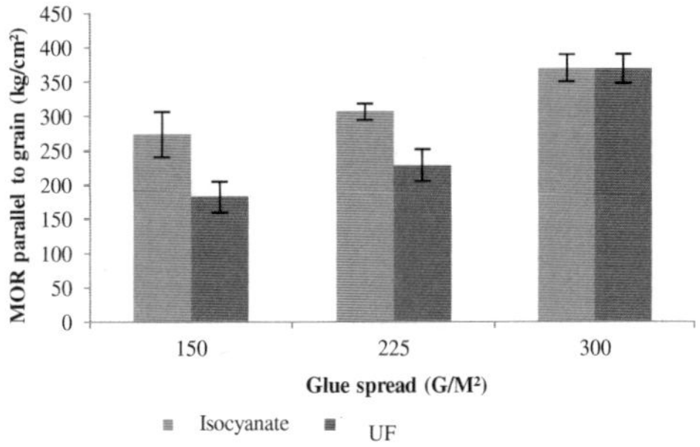

Figure 9 Histogram of composite boards MOR parallel to the grain.
Figure 10. Histogram of composite boards MOR perpendicular to the grain.

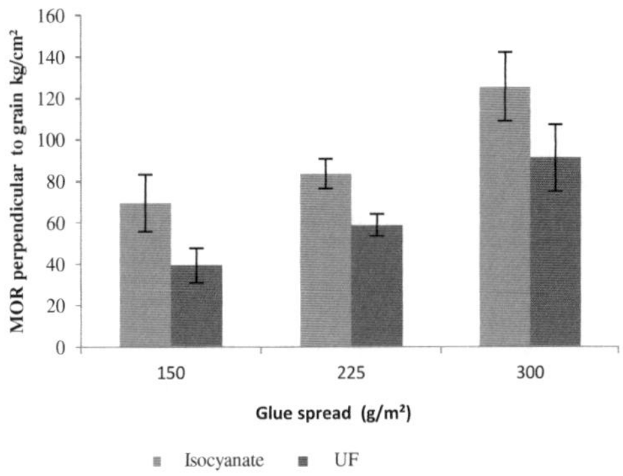

Modulus of Rupture (MOR) is the maximum load that could be resisting until damaged. Figures 9 and 10 showed that the MOR values parallel and perpendicular to the grain of composite boards bonded by isocyanate adhesive ranged from 274 to 371 kg/cm² and 69 to 125 kg/cm², respectively. While those of bonded by urea formaldehyde adhesives ranged from 182 to 369 kg/cm² and 39 to 91 kg/cm², respectively. The composite boards bonded by isocyanate adhesive fulfilled the JIS A 5908 (2003) standard, while those of bonded by urea formaldehyde fulfilled the standard at glue spread of 300 g/m². The composite board bonded by isocyanate resulted higher MOR compared to those of bonded by urea formaldehyde, and MOR parallel to the grain were higher compared to those of perpendicular to the grain. This phenomenon was influenced by the testing positions that the applied force across the fibers length in face and back layers at parallel to the grain. This results also higher compared to solid oil palm trunk. This phenomenon proved that compaction with high temperatures increased the MOR value of oil palm trunks [22]. The MOR value increased following increasing of glue spread used in composite board production. MOR parallel to the grain of solid oil palm trunk ranged from 97-151 kg/cm² [10]. Statistical analysis indicated that glue spreads and adhesive types had significantly affected the MOR of composite boards made from oil palm trunk waste.

4. Conclusions

Based on the test results could be concluded that : (1) composite boards bonded by isocyanate performed better physical and mechanical properties compared to those of bonded by urea formaldehyde, (2) utilization of higher glue spread level would improve the physical and mechanical properties of the composite boards. (3) composite boards bonded by isocyanate and urea formaldehyde adhesives at glue spread of 225 g/m²,300 g/m², respectively were enough to fulfill the JIS A 5908 (2003) standard.

Acknowledgements

The authors gratefully acknowledge financial support from the Directorate of Higher Education, Ministry of Research, Technology and Higher Education of the Republic of Indonesia. The authors also thank for Bogor Agricultural University (IPB) for supporting research facilities.

References

[1] [BPS] Badan Pusat Statistik. 2015. *Statistik Indonesia : Statistical Yearbook of Indonesia* 2015. Jakarta (ID): Badan Pusat Statistik Republik Indonesia.
[2] Hartono *et al.* 2012. Peningkatan Kualitas Batang Kelapa Sawit Bagian Dalam dengan Metode Close System Compression dan Kompregnasi Fenol Formaldehida. [Tesis]. Program Pasca Sarjana, Institut Pertanian Bogor. Bogor.
[3] Dungani R, Islam MN, Abdul Khalid HPS, Davoudpour Y dan A Rumidatul. 2014. Modification of the inner part of the oil palm trunk (OPT) with oil palm shell (OPS) nanoparticles and phenol formaldehyde (PF) resin: physical,mechanical,and thermal properties. *J. Bioresources.* 9 (1) : 455-47.
[4] Bakar ES, Rachman O, Hermawan D, Karlinasari L dan Rosdiana N. 1998. Pemanfaatan Batang Kelapa Sawit (*Elaeis guineensis* Jacq.) Sebagai Bahan Bangunan dan Furniture (I) : Sifat Fisis, Kimia dan Keawetan Alami Kayu Kelapa Sawit. *Jurnal Teknologi Hasil Hutan* Vol. XI (1): 1-11.
[5] Bakar ES, Rachman O, Darmawan W, Hidayat I. 1999a. Pemanfaatan Batang Kelapa Sawit (*Elaeis guineensis* Jacq.) Sebagai Bahan Bangunan dan Furniture (II): Sifat Mekanis Kayu Kelapa Sawit. *Jurnal Teknologi Hasil Hutan* Vol. XII (1) : 10-20.
[6] Bakar ES, Massijaya Y, Tobing TL dan Ma'mur A. 1999b. Pemanfaatan Batang Kelapa Sawit (*Elaeis guineensis* Jacq.) Sebagai Bahan Bangunan dan Meubel (III) : Sifat Keterawetan Kayu Sawit dengan Basilit-CFK dan Impralit-BI. *Jurnal Teknologi Hasil Hutan* Vol. XII (2): 13–20.
[7] Bakar ES., MY Massijaya, O Rahman dan Bahruni. 2000. Pemanfaatan batang kelapa sawit (*Elaensis guineensis*. Jacq) sebagai bahan bangunan dan furniture. Laporan Penelitian Hibah Bersaing VI Dirjen Dikti. Jurusan Teknologi Hasil Hutan. Fakultas Kehutanan IPB. Bogor.
[8] Bakar ES, Febrianto F, Wahyudi I, Ashaari Z. 2006. Polygon sawing: an optimum sawing pattern for oil palm stems. *J Biol Sci.* 6 (4) : 744 - 749.
[9] Younesi M and Bahrololoom M E. 2009. Effect of temperature and pressure of hot pressing on the mechanical properties of PP-HA bio-composite. *Mater.Des.* (30) : 3482-3488.
[10] Sulistyono. 2003. Studi Rekayasa Teknis, Sifat Fisis, Sifat Mekanis dan Keandalan Konstruksi Kayu Agatis (*Agathis loranthifolia* Salisb.) Terpadatkan. [Thesis]. Program Pascasarjana, Institut Pertanian Bogor. Bogor.
[11] Massijaya MY, Kusumah SS. 2005. Analisis kelayakan teknis papan komposit dari limbah kayu dan karton gelombang untuk bahan bangunan dan mebel. *Jurnal Teknologi Hasil Hutan.* 18(2).
[12] Berglund L, Rowell RM. 2005. *Wood Composites, Handbook of Wood Chemistry and Wood Composites.* Boca Raton : CRC Press.
[13] [JIS] Japanese Industrial Standard 2003. *Japanese Industrial Standard for Particle Board* 5908. Tokyo (JP). Japanese Standard Association.
[14] Tsoumis G. 1991. *Science and Technology of wood* (*Structure, Properties, Utilization*). New York (US): Van Nostrand Reinhold.
[15] Iswanto AH, Sucipto T, Coto Z, F Febrianto. 2010. Sifat fisis dan mekanis batang kelapa sawit (*Elaeis guineensis* Jacq.) asal kebun Aek Pancur-Sumatera Utara. *JTHH.* 3(1) : 1-7.

IOP Conf. Series: Earth and Environmental Science **141** (2018) 012012 doi:10.1088/1755-1315/141/1/012012

[16] Bowyer JL, Smulsuky R, Haygreen JG. 2003. *Forest Product and Wood Science an Introduction*. Iowa (US) : IOWA State University Press.
[17] Putra E. 2011. Kualitas papan partikel batang bawah, batang bawah dan cabang kayu jabon (*Anthocephalus cadamba* Miq.) [Skripsi]. Institut Pertanian Bogor. Bogor.
[18] Massijaya YM, YS Hadi, B Tambunan, ES Bakar, I Sunarni. 1999. Studi pembuatan papan partikel dari limbah kayu dan plastic polystirena. Bogor. *JTHH*. 1(2):19-29.
[19] Marra AA.1992. *Technology of Wood Bonding Principles in Practices*. New York (US) : Van Nostrand Reinhold.
[20] Alghiffari AF. 2008. Pengaruh kadar resin perekat urea formaldehida terhadap sifat papan partikel dari ampas tebu. [skripsi]. Institut Pertanian Bogor. Bogor.
[21] Darwis A. Massijaya MY, Nugroho N, EM Alamsyah. 2014. Karakteristik papan laminasi dari batang kelapa sawit. *J. Ilmu Tekonologi Kayu Lapis*. 12 (2) : 157-158.
[22] Salim *et al.* 2013. Effect of steaming on some properties of compressed oil palm trunk lumber. *BioResources*.Part B. 8(2) : 2310-2324.

Lampung natural zeolite filled cellulose acetate membrane for pervaporation of ethanol-water mixtures

D A Iryani[1,3,*], N F Wulandari[1], AW Cindradewi[1], S Br Ginting[1,3], E Ernawati[2], U Hasanudin[3,4]

[1]Department of Chemical Engineering, Engineering Faculty, University of Lampung
[2]Department of Chemistry, Mathemathic and Science, University of Padjajaran Bandung
[3]Research and Development Center for Tropical Biomass, University of Lampung
[4]Department of Agroindustrial Technology, Agriculture Faculty, University of Lampung

*Corresponding Author: dewi.agustina@eng.unila.ac.id

Abstract. Pervaporation of ethanol–water can be cost-competitive in the production of renewable biomass ethanol. For the purpose of improving the pervaporation performance of polymeric membranes, we prepared cellulose acetate (CA) filled Lampung Natural Zeolite (LNZ) membranes by incorporating LNZ into CA for pervaporation separation of ethanol-water mixtures. The characteristics and performance of these filled membranes in the varied ratio of CA:LNZ (30:0, 30:5, 30:10, 30: 20, 20:20 and 40:10) wt% were investigated. The prepared membranes were characterized for pervaporation membrane performance such as %water content and membrane swelling degree. Further, the permeation flux and selectivity of membrane were also observed. The results of investigation show that water content of membrane tends to increase with increase of LNZ content. However, the swelling degree of membrane decrease compared than that of CA control membrane. The permeation flux and the selectivity of membranes tend to increase continuously. The CA membrane with ratio of CA:LNZ 30:20 shows the highest selectivity of 80.42 with a permeation flux of 0.986 kg/(m^2 h) and ethanol concentration of 99.08 wt%.

Keywords: Cellulose acetate membrane; Lampung natural zeolite, pervaporation

1. Introduction

Bioethanol is one of the promising renewable biofuels that continue to be developed as an alternative fuel. However, in order to be used as an alternative fuel, ethanol must have high purity (less than 0.5 wt% water). To achieve high purity, separation techniques such as extractive distillation or azeotropic distillation are required because the ethanol-water mixture forms an azeotrope. However, the ethanol purification process by using conventional distillation followed by extractive distillation is disadvantageous due to the high cost of service [1]. Therefore, a cheaper alternative of purification process is needed to make ethanol to be applied as fuel.

Currently, the pervaporation technique to purification of bioethanol has attracted considerable attention since its high efficiency and low cost operation. Pervaporation is one of purification method by using membrane like material for separation of mixture of liquids by partial vaporization. Pervaporation has several advantages compared with conventional distillation, such as able to separate the azeotropic mixture, no additives required and relatively low energy consume [2]. Membrane pervaporation technique is used to separate a liquid mixture by partly vaporizing it through a nonporous permselective membrane [3]. There are two types of membrane can be applied for ethanol production, namely organophilic membranes for recovery of ethanol from fermentation broth and hydrophilic membranes for dehydration of ethanol near azeotrope point in a hybrid distillation-pervaporation process [1].

There are three types of hydrophilic membranes are often used on pervaporation of ethanol-water, such as: (1) polymeric membranes which are based on organic polymer chains that are cross-linked together, (2) inorganic membranes, fabricated from ceramics or zeolites, and (3) composite membranes that are based on an organic polymeric membrane but have inorganic particles dispersed throughout the polymeric structure, these are commonly termed mixed matrix membranes [2]. One type of polymeric membrane is widely applied to separate substances in liquid and gas phase is cellulose acetate (CA) [4], [5]. CA is widely used as a membrane material because it has several advantages, such as easy to produce and derived from natural renewable substances, low price, moderate chlorine resistance, and it has good biocompatibility [6].

However, organic polymer such as CA is not suitable for more aggressive cleaning as it has easily oxidized and low chemical resistances and mechanical strength CA membrane also has several disadvantages, such as its easily oxidized, low chemical and mechanical resistance [4]. Therefore, a special treatment is required to improve performance by improving the deficiency of CA properties [7],[4],[5]. Membrane modification is one of the special treatments that can be performed to correct its deficiency properties resulting in increased membrane performance by adding additives such as alumina and silica as fillers to improve the characteristics and improve membrane performance [8].

As above mentioned, the performance of the CA membranes can be improved by mixing the additive substance into the membranes. Several researchers have investigated the effect of mixing many CA as organic polymers with inorganic materials such as alumina, silica, titania, and etc [6] and [5]. The presence of inorganic material able to improve mechanical stability, increasing fouling resistance ability, surface porosity, permeability properties and membrane selectivity in separation processes [6], [9], [5]. Recently, few researchers have investigated that the blend of zeolite with organic polymer able to improve pervaporation performance of ethanol-water. [10] had studied the pervaporation separation of ethanol-water mixture using PVA zeolite-clay membranes to investigate the effect of ethanol feed concentration (10%, 30%, 50%, 70% dan 90%) and operation temperature (30, 40, 50, 60, and 70 ^{o}C). Their investigation shows that, when the ethanol concentrations of feed and operation temperature increased, the membrane selectivity increased, on the contrary, the absorption (flux) decreased. Similarly, [11] conducted the investigation of thin-film zeolite-filled silicone/PVDF composite membranes which were fabricated by incorporating zeolite particles into PDMS (poly(dimethylsiloxane)) membranes. The result showed that the increase of zeolite loading from 10% to 30% was increasing the total flux with significantly from 265.0 g/(m^{2}h) into 820.7 g/(m^{2}h) with 5 wt% ethanol feed concentration at 50°C, and the separation factor from 11.3 to 13.7.

In the present study, Lampung Natural Zeolite (LNZ) was added from in 5 to 30 wt % into the casting solution to prepare the new CA/LNZ hybrid membranes. The effects of LNZ addition on prepared membranes were investigated by swelling degree test, and water content. In addition, the pervaporation performance of the membranes such as permeability and selectivity were also evaluated for dehydration of ethanol/water mixture in order to produce ethanol according to fuel specifications.

2. Materials and Methods

2.1 Materials

Commercial grade cellulose acetate (Sigma Aldrich, USA) with acetyl content 39 %wt - 41 %wt and Mw 30,000 was used as the matrix polymer membrane. Lampung natural zeolite (LNZ) (200 mesh) which purchased from CV. Minatama, Lampung,Indonesia was used as a filler and N_2 liquid was used as cold trap permeate on pervaporation process.

2.2 Solution Blending and Hybrid Membranes Fabrication

CA filled zeolite membrane were fabricated using the casting solution method by following procedure of [5]. In this case, CA dissolved in 100 ml of acetone (J.T baker, USA) and stirred until homogen. Afterward, the calculated amounts of LNZ were added into CA solution to making the various ratio of CA : LNZ (30 : 0 ; 30 : 5 ; 30 : 10; 30 : 20; 20 : 20) wt%. The resulting of CA/LNZ slurry was stirred for 24 hours. The resulting CA slurry was set aside for 2 hours to remove air bubbles after 24 hours of stirring. Thus, the CA slurry was casted on a glass plate. The film was then placed in air for 20 seconds to evaporate the solvent, and then dipped in ice water for 20 min to complete the phase separation process. Further, the ice water was replaced with hot water (80 °C) to separate the membranes from the glass surface. For drying purpose, the hybrid membranes underwent to solvent exchange method. Finally, the hybrid membranes were placed in an oven that was maintained at 60 °C for 2 hours.

2.3 Pervaporation test of membrane

The pervaporation experiments test were conducted by using an apparatus as shown in Figure 1. The prepared CA/LNZ membrane was setted into the membrane cell module in pervaporation apparatus. The diameter of effective membrane area in contact with feed was 12.56 cm^2. First of all, the pervaporation apparatus was setting and tested by run the machine using distilled water to remove unwanted material in the machine. Then, the feed composition was running up to free machine from distilled water. The fed solution with 96 wt-% of ethanol/water mixture was stored in the feed flask and heated into 40 °C using heater. A perilstatic pump was used to circulate the feed solution from the feed flask into the membrane cell module. The circulation rate was kept high to minimize concentration polarization in the permeation cell and maximize mixing of solution in the feed tank. The vacuum in the downstream side was maintained at about 0,3 mBar using a vacuum pump, and permeate was collected in liquid nitrogen-cold traps. The compositions of the feed remain in the flask and permeate composition were measured using picnometer and gas chromatography to determine the separation factor and total flux. The permeation flux and the separation factor of membrane were determined as equation as follows:

$$J = \frac{1}{A} \text{ x dm/dt} \tag{1}$$

Where J is permeate flux (/m^2 h), m is total mass of permeate (gram) collected in t (hours), and A (m^2) denotes the effective area of the membrane. The dm/dt is slope which was derived from graph from relations between mass of permeate and time of pervaporation.

$$\alpha = \frac{(Y_W / Y_{eth})}{(X_W / X_{eth})} \tag{2}$$

Where X_w and X_{eth} are the weight fraction of ethanol and water (wt%) in the feed solution respectively, and Y_w and Y_{eth} represent the weight fraction of ethanol and water (wt%) in the permeate.

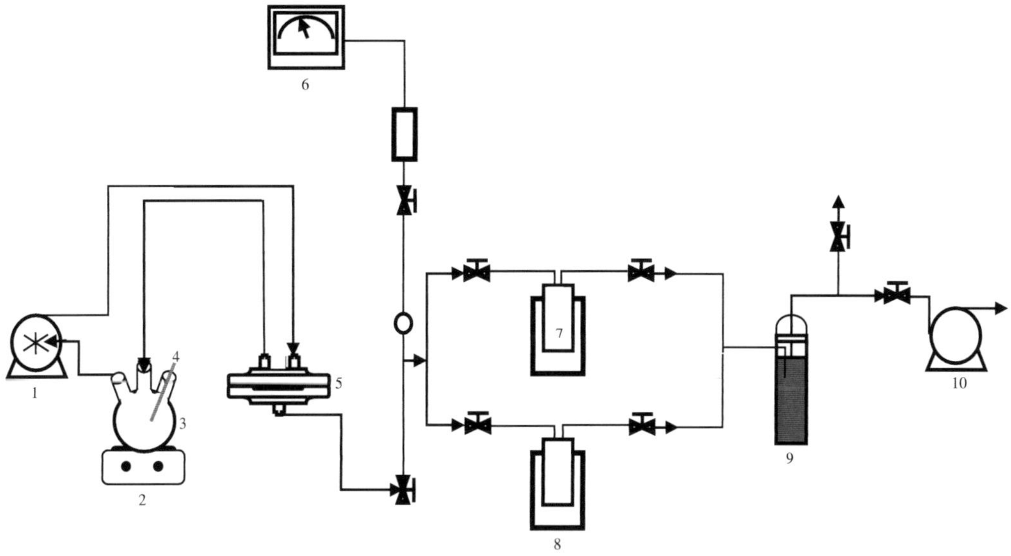

1. Peristaltic Pump, 2. Heater, 3. Three neck glass, 4. Thermometer, 5. Module membrane, 6. Manometers, 7. Cold Trap, 8. Condenser, 9. Silica gel, 10. Vacuum Pump

Figure 1. Scematic of pervaporation experiment apparatus

2.4 Membrane characterization test

2.4.1 Swelling Degree

The swelling degree of the membranes was obtained after soaking a known weight of samples in fed solution (ethanol 96 wt%) for 48 h (until equilibrium reached) at room temperature, then removed. The surface of membrane was on swollen samples was blotted by pressing between the folds of filter paper and quickly weighed. The percentage of swelling degree was determined as follows:

$$\text{Swelling Degree (\%)} = \frac{\text{wet sample weight - dry sample weight}}{\text{dry sample weight}} \text{ x } 100\ \% \quad (3)$$

2.4.2 Water content analysis

The procedure for water content analysis was conducted by following the method of [7]. Water content of the membranes was obtained after soaking a known weight of samples in distilled water for 24 h and the membranes were weighed followed by mopping it with tissue paper. The wet membranes were then dried in vacuum drier at 75 °C for 48 h and the dryweights of the membranes were determined. The percent water content was determined by using the equation as follows:

$$\text{\%Water Content} = \frac{\text{(wet sample weight - dry sample weight)}}{\text{wet sample weight}}\ x\,100\ \% \quad (4)$$

3. Results and Discussions

3.1 Effect of zeolite addition to swelling degree

The ideal pervaporation membrane should have not easy to swell excessively to maintain its structural selectivity and stability [12]. It is observed from Figure 2 that the presence of LNZ as the additive to casting solution decrease membrane swelling degree. The pure 100 wt.% CA membrane was compared with the CA/LNZ ratio of 30:5, 30:10, 30:20, 20:20, and 40:10 wt.%. It was observed that 100 wt.% CA membrane shows almost similar value of swelling degree percentage. In the presence of 10 - 20 wt.% LNZ, the swelling degree percentage showed a lower value than 100 wt % CA membrane. The LNZ presence on casting solution can serve to limit the molecular motions of the polymer chains. Limited molecular motions and a favorable increase in the mean distance between chains can lead to a simultaneous improvement of membrane stability or rigidity and porosity. Thus, it can be concluded that the addition of LNZ into CA gives a significant effect on the swelling degree of the membrane.

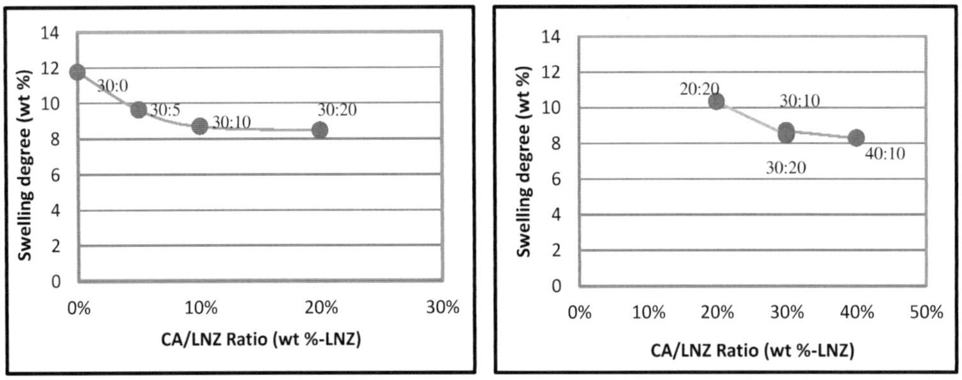

Figure 2. Effect of addition of LNZ into CA solution in varied ratio of CA/LNZ to percentage of swelling degree

3.2 Water Content analysis

Figure 3 shows the effect of LNZ addition into water content of membranes. As can be seen from Figure 3, the percentage water content of 100 wt.% CA membranes are lower than that of the membrane prepared by addition of LNZ. Hydrophilicity properties of the membrane relate to its water content percentage [13]. The hydrophilicity of the membrane affects the percentage of water content due to water absorbed into the membrane matrix. Similar results were also demonstrated by [7]. In their study, it was observed that an addition of polyethylene glycol (PEG 600) and TiO_2 able to increase the membrane water content. The increase of water content possibly due to a presence of TiO_2 as inorganic phase was able to create the spaces in the polymer matrix and make an increase in the mean distance (free volume) between the polymer chains which leads to increasing the water content [14]. Moreover, the occurrence of "leached out" or dissolved during the process of gelation leading to the formation of pores, thus, the water molecule is trapped in the pores of the membrane matrix.

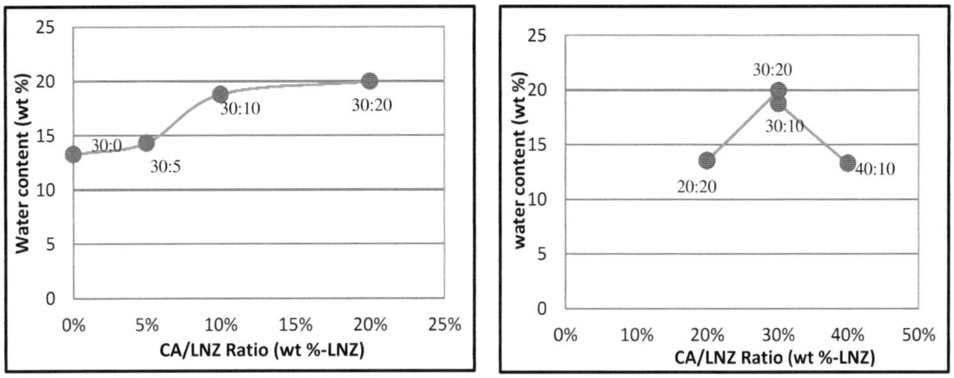

Figure 3. Effect of addition of LNZ into CA solution in varied ratio of CA/LNZ to percentage of water content

3.3 Permeation and selectivity (Separation Factor) of membranes

The results of permeation fluxs test and selectivity of membrane shows in Figure 4 and Table 1. An increase of LNZ addition from 5 to10% resulted an increase of membrane total flux. In contrast to LNZ addition to 20% in the casting solution decreases membrane flux. Similar results were obtain in the ratio of CA/LNZ 20:20 and 40:10. This result occurs due to the thickness of the synthesized membrane were reduced by increasing CA concentration. Increasing CA concentration from 20% to 40% and LNZ loading from 10% to 20% intensifies thermodynamic instability of the cast film solution and thus demixing of this concentrated solution can be performed with less amount of nonsolvent. It also observed from the experiment, viscosity values of the casting solutions are too high and thus it affects diffusivities of ingredients. Increase in the CA concentration 30% to 40% increase in viscosity values and intensive reduction of mutual diffusivities between the nonsolvent (water) and the solvent (acetone) in the system during solidification of the casting solution. Thus, it influence the precipitation process and this leads to preparation of thinner and denser membranes.

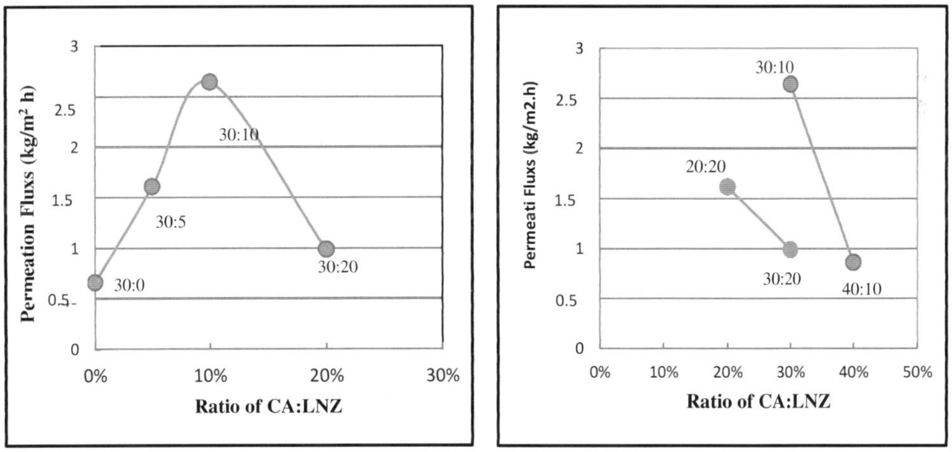

Figure 4. Effect of addition of LNZ into CA solution in varied ratio of CA/LNZ to pervaporation performance of permeation fluxs

Differ with flux permeation value, the selectivity of membrane in ratio of CA/LNZ 30:10 decreases and it was decreasing the dehydration performance of ethanol (97.67%). However, when the ratio of CA/LNZ 30:20 the selectivity of membrane was higher than the separation factor of pure CA and other CA:LNZ ratio. Based on the results, it was observed that increase of LNZ loading into membrane casting solution made hydrophilicity of the membrane changes the membrane morphology in a manner that the flux permeation increase. This can be explained that the transport of ethanol molecules through the CA filled LNZ membranes took place in a straight path through the zeolite pores and led to increase of the separation factor.

Increase in separation factor has achieved when the LNZ content was 20 wt%. The increase in separation factor may have been attributed to the pore size effect of LNZ and a close contact between the polymer and LNZ particles.When excess LNZ was added, more interfacial defects could be generated. Consequently, more ethanol molecules got the opportunity to pass through the membrane. However, the data are shown in the CA / LNZ ratio of 30:10 disagree with the theory or the trend shown by the other data. This is predicted when LNZ mixing in the casting solution causes the viscosity was too high, thus, resulting in an inhomogeneous mixing between CA and LNZ.

Tabel 1. Effect of CA/LNZ Ratio to selectivity and ethanol concentration

Ratio of CA/LNZ	Selectivity (%)	Ethanol Conc. (%)
30 : 0	20,68	97,85
30 : 5	22,15	97,92
30 : 10	17,27	97,67
30 : 20	80,42	99,08
40 : 10	46,88	98,65
20 : 20	31,77	98,28

4. Conclusions

Ratio of CA:NZL affects membrane characteristics and performance. However, the portion of the LNZ added as filler should not be too high, as it may decrease the efficiency of the separation. The best membrane performance was obtained at CA/ZAL ratio 30:20 with flux value 0.986 kg/ m^2.jam, selectivity 80.42 and obtained highest ethanol purity that is equal to 99.081%. Addition of ZAL can improve the hydrophilicity properties of CA membranes.

5. Acknowledgement

The authors gratefully acknowledge the financial support of the Directorate General of Higher Education, National Ministry of Republic Indonesia for research grant of "Hibah Fundamental Dikti 2016". This study was successfully done by the facilities support of Integrated Laboratory for Biomassa University of Lampung.

6. References

[1] Samanta HS and Ray SK 2014 Separation of ethanol from water by pervaporation using mixed matrix copolymer membranes, Separation Purification Technology Journal doi: http://dx.doi.org/10.1016/j.seppur.2015.03.006

[2] Chapman PD, Oliveirab T, Livingstona AG, Li K 2008 Membranes for the Dehydration of Solvents by Pervaporation, Journal of Membrane Science 318 5–37

[3] Kanse NG and Dawande SD 2015 A Review Of Pervaporation Membrane System For The Separation Of Ethanol/Water (Azeotropic Mixture), International Journal Engineering Scince & Research Techology ISSN: 2277-9655 http: // www.ijesrt.com

[4] Kanagaraj P, Neelakandan S, Nagendran A 2014 Preparation,Characterization and Performance of Cellulose Acetate Ultrafiltration Membrane Modified by Charged Surface Modifying Macromolecule, India : Korean Journal of Chemical Engineering 31:6 1057-1064.

[5] Minhas FT, Farrukh S, Hussain A, Mujahid M 2015 Comparison of Silica and Novel Functionalized Silica-Based Cellulose Acetate Hybrid Membrans in Gas Permeation Study, Journal of Polymer Research 22: 63

[6] Arthanareeswaran G, Thanikaivelan P, Srinivasn K, Mohan D, Rajendran M 2004 Synthesis, Characterization and Thermal Studies on Cellulose Acetate Membranes with Additive, European Polymer Journal 40 2153–2159.

[7]Arthanareeswaran G, Sriyamuna Devi T.K, Raajenthiren M 2008 Effect of Silica Particles on Cellulose Acetate Blend Ultrafiltration Membranes: *Part I*, Separation and Purification Technology 64 38–47

[8] Huang HJ, Ramaswamya S, Tschirner UW, Ramaraob BV 2008 A Review of Separation Technologies in Current and Future Biorefineries, Separation and Purification Technology 62 1–21.

[9] Dogan, H and Hilmioglu N D 2010 Chitosan Coated Zeolite Filled Regenerated Cellulose Membrane for Dehydration of Ethylene Glycol/Water Mixtures by Pervaporation, Desalination 258 120–127.

[10] Ling LK, Ghazali M. Nawawi M, Sadikin AZ 2008 Pervaporation of Ethanol-Water Mixture Using PVA Zeolite-Clay Membranes, Journal Teknologi Universiti Teknologi Malaysia, 49(F) 167–177.

[11] Zhan X, Li JD, Chen J, Huang J 2009 Pervaporation of Ethanol/Water Mixtures with High Flux by Zeolite-Filled PDMS/PVDF Composite Membranes, Chinese Journal of Polymer Science 27:6 771–780.

[12] Pabby, Anil K., Syed SH, Rizvi, Requena AM 2015 *Handbook of Membran Separations: Chemical, Pharmaceutical, Food, and Biotechnological Applications, 2nd edition CRC Press, Taylor & Francis Group.*

[13] Sivakumar, Muthusamy, Mohan D R, Rangarajan R 2006 Studies on Cellulose Acetate Polysulfone Ultrafiltration Membranes II: Effect of Additive Concentration, Journal of Membran Science Volume 268; 2 208–219.

[14]Wang Y, Yang L, Luo G, Dai Y 2009 Preparation of Cellulose Acetate Membrane Filled With Metal Oxide Particles for the Pervaporation Separation of Methanol/Methyl Tert-Butyl Ether Mixtures, Chemical Engineering Journal 146 6–10

Ampel Bamboo Leaves Silicon Dioxide (SiO_2) Extraction

Irzaman[1,2], Novi Oktaviani[1], Irmansyah[1]

[1]Physics Department, FMIPA Bogor Agricultural University, Kampus IPB Dramaga Bogor, Indonesia
[2]SBRC Researcher, Bogor Agricultural University, Kampus IPB Baranangsiang Bogor, Indonesia
Email: irzaman@apps.ipb.ac.id

Abstract. The bamboo tree trunk was the most commonly used part of daily life. Bamboo leaves often wereconsidered waste by the community, and bamboo leaves contain Silicon dioxide (SiO2). We have developed and compare two silicon dioxide method, using combustion to washing method (A) and washing to combustion method (B). Atom purity of either method was 99.9 %, with the tetragonal crystal structure. Mg, Au, Ca, and K impurities were found in Method A sample, andnot in Method B.
Keywords: Silicon dioxide, Ampel bamboo leaves, amorphous.

1. Introduction

Indonesia was a tropical country rich in flora and fauna with variations and varieties. One of them is bamboo. There were 143 species of bamboo in Indonesia, and 60 species are on the island of Java. Bamboo leaves cloud to grow in wet climates to dry climates, plateau or low [1].

Bamboo rod was the most common part used by the community to make paper, handicrafts and medicines. Utilization of other bamboo parts such as the roots, leaves, and branches of bamboo plants have not been fully utilized [1]. Bamboo leaves were often considered garbage by the public and received less attention. It was very regrettable because in bamboo leaves there are still compounds that can be utilized, namely silicon dioxide or silica [2].

Silicon oxide (silica) was the basic material for the manufacture of glass, ceramics, and refractory industries. Silica has been used for essential raw materials for the production of soluble silicates and silicon carbide [3]. The price of mineral silica was currently quite expensive, so the use of silica from organic materials has become another alternative in silicon chip industry [4]. The average weight of silica in the earth's crust was 27.7% [5].

In 2015, Aminullah has extracted silica from bamboo leaves with 3% HCl done after burning. The EDX test has shown that the purity of silica obtained is 65.85-74.49% with the impurities of $A_{l2}O_3$, Rb_2O, and Fe_2O_3 [6]. In 2016, Sa'diyah has extracted the silica from bamboo leaves by changing the phase in the extraction method, i.e. by transferring the washing process to before the combustion with the concentration of HCl to wash the bamboo leaves was 3%. The purity has been gained by changing this phase is 67.62% with one impurity, that was carbon. Therefore, it was necessary to optimize HCl concentration to obtain silica with higher purity and free of impurities. The objective was to calculate atomic purity and analyze the effect of HCl concentration (1%, 3%, and 5%) with two different extraction phases, and calculates the elemental composition and crystalline silica of Ampel bamboo leaves.

Bamboo has been known to contains high enough silicon oxide [7]. Based on the results of the research, bamboo has a cellulose content ranging from 42.4% -53.6%, lignin content ranges from 19.8% -26.6%, powdered 1.24% -3.77%, ash content 1.24% -3.77%, silica 0.10% 1.28%, extractive levels of 0.9% -6.9% and bamboo containing holocellulose (cellulose and hemicellulose) ranged from 73.32% -83.80% [8]. Due to the silica content owned by bamboo, it was possible to utilize bamboo to be more optimal by extracting the silica content from the bamboo. Silica was a chemical that utilization and its application is quite wide in various fields [2][9]. Table 1 hasshown the Chemical Composition of Bamboo Leaf Ash, whereas Table 2 have shown the Chemical Composition of Abu Husam Rice [10].

Table 1. Chemical Composition of Bamboo Leaf Ash [12]

Element	Mass (%)
SiO_2	75.90
Al_2O_3	4.13
Fe_2O_3	1.22
CaO	7.47
MgO	1.85
K_2O	5.62
Na_2O	0.21
TiO_2	0.20
SO_3	1.06

Table 2. Chemical composition from rice husk ash [12]

Element	Mass (%)
Fe_2O_3	0.95
SiO_2	67.30
CaO	1.36
Al_2O_3	4.90
MgO	1.81
L.O.I	17.78

Silicon dioxide was a compound that can be found in everyday life and often has been used as raw materials of the electronics industry [4]. Silicon atoms could form four bonds simultaneously and arrange tetrahedrally. In silica, each Si atom was bonded to four O atoms and each O atom was bonded to two sides of Si atom (Figure 1). The oxygen atom was electromagnetic and the atomic density of silica has been transferred to the oxygen atom [6]. Table 3 has shown thecharacteristics of SiO_2 [11].

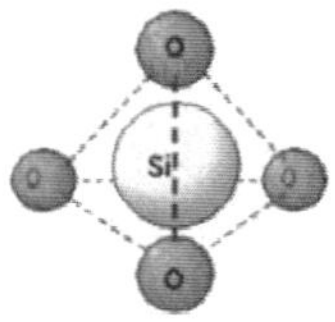

Figure 1. Crystal Structure Silica[2]

Tabel 3. SiO_2 Characteristics[13]

Crystal Structure	Amorf
Atomic Weight	60.08 g/mol
Density	2.27/2.18 g/cm^3
Molecule	2.3 x 10^{22}/cm^3
Specific heat	1.0 J/g-K
Melting Point	1700 ^{0}C
Thermal expansion coefficient	5.6 x 10^{-7}/K
Young Modulus	6.6 x 10^{10}N/m^2
Poisson Ratio	0.17
Thermal conductivity	1.1 W/m-K – 1.4 W/m-K
Relative dielectric constant	3.7-3.9
Dielectric strength	10^{7}V/cm
Energybandgap	8.9 eV
DC resistivity	10^{17}Ωcm

2. Materials and Methods

This research was consists of two methods. Method A has started from the burning of bamboo leaves, bamboo charcoal stalking, silica extraction, and analysis of the composition of crystal structure elements, and electrical properties. The result of extraction from method A was called silica A. Method B has been started from leaching (extraction process), bamboo leaves burning, bamboo charcoal leafing and analysis of elemental composition, crystal structure. The result of extraction from method B was called Silica B.

Bamboo Leaf Burning

Bamboo leaves weredried with the help of sunlight and weighed as much as 2 kg. Dried bamboo leaves have been burned in open space without additional fuel. The resulting bamboo charcoal then was weighed with an analytical balance [12].

Bamboo Coal Billing

The milling process was carried out by adopting several studies on the process of rice husk fertilization [4][7][12]. The bamboo charcoal was weighed 10 grams, then put in a porcelain dish and arranged so that it has the same thickness and burned in a furnace with an initial temperature of room temperature, then raised to 400 ^oC with 2 hours hold time. The next heating was continued with a temperature of 950 ^oC with a holding time of 1 hour. The temperature rise rate was set at 1 ^oC / min. Furthermore, the temperature was lowered to room temperature.

Silica Extraction Process A

After the probe process, bamboo ash leaves were weighed and washed using 1%, 3%, and 5% HCl. The washing process was aimed to reduce the impurities present in the ash of bamboo leaves other than silica. The washing process was carried out as follows: first ash was inserted into a cup glass, then mixed with 1% HCl, 3%, and 5% technical (12 mL 1% HCl, 3%, and 5% technical for 1 gram of bamboo ash) , then was heated over hotplate, temperature control knob on the hot plate temperature to show the temperature scale of 200 oC and stirred with a magnetic stirrer at 240 rpm for 2 hours. Afterwards, the sample was washed with repeatedly acid-free double distilled water (aquabides) (tested using indicator pH), then filtered with ash-free filter paper. The result of the filtration was heated in a kiln at a temperature of 1000 ° C for 1 hour.

Silica B Extraction Process (Bamboo Leaching)

The washing process was aimed to remove the existing impurities in the ash of bamboo leaves other than silica. Dry bamboo leaves have been soaked in 1%, 3%, and 5% technical HCl (i.e. 12 mL HCl 1%, 3%, and 5% technical for 1 gram of bamboo leaves). The sample was heated on top of the hotplate, the temperature setting button on the hotplate is set so that it shows a 200 °C temperature scale for 2 hours. Afterwards, the sample was washed using aquabides repeatedly until it was acid-free (tested using indicator pH). Furthermore, the sample was filtered and dried. After the washing process, then bamboo burning was done then the process of making bamboo charcoal briquettecontinues. The bamboo leaves have been heated in a kiln with a temperature of 1000 °C for 1 hour.

SEM/EDX

The resulting silica of all treatments was analyzed using SEM / EDX. The EDXini test is performed to identify the composition of the ingredients contained in the sample so as to determine the purity of the silica [13]. The morphology of the silica surface was analyzed using Scanning Electron Microscopy (SEM).

X-Ray Diffraction

X-ray diffraction X-ray spectroscopy (XRD) was used to identify the crystalline phase in the material by determining the lattice structure parameters. XRD has been used the principle of X-ray diffraction that undergoes scattering. The resulting diffraction pattern was representing the crystal structure. From the analysis of diffraction patterns, we cloud be determined lattice parameters, crystal size, crystalline phase identification. The type of material could be determined by comparing the XRD results with the catalogue of results diffused by various materials. The XRD randomized detector used was the SHIMADZU XRD 7000 X-RAY MAXima Diffractometer, the target source using CuKα1 (λ = 1.54060 Å). Samples wereprepared as much as 2 grams, then inserted into a holder of 2 × 2 cm in the diffractometer. The initial angle was taken at 10 ° and the end angle at 80 ° with a reading rate of 0.02 ° per minute. This analysis was conducted to find out the degree of sample crystallinity and calculate the lattice parameter value by using Cohan method. The results have been compared with the Joint Committee on Powder Diffraction Standards database (JCPDS).

3. Results and Discussions

This research has resulted in silica derived from Ampel bamboo leaves waste. Silica extraction in this study was obtained by using two methods, namely method A and method B. Method A began with the burning of bamboo leaves until the ampere of bamboo leaves, the bamboo leaves have been heated using a blast furnace to produce ash. The bamboo leaves of this ample were then reheated in the furnace to obtain silica. The extraction silica using method A was called silica A. The bamboo charcoal of 40 amps of bamboo leaves was yielding 22.63 g of silica A. Method B has been started with the washing of bamboo leaves of Ampel. Bamboo leaves that have been washed and neutralized then burned to charcoal. Charcoal bamboo charcoal leaves were heated into the furnace until silica is obtained.

The extraction silica using method B was called silica B. The leaves of bamboo amplewas 80 g, which gives silica 8.15 g. Mass shrinkage for silica B extraction was more than silica A. The purity of Silica A was higher but has many impurities. This Impurity has been allowing for inaccuracies of the data being analyzed because when testing the sample, the probable probability was the material other on impurities, so only one silica A sample could be tested SEM / EDX and XRD, while Silica B should be tested SEM / EDX and XDR for each washing variation.

Silica Purity based on EDX Test

EDX test has been aimed to determine the chemical content contained in the sample. The EDX test results for method A and method B has been shown in Table 4.

Tabel 4. Percentage (%) of sample atoms of EDX test results

Elemental Name	Atom (%)			
	SilicaA	SilicaB		
	1%	1%	3%	5%
Oxygen	38.01	47.6	65.19	37.71
Silicon	54.89	52.29	34.81	60.19
Magnesium	0.98	-	-	-
Aurum	3.97	1.1	-	2.1
Calsium	1.13	-	-	-
Kalium	1.01	-	-	-
Silica Purity	**99.99**	**99.99**	**99.99**	**99.99**
Silicon Purity	**39.43**	**28.04**	**2.215**	**42.435**

Silica A was the silica obtained by washing process after combustion. Previous research has shown that silica A still contains some impurities, but the high purity of silica. So for silica A only sample by washing using 1% technical HCl hastested its chemical compositions because it has the best conductivity value. While Silica B was performed washing before burning, based on the calculation of the purity of silica atoms. Silica A has an atomic purity value of 99.99% with four impurities, namely magnesium, aurum, calcium, and potassium. When burning, the composition of the ampere bamboo leaves has undergone a chemical reaction so that there was a breaking of the bond. This breaking of the bonds has caused reactions that occur when leaching of bamboo charcoal using HCl cannot bind to the compound and the impurities found as shown in Table 4. The HCl concentration during washing was affecting the compound successfully attached by the HCl. Reactions that may occur when washing the bamboo leaves of the ample:

$$K_2O_{(s)} + 2\,HCl_{(aq)} \rightarrow 2\,KCl_{(aq)} + H_2O_{(l)} \quad (1)$$

The higher the technical HCl concentration, the reaction will run toward the product. In the reaction was written in equation (1), the resulting product was KCl and H_2O. KCl was lost when washed with aquabides. While H_2O or water evaporates during heat treatment. The purity of silica B atoms based on the EDX test result was 99.99%. Silica B washed using 1% and 5% HCl concentrations have one impurity, ie aurum. While silica B washed using 3% HCl concentration did not have any impurities. Aurum was a layer used during testing to make silica more conductive. Based on table 4, there was no aurum content in bamboo leaves. Thus the aurum present in the sample was a deliberately coated impurity on the sample at the time of testing. Silica B with high purity present in Silica B by washing using 3% HCl was technically preferable because the percentage of atoms was close to the required value for obtaining pure SiO_2. In addition to silica, there was pure silicon in the sample. So if the silicon extraction could be done from the sample, the process will be easier.

Silica surface morphology based on SEM Test

Figure 2. Morphology of Silica A by washing using HCl 1% technical magnification 5000x

Figure 3. Silica B morphology with washing using HCl 1% technical 5000x magnification

Figure 4. Morphology of Silica B by washing using HCl 3% technical magnification 5000x

Figure 5. Silica B morphology with washing using 5% HCl technical 5000x magnification

The magnification performed during SEM testing was 5000x for each sample. The morphology of silica A and silica B surfaces was not homogeneous and there are elliptical clumps that look like colonies. In silica A, there are crystalline grains that have been dispersed although not homogeneous (Fig. 2). There are very few crystals was present in silica B (Fig. 3, 4, 5). The distribution of crystal grains that was visible on the surface morphology may affect the X-ray diffraction pattern on the XRD test. Silica B has produced images with more pores. This because silica B was more dominated by amorphous structures.

Grid Parameter Value and Degree of Silica Crystal based on XRD Test

The XRD characterization test yields the diffraction pattern shown in Figure 6. From the diffraction pattern shows prominent peaks. The peak then produces a lattice parameter of silica.

Table 5. Crystal and Silica Crystal Lattice Parameters based on XRD data

Silica Samples		Grid Parameters (Å)		Structure
		a=b	c	
Silica A	1 %	5.2798	7.5730	Tetragonal
Silica B	1 %	2.4471	8.4543	Tetragonal
	3 %	2.8103	4.2831	Tetragonal
	5 %	2.2946	8.5860	Tetragonal
JCPDS		4.9730	6.9230	Tetragonal

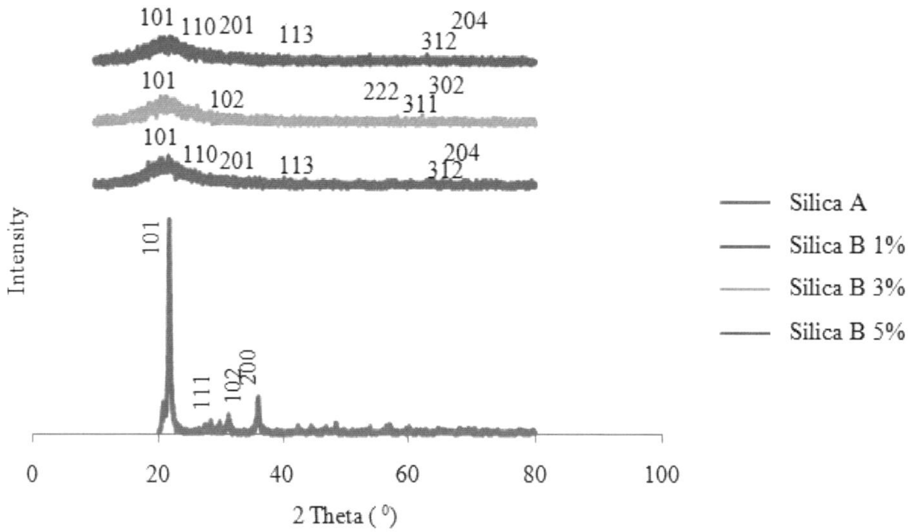

Figure 6. Pattern of X-Ray Diffraction on Silica A and Silica B

Table 5 has shown the values of the silica lattice parameters A and silica B compared with the lattice parameter values in JCPDS ICDD 1997. The values of the silica lattice parameter A was closer to JCPDS data than the values of the silica B lattice parameters. According to the XRD test, silica A has a higher crystallinity value than Silica B. The resultant peak based on the XRD test has shown that the intensity of silica A was higher and was found in four angles. The heat treatment has caused the amorphous crystal structure changes to either the crystal or the tridymite. In the combustion phase prior to washing it has caused the crystalline structure of Silica A to turn into a crystalline. This cristobalite structure has caused an increase in intensity when fired X-ray. High-intensity values have shown constructive interference. Constructive interference has occurred because of the similarity of phase, the same atomic orientation. Silica B was a silica with an amorphous phase so that the resulting peak was more gentle and steep with high intensity hard to find. The intensity was produced when X-ray shooting on silica B was not high so that the peaks do not stand out. Washing on Silica B does not change the amorphous structure. The peaks formed on silica A and silica B did not correspond to the 2nd angle of the JCPDS database.

4. Conclusions

High purity silica has been obtained with two different leaching phases. Silica A has a purity of 99.99% but has many impurities, while Silica B has a purity of 99.99% with one impurity for silica washed using 1% HCl and 5% technical. The impurity was the gold obtained by the coating process during the SEM test. Silica B with 3% leaching was 99.99% purity without impurities. Silica A with washing using 1% HCl and silica B with a 1%, 3%, and 5% HCl washers was containing pure silicone content other than silica. With the existence of pure silicon content other than silica, then the silicon extraction process will be easier to do.

The effect of HCl concentration during washing was the binding of the compound performed by the HCl so that the purity and impurities of silica depend on the concentration of HCl. Silica has a tetragonal crystal structure. Silica A has a crystalline phase with a high degree of crystallinity. While silica B has an amorphous phase with a low degree of crystallinity. Silica A was a semiconductor for high frequency and isolator at low freq. This shows that silica A was not a good electrical conductor. While silica B was a semiconductor both at high frequency and low frequency.

Acknowledgements

This work was funded by the Grant of International Cooperation and International Publication, Ministry of Research, Technology and Higher Education, the Republic of Indonesia under contract No. 011 / SP2H / LT / DRPM / IV / 2017 and Grant Mandate Division of Ministry of Research, Technology and Higher Education, the Republic of Indonesia under contract No. 011 / SP2H / LT / DRPM / IV / 2017.

References

1. Priyanto A. Sintesis dan aplikasi silika dari abu daun Bambu Petung (Dendrocalamus Asper (Schult F.) Backer Ex Heyne) untuk mengurangi kadar ammonium dan nitrat pada limbah cair tahu. UIN Walisongo; 2015.
2. Amu OO., Adetuberu AA. Characteristics of bamboo leaf ash stabilization on lateritic soil in highway construction. Int J Eng Technol. 2010;2(4):212–219.
3. Kurniati E. Ekstraksi Silica White Powder dari Limbah Padat Pembangkit Listrik Tenaga Panas Bumi Dieng. UPN Press. Surabaya; 2009.
4. Irzaman DH, Alatas H, Irmansyah HA, Indro M, Hardhienata H, Abdullah K, et al. Optimization of thermal efisiency of cooking stove with rice-husk fuel in supporting the proliferation of altenative enrgy in Indonesia. In: Procceding Symposium on Advanced Technological Development of Biomass Utilization in Southeast Asia. 2009. p. 20.
5. Tanah BP. Mengenal Silika sebagai Unsur Hara. War Penelit Dan Pengemb Pertan. 2010;32.
6. Aminullah, Rohaeti E, Irzaman. Reduction of High Purity Silicon from Bamboo Leaf as Basic Material in Development of Sensors Manufacture in Satellite Technology. Procedia Environ Sci. 2015;24:308–316.
7. Suhardiman M. Kajian Pengaruh Penambahan Serat Bambu Ori terhadap Kuat Tekan dan Kuat Tarik Beton. J Tek. 2011;1(2).
8. Noverliana N, Asmi D. Sintesis Keramik Silika dari Daun Bambu dengan Teknik Sol-Gel dan Karakterisasi pada Suhu Kalsinasi 500 OC, 600 OC, dan 700 OC. J Teori Dan Apl Fis. 2017;3(1).
9. Krisdianto GS, Ismanto A. Sari hasil penelitian bambu. Pus Penelit Has Hutan Badan Litbang Kehutan Dan Perkeb Bogor Hlm. 2000;29.
10. Dwivedia V, Singhb N, Dasa S, Singha N. A new pozzolanic material for cement industry: Bamboo leaf ash. Int J Phys Sci. 2006;1(3):106–111.
11. Filipovic L. Topography simulation of novel processing techniques. na; 2012.
12. Faiz MA, others. Pengembangan Ekstraksi Silikon Sekam Padi untuk Semikonduktor. 2013;
13. PADI SDSDS. OPTIMASI KELAJUAN PEMANASAN PADA EKSTRAKSI. J Biofisika. 2013;9(2):13–20.

IOP Conf. Series: Earth and Environmental Science **141** (2018) 012015 doi:10.1088/1755-1315/141/1/012015

Lipid and fatty acid composition microalgae *Chlorella vulgaris* using photobioreactor and open pond

M I Jay[1], M Kawaroe[1,2], H Effendi[3,4]

[1]Department of Marine Science and Technology, Faculty of Fisheries and Marine Science, Bogor Agricultural University, Dramaga Campus, Bogor 16680, Indonesia
[2]Surfactant and Bioenergy Research Centre, Bogor Agricultural University, Indonesia
[3]Department of Aquatic Resources Management, Faculty of Fisheries and Marine Science, Bogor Agricultural University, Indonesia
[4]Center for Environmental Research, Bogor Agricultural University, Indonesia
ismetjay@gmail.com

Abstract. Microalgae contain lipids and fatty acids that can be the raw materials of biofuel. Previous studies have been known of using cultivation systems to obtain biomass of *C. vulgaris* which can be extracted to obtain lipid and fatty acid content. The observational step was observed ten days in photobioreactor and open pond for harvesting biomass using NaOH, lipid extraction using hexane and methanol, and fatty acid analysis using Gas Chromatography. Lipid content of microalgae biomass in photobioreactor and open pond was 2.26 ± 0.51% and 3.18 ± 0.80%, respectively. Fatty acid content ranged between 0.7-22.8% and 0.9-22.6% and the dominant fatty acids in both cultivating system was palmitic acid.

1. Introduction

Microalgae are microscopic organisms that live in fresh and marine waters and produce their own food through photosynthesis [1]. Microalgae contain lipids and fatty acids that can be the basic ingredients of biofuel [2]. Another advantage of microalgae does not compete with other crops in land use, otherwise it produces oil in larger quantities when compared to land plants in the same area [1]. The genus *Chlorella* is one of the most abundant types of microalgae in aquatic habitat, with green features, spherical structure, can grow well in the salinity range 0-35 psu, and the optimum temperature is 25-30°C [1].

Various advantages of microalgae make the organism potentially as industrial raw material, so much conducted the development of cultivation system and lipid extraction. The microalgae cultivation system influences the cultivation outcome of biomass, further extracting lipid and fatty acids [1]. A good cultivation system is needed to support the growth of microalgae so the optimum biomass can be known. In general, microalgae can be cultivated using photobioreactor and open pond.

Photobioreactor is a closed harvesting system that can be conducted indoor or outdoor. For indoor culture, fluorescent lamps is used as a light source, for photobioreactor that using sunlight as light source should be conducted in outdoor[1]. Photobioreactor is made using glass to facilitate light penetration. The application of photobioreactor has increased and developed for application in the pharmaceuticals, cosmetics, aquaculture, and agriculture. Open ponds used to cultivate microalgae can be artificial ponds or natural ponds such as lakes. Decreased media due to evaporation, atmospheric CO_2 diffusion, and high risk of exposure to contaminants are a major problem in microalgae cultivation using open pond [3].

Lipid and fatty acid from biomass microalgae have similar chemical content to vegetable oil and considered as potential biofuel feedstock [4]. Microalgae lipid content has form of glycerol and fatty acids. Under certain conditions, microalgae are known have 85% of the dry weight lipids, this amount exceeds the amount of lipids if compared with the other land plant [1]. The purpose of this study is to analyze the biomass, lipid, and fatty acid content of microalgae *C. vulgaris* using two different cultivation system, the photobioreactor and open pond.

2. Materials and Methods

The study was conducted from August 2016 to January 2017. Microalgae culture was conducted in the laboratory of the Marine Bioprospecting and Biomaterial Department (MBB) of Marine Science and Technology and the laboratory of Surfactant and Bioenergy Research Center (SBRC). Microalgae seeds *C. vulgaris* was obtained from Surfactant and Bioenergy Research Center.

2.1. Microalgae cultivation procedure

The cultivation on the laboratory scale was conducted to multiply the microalgae seeds to reach the required amount for the next stage. Lab scale cultivation was carried out to the size of 1-5 liters per container. Walne fertilizer was given and microalgae placed under a TL lamp (1000 lux) for cultivation at this stage. Semi-mass cultivation is carried out from volume of 5 liters to 50 liters. The fertilizers used at this stage were replaced by the technical fertilizers referring to [1]. At the stage of mass scale cultivation, the process is conducted using photobioreactor and open pond. The light obtained comes from sunlight because mass cultivation was conducted outdoors. Mass cultivation was performed on 100 liters volumes for 10 days with the same volume in photobioreactors and open pond. Cultivation was conducted with twice repetitions based on time. The fertilizer used was TSP (CaH2PO4) 30 ppm, ZA (ammonium sulphate) 60 ppm and 60 ppm Urea which refers to [5]. While the comparison between microalgae and medium was 1: 10 [5].

2.2. Calculation of biomass weight

Biomass weight calculations are performed daily using gravimetric method. Whattman filter paper of 90mm diameter was inserted into an oven with 60°C temperature for 3 hours to remove the moisture present in the filter paper. Then cooled for 10 minutes and recorded the empty weight (b_0) on the digital scale.

After filter paper used for microalgae biomass, put it back into the oven at the same temperature and time. After that it is cooled in the desiccator at the same time then weighed again and recorded the weight of its contents (b_1). The difference between (b_0) and (b_1) is the biomass of microalgae (g/L). Then put into equations referring to [6]:

$$\text{Biomass (g/l)} = (b_1\text{-}b_0)\frac{1000}{ml\ sample} \qquad (1)$$

Information :
b_0 = filter paper weight before filtering
b_1 = filter paper weight after filtering

2.3. Analysis of lipid and fatty acid

The lipid extraction and calculation formula in this study refers to [7]. After dry biomass is obtained, the biomass was dissolved into hexane and metanol at a ratio of 1: 1, as much as 200 mL in erlenmeyer. The next stage was sonication 38 kHz for 20 minutes. The mixture was homogenized with a magnetic stirrer with a temperature 25°C for 24 hours. The extraction results were centrifuged at 4000 rpm for 10 min. The results obtained are supernatant and natant (pellet). Then the supernatant evaporated on a rotary evaporator with a temperature 50°C to obtain lipid from microalgae biomass. Then used extracted lipid microalgae for analyzed fatty acid composition based on methods issued by Environmental Protecting Agency (EPA 1996) with GC-MS [1]. Lipid from microalgae biomass calculated using formula [7] :

$$L = \frac{M_l}{M_s} \times 100 \qquad (2)$$

Information :

L = lipid content (%)

M_l = lipid mass (g)

M_s = sample mass (g)

3. Results and Discussions

3.1. Microalgae Biomass C. vulgaris

The current photobioreactor biomass (h0) is 0.43 ± 0.29 g/L. The biomass at open pond at (h0) is 0.41 ± 0.15 g/L (Figure 1). There is a fluctuation of microalgae biomass that is cultivated in both systems according to their life phase.

Photobioreactor biomass at (h0) or lag phase of 0.43 ± 0.29 g/L then increases to the highest value at (h4) or stationary phase is 1.27 ± 0.44 g / L. The current photobioreactor biomass (h5) weighing 0.93 ± 0.58 g / L then decreases to the end of the observation or when the microalgae reaches the death phase at (h10) weighing 0.34 ± 0.33 g / L. The open pond biomass at (h0) or lag phase is 0.41 ± 0.15 g/ L then increases to the highest value at (h5) or when the stationary phase is 0.94 ± 0.11 g / L. Biomass continues to decline to the end of observation when (h10) or when it reaches the death phase to be 0.49 ± 0.13 g / L.

There is an increase and decrease in biomass from the beginning of observation. Increase biomass (h0) to (h4) on photobioreactor, in open pond there is an increase in biomass from (h0) to (h5). This increase is a result of the availability of nutrients in the media so the biomass is still increased, while the decline in the value of biomass due to nutrients in the media is reduced. This is in accordance with the statement of [8] and [9], that biomass is influenced by nutrients in the media. [10] stated a decline in growth and biomass due to the number of cells that have a lot of competition in obtaining nutrients.

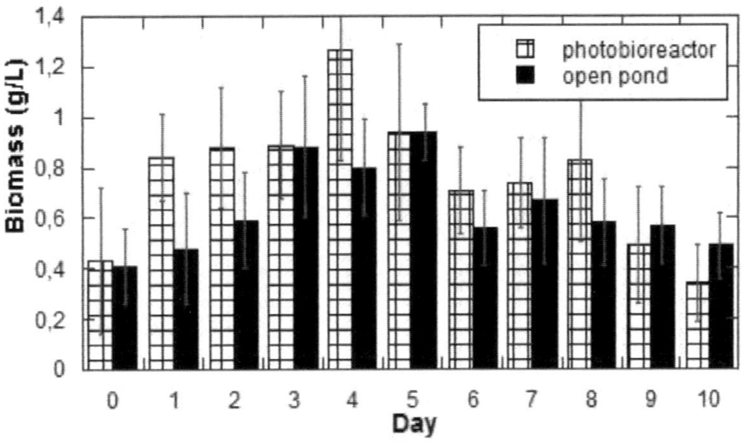

Figure 1. Biomass microalgae from photobioreactor and open pond

3.2. Lipid Content

Photobioreactor had a lipid percentage of 2.26 ± 0.51%, while the open pond had a lipid percentage of 3.18 ± 0.80% (Figure 2). In addition, differences in lipid content caused by environmental stress. Open pond has a wider surface than photobioreactor, so open pond has greater risk of evaporation. The occurrence of evaporation due to rising temperatures in the open by the sunlight. Another research showed that there is an increased in lipid content of the microalgae when the temperature rised from 25°C-30°C, this is because the microalgae accumulated high lipid levels under stressful conditions and stored them in triaglyceride (TAG) form in the lipid body [11].

The biomass of the *C. vulgaris* microalgae obtained from both cultivation systems then extracted using hexane and methanol chemical solvents to obtain lipid from microalgae. The microalgae did not produce large amounts of lipids in the lag phase, because in the first 3-4 days the microalgae were still active dividing, leaving energy and carbon stored in protein [12]. The decrease in nutrients in stationary phase causes the cell division to decrease gradually and cell begin to store their products in the lipid form [13]. According to [1] the lipid content of microalgae is usually present in the form of glycerol and fatty acids.

The microalgae *C. vulgaris* biomass from both cultivation systems was extracted using chemical solvents of n-hexane and methanol to obtain lipids. The solvent interacts with the microalgae biomass by penetrated the cell membrane and entering the cytoplasm, and the lipid is extracted out of the microalgae cell. [14] stated that the use of n-hexane or non-polar solvents is not strong enough to break the protein bonds with lipids, so a mixture of methanol or polar solvent is used to form hydrogen bonds that can break the lipid bond with proteins.

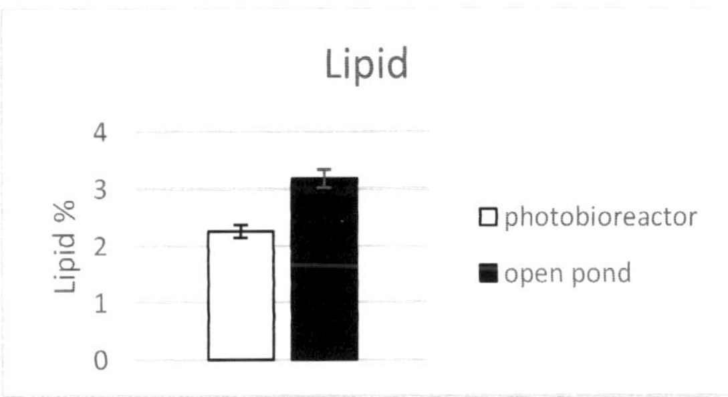

Figure 2 Lipid content *C. vulgaris* from photobioreactor and open pond

3.3. Content of C. vulgaris fatty acids

The content of the fatty acid microalgae is an important component for the formation of industrial raw materials because fatty acids become the compositions of the lipids, the highest fatty acid cultivation techniques are potentially used for larger cultivation scales [1].

The highest fatty acid analysis on photobioreactor is palmitic acid (C16: 0) 22.8% (Table 1). Stearic acid (C18: 4) 21.6%. Other fatty acids have a smaller percentage, such as linolenic acid (C18: 3) 14.9%, palmitoleic acid (C16: 1) 10.2%, oleic acid (C18: 1) 6.9%, linoleic acid (C18: 2) 6.9%, myristic acid (C14: 0) 6.3%, eicosapentaenoic acid (C20: 5) 6.2%, arachidoric acid (C20: 4) 2.0%, lauric acid (C12: 0) 0.7%. Open pond has a content of palmitic acid (C16: 0) 22.6%. Stearic acid (C18: 4) 21.4%. Other fatty acid components have a smaller percentage, such as linolenic acid (C18: 3) 14.3%, palmitoleic acid (C16: 1) 10.4%, oleic acid (C18: 1) 6.9%, linoleic acid (C18: 2) 6.6% , myristic acid (C14: 0) 6.1%, eicosapentaenoic acid (C20: 5) 6.0%, arachidoric acid (C20: 4) 2.3%, lauric acid (C12: 0) 0.9%.

The results of the study of *C. vulgaris* microalgae by [15] using wastewater had palmitoleic acid (C16: 1) of 30.54%, higher when compared with 10.2% photobioreactor and 10.4% open pond. Stearic acid (C18: 4) was 13.65% lower when compared with photobioreactor of 21.6% and open pond by 21.4%. There is no linolenic acid (C18: 3), when linolenic acid [18:3] at photobioreactor 14.9% and open pond 14.3%.

[16] used *Nannochloropsis gaditana* microalgae has a content of palmitic acid (C16: 0) of 53.4% higher than photobioreactor of 22.8% and open pond 22.6%. The content of linolenic acid (C18: 3) was 0.8% smaller than this study, with photobioreactor of 14.9% and open pond at 14.3%. The results of [16] did not find the content of stearic fatty acid (C18: 4), while in photobioreactor 21.6% and open pond 21.4%.

Previous research has shown that microalgae of different species has different fatty acid content, while microalgae with the same species with different cultivation conditions has different fatty acid content. [2] stated that differences in fatty acid content in microalgae due to differences used in initial conditions, techniques, and cultivation media.

The saturated fatty acids content *C. vulgaris* cultivated using photobioreactor and open pond is quite high when compared with unsaturated fatty acids, thus having the potential to be biodiesel. According to [2], high levels of unsaturated fatty acids are not suitable for bio-fuels because at the time of the production process still have to do the fatty acid oxidation balance. Production of biodiesel containing high unsaturated fatty acids is more susceptible to oxidation and will cause rancidity when compared to saturated fatty acids.

Fatty acid analysis of photobioreactor and open pond showed high content of palmitic acid (C16: 0), stearic acid (C18: 0), and linoleic acid (C18: 2). According to [17] fatty acid components that support for biodiesel are palmitic acid (C16: 0), stearic acid (C18: 0), and linoleic acid (C18: 2). [18] states that high palmitic acid and oleic acid can help the formation of biodiesel with better quality.

Table 1 *C. vulgaris* microalgae fatty acid composition

Fatty acid (content)	**SFA/MUFA/PUFA**	**photobioreactor**	**open pond**
Lauric acid (C 12:0)	Saturated	0.7%	0.9%
Myristic acid (C 14:0)	Saturated	6.3%	6.1%
Palmitic acid (C 16:0)	Saturated	22.8%	22.6%
Stearic acid (C 18:0)	Saturated	21.6%	21.4%
Palmitoleic acid(C 16:1)	Unsaturated	10.2%	10.4%
Oleic acid (C 18:1)	Unsaturated	6.9%	6.9%
Linoleic acid (C 18:2)	Unsaturated	6.9%	6.6%
Linolenic acid (C 18:3)	Unsaturated	14.9%	14.3%
γ- linolenic acid (C 18:3n6)	Unsaturated	1.6%	1.8%
Arachidonic acid (C20 : 4)	Unsaturated	2.0%	2.3%
Eicosapentaenoic acid (C20 :5)	Unsaturated	6.2%	6.0%
Total fatty acid		**100.1%**	**99.3%**

4. Conclusions

The highest biomass in photobioreactor at (h4) 1.27 ± 0.44 g / L while at open pond (h5) 0.94 ± 0.11 g / L. The biomass from both cultivation systems has the potential to produce lipids and fatty acid. The lipid content of open pond is higher when compared to photobioreactor. Photobioreactor had a lipid percentage of 2.26 ± 0.51%, while the open pond had a lipid percentage of 3.18 ± 0.80%. Photobioreactor had total percentage fatty acid 100.1% with the highest fatty acid analysis is palmitic acid (C16: 0) of 22.8%, open pond had total percentage fatty acid 99.3% % with the highest fatty acid analysis is palmitic acid (C16: 0) of 22.6%.

Acknowledgments

Acknowledgments are addressed to the Marine Bioprospecting and Biomaterial laboratory of the Marine Science and Technology Department FPIK IPB and the laboratory of the Surfactant and Bioenergy Research Center (SBRC).

References

[1] Kawaroe M, Prartono T, Rachmat A, Sari D, and Augustine D 2010 *Mikroalga Potensi dan Pemanfaatannya untuk Produksi Bio Bahan Bakar* (Bogor) IPB Press

[2] Kawaroe M, Prartono T, Rachmat A, Sari D, and Augustine D 2012 Laju pertumbuhan spesifik dan kandungan asam lemak pada mikroalga *Spirulina platensis* dan *Porphyridium cruentum Ilmu Kelautan* **17** 125-131

[3] Pulz O Photobioreactor : Production system for phototrophic microorganism *Appl Microbiol Biotechnol.* **57** 287-293

[4] Saadudin E, Fitri S R and Wargadalam J V 2011 Karakteristik asam lemak mikroalga untuk produksi biodiesel *Ketjurnal* **10** 131-140

[5] Kawaroe M, Prartono T, Sunuddin A and Saputra D 2016 Marine microalgae *Tetraselmis suecica* as flocculant agent of bio-flocculatinon method *HAYATI J Biosci* **23** 62-66

[6] Lin Q, Gu N, Li G, Lin J, Huang J and Tan L 2012 Effect of inorganic carbon concentration on carbon formation, nitrate utilization, biomass and oil accumulation of *Nannochloropsis oculata* CS 179 *Bioresource Technol* **111** 353-359

[7] Malekxadeh M, Najafahadi H A, Hakim M, Feiliszadeh M, Vossoughi M and Rashtchian D 2016 Experimental study and thermodynamic modeling for determining effect of non-polar solvent (hexane)/polar solvent (methanol) ratio and moisture content on the lipid extraction efficiency

from *Chlorella vulgaris Bioresource Technol* **201** 304-311

[8] Ashokkumar V and Rengasamy R 2012 Mass culture of *Botryococcus braunii* under open raceway pond for biofuel production *Bioresource Technol* **104** 394-399

[9] Kumari A, Sharma V, Pathak K and Guria C 2014 Cultivation of *Spirulina platensis* using NPK-10:26:26 complex fertilizer and stimulated flue gas in sintered disk chromatographic glass bubble column *Journal Env Chem Eng* **2** 1859-1869

[10] Daniyanti R, Yudoyon9o G, Rubiyanto A 2012 Desain closed photobioreactor *Chlorella vulgaris* sebagai mitigasi emisi CO_2 *JSS ITS* **1**

[11] Subhash GV, Rohit MV, Devi PM, Swamy YV, Mohan SV 2014 Temperature induced stress influence on biodiesel productivity during mixotrophic microalgae cultivation cultivation with wastewater *Bioresource Technol* **169**:789-793

[12] Jiang Y, Yoshida T, Quigg A 2012 Photosynthetic performance, lipid production and biomass composition in response to nitrogen limitation in marine microalgae *Plant Physiol Biochem* 54 70-77

[13] Widianingsih, Hartati R, Hendrawati A, Yudirti E and Iriani V R 2011 Pengaruh pengurangan konsentrasi nutrien fosfat dan nitrat terhadap kandungan lipid total *Nannochloropsis oculata IJMS* **16** 24-29

[14] Halim R, Danquah MK, Webley PA. 2012. Extraction from oil from microalgae for biodiesel production: A Review. *Biotechnol Adv*. **30** : 709-732

[15] Lam MK, Tyusoff MI, Uemura Y, Lim JW, Khoo CG, Lee KT, Ong HC 2017 Cultivation of *Chlorella vulgaris* using nutrients source from domestic wastewater for biodiesel production : growth condition and kinetic studies *Renew Eng* **103** : 197-207

[16] Augustine D, Kawaroe M, Sudrajat AO 2014 Effect of ethyl methanesulfonate on fatty acid characteristic of *Nannochloropsis sp*. *Int J Envi Bio* **9**(3):196-204.

[17] Kao CM, Chen TY, Lin TH, Kao CY, Lai JT, Chang JS, Lin CS 2012 Cultivation of *Chlorella vulgaris* GD using piggery wastewater for biomass and lipid production. *Bioresource Technol*

[18] Yang IS, Salama ES, Kim JO, Govindwar SP, Kurade MB, Lee M, Roh HS, Jeon B.H. 2016. Cultivation and harvesting of microalgae in photobioreactor for biodiesel production and simultaneous nutrient removal. *Energ Conver Manage*. **117**:54–62.

Kinetic study of *Escherichia coli* BPPTCC-EgRK2 to produce recombinant cellulase for ethanol production from oil palm empty fruit bunch

S Limoes[1], S F Rahman[1], S Setyahadi[2], M Gozan[1]
[1]Chemical Engineering Department, Faculty of Engineering, University of Indonesia, Depok 16424, Indonesia
[2]Center for Bioindustrial Technology, Agency for Assessment and Application of Technology, 15314 Serpong, Indonesia
misrigozan@gmail.com

Abstract. Oil Palm Empty Fruit Bunch (OPEFB) is an abundant biomass resource in Indonesia, which contains 46,77% (w/w) of cellulose. The high cellulose content of OPEFB can be used as a substrate for bacteria cultivation to produce cellulase. By using OPEFB as an alternative substrate, the production cost of cellulase in industrial scale can be suppressed. However, currently there are no available research that simulate a cellulase production plant design. Prior to simulating the cellulase plant design, kinetic studies of bacteria used in cultivation are needed to create an accurate simulation. In this research, kinetic studies of E. coli BPPTCC-EgRK2 growth were examined with the Monod approach to get the Monod constant (Ks) and maximum specific growth rate (μmax). This study found that E. coli BPPTCC-EgRK2 have μmax and Ks of 1.581 and 0.0709 respectively. BPPTCC-EgRK2 produced intracellular cellulase, thus gave linear correlation between cell concentration and cellulase production.

1. Introduction

As a tropical country, Indonesia is rich with potential biomass that can be utilized as bioenergy feedstock. Bioenergy can be used as fuel in the form of liquid (biodiesel, bioetanol) or gas (biogas). Indonesia Ministry of Energy and Mineral Resources stated that Indonesia's bioenergy potential derived from biomass waste is estimated to reach 49,810 MW. However, bioenergy utilization until today has only reached 1,618 MW or about only 3.25% of the total potential. One form of alternative energy that is environmentally friendly as well as promising in the future is bioethanol. The demand for bioethanol in the world continues to increase from year to year, where the rate of global bioethanol production from 13.2 billion gallons in 2007 is increased to 25.7 billion gallons in 2015 [1].

To produce bioethanol from biomass, there are some pre-treatment that has to be done on the biomass before converting it to bioethanol with fermentation. One of the hydrolysis agents topre-treat the lignocellulosic biomass is cellulase enzyme. As the bioethanol needs increases, the needs for cellulase will also increase. This can be seen from the market revenue of the world biofuel enzymes (cellulase and amylase), which keeps increasing from 2010 to projection year 2018 [2]. Not only the global level, but also domestic cellulase demand are increasing. Enzyme consumption for local industries is estimated to reach around 2,500 tons in 2015 with an import value of Rp187.5 billion [3]. However, according to Juniarto as quoted by Agency for Assessment and Application of Technology in 2013, 99% of the enzymes used for industries in Indonesia are imported from other countries. The import value of the cellulose enzyme is estimated to reach 121.85 billion rupiah and will continue to increase with a rate of increase of 6.67% per annum. This is unfortunate, as there is enormous potential in Indonesia's natural resources that should be able to meet those needs.

One example of Indonesia's natural resource potential is oil palm. The processing of palm leads to the formation of several by-products and residues that have economic potential. Empty fruit bunches (EFBs) are solid residue and they have a high cellulose content, at 46,77% of EFB dry weight [4]. In this case, the high cellulose content of EFB is used as a substrate for bacteria cultivation to produce cellulase. By using EFB as an alternative substrate, the production cost of cellulase on an industrial scale can be suppressed.

Currently, most cellulase enzymes are produced from saprophytic microorganisms [5]. However, industrial applications often require a cost-effective production of enzymes on a large scale, which is a common bottleneck. In producing cellulase enzymes, bacteria have several advantages over fungi, which are faster in growth rates and easily genetically engineered, although the number of enzymes produced is not as much as mushrooms [5]. In this study, bacteria were used to produce cellulase from the genetically engineered *Escherichia coli* BPPT-CC EgRK2 isolates.

Prior to simulating the cellulase production in large scale, kinetic studies of bacteria used in cultivation are needed to create an accurate simulation. In this research, kinetic studies of E. coli BPPTCC-EgRK2 growth are examined with the Monod approach to get the Monod constant (Ks) and maximum specific growth rate (μm). After determining the growth kinetics parameter of *E. coli* BPPTCC-EgRK2, one can design the shape and size of the fermentor to produce cellulase enzyme by *Escherichia coli* BPPT-CC EgRK2 on an industrial scale.

2. Materials and Methods

This research is divided into 3 stages: the preparation stage, the making of growth curve, the last is the stages are determining kinetic parameters and simulation. The preparatory stage includes a literature study of theory as well as previous research related to the study of *Escherichia coli* growth kinetics. After preparation of literature, tools and materials finished, medium of Luria Bertani (LB) for fermentation is prepared and *Escherichia coli* BPPT-CC EgRK2 is rejuvenated. At the stage of making growth curves, Escherichia coli BPPT-CC EgRK2 is cultured at optimum conditions with glucose substrate. Glucose substrate is varied with 1 $g.L^{-1}$ and 10 $g.L^{-1}$ concentration. While the fermentation process is going, the samples is analysed to get the optical density, dry cell weight, and glucose concentration data. In the final stage, microbial growth analysis is done through Monod approach and data validation is performed. In addition, a review of batch production method and fed batch is done by using SuperPro Designer 9.0 application.

The culture was obtained from BPPT (Indonesia's Agency for Assessment and Application of Technology), Serpong, Indonesia. Growth and culture conditions were referred to the previous research [6]. 1-2 loopful of *E. coli* BPPTCC Eg-RK2 was inoculated into a 100 ml of LB Erlenmeyer flask containing 25 mL of Luria Bertani medium made from ($g.L^{-1}$): yeast extract, 5; peptone, 10; and NaCl, 10. The pH of each medium was adjusted to 7.0 with 0.1N NaOH. Sterilization was done for 15 min at 121 °C, 1.2 atm. This seed culture was grown for 6 hours at 37°C with agitation at 150 rpm. The 10 $g.L^{-1}$ glucose condensed solution (25 mg glucose + 20 mL H_2O) was sterilized for 15 min at 121 °C and was added to the production medium before inoculation. After 6 h, the seed culture was fed into the 225 mL production medium, so that the production medium contains 10% inoculum. The cultures were incubated at 150 rpm at 37°C. Several samples were withdrawn from each culture.

Sample was measured every one hour to check the cell growth and glucose concentration. Cell growth was monitored by measuring the optical density of sample at 600 nm (OD600). The glucose concentration, Cs ($g.L^{-1}$), was assayed using the 3,5-dinitrosalicylic acid method [7]. The maximum specific growth rate (μ), Monod constant (Ks), and maximum specific growth rate were solved using the linearization of Monod's formula, which was known as Lineweaver Burk formula [8].

After kinetic parameters are obtained, simulation of cellulase production through batch fermentation method and fed batch fermentation method is done with simulation software SuperPro Designer v9.0. Fermentor design and sizing is made with the same simulation software and manual calculation. Manual calculations are performed using reactor design conditions that are in accordance with the provisions in the literature [9, 10]

3. Results and Discussions

3.1. Determination of Escherichia coli BPPT-CC EgRK2 growth kinetic parameter
Fermentation results can be seen from Figure 1.

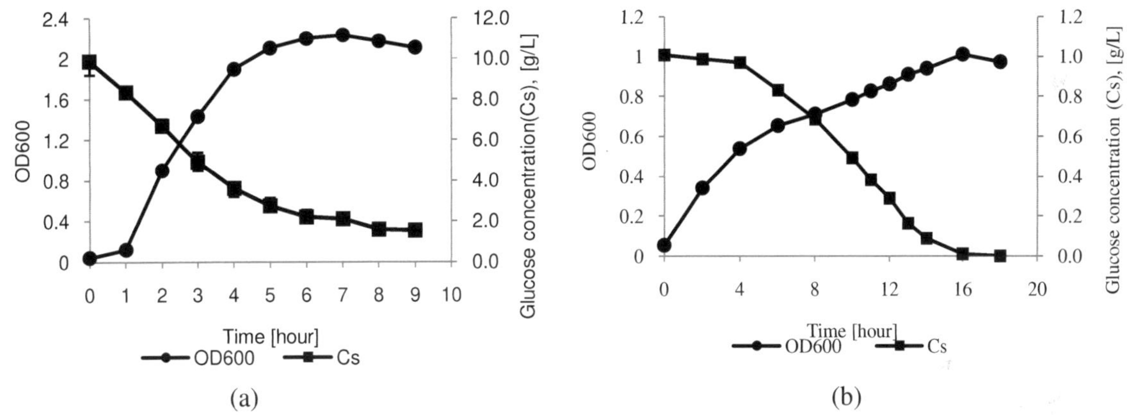

Figure 1. Growth curve of *Escherichia coli* BPPT-CC EgRK2 in different initial glucose concentration. (a) Cs0 10 $g.L^{-1}$; (b) Cs0 1 $g.L^{-1}$

Figure 1 shows that bacteria growth in study A (initial concentration of glucose 10 $g.L^{-1}$) was much faster and higher than study B (initial glucose concentration 1 $g.L^{-1}$). This is happened due to the presence of adequate glucose content in study A. The addition of glucose is known to produce greater cell growth and biomass over time than not giving it at all [11]. From Figure 1, it is seen that the optical density of the cell indicates bacterial growth consisting of several phases, which are lag phase, exponential, stationary and death phase [12]. From Figure 1 it can be seen that at the 9th hour the bacteria growth is almost stopped because the glucose substrate concentrations are getting smaller and almost exhausted. This condition is true according to study that has been done by Monod, that glucose is the growth limiting substrate in bacteria fermentation [13].

Determination of growth kinetics parameter is done by the linearization of Monod equation, which is known as the Lineweaver-Burk equation [8]. The results of Lineweaver- Burk plot at initial glucose concentration 10$g.L^{-1}$ are illustrated by Figure 2.

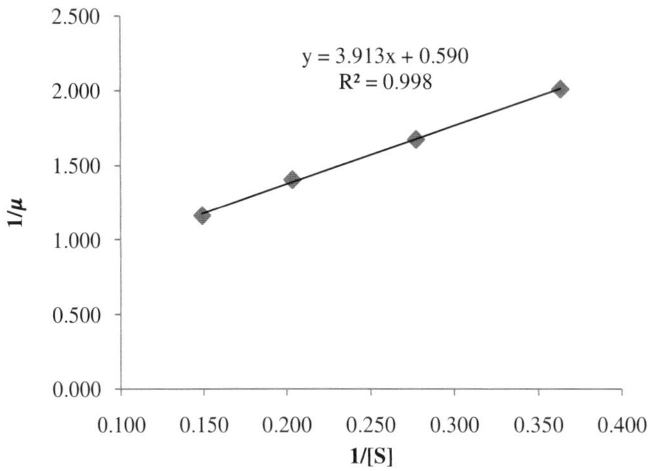

Figure 2. Lineweaver-burk plot to estimate Monod constants on *Escherichia coli* BPPT-CC EgRK2

From calculation of Figure 2, we get the results of μ_m and KS in this study are 1,694h^{-1} and 6,629 g.L^{-1}. The results obtained are quite high compared to kinetic parameters of E. coli on the glucose substrate found in other studies. Kinetic study of E. coli with Monod's equation was first done by Monod himself, and from the data interpolation, it was found that μ_m and KS are 1.35h^{-1} and 0.22 x 10^{-4} M (Monod, 1949). Fogler in his book also stated that the E. coli kinetic parameters on glucose substrate that generally used are according to Monod's study [8]. Poccia et al also estimated the growth model of E. coli JM 109 and BL 21 on multi-substrate media, and they obtained μ_m for JM 109 0.53 h^{-1}, while for BL 21 0.76 h^{-1}[14]. Study conducted by Kovarova showed E. coli ML30 on 500 mg.L^{-1} glucose substrate have kinetic parameters μ_m 0.76 h^{-1} dan Ks 0.0328 mg.L^{-1} [15].

From the above data it can be seen that μ_m and K_S for *Escherichia coli* BPPT-CC EgRK2 on glucose substrate is much higher than *Escherichia coli*strain in general. This is due to the genetic recombination that has been done on *Escherichia coli*BPPT-CC EgRK2 which ultimately leads to high substrate consumption. In addition, the high value of $\mu_m$may be due to the glucose content in fermentation medium. The article written by Schaechter states that *E. coli* grows faster on glucose media than other carbon and energy sources and can achieve doubling time of 20 minutes at 37 ° C [16].

The value of Ks obtained is 6,629 g.L^{-1}. High Ks value indicates that in order to achieve its maximum growth rate, bacteria need a large amount of substrate, and this is not desirable in industrial production. Compared to other studies in literatures, this value is very high and abnormal, since Ks values normally reside in the micromolar range, making it difficult to determine experimentally [17]. The high Ks value also shows that bacteria are not efficient in using the substrate. There is also a possibility that substrate concentration used in this study is not optimum for *Escherichia coli* BPPT-CC EgRK2 growth, so that the Ks value is too high; therefore it is necessary to do further research on optimum substrate concentration for growth of *Escherichia coli* BPPT-CC EgRK2.

3.2. Fermentor Design in Cellulase Production

In this study, a review of two types of fermentors is done, which are batch fermentor and fed batch fermentor. The operating conditions used in the fermentation process are in accordance with the optimum condition of *Escherichia coli* BPPT-CC EgRK2, which are temperature at 37°C and pH 7.0 [6]. Input and output balance data were obtained using SuperPro Designer v9.0 simulation. Calculation of the fermentor size was performed using the condition obtained from the literature [9, 10]. Here is the detail of batch fermentor specification used in cellulase production.

Table 1. Batch Fermentor Specification

Fermentor *Batch*			
Identification	Type	*Batch*	
	Amount	1	
Function	Vessel where the bacteria growth take place		
Handled material	Cell and fermentation media		
Fluid volume (m^3)	81.96		
Fermentor volume (m^3)	102.44		
Mass Balance	**Component**	**Input (kg)**	**Output (kg)**
	Ash	18.63	18.65
	Biomass	98.78	3951.14
	Cellulase	8.00	8.00
	Cellulose	455.94	455.91
	Fish powder	2483.10	380.90
	Glucose	2266.09	0.00
	Hemicelullose	874.07	874.07
	Lignin	389.85	389.85
	Sodium Hidroxide	1.42	1.45
	Sodium Sulfate	72.32	72.34
	Sulphuric Acid	1.74	1.72
	Amonium Hidroxide	402.67	68.36
	Water	59039.78	59566.97
	Air	13753.00	13719.06
	Carbondioxide	0.00	356.98
	Total amount	79865.39	79865.40
Design Specification	*Pressure (atm)*		1
	Temperature (°C)		37
	Construction material		Stainless steel S316
	Vessel diameter (m)		3.5
	Cylinder height (m)		10.5
	Ellipsoidal height (m)		0.6
	Cylinder thickness (in)		1.2
	Ellipsoidal thickness (in)		0.8
	Total fermentor height (m)		11.7
	Hydrostatic pressure (bar)		7.9
	Design pressure (bar)		9.9
	Impeller type		3 Rushton turbin
	Agitation speed (rpm)		150
	Impeller diameter (m)		1.2
	Impeller width (m)		0.2

	Distance of impeller to base (m)	1.2
	Length of impeller (m)	0.3
	Baffle width (m)	0.3
	Amount of baffle	4
Utility	Agitation power (HP)	37.3

In this research, fed batch fermentation simulation is also done on SuperPro Designer v9.0. SuperPro Designer provides the Fed-batch Fermentation feature as an option on batch fermentation if there is a simultaneous addition of material during the fermentation process. In this feature, the feed flow rate can be determined by specifying the flow rate of the stream (mass/ volume) or determining the amount of material fed per cycle (mass/ cycle). Nutritional feed on fed batch fermentation is composed of glucose as the main carbon source, fish powder (yeast extract), and water. The calculation of the feed flow rate was determined by the flow rate of streams on the Fed-batch tab. The constant volume flow rate of the feed is 320 $L.h^{-1}$ and begins to be added after fermentation runs 6 hours. Table 2lists the detail of fed batch fermentor specification used in cellulase production.

Table 2. Fed batch Fermentor Specification

Fermentor *Fed-batch*			
Identification	Type	*Fed Batch*	
	Amount	1	
Function	Vessel where the bacteria growth take place		
Handled material	Cell and fermentation media		
Fluid volume (m^3)	84.188		
Fermentor volume (m^3)	105.24		
Composition (%)	**Component**	**Input (kg)**	**Output (kg)**
	Ash	18.6	18.6
	Biomass	98.8	4630.5
	Cellulase	8.0	8.0
	Cellulose	455.9	455.9
	Fish powder	2910.1	412.3
	Glucose	2693.1	0.5
	Hemicelullose	874.1	874.1
	Lignin	389.8	389.8
	Sodium Hidroxide	402.7	1.4
	Sodium Sulfate	3.2	72.3
	Sulphuric Acid	70.5	1.7
	Amonium Hidroxide	1.7	5.6
	Water	59894.1	60520.5
	Air	13753	13712.2
	Carbondioxide	0	424.2
	Total Amount	81573	81573
Design specification	*Temperature (°C)*		37
	Construction material		Stainless steel S316L

	Vessel diameter (m)	3.5
	Cylinder height (m)	10.6
	Ellipsoidal height (m)	0.6
	Cylinder thickness (in)	1.0
	Ellipsoidal thickness (in)	1.0
	Total fermentor height (m)	11.8
	Hydrostatic pressure (bar)	157.1
	Design pressure (bar)	188.5
	Impeller type	3 Rushton turbin
	Agitation speed (rpm)	150
	Impeller diameter (m)	1.2
	Impeller width (m)	0.2
	Distance of impeller to base (m)	1.2
	Length of impeller (m)	0.3
	Baffle width (m)	0.3
Utility	Agitation power (kW)	6.55

From Table 1, it is known that the batch fermentation method produces 3,951.14 kg of biomass. On the other hand, the fermentation by fed batch method is capable of producing 4,630.5 kg of biomass. It can be concluded that by using simulations, the fed batch fermentation method produces 18% more biomass (cell) than in batch fermentation

This addition occurs because basically the main advantage of fed batch fermentation method is to add nutrients when the amount of nutrients in the media starts to run out. In addition, fed batch fermentation can extend the long log phase of bacterial growth, resulting in rapid cell production over long periods of time. Because of this, the fed batch fermentation method is suitable for microbial production using Escherichia coli, since Escherichia coli fermentation time is short due to its rapid doubling time, which is about 15-20 minutes at 37 ° C [18].

From the specifications shown above, it can be seen that the two methods of fermentation do not have much effect on the volume and size of the fermentor. In the batch fermentation method, the fermentor volume required with 20% safety factor is 102.44 m^3; while for fed batch fermentation method, the required fermentor volume with a 20% safety factor is 105.2 m^3. In other words, the two methods of fermentation have a difference in volume size of approximately 3 m^3. This volume difference occurs because in the fed batch fermentation method there is constant addition of nutrient feed. This contributes to the addition of fluid volume in the fermenter. It is for this reason that the fermentor size determination should also take into account the possible increase in volume in order for the fermentor to avoid overflow in fermentor.

In the fermentor size calculation, the size of the fermentor tank and its stirrer must be adjusted to the process and type of organism handled inside [9]. In both fermentors, the ratio of fermentor diameter to height is 1: 3, which is the optimum ratio for microbial applications on an industrial scale [9]. This ratio provides good aeration and a large area for heat exchange within the media. Both fermenters also given impellers to provide better agitation. This is happened because the vessel high can become very critical on the large fermentor vessel. The high size of the vessel will cause high hydrostatic pressure at the bottom of the vessel, and it can affect the life of cells and products it produces.

4. Conclusions

Monod growth kinetic parameters of *Escherichia coli* BPPT-CC EgRK2 in glucose substrate on this study are 1.694 h^{-1} and 6.629 $g.L^{-1}$. Total biomass amount produced in batch fermentation is 3951.14 kg/batch and total biomass amount produced in fed batch fermentation is 4676.48 kg/batch. In large scale cellulase production, fermentor volume needed is 102. 44 m^3 for batch and 105.23 m^3 for fed batch. Fed batch fermentation can produce more cellulase than usual batch fermentation. Therefore, fed batch fermentation is the better fermentation method in producing cellulase in a large scale.

Acknowledgement
The authors would like to thank PITTA Ristekdikti funds, Universitas Indonesia to provide the funding for this research.

5. References

[1] Renewable Fuel Association 2016 World Fuel Ethanol Production Accesed 5 Juni 2016. http://www.ethanolrfa.org/resources/industri/statistics/#1460747648251-1256ef3d-12da

[2] Transparency Market Research 2014, 28 Juli Biofuel Enzymes Market - Global Industry Size, Market Share, Trends, Analysis And Forecast, 2012 - 2018. Accesed 5 September 2016. http://www.transparencymarketresearch.com/biofuel-enzymes-market.html

[3] BPPT 2014, 11 Juli Kepala BPPT dan Menristek Resmikan Pembangunan Bio Center Plant dan Ground Breaking Unit Produksi Enzim di Gresik. Accesed 5 September 2016. http://www.bppt.go.id/teknologi-agroindustri-dan-bioteknologi/2017-kepala-bppt-dan-menristek-resmikan-pembangunan-bio-center-plant-dan-ground-breaking-unit-produksi- enzim-di-gresik

[4] Piarpuzán D, Quintero J A and Cardona C A 2011 Empty fruit bunches from oil palm as a potential raw material for fuel ethanol production *Biomass Bioenergy* **35(3)** 1130

[5] Sadhu S and Maiti T K 2013 Cellulase production by bacteria: a review Br. *Microbiol. Res. J.***3(3)** 235

[6] Ariyani L P 2013 Produksi Selulase dengan Bacillus amyloliquefaciens BPPTCC-RK2 dan Rekombinan EgRK2 Berbasis Tandan Kosong Kelapa Sawit: Pengaruh pH dan Suhu Serta Uji Aktivitas. Skripsi, Program Sarjana (Depok, Indonesia: Universitas Indonesia)

[7] Miller G L 1959 Use of dinitrosalicylic acid reagent for determination of reducing sugar *Anal. Chem.* **31(3)** 426

[8] Fogler S H 2001 Elements of Chemical Reaction Engineering 4th ed (New Jersey: Prentice Hall International Series in the Physical and Chemical Engineering Sciences)

[9] Meyer H, Minas W dan Schmidhalter D 2017 Industrial Biotechnology: Products and Processes, First Edition (Wiley-VCH Verlag GmbH & Co)

[10] Najafpour, Ghasem D 2007 Biochemical Engineering and Biotechnology (Oxford: Elsevier)

[11] Shehata T E and Marr A G 1971 Effect of nutrient concentration on the growth of Escherichia coli *J. Bacteriol.***107(1)** 210

[12] Doran P M 1995 Bioprocess Engineering Principles 1st ed. (London: Academic Press)

[13] Monod J 1949 The growth of bacterial cultures. *Annu. Rev. Microbiol.* **3** 371-39

[14] Poccia M E, Beccaria A J, Dondo R G 2014 Modeling the microbial growth of two Escherichia coli strains in a multi-substrate environment Braz. *J. Chem. Eng.***31(2)** 347

[15] Kovarova K, Zehnder A J B, Egli T 1996 Temperature-Dependent Growth Kinetics of *Escherichia coli* ML30 in Glucose-Limited Continuous Culture. *J. Bacteriol***8** 4530.

[16] Schaechter M 2009 Encyclopedia of Microbiology. 1st Ed. (Amsterdam: Elsevier)

[17] El-Mansi E M T, Bryce C F A, Demain A L and Allman A R 2011 Fermentation Microbiology and Biotechnolog 2nd ed (London: CRC Press)

[18] Sezonov G, Joseleau-Petit D and D'Ari R 2007 *Escherichia coli* Physiology in Luria-Bertani Broth. *J. Bacteriol***189(23)** 8746

Assisted of electromagnetic fields in glucose production from cassava stems

Lia Lismeri[1,3], Sri Haryati[2], M Djoni Bustan[2], Yuli Darni[1,3].

[1]Department of Chemical Engineering, Lampung University, Bandar Lampung-Lampung, Indonesia
[2]Department of Chemical Engineering, Sriwijaya University, Palembang, South Sumatera, Indonesia
[3] Tropical Biomass R & D Centers of Lampung University, Bandar Lampung, Lampung, Indonesia
*Corresponding author: E-mail: lismeri@yahoo.co.id

Abstract. Decrease in fossil fuel reserves that led to high price has become major problem in manycountries around the world. To acquire the sustainability of energy reserves, the renewable energies obtained from plant biomass will therefore have to play an increasing role in fulfilling energy demand throughout the century. Renewable energy source must be explored by innovative techniques which is safe to the environment and low in energy consumptions. This research conducted to produce glucosefrom cassava stems assisted by electromagnetic field inductions process. The parameters used in this research were pretreatment solvent, concentration, temperature and electrical currents. The electromagnetic field inductions could be applied to increase glucose productivity with the maximum yield of glucose was 47.43%.

Keywords:*cassava stems, electrical current, electromagnetic fields and glucose.*

1. Introduction

The energy crisis during the 21st century is indicated by the increasing prices of fossil and has been faced by all countries on the world. The fossil fuel is not renewable and the availability isdecreasing every year. Production of crude oil comes down over the year, from the present 25 billion barrels to about 5 billion barrels in 2050 [1]. National energy consumption in 2000-2011 was dominated by fuel (fuel oil). The projection gasoline consumption in 2030 is not expected changed a lot which is about 45.4%. Nonetheless, fuel consumption in 2030 will rise by 154% compared to the consumption in 2011 to around 107 million kl/year. In the 2030 national fuel consumption was amounted to 55.64% and met by imports. Production of domestic crude oil supply for the year is estimated at 10 % and the rest is met by imports. From that situation this makes it is imperative to develop new economical and energy-efficient process for the fuel Production from renewable resources. The potential renewable energy source that can substitute fossil energy is biomass [2]. Our government has made a policy to support efforts in the use of alternative energy sources such as biofuels which is a biomass-based energy that is considered reliable in terms of technical, economic, environmental, and renewable fuels.

As an agricultural country which is rich in biodiversity of flora, Indonesia became the largest cassava producer countriesafterNigeria, Brazil, and Thailand [3]. Cassava is a plant widely cultivated in Indonesia and nearly all parts of this plantareuseful either for humans or animals. The roots (tubers) are used as food and bio-ethanol feedstock and the leaves are generally used as a source of food and medicine. The stems which are produced from the harvest has not been used optimally and became crop residue. It'susually end up on open burning activity which increase negative contribute to the air pollution and dangerouseffect on human respiration system.This means that the chances of utilization of cassava stems for the fulfillment purposes of new and renewable energy sources is still very widely open and also can increase the economicvalue of this plant. This study tried to utilize the cassava stemsby converting it into glucoseas an intermediate in biofuel production by two main step; the chemical pretreatment process and hydrolysis.

IOP Conf. Series: Earth and Environmental Science **141** (2018) 012017 doi:10.1088/1755-1315/141/1/012017

Cellulose, hemicellulose, and lignin are the major component of lignocellulose. In nature, cellulose is usually associated with other polysaccharides such as xylan and lignin [4]. Cellulose is the skeletal basis of plant cell walls. Lignin is a highly cross-linked phenylpropylene polymer [5]. Lignin plays an important role in cell wall structure as a pretreatment bonding agent among plant cells. There are chemical bonds between lignin and hemicellulose and even cellulose [6]. As cell walls in biomass feedstockdiffer in structure and chemical composition, one pretreatment method will not necessarily fit all applications. Therefore, developing a pretreatment technology that is effective over a wide range of biomass materials is important.Many pretreatment methods have been reported and several detailed review papers have been published [7]. An ideal pretreatment should be cheap, as much as removal lignin, effective for various lignocellulosic substrates, minimal glucan loss and inhibitorgeneration,safe to environment and low in energy consumptions. According to [8], dilute sulfuric acid pretreatment with microwave-assisted heating can save amount of energy in disruption of sugarcane bagasse lignocellulose. The combination of ionic liquid and microwave heating also done [9]. Consequently, there is currently no single pretreatment technology that is potentially acceptable for the multiple biomass conversion.The pretreatment process indicates positive impact on the cellulose hydrolysis and consequently the glucose yields. The purpose of the pretreatment is to separate lignin and hemicelluloses from cellulose, reduce cellulose crystallinityand increase the porosity of the lignocellulose so that cellulose hydrolysis can be improved significantly [10].

Hydrolyzing the cellulose into glucose is performed by using specific enzyme in certain conditions such as temperature, the concentration of enzyme, the quantity of cellulose, pH and time so Acid hydrolysis occurs rapidly and is more cost effective than enzymatic hydrolysis [11].Direct hydrolysis by using ultra high temperature and pressure steam explosion were very effective [12]. The electromagnetic field application for producing bio-gasoline from CPO has been done and shows positive impact on the bio-gasoline properties [13]. Based on the previous study, came the idea to conduct research related to the effect of the dilute acid, electromagnetic and steam combination reaction in the pretreatment and hydrolysis of cassava stems.

2. Materials and Methods

To convert cassava stems into glucose, there are two main step processwere done in this research: Swelling of cassava stemsby using acid solvent as lignocellulosepretreatment and hydrolysis of cellulose in electromagnetic reactor. Cassava stems used in this research was taken from PT. Budi Acid Jaya in Central Lampung, Lampung, Indonesia. The chemical used in this research were: phosphoric acid (H_3PO_4), acetic acid (CH_3COOH), hydrochloric acid (HCl), hypochlorous acid (HOCl), distilled water and steam.

2.1 Cassava Stems Preparation.

Cassava Stems was reduced in size by using ball mill then sieved to the particle size of 60 meshes. To remove the moisture content, raw material then dried in 80 $^{\circ}$C for 4 hour until the weightwas constant as in figure 1. Thereafter, the prepared raw cassava stems was placed in sealed plastics bags and stored at room temperature until the experiments were carried out. Before experiment conducted, the sample was analyzed to determine density and the composition of cassava stems (cellulose, hemicellulose and lignin content).

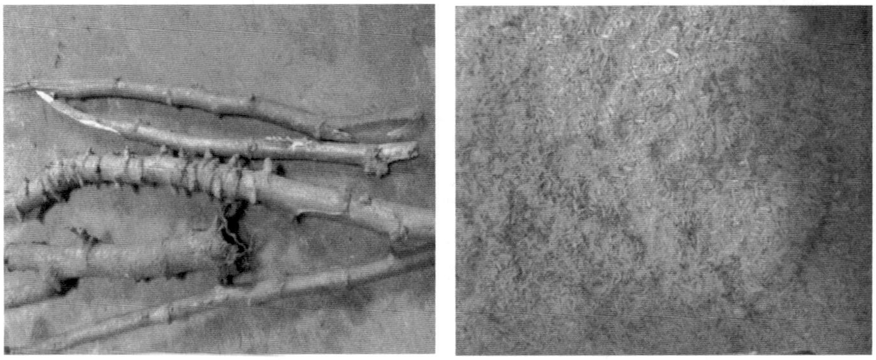

a) b)

Figure 1 .Cassava stems : (a) Raw material, (b) after size reduction

2.2 Solvent Preparation.

Phosphoric acid, acetic acid, and hydrochloric acid were the solvent used in the pretreatment step, the concentration of solvent used in this research were 0%, 1%, 3 %, and 5%, respectively. Hydrolysis of cellulose into glucose occurred in a set of electromagnetic reactor equipped with boiler, preheater, steam drum and condenser shown in figure 2. 3%hypochlorous acid wasused in this step, made from calcium hypochlorite and water based on reaction below:

$$Ca\,(ClO)_{2(s)} + \quad 2\,H_2O_{(l)} \longrightarrow 2\,HOCl_{(l)} + \quad Ca(OH)_{2(s)}$$

Calcium hypochlorite Water Hypochlorous acid calcium hydroxide

This research was conducted to determine the effect of electromagnetic field and reaction temperature of glucose production from cassava stems. The electromagnetic field generated from the solenoid tool that electrified. Electrical current were 1, 2, 3 and 4 Ampere. Temperature variation were 100, 120, 140 and 160 $^{\circ}$ C. First steps, swelling pretreatment by phosphoric acid (H_3PO_4), acetic acid (CH_3COOH), hydrochloric acid (HCl) for 1 hour at 90° C temperature process. Hydrolysis process took place in electromagnetic reactor using acid solvent in 50 minutes. The cassava stems residue and solvent residual were analyses to identify the lignocellulose content.

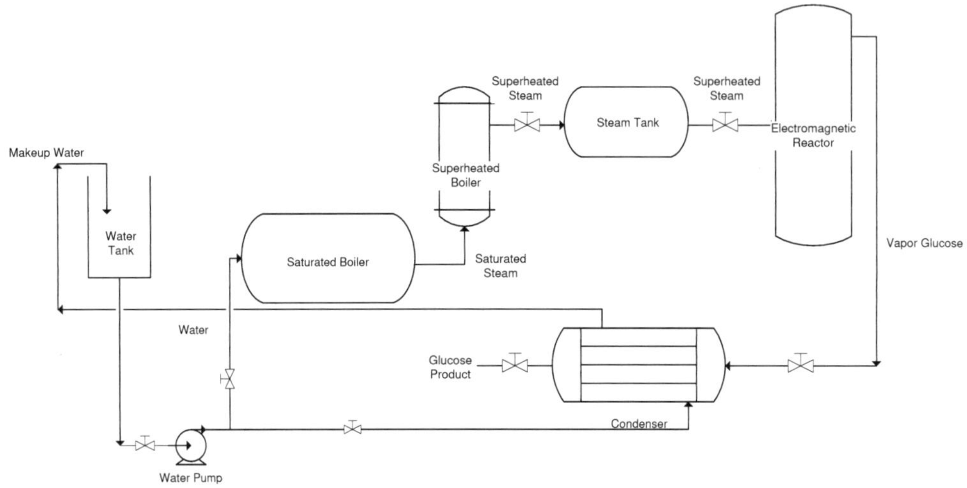

Figure 2. Schematic design of glucose production from cassava stems.

2.3 Acid Pretreatment

Determinations of solvent, concentration, time and temperature process are the main variables in pretreatment. In this step, the phosphoric acid (H_3PO_4), acetic acid (CH_3COOH) and hydrochloric acid (HCl) were used as solvent. The acid pretreatment took placein the reflux apparatus. A total of 50 grams of cassava stem heated in reflux with distilled water and acid solution. The ratio of the samplesand cooking liquor was 1: 6 gr/ml. Variable concentrations of solvents were 1, 3, and 5%. After swelling wasdone for 1 hour and at 90 °C, the samplewas separated from the liquid cookers, filtered, washed with distilled water to neutral pH, and dried using an oven at 80°C. To identify the effect of swelling process to the samples several chemical and physical analysis done such as density, composition, SEM and FTIR test.

2.4 Hydrolysis

Pretreated cassava stemsafter swelling then used in second step to produce glucose. To perform this process an experimental device which combined the steaming process with electromagnetic fields has been designed as shown in figure 3. 40 grams of pretreated cassava stemwas poured into sample bucket until its cover the solenoid which placed in the middle of reactor. Then 800 ml solvent placed at the bottom of a reactor. Sample and solvent heated for 30 min to evaporate the solvent at varied temperature. The solvent vapor cause the pressure change inside reactor and the steaminjection was assisted in the reactor to maintain theatmospheric pressure at the reactor. When all operation condition achieve the electromagnetic ampere turned on which varied 1,2,3 and 4 ampere for 50 minutes reaction time. After the time is up, amperemeter and temperature reactor control turned off and glucose product condensed would be analyzed by spectrophotometry.

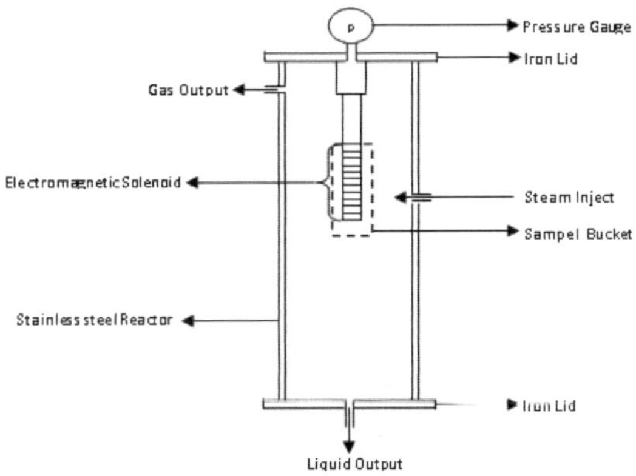

Figure 3. Electromagnetic Reactor Design

2.5 Reduction sugar analysis

Concentration ofglucose was predictedby JENWAY 6305 spectrophotometry. First step to do was making glucose standard solution which material analysis composed of glucose solution, Nelson reagent, distilled water, and Arsenic reagent. Glucose standard solution made of 10 mg of pure glucose dissolved in 100 ml distilled water. Then, glucose solution was taken as 0-4 ml in test tube. Each sample is was added with distilled wateruntil 10 ml. Add solution with 1 ml Nelson reagent heated in boiling water during 20 minute and cooled till room temperature. Mix each solution with 1 ml Arsenic reagent and put it to spectrophotometry cuvette to test glucose content.Perform these treatments on all samples.

2.6 Cellulose analysis

The method to measure the content of cellulose and lignin based Datta method proposed by Chesson (1981). 1 gr of dry sample (weight a) was added to 150 mL H_2O or alcohol-benzene and refluxed at 100°C with a water bath for 1 hour. The resultswere filtered, the residuewas washed with 300 mL of hot water. The residue then dried in an oven until constant in weight (weight b). Residues and 150 mL of 1 N H_2SO_4, then refluxed with a water bath for 1 hour at a 100° C. The results were filtered and neutral (300 mL) and dried to constant weight(weight c). Dried residue was added with 10 mL of 72% H_2SO_4 and soaked at room temperature for 4 hours. Add 150 mL of 1 N H_2SO_4 and refluxed at 100°C with a water bath for 1 hour in a cooler.Residuewas filtered and washed with H_2O until neutral (400 mL). Then the residue was heated in an oven with a temperature of 105°C until constant weight and weighed (weight d). Calculation of cellulose content using the following formula:

Cellulose content = (c-d) / a x 100%

3. Results and Discussions

Cassava stems as lignocellulosematerial used in this study was analyzed and has a composition such as Cellulose 39.29 %, Hemicellulose 24.35% and Lignin 13.42%, respectively.

a. Effect of variation and concentration of solvent to the lignocellulose density of pretreatment process.

The pretreatment purpose is to reduce lignocellulose material recalcitrance by altering cell wall structural. By this process, the polysaccharide fractions (mainly cellulose) locked in the intricacy of plant cell walls will be more accessible and improve the sugar yields. The pretreatment of lignocelluloses impactsthe recalcitrance of biomass including the resulting biomass constituents, cellulose crystallinityand ultrastructure, lignin/hemicellulose structures, cellulose degree of polymerization, and accessibility.Hemicellulose fills the gap between lignin and cellulose. The solubility of hemicellulosesis directly linked increase in biomass porosity [14].In this study, pretreatment by using acid solvent in various concentrations as swelling process was indicated by the change in pore volume of biomass. The change of biomass pore volume means the change in its density. The density of pretreatment cassava stems and raw material show the various change in each conditionsand shown in figure 4.

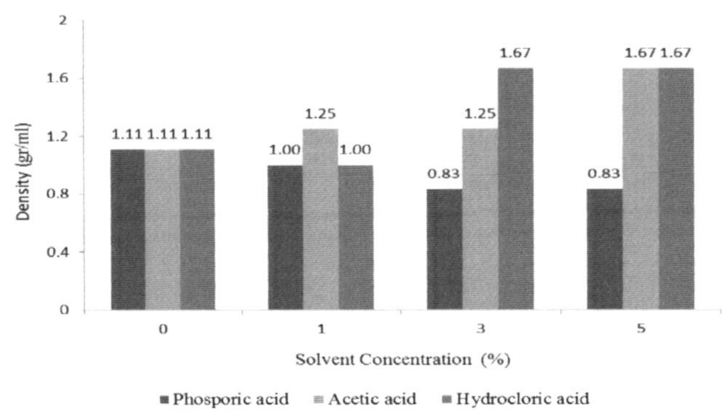

Figure 4.Effect of variation and concentration of solvent to cassava stems density.

At this swelling process, variation and concentrations solvents were used to determine the best density of the sample. The best density is the smallest density value, because the swelling goal is to break the structure of cellulose from hemicellulose and lignin bonding, damage the crystal structure of cellulose and increase the volume of material. By the swelling, cellulose volume will be greater than the initial volume prior to thetreatment. Dividing the mass of the lignocellulose to obtain the volume after treatment gives the density value of sample.

Based on Figure 4, the use of phosphoric acid shows the greater the concentration of the solvent, the smaller the density obtained. However, the use of other solvent sulfuric acid and hydrochloric acid, show the greater the concentration of the solvent impact to the greater of the density values. It can be concluded that the best solvent for swelling step is phosphoric acid with the concentration of 3%, the smallest result of the density value achieve in this step is 0.833 g/ml.

b. SEM (Scanning Electron Microscopy) images

SEM analysis was conducted to determine the effect of acid pretreatment on the material structure. The SEM images presented the surface of raw and treated cassavastems in best conditions of the step. The SEM images clearly show that raw material of cassava stemshas smooth and continuous surface. Figure 5 (a) Shows basic and compact of fiber surface structure. By swelling in 3% H_3PO_4 for 1 hours and 90°C, the gradual breakdown of fiber cell wall and the reduction in fiber dimensions were apparent on the surface of solid product. Under this condition, the high levels of residual lignin an a little number of hemicellulose removalwas observed by structural rapture and pores which present at the surface [15]. The accessibility of cellulose fibrils to separate was enhanced by the pretreatment of lignocellulose biomass. These results indicate that dilute acid pretreatment induces severe morphological changes in the plant cell walls studies and are in general agreement with significant lignin removal observed.

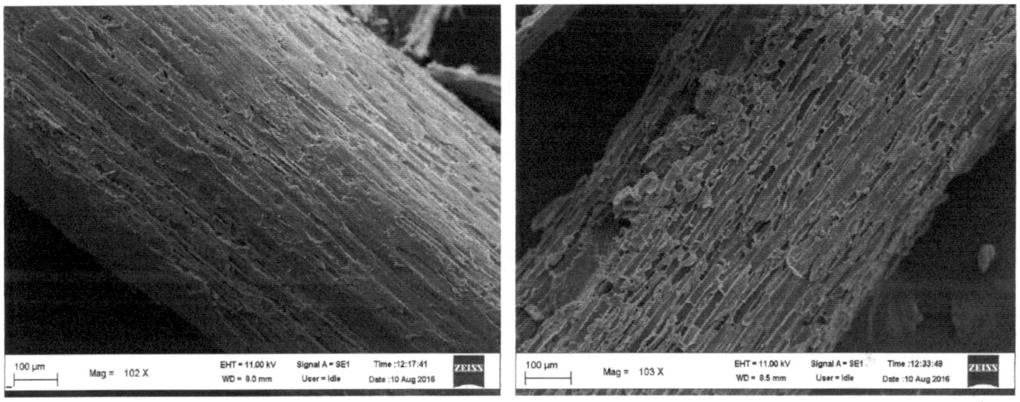

(a) (b)

Figure 5. SEM images of raw (a) and treated (b) cassava stems at 3% H_3PO_4

c. FT-IR analysis of functional group of treated cassava stems

FT-IR spectroscopy is frequently used to investigate the structure of constituents and the chemical changes in the lignocellulose biomass during pretreatment. The representative chemical changes related to lignin removal based on FT-IR analysis are shown in figure 6. According to [16], the peak approximately recorded at 3300 $cm^{-1}$was attributed to –OH groups indicatingthat hydroxyl group or watermolecules was decreased within the solids after pretreatment. The water gradually released and the feed material wasdehydrating.The characteristics bondf C=C stretching aromatic skeletal mode of lignin, observed in raw cassava stems around 1900 cm^{-1}, which not present after acid pretreatment indicates that lignin was removed after pretreatment process. The band at 1740 $cm^{-1}$was related to the C=O stretching of carbonyl and acetyl groups attributed mainly to a presence of hemicelluloses, and thedecrease of this band inacid pretreatmentwas the result of reduction of hemicellulose. The absorption band of cellulose at 879 cm^{-1}showed a slight decrease as compared with raw material of cassava stems, corresponding to the degradation of cellulose in acid pretreatment.

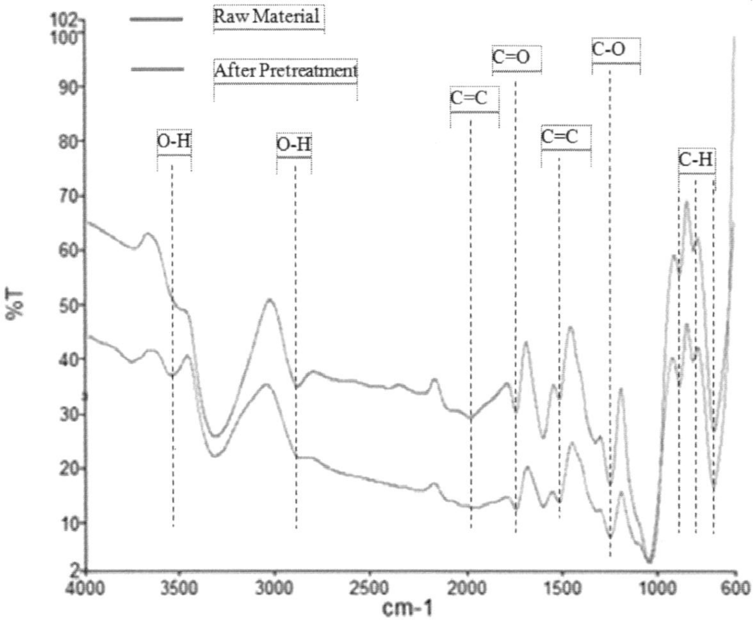

Figure 6. FT-IR spectra of raw and solid products pretreated in 3% of Phosphoric acid

d. Effect of Temperature to Glucose Yield at Different Electrical Current

The glucose yield was increasing when the temperature rises as shown in figure 7. The best value of the glucose yield was 47.76 % in 160 °C with4 ampere electrical current.

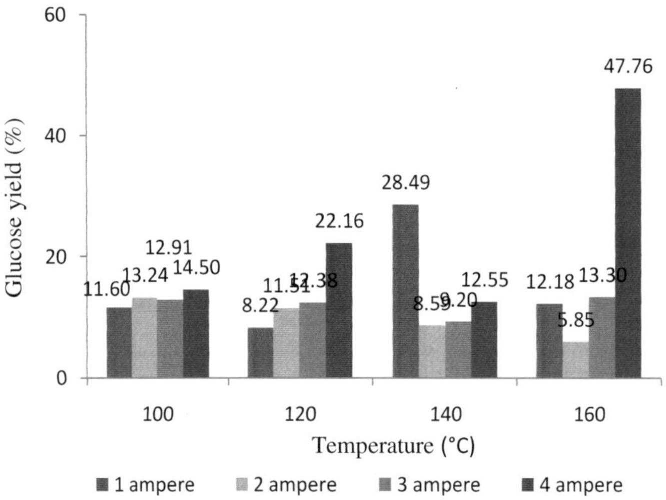

Figure 7. Effect of Temperature to Glucose Yield

This means that in higher temperature, the molecules react with enough force to create a reaction, based on collision theory. The higher temperature produces the higher energy to cause the molecules to collide, so that the activation energy which is the energy kinetic required for the collision to yield a successful reaction become minimum. By the kinetic theory, increasing the temperature makes the molecules accelerate thus there will be more chance of a collision. Since the molecules will move faster than average at high temperatures the amount of energy generated on impact is more than enough to sustain a successful product. On the contrary, if the reaction occurs at low temperatures, there will be less chance for the molecules to collide due to decreased acceleration.

e. ***Effect of Electromagnetic Field to Glucose Yield.***

The electromagnetic field generated by solenoids which was electrified [17]. A solenoid coil made of conductor wire that is wrapped tightly. The strength of the magnetic field generated from the solenoid is affected by strong currents that flow in the solenoid, the number of windings and length of wire solenoids. When electrical current transmitted to the solenoids, magnetic fields aregenerated and give an induction to the spin electron of the molecules[18].The higher the electrical current transmitted, the higherinduction or electromagnetic fields were produced as shown in figure 8.

The electromagnetic field gives impact to the yield of glucose as shown in the figure 9.At higher electromagnetic field, theyieldincreased to the value 47.76 % with 4 ampereelectrical current. There is a significant impact resulting from external magnetic fields on the reaction. Under certain conditions, the magnetic field that changes the rate constant of reactions involving radicals can be interrupted its steady state and move toother system conditions with differentinitial conditions [19].The atomic particles of the molecules which are affected by group magnetic fields will become more active and in the same direction neatly correspond to the magnetic field direction. Molecular events increase due to the magnetic field will cause the molecular group becomes fragmented.The induction of electromagnetic field causes the polarization effect of the molecule. The polarization will arrange the spin of unpairs electron from radical (intermediate) to excite to the molecules orbital and form the product.The Zeeman interaction between the unpaired electron spins on each radical and the magnetic field obviously plays an essential role in the product formation [20].

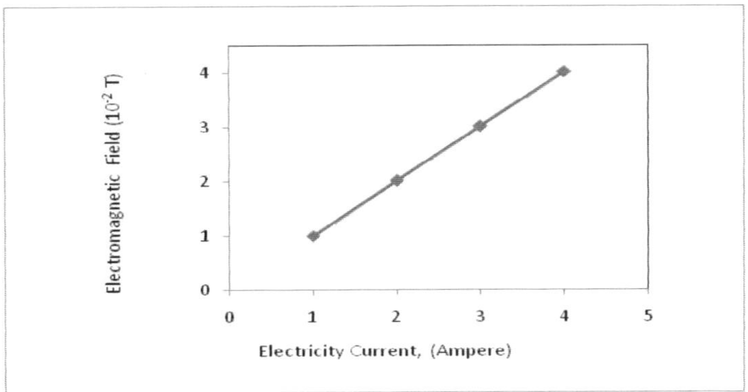

Figure 8. Plot of Electromagnetic Field and Electrical Current

IOP Conf. Series: Earth and Environmental Science **141** (2018) 012017 doi:10.1088/1755-1315/141/1/012017

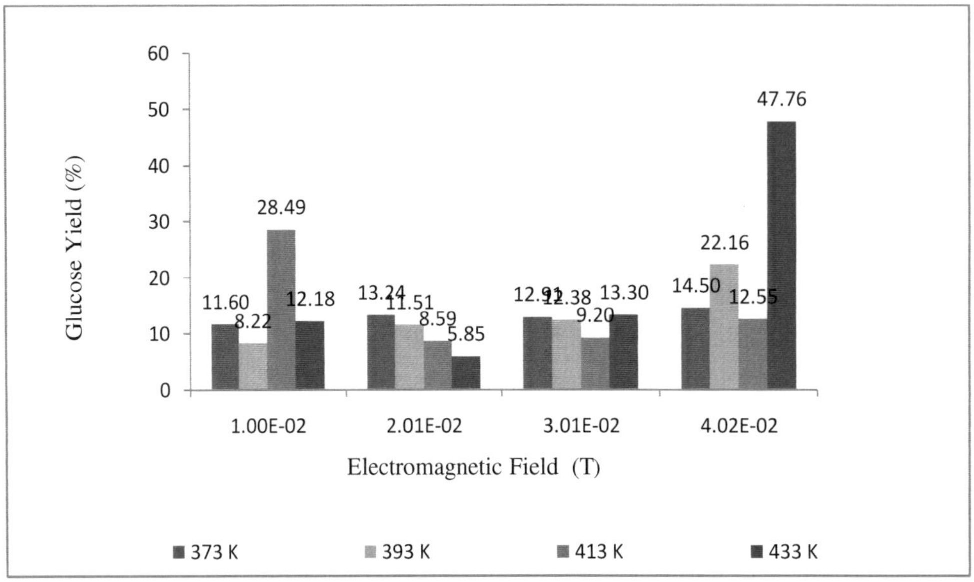

Figure 9. Effect of Electromagnetic Field to GlucoseYield

f. ***Effect of Energy Supply to Glucose Yield at Different Temperature.***

The sum of solventand steam energycalculated as energy supply to the hydrolysis reaction. Energy supply gives the significant effect to percent yield of glucose as shown in figure 10.

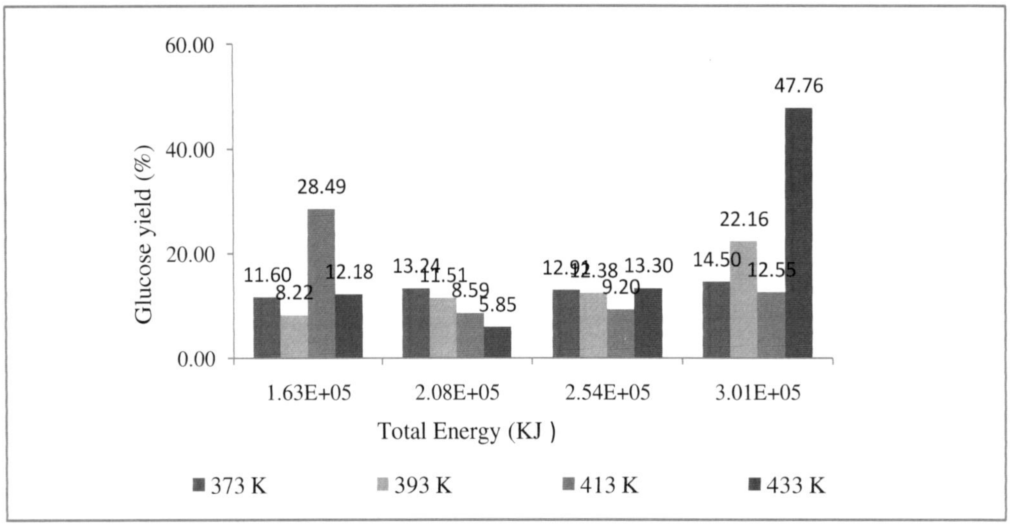

Figure 10. Effect of Energy Supply to Glucose Yield

The higherenergy supplythe larger amount of energy produced to break the chemical bond of lignocellulose polymer, where one of the bonds is the glycosidelinkage's which is consist of oxygen and carbon linkages with 356 kJbond energy. Breaking up the carbon – oxygen linkages of glycoside bond will produce two monomer of glucose. Thus the higher energy supply, the higher yield of reaction form, until itreaches the point that the higher energy supply will decrease the glucose yieldas the cause of polymer degradation to reform and produce another product.

4. Conclusions

Based on this research, it can be concluded that the optimum result of acid pretreatment process was using3% phosphoric acid with 0.833 gr/ml density value. The highest glucose yield was 47.76% occuredat temperature 160°C, 4 x 10^{-2} T electromagnetic field in 50 minutes reaction time.

5. Acknowledgement

On this occasion, the authors would like to acknowledge my advisor at Chemical Engineering Graduate ProgramSriwijayaUniversity, for their enlightening suggestions, guidance, and kind encouragement throughout this study. The authors also would like acknowledge all people who have helped in completion of this research.

6. Reference

[1] Colin J. Campbell and Jean H. Laherrère. 1998. The End of Cheap Oil. Scientific American
[2] George W,. 2006. An overview of aqueos-phase catalytic processes for production of hydrogen and alkanes in a biorefinery. Elsevier.
[3] DepartemenPertanian republic Indonesia. 2009. Basis Data StatistikPertanian http://database.deptan.go.id/bdsp/index.asp.
[4] Hanwu Lei, Iwonacybulska, James julson. 2013. Hydrothermal Pretreatment of lignocellulosic Biomass and Kinetics. Journal of sustainable bioenergy Systems, page 250-259
[5] Agbor VB, Cicek N, Sparling R, Berlin A, Levin DB. 2011. Biomass pretreatment: fundamentals toward application. Biotechnol Adv. 2011;29(6):675–685. doi: 10.1016/j.biotechadv.2011.05.005.
[6] Harmsen P. 2010. Literature review of physical and chemical pretreatment processes for lignocellulosic biomass. Biosynergy.
[7] Kumar, R., Wyman, C.E., 2013. Physical and chemical features of pretreated biomass that influence macro-/micro-accessibility and biological processing. In: Wyman, C.E. (Ed.), Aqueous Pretreatment of Plant Biomass for Biological and Chemical Conversion to Fuels and Chemicals. John Wiley & Sons, pp. 281–310.
[8] Wei-Hsin Chen. 2011. Disruption of sugarcane Bagasse lignocellulosic structure by means of dilute sulfuric acid pretreatment with microwave-assisted heating. Elsevier.
[9] DU jun. 2010.Fast pyrolysis of biomass for bio-oil with ionic liquid and microwave irradiation. Journal of fuel and technology, 38(5) : 554-559.
[10] Kuo, C.H., Lee, C.K. 2009. Enhancement of enzymatic saccharification of cellulose by cellulose dissolution pretreatments. Carbohydrate Polymers.
[11] Minhee Han, Yule Kim. 2011. Bioethanol production from optimized pretreatment of cassava stems. Changhae Institute of Cassava and Ethanol Research, Changhae Ethanol Co., Ltd., Jeonju 561-203, Korea.
[12] Sasaki. 2012. Direct hydrolysis of cellulose to glucose using ultra-high temperature and pressure steam explosion. Elsevier.
[13] Nina Haryani1, M. DjoniBustan, Sri Haryati. 2014. Studipengaruhkonsentrasilarutanelektrolitkoh, voltaseelektrolisadanmedanelektromagnetik, sertarasiocpo/kataliszeolitalam yang diaktifkanterhadapkonversitrigliseridacpomenjadibiogasolin. Seminar Nasional Added Value of Energy Resources (AVoER) Ke-6.Palembang.Indonesia.
[14] Izshikawa C.I., Davis M.F., Schell D.F., Johnson D.K. 2007. Porosity and its effect on the digestibility of dilute sulfuric acid pretreated corn stover. J. Agri. and Food Chem., 55, 2575-2581.

[15] Wi et al. 2015. Lignocellulose conversion for biofuel: a new pretreatment greatly improves downstream biocatalytic hydrolysis of various lignocellulosicmaterialsBiotechnol Biofuels (2015) 8:228 DOI 10.1186/s13068-015-0419-4
[16] Yang H, Yan R, Chen H, Lee DH, Zheng C. 2007.Characteristics of hemicellulose,cellulose and lignin pyrolysis. Fuel; 86: 1781-1788.
[17] Gillesberg, Tom. 2009. 21st Century Science &Technology : Flu Dynamics in the Noösphere, 22-33.
[18] Cullity, B. D., and Graham, C. D., 2009. Introduction to Magnetic Materials.2nd Edition.A john Wiley & Sons, Inc.
[19] Kipriyanov, A.A., and Purtov, P.A. 2012. Prediction of a Strong of a Weak Magnetic Field on Diffusion Assisted Reaction in Non Equilibrium Condition. Bull. Korean Chem. Soc. Vol 33, No. 3, pp. 1009-1014.
[20] Schwarz, C., and Trappe, M.I. 2006. Normal Zeeman Effect.

IOP Conf. Series: Earth and Environmental Science **141** (2018) 012018 doi:10.1088/1755-1315/141/1/012018

Optimization of moistening solution concentration on xylanase activity in solid state fermentation from oil palm empty fruit bunches

Efri Mardawati[1(a)], Parlan[2], Tita Rialita[2] and Bambang Nurhadi[2]

[1]Departement of Agro-Industrial Technology, Faculty of Agro-industrial Technology, Universitas Padjadjaran, Jl. Raya Bandung-Sumedang Km 21. Jatinangor, Sumedang, 40600, Indonesia
[2]Departement of Food Industrial Technology, Faculty of Agro-industrial Technology, Universitas Padjadjaran, Jl. Raya Bandung-Sumedang Km 21. Jatinangor, Sumedang, 40600, Indonesia
[(a)]Corresponding author: efri.mardawati@unpad.ac.id

Abstract. Xylanase is an enzyme used in the industrial world, including food industry. Xylanase can be utilized as a 1,4-β-xylosidic endo-hydrolysis catalyst of xylanase, a hemicellulose component for obtaining a xylose monomer. This study aims to determine the optimum concentration of the fermentation medium using Response Surface Method (RSM) in the production of xylanase enzyme from oil palm empty fruit bunches (OPEFB) through solid state fermentation process. The variables varied in this study used factor A (ammonium sulphate concentration 1.0-2.0 g/L), B (concentration of potassium dihydrogen phosphate 1.5-2.5 g/L) and C (urea concentration 0.2 – 0.5 g/L). The data was analysed by using Design Expert version 10.0.1.0 especially CCD with total 17 running including 3 times replicated of canter point. *Trichoderma viride* was used for the process production of xylanase enzyme. The ratio between substrate and moistening solution used was 0.63 g / mL with temperature of 32.8^{0}C, 60 h incubation time. The analysis of enzyme activity was done by DNS method with 1% xylan as substrate. Analysis of protein content in enzyme was done by Bradford method. The optimum of moistening solution concentration in this fermentation was obtained. They are, the ammonium sulphate concentration of 1.5 g/L, potassium dihydrogen phosphate 2.0 g/L and urea 0.35 g/L with activity of 684.70 U/mL, specific activity enzyme xylanase 6261.58 U/mg, protein content 0.1093 U/mg, the model was validated using experiment design with perfect reliability value 0.96.

1. Introduction

The usage of enzyme in Indonesia reaches 2500 tons with an import value of $ 14.0 millions. Many food and non-food industries have used enzyme. One of the enzyme that has high economic value in the industry is xylanase that can be used for bioleaching in paper industry, cleansing and improving aroma of juice and wine, improving the quality of bread and animal feed, waste processing and composting

Xylanase is a group of enzymes that have the ability to hydrolysed hemicellulose in this case is xylene or polymer of xylose and xylo-oligosaccharide [1]. Xylan which is a substrate for xylanase enzymes is found in many annual crops and especially agricultural wastes such as corncobs, bagasse, rice straw, wheat bran, and cottonseed. Oil palm empty fruit bunches (OPEFB) are the main waste of the palm oil processing industry. The percentage of OPEFB waste is 23% of fresh fruit bunches, while the percentage of fibre and shell of kernels are 13% and 5.5% respectively of fresh fruit bunches [2]. The main components of palm solid waste are cellulose and lignin so that this waste is also called lignocellulosic waste. The results also showed similar results, OPEFB containing cellulose 43-43.7%, hemicellulose 22.93-23.67% and lignin 21.28-22.10% [3].

The high content of hemicellulose in OPEFB has the potential to produce xylanase by microorganisms that produce xylanase by utilizing xylan. Xylanase can be produced from microorganisms such as fungi and bacteria, one of them is Trichoderma sp. through the fermentation process [3]. Also reported that the best activity of xylanase was obtained from *Trichoderma viridae* incubated at 32.8°C compared with *Aspergilus niger* and Penicillium sp. on OPEFB substrate [4].

In the solid-state fermentation, the moisture obtained from the addition of moistening solution, the addition of moistening solution is intended to increase the number of cells and assist cell adaptation to new environmental conditions [5]. Metabolism of microorganisms requires macronutrients such as nitrogen, phosphorus, potassium, sulphur, and micronutrients such as chromium, cobalt, selenium, tungsten, magnesium, iron, zinc, nickel, molybdenum, etc. [6]. Mineral salt solutions are important for enzyme production because some enzymes such as xylanase, cellulase and proteinase are very sensitive to the moistening solutions used [7].

Salt solutions containing a mixture of essential minerals such as Mg, K, and P need to be added with a certain concentration into the fermentation medium because of the addition of one or several combinations of mineral solution concentration to the fermentation process and metabolite products produced by microbes. Based on the background, it is necessary to conduct further study to find out the optimum concentration of moistening solution.

2. Materials and Methods

2.1 Raw Material Preparation

OPEFB was collected from PTPN VIII Cikasungka, Bogor, Indonesia. OPEFB fibre was washed using tap water and then sun dried for 12 h and oven dried at 105^0C overnight. Dry OPEFB was grinded using a disc mill and sieved using Tyler 60 mesh.

2.2 Fungal Preservation

The fungal strain was tested during the experiment is *T. viridae* ITB CC L.67 was preserved in potato dextrose agar (PDA) slants. The spore of *T. viridae* were preserved as immobilized spores on dried rice pellets. Inoculum was prepared by suspending the fungal spores in sterile distilled water to give a final spore count of 1×10^6 spores/mL.

2.3 Preparation of Cultivation Medium

The media for solid state fungal cultivation was prepared by mixing dried OPEFB as the carbon source for the fungal cultivation with liquid moistening solution which was adapted from Aves-Prado et al., (2010). The moistening solution contained $(NH_4)_2SO_4$, KH_2PO_4, urea, $CaCl_2.2H_2O$ and $MgSO_4.7H_2O$. The media was first sterilized in an autoclave at 121^0C for 15 minutes before use.

2.4 Solid State Fermentation

2.4.1 Evaluation of cultivation time

In the process of determining the optimal cultivation time for each of the *T. viridae*, 6.3 g OPEFB was mixed with 10 mL of mineral medium (moistening solution) to give the solid to liquid substrate ratio of 0.63 g substrate/mL moistening solution. The fungal cultivation of 1×10^6 spores/mL were carried out in 250 mL shake flasks at 32.8^0C and were performed over a duration of 60 h. Sampling was done 60 h incubation and the obtained samples were used for the analysis of xylanase activity, protein content and xylanase activity specific. Each run was performed in Duplo.

2.4.2 Evaluation of OPEFB particle size

The drying process was done before using the raw material of palm stem, the reduction and uniformity of the size to 40-60 mesh to increase the surface area and increase the solubility in water. The reaction rate increases with time to particle size. The smaller particle size gives an increasingly fast cooking time.

2.4.3 Experimental design for optimization of parameters using RSM

The production of xylanase was optimized using the method of response surface methodology, which is the Central Composite Design (CCD) experimental design with application Design Expert version 10.1.0.0, on three variables: ammonium sulphate concentration (1.0–2.0 g/L), calcium dihydrogen phosphate concentration (1.5–2.5 g/L), and urea concentration (0.1–0.5 g/L). The cultivation temperature was varied at 32.8^{0}C, whereas the substrate ratio applied were in 0.63 (g OPEFB/mL moistening solution).

The total amount of treatment variation is 17 running including 3 times the repetition of the value of centre (central point). Furthermore, in a near-response state, two or more order models are usually required to approximate the response because of the curvature on its surface expressed by (1):

$$Y = \beta_0 + \sum_{i=1}^{k} \beta_i X_i + + \sum_{i=1}^{k} \beta_{ii} X_i^2 + + \sum\sum_{i<j}^{k} \beta_{ij} X_i X_j + \varepsilon \qquad (1)$$

where Y is the predicted response, x_i, x_j,. .., x k are the input variables, β_0 is the intercept state, β_i (i = 1, 2,. . ., k) is the influence of linear, β_{ii} (i = 1, 2,. . ., k) is the influence of squared, X_{ij} (i = 1, 2,. . ., k; j = 1, 2,. . ., k) is the influence of interaction and ε is a random transgression.

2.5 Harvesting of Enzymes

Enzyme harvesting process was conducted according to [8]. Enzymes were extracted from the fungal cultivation by adding 40 mL distilled water (four times volume of liquid medium) to the cultivation solution. Subsequently the solution was stirred with a sterile glass stick and then shaker at 100 rpm for 1 hour at room temperature (25^{0}C). The solution was then vacuum filtrated to remove solid and particulates. The filtered solution was further centrifuged at 10.000 rpm for 12 minutes at 4°C. The obtained supernatant (the crude enzyme) could be used for enzyme activity analysis.

2.6 Analytical Methods

2.6.1 Lignocellulosic material composition

Analysis of chemical components of raw materials aims to determine the chemical composition contained in the raw material, consisting of moisture content (b/b) gravimetric method, ash content and lignocellulosic levels.

2.6.2 Enzyme Activity

The xylanase activity was determined using larch wood xylan (Sigma Co., USA) as substrate [9]. In brief, the reactant containing 0.5 mL of crude enzyme and 0.5 mL of 1% (w/v) larch wood xylan (pH 5.0 acetate buffer) were incubated in a water bath at 40°C for 15 min. The reaction was stopped by adding 1.5 mL of dinitro salicylic (DNS) acid solution. After incubation in a boiling bath for 5 min, the liberated reducing sugars were measured with spectrophotometer at 540 nm wavelength. The reducing sugars produced were quantified by the dinitro salicylic acid method using D-xylose as standard. One unit (U) of xylanase activity was defined as the amount of enzyme that released 1 μmol of xylose per minute under the assay conditions.

2.6.3 Protein Content

IOP Conf. Series: Earth and Environmental Science **141** (2018) 012018 doi:10.1088/1755-1315/141/1/012018

The measurement of protein content in the sample was determined by Bradford method with Bovine Serum Albumin (BSA) as the standard. A total of 100 µL filtrate (rough xylanase enzyme) was added with 2 mL of the Bradford reagent inserted into the test tube. The reaction tube was vortex and incubated at room temperature (25°C) for 5 minutes and the measured absorbance at 595 nm. The protein content of the sample was determined from the standard BSA solution curve with a concentration of 0.2 - 1.0 mg/L.

2.6.4 Enzyme Specific Activity

The specific activity of crude extract of xylanase enzyme states a number of µmol substrates which can be converted into products within a minute by 1 milligram of enzyme under optimum conditions. The higher the value of the specific activity of the enzyme, the better the ability of xylanase to utilize the substrate. Calculation of specific activity of crude extract of xylanase enzyme was done by comparing the value of activity of crude extract of xylanase enzyme with protein content.

3. Results and Discussions

3.1 Characterization of OPEFB

Characterization of chemical composition of oil palm empty bunches carried out include testing the levels of cellulose, hemicellulose, lignin, water, and ash. The results are presented in Table 1.

Table 1. Composition of OPFEB

Component	Percentage (5)
Hemicellulose (db)	17.31 ± 0.5586
Cellulose (db)	39.99 ± 0.7354
Lignin (db)	23.96 ± 0.9970
Water (wb)	4.60 ± 0.1131
Ash (db)	4.78 ± 0.1626

Hemicellulose levels need to be analysed to find out how big the potential of empty bunches of palm oil to be used as raw materials in the production of xylanase enzymes. Hemicellulose level in OPEFB used was 17.31%. The main component of hemicellulose is needed, xylan contained in hemicellulose. Xylan acts as an inducer for xylanase production in the hope that xylanase production will increase.

In addition, OPEFB contained quite high carbon because they contain cellulose of 39.99%. The component of cellulose as a carbon source for microorganisms. Lignin prevents the entry of enzymes in breaking down polysaccharides into monosaccharides [10]. Based on the results of analysis known that the lignin content of 23.96%. This indicates that empty bunches of palm oil can be used as substrate because it contains lignin ≤25% [11].

Water content needs to be analysed to determine whether OPEFB have long or no shelf life. High water levels cause rapid OPEFB damage due to the growth of microorganisms. If the water content is known then it can be adjusted raw material storage conditions to minimize the possibility of the growth of microorganisms. In addition, knowing the initial moisture content of the material can be used as a reference to the addition of a fermentation medium (moistening solution) so that the water content in the material (substrate) is not more than 20% so it can be referred to as solid state fermentation (SSF).

Ash content indicates the content of organic substances or mineral elements of a material. Ash content is influenced by type, age of material, and others. Based on the analysis, OPEFB have ash content of 4.78%.

3.2 Optimization production of enzyme xylanase

Parameters for SSF optimization to be considered are carbon and nitrogen sources, adaptation of inoculum, pH, temperature, incubation time, water content [12]. In addition, the fermentation medium is also an important factor in the state fermentation process for enzyme production because some enzymes such as xylanase, cellulase and proteinase are very sensitive to the composition of the fermentation medium (moistening solution). Response to each of the moistening solution concentrations is presented in Table 2 and the analysis of ANOVA is presented in table 3.

Table 2. Response on Overall Condition of Optimization

Run	Concentration (g/L)			Response		
	$(NH_4)_2SO_4$	KH_2PO_4	Urea	Xylanase Activity (U/mL)	Protein content (mg/mL)	Xylanase Activity Specific (U/mg)
1	1.50	2.00	0.35	712.71 ± 43.21	0.1076 ± 0.0017	6624.53
2	1.50	2.00	0.60	562.46 ± 97.23	0.0506 ± 0.0025	11114.87
3	2.00	2.50	0.50	554.82 ± 21.60	0.0718 ± 0.0008	7732.44
4	2.00	2.50	0.20	535.53 ± 2.88	0.1123 ± 0.0133	4769.37
5	2.34	2.00	0.35	450.10 ± 5.76	0.1011 ± 0.0025	4450.98
6	1.00	1.50	0.50	542.09 ± 32.41	0.0671 ± 0.0025	8084.49
7	2.00	1.50	0.20	524.26 ± 43.21	0.0923 ± 0.0033	5679.17
8	1.00	2.50	0.20	590.47 ± 0.00	0.0571 ± 0.0017	10347.16
9	1.00	2.50	0.50	593.02 ± 68.43	0.1440 ± 0.0017	3905.75
10	1.50	2.00	0.35	733.08 ± 7.20	0.0988 ± 0.0008	7421.75
11	1.50	2.84	0.35	598.11 ± 3.60	0.0970 ± 0.0017	6165.31
12	1.00	1.50	0.20	565.01 ± 14.40	0.1229 ± 0.0017	4598.78
13	1.50	2.00	0.10	577.74 ± 32.41	0.0876 ± 0.0050	6594.20
14	2.00	1.50	0.50	536.99 ± 90.03	0.0923 ± 0.0050	5817.11
15	1.50	1.16	0.35	562.46 ± 82.83	0.0588 ± 0.0008	9561.01
16	1.50	2.00	0.35	692.34 ± 0.00	0.1187 ± 0.0008	5830.30
17	0.66	2.00	0.35	557.37 ± 104.44	0.0894 ± 0.0042	6236.22

Table 3. ANOVA for Optimisation of Xylanase Enzyme Activity Resapons

Source	Sum of Squares	df	Mean Square	F Value	p-value Prob > F	
Model	79047.49	9	8783.05	25.47	0.0002	*significant*
A-*Ammonium sulphate*	7467.38	1	7467.38	21.65	0.0023	
B-*Potassium dihydrogen phosphate*	2004.51	1	2004.51	5.81	0.0467	
C-*Urea*	14.08	1	14.08	0.041	0.8456	
AB	279.54	1	279.54	0.81	0.3979	
AC	343.09	1	343.09	0.99	0.3518	
BC	128.24	1	128.24	0.37	0.5613	
$\mathbf{A^2}$	56340.70	1	56340.70	163.36	< 0.0001	
$\mathbf{B^2}$	21411.40	1	21411.40	62.08	0.0001	
$\mathbf{C^2}$	25181.38	1	25181.38	73.01	< 0.0001	
Residual	2370,08	7	338,58			
Lack of Fit	1584.31	5	316.86	0.76	0,6596	*not significant*
Pure Error	829.87	2	414.94			
R^2	0,9704					

3.2.1 Response to Xylanase Enzyme Activity

The ANOVA analysis showed that the F-values of 25.47 and p-values (Prob > F) were less than 0.05 (0.0002), indicating that the (quadratic) model was significant at 95% confidence level (Table 3). This means that the independent variable (X) in the model has a significant effect on the response (Y)

xylanase enzyme activity. In addition, non-significant Lack of Fit values also indicate that the model is good to implement. Lack of Fit is a model that has not been exact or there is no match Between the data and the model.

Based on Table 3, it is known that the F_{count} value is more than 0.05 at the $\alpha = 5\%$ level so that the model is suitable or the suitability between the model and the data is appropriate. In Table 3, it can also be known the value of coefficient of determination or also known as R^2 of 0.9710 which means the influence of independent variables on the change of dependent variable (xylanase enzyme activity) is 97.10% while the rest of 2.90% is affected by other factors outside the independent variables used in this study. In general, a regression that has a coefficient value of determination over 0.90 indicates a very strong correlation. The value of the adjusted determination coefficient (Adjusted R-Squared) is also fairly high (0.9336). This is sufficient to prove that the equation model is significant.

Checking the suitability of the model can also be done using a normal probability plot of residual and residual plots on the response prediction. Plot The normal probability to residual is considered to be satisfied when the residual value (the difference between the actual value and the response prediction value) is in the vicinity of a straight line. The residual value residing in the straight line on the curve indicates that the model's non-conformance to the response data is still normally distributed. Meanwhile, the residual plot of the response prediction is considered to be satisfactory when the model in the sample is drawn at random and the result is around a straight line.

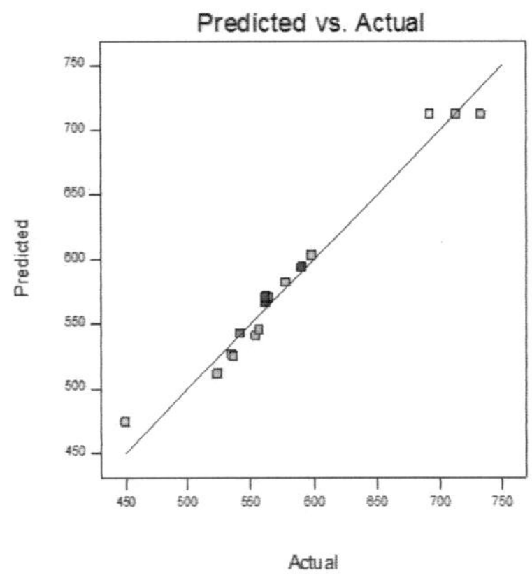

Figure 1. Prediction vs Actual Curves for Xylanase Enzyme Activity Response

The predictive and actual plot shows that residuals are drawn at random but there are some points far enough away from the straight line. The point that is quite far from the line still can be said to meet the model if it is still inside the graph area (not missing from the picture). The optimization equation in the form of the actual value to predict the response of the xylanase enzyme activity of factor A (Ammonium sulphate), B (potassium dihydrogen phosphate) and C (urea) where Y is the predicted response activity xylanase enzyme, can be seen in equation (2):

$$Y = -846.85 + 818.55\mathbf{A} + 738.55\mathbf{B} + 1246.12\mathbf{C} - 23.65\mathbf{AB} + 87.32\mathbf{AC} + 53.38\mathbf{BC} - 282.87\mathbf{A}^2 - 174.38\mathbf{B}^2 - 2129.51\mathbf{C}^2 \quad (2)$$

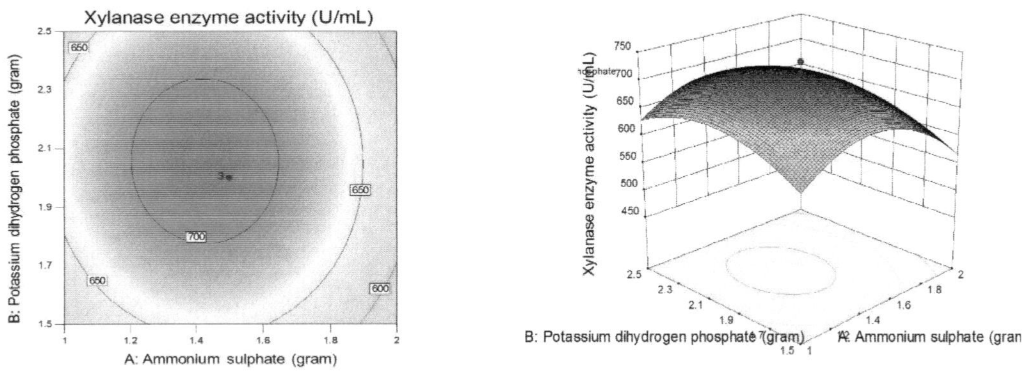

(a) (b)

Figure 2. Contour (a) and 3D plots (b) Effect of interaction between Ammonium sulphate and Potassium dihydrogen phosphate to Xylanase Enzyme Activity

Fig.2a contour shows the condition of both factors where optimum response value at ammonium sulphate concentration 1.5 g/L, concentration of potassium dihydrogen phosphate 2.0 g/L and the region is marked with red colour. While in Fig. 2b (3D surface) can be seen more clearly form model that resembles parabola, this is considered good because this graph shows existence of maximal value before finally declining. Based on this research, it can be seen that optimum xylanase enzyme activity at ammonium sulphate concentration 1.5 g/L and concentration of potassium dihydrogen phosphate 2.0 g/L while urea concentration 0.35 g/L.

Studies stated that higher xylanase enzyme activity uses ammonium sulphate ($(NH_4)_2SO_4$) as an inorganic nitrogen source at 0.14% (1.4 g/L) concentration of fermentation medium compared to sodium nitrate ($NaNO_3$) and ammonium nitrate (NH_4NO_3) [13]. Also stated that the highest enzyme activity of xylanase used ammonium sulphate ($(NH_4)_2SO_4$) at 0.12% - 0.15% (1.2 – 1.5 g/L) concentrations [14]. The interaction between ammonium sulphate and urea can be seen in Figure 3.

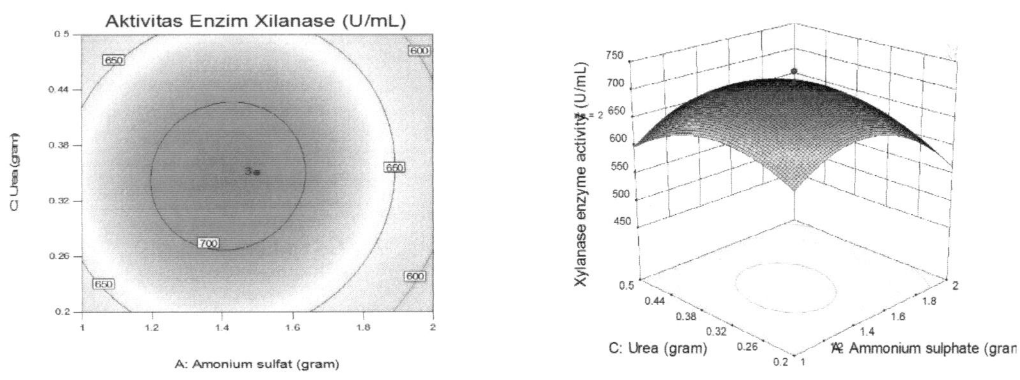

(a) (b)

Figure 3. Contour (a) and 3D plots (b) effect of interaction between ammonium sulphate and urea on xylanase enzyme activity

As has been reported [13] the activity of xylanase enzyme is higher with activity of 760.0 U/mg in fermentation medium which added urea compared with medium without urea with activity of 711.5 U / mg. Also that the highest production of xylanase enzyme from T. reesei RUT-C30 is on urea containing medium and producing xylanase enzyme from Melanocarpus albomyces states that the highest activity occurs on urea containing medium [8]. The produced of xylanase enzyme from Aspergillus niger XY-1 increases with increasing ammonium sulphate and urea concentrations. The interaction between potassium dihydrogen and urea can be seen in Figure 4.

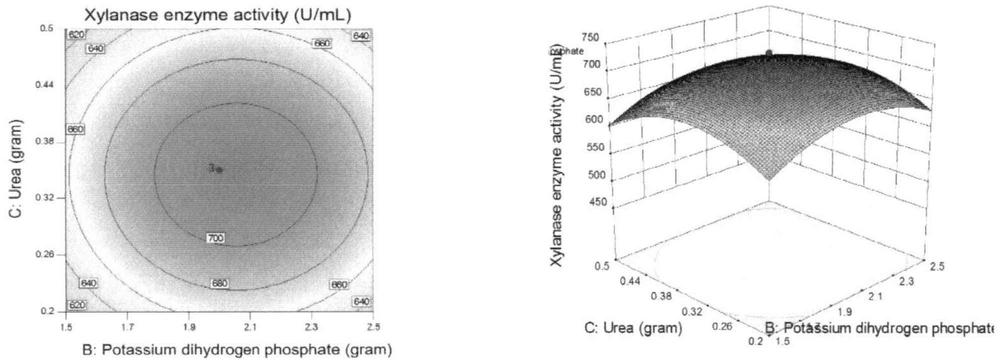

(a) (b)

Figure 4. Contour (a) and 3D plots (b) effect of interaction between Potassium dihydrogen phosphate and urea on xylanase enzyme activity

The graphs in Fig. 4a and 4b show the relationship between potassium dihydrogen and urea. The contour graph displays the conditions of the two factors where the optimal response value, the region is marked with red colour. While the 3D surface can be seen more clearly form a model that resembles a parabola, this is considered good because this graph shows the maximum value before it eventually declines. Based on the result of the research, it can be seen that the optimum xylanase enzyme activity is obtained at the concentration of 2.0 g/L potassium dihydrogen phosphate and urea concentration 0.35 g / L.

Potassium dihydrogen phosphate (KH_2PO_4) is one of the largest components present in the growth medium of T. viridae. In the solution potassium dihydrogen phosphate will decompose into potassium and phosphorus is needed microbe for metabolism. Phosphorus (P) required microbes as part of the formation of nucleic acids, phospholipids and coenzymes. Potassium acts as an inorganic cation in the cell and as a cofactor of some enzymes.

3.2.2 Protein content and specific activity xylanase enzyme

High protein levels allegedly indicate a high enzyme as well, but the enzyme formed cannot be ascertained is xylanase. The protein of the xylanase can be determined by knowing the activity of the xylanase first. The relationship between xylanase activity and the resulting protein is expressed by specific activity. Based on Table 4, the measurement of xylanase activity in some treatments showed that the highest enzyme activity was 711.90 U/mL. High enzyme activity is not always followed by high specific activity, and vice versa.

Enzymes are the most common group of proteins in living cells and have an important function as catalysts of biochemical reactions that collectively form metabolisms-intermediates from cells. The protein content is used as one of the tests because the solid state fermentation causes the cells to mix with the substrate so it is difficult to calculate the number of dry weight cells. The protein testing process, similar to enzyme activity, was applied to T. viridae with a temperature of 32.8^{0}C and a substrate ratio of OPEFB with a moistening solution of 0.63 g substrate/mL moistening solution.

3.2.3 Optimization of conditions and response

The optimum value (maximization) of the xylanase enzyme activity response through solution provided by Design Expert software version 10.0.1.0. The solution provided is proven by way of experimenting again with the concentration of each factor that has been given. The results of confirmation of the equations and solutions can be seen in Table 4.

Table 4. Concentration Optimization Solutions and Prediction Response

Factor Optimization Solution	Factor Concentration (g/L)	Xylanase Enzyme Activity		
		Prediction (U/mL)	Results (U/mL)	Validation (%)
Ammonium sulphate	1.50	711.91	684.70	96.18
Postassium dihydrogen phosphate	2.00			
Urea	0.35			

The accuracy of an optimization prediction model to predict the actual value can be measured by the degree of reliability. Software Design Expert version 10.0.1.0 defines that:

- Comparison validation> 0.90 then reliability is perfect
- Comparison of validation between 0.70 - 0.90 then high reliability
- Comparison of validation between 0.50 - 0.70 then moderate reliability
- Comparison of validation <0.50 then reliability is low

Based on RSM model validation data in Table 4, it can be seen that the accuracy of RSM model can be said perfect because the value of validation comparison of experimental results with optimization model prediction value on all responses above 90%.

4. Conclusions

The optimum concentration of moistening solution for the production of xylanase enzyme was at ammonium sulphate concentration of 1.5 g/L, concentration of potassium dihydrogen phosphate 2.0 g/L and urea concentration 0.35 g/L with perfect reliability value (0.96). The optimum condition of fermentation using oil palm empty fruit bunches as substrate were autoclave pretreatment with substrate ratio 0.63 g substrate/mL moistening solution, temperature 32.8°C with fermentation time for 60 h.

References

[1] Dekker, R.F.H. 2001. Bioconversion of Hemicellulose: Aspect of Hemicellulose Production by Trichoderma reesei QM 9414 and Enzymic Scarification of Hemicellulose. Available at: http://onlinelibrary.wiley.com/doi/10.1002/bit.260250419/epdf?r3_referer=wol&tracking_action=preview_click&show_checkout=1&purchase_referrer=onlinelibrary.wiley.com&purchase_site_license=LICENSE_DENIED. CSR Limited, Sugar Division, Sydney. NWS. Asutralia. Biotechnol. Bioeng. 25:1127-1146.

[2] Peni S P. 1995. Palm Bunches For Kraft Paper. Trubus Vol.311 : 52-54.

[3] Mardawati E, Wira DW, Kresnowati M, Purwadi R, Setiadi T. 2015. Microbial Production of Xylitol from Oil Palm Empty Fruit Bunches Hydrolysate : The Effect of Glucose Concentration. J Japan Inst Energy. 2015;94:769-774.

[4] Richana N., Irawadi, T., Nur, A., dan Syamsu, K. 2008. Isolation of Identification of Xylanase Producing Bacteria and their Characterization. Jurnal AgroBiogen 4(1):24-34.
[5] Alves-Prado, H.F., Pavezzi, F.C., Leite, R.S.R., de Oliveira, V.M., Sette, L.D., Dasilva, R., 2010. Screening and Production Study of Microbial Xylanase Producers from Brazilian Cerrado. Appl. Biochem. Biotechnol. 161, 333–346. doi:10.1007/s12010-009-8823-5.
[6] Yang, S.-T., El-Ensashy, H., Thongchul, N., 2013. Bioprocessing Technologies in Biorefinery for Sustainable Production of Fuels, Chemicals, and Polymers. John Wiley & Sons.
[7] Shah, A. R and Datta Madamwar. 2005. Xylanase Production under Solid State Fermentation and its Characterization by an Isolated Strain of Aspergillus foetidus in India. World Journal of Microbiology & Biotechnology, 21: 233–243.
[8] Xu, Y., Li, Y., Xu, S., Liu, Y., Wang, X., & Tang, J. 2008. Improvement of Xylanase Production by Aspergillus niger XY-1 Using Response Surface Methodology for Optimizing the Medium Composition. Journal of Zhejiang University SCIENCE B. Vol. 9 (7): 558–566. https://doi.org/10.1631/jzus.B0820038.
[9] Mardawati, E., Werner, A., Bley, T., Kresnowati, M., & Setiadi, T. 2014. The Enzymatic Hydrolysis of Oil Palm Empty Fruit Bunches to Xylose. Journal of Japan Institute of Energy, 93, 973-978. http://dx.doi.org/10.3775/jie.93.973.
[10] Fadilah dan Distantina, S., 2012. Delignification of Aren Trunk: Benchmarking the Effect of Glucose Addition by Adding Drops. Jur. Tek. Kim. Fak. Tek. Univ. Sebel. Maret Surak. 19–25.
[11] Cirakovic, J., Diner, B.A., 2010. Ozone Treatment of Biomass to Enhance Enzymatic Scarification. US20100159521 A1.
[12] Lio. Y. J. 2012. Using Soybean and Corn Processing Co-Products in Different Fermentation Systems. Graduate Theses and Dissertations. Paper 12767. Iowa State University.
[13] Seyis, I. and Aksoz, N. 2005. Effect of Carbon and Nitrogen Sources on Xylanase Production by Trichoderma harzianum 1073 D3. International Biodeterioration and Biodegradation. 55(2): 115-119.
[14] Gomes, J., Gomes, I., Steiner,W., Esterbauer, H. 1992. Production of Cellulase and Xylanase by a Wild Strain of Trichoderma viride. Applied Microbiology and Biotechnology. Vol. 36 : 701–707.

IOP Conf. Series: Earth and Environmental Science **141** (2018) 012019 doi:10.1088/1755-1315/141/1/012019

Production of xylitol from corn cob hydrolysate through acid and enzymatic hydrolysis by yeast

Efri Mardawati [1], R Andoyo [1], K A Syukra [1], MTAP Kresnowati [2], Y Bindar [2]

[1]Departement of Agroindustrial Technology, Universitas Padjadjaran, Jatinangor 40600, Indonesia
[2]Departement of Chemical Engineering, Institut Teknologi Bandung, Bandung 40132, Indonesia
efri.mardawati@unpad.ac.id

Abstract. The abundance of corn production in Indonesia offers the potential for its application as the raw material for biorefinery process. The hemicellulose content in corn cobs can be considered to be used as a raw material for xylitol production. The purpose of this research was to study the effect of hydrolysis methods for xylitol production and the effect of the hydrolyzed corn cobs to produce xylitol through fermentation. Hydrolysis methods that would be evaluated were acid and enzymatic hydrolysis. The result showed that the xylitol yield of fermented solution using enzymatic hydrolysates was 0.216 g-xylitol/g-xylose, which was higher than the one that used acid hydrolysates, which was 0.100 g-xylitol/g-xylose. Moreover, the specific growth rate of biomass in fermentation using enzymatic hydrolysates was also higher than the one that used acid hydrolysates, 0.039/h compared to 0.0056/h.

Introduction

Xylitol is a five-carbon polyalcohol sugar with a similar level of sweetness as that of sucrose, but contains only 2.4 cals/g, while sucrose 4 cal/g. Xylitol is very suitable to be consumed by people with diabetes because it has a low glycemic index [1]. Xylitol can be produced from xylose, either through fermentation or chemical processes. A monosaccharide xylose is produced from hydrolysis of hemicellulose with the main content in the form of xylan. Xylitol is also known as wood sugar because it is commonly contained in the woods that are rich with hemicellulose.

In the industrial scale, xylitol is produced chemically by reducing the pure xylose by Hydrogen gas at a temperature of 80°C-140°C used metal catalysts [2] . This method is costly because xylose derived from hemicellulose hydrolysis needs to be purified beforehand [3]. Nowadays, the world's production of xylitol is about 20,000 to 40,000 tons per year with a value of 40 to 80 million euros [4]. Whereas through biotechnological methods, xylitol production process is done by utilizing catalyst of microorganisms or enzymes. Microorganisms that can be used are the types of bacteria, fungi, and yeasts [2]. However, several studies have shown that xylitol production by using yeast produces xylitol yield more than the production of xylitol using bacteria and fungi [2].The advantages of producing xylitol with the bioprocess, among others, are: it does not cause environmental problems because it does not produce metal waste; it has a relatively low cost because the process takes place at room temperature and pressure of 1 atm; and there are many types of microorganisms capable of producing the xylose reductase enzyme which can turn xylose into xylitol and utilize hemicellulose hydrolysate contained in plants.

Hemicellulose is obtained from lignocellulosic waste that mainly contains 35-50% cellulose, 20-35% hemicellulose and 10-25% lignin [5,6]. Lignocellulosic wastes that are abundant in Indonesia are rice husks, tapioca pulp, empty fruit-bunches of oil-palm, and corncobs.

Corn is one of the agricultural products that are commonly produced in Indonesia. The national maize production amounted to 19.03 million tons, an increase of 2.81% compared to 2013. The plantation area in 2014 reached 3.91 million ha, while the harvesting area in 2014 amounted to 3.83 million hectares (ha) with the productivity of 4.9 tons per ha. Corn utilization is currently very diverse, ranging from food to biofuels. The fruit consists of 30% biomass waste in the form of corn cobs [7]. However, utilization of waste corn cobs in Indonesia is still limited when the content of hemicellulose in corn cobs is quite high, namely at 36%. Hydrolysis of hemicellulose will result in a constituent component which is generally a monosaccharide that is composed of the five-carbon chain. Hydrolysis is a chemical process of adding water molecules in a chemical compound that causes the termination of a chemical bond in the compound. This reaction is used to break down a polymer into its monomers. One of the results of the hydrolysis of hemicellulose is xylose which can be processed further into products of high economic value such as xylitol [8].

This study carried out two different types of hydrolysis process, that were acid hydrolysis and enzymatic hydrolysis on corn cobs to produce xylose. Hydrolysis by acid will yield a mixed product [2]. Therefore, in acid hydrolysis of corn cob, it is not only hemicellulose that will be hydrolyzed but also the cellulose whereas the monosaccharide can be further converted into another substance [5]. The resulted products are a mixture of compounds among other containing xylose, arabinose, and glucose. In addition, it will also produce byproducts in the form of oligomers, furfural, and acetic acid. Furfural compounds are toxic to some microorganisms and should be removed from the hydrolysate for it can be inhibitors in the fermentation process which may reduce the concentration of xylitol produced. The remaining acid from the hydrolysis process must also be neutralized because it can inhibit the fermentation reaction. On the contrary, in the process of the hydrolysis with enzymes, xylose will be produced more specifically because enzymatic hydrolysis does not produce mixed products or cause any product decomposition [9]. Therefore, it is important of this research to assess the effect of the acid hydrolysate and enzymatic hydrolysate in the production of xylitol by yeasts. Overall, this study was intended to turn corn cob biowaste into xylitol. Specifically, the purpose of this research was to study the effect of types of hydrolysis method on the production of xylitol via fermentation.

Materials and Methods

Corn cobs obtained from the Faculty of Agriculture, Universitas Padjadjaran, is the type of Unpad hybrid F1 originated from the crossover between Unpad A#001 x Unpad#002 that were cut, cleaned and then washed repeatedly with clean water. It was obtained from Arjasari, Bandung regency, with harvest age of 15 weeks of the growing season [10]. Afterward, the corncobs were dried in an oven at 50°C for 24 hours. Furthermore, it was crushed using a disk mill and screened with a sieve of 60 mesh in order to produce more refined corn cob powder. The composition of corn cob used was tested for its content: lignin, cellulose, and hemicellulose [11], and the morphology of the surface of the material using SEM (Scanning Electron Microscope). The tests were conducted to determine the level of hemicellulose contained in the corncobs which will then be used as a raw material in the manufacture of xylitol.

Microbe used for the fermentation was *Debaryomyces hansenii* ITBCC R85, obtained from Microbiology and Bioprocess Laboratorium, Chemical Engineering Institute of Technology Bandung. The yeast was maintained by swiping the old yeast cells in a medium of slant GYE (Glucose-Yeast Extract) agar and incubating at 30°C for 48 hours.

1.1. Hydrolysis Process

Acid hydrolysis was conducted based using sulfuric acid concentration of 6%, at the temperature of 120°C for 15 min [6]. Solid loading applied was 1 gram of corn cob fiber on the dry basis in 8 g of solution.

Enzymatic hydrolysis was performed which solid was pretreated with autohydrolysis at 121°C for 15 minutes in an autoclave. Then, samples were hydrolysis using xylanase at 60 °C, pH 5 and solid loading 15% [8,12].

1.2. Fermentation
The hydrolysates were further used as the substrate for the fermentation process. The fermentation process conducted using *D.hansenii* inoculum in a batch process at 30°C and at the anaerobic condition for 96 hours in Bioreactor. The solution was sampling at the beginning of the fermentation, at 24 hours fermentation, and at the end of fermentation [7]

1.3. Analysis
Samples of the fermentation liquid were analyzed by using a spectrophotometer and HPLC. From the analysis by a spectrophotometer, the cell growth curve was obtained, and from the HPLC analysis, the composition of the compound in the fermentation broth was observed.

Pulps from hydrolysis process were also analyzed using SEM (Scanning Electron Microscope) serves to determine the morphology of the surface of the material. Images that were obtained from SEM analysis were then processed using ImageJ 1.4 software to determine the average size of the particles and the pores of corn cob substrate.

The obtained data were further processed to parameterize the process performance following the growth of yeast (measured as specific growth rate), the product yield, and the utilization of xylose substrate. The specific growth rate (μ) was calculated using Equation 1 [12] :

$$\mu = \frac{1}{X}\frac{dX}{dt} \qquad (1)$$

Yield of biomass ($Y_{X/S}$) (g / g) is the biomass yield which is determined using the results of the growth curve of cells (dry cells), namely the ratio of the concentration of cells with the substrate concentration used includes xylose and glucose and expressed by Equation 2 [12]:

$$Y_{X/S} = -\frac{\Delta X}{\Delta S} = \frac{X - X_o}{S_o - S} \qquad (2)$$

Product yield ($Y_{P/S}$) (g/g) of xylitol products and other metabolites was calculated by Equation 3 [12]:

$$Y_{P/S} = -\frac{\Delta P}{\Delta S} = \frac{P - P_o}{S_o - S} \qquad (3)$$

Product yield per cell biomass (($Y_{P/X}$) (g/g) is the ratio between the number of xylitol products obtained by biomass produced at the end of fermentation (Equation 4) [12]:

$$Y_{P/X} = -\frac{\Delta P}{\Delta X} = \frac{P - P}{X - X_o} \qquad (4)$$

Xylose utilization (%) is the amount of xylose used by the yeast in the production of biomass and products, obtained by comparing the amount of xylose that was used during the fermentation with the initial xylose concentration.

Results and Discussions

1.4. Characterization of Corn Cob
Before being used as a substrate hydrolysate in the research, characterization on corn cob fiber was performed. Corn cob fiber characterization performed included the test of cellulose, hemicellulose, and lignin (Table 1).

Table 1. Corn cob fiber composition.

Fiber Components	% dry basis	
	Unpad Hybrid Corn Cobs	Reference[a]
Hemicellulose	41.170 ± 0.930	36.210
Cellulose	20.890 ± 0.220	34.400
Lignin	16.260 ± 0.210	18.800

[a]Koswara, 1991

Hemicellulose concentration of Unpad hybrid corn cobs is 41.17 (% DB) where the value is greater than the corn cobs, in general, that is 36.21 (% db) (Table 1). Hemicellulose is sugar polymer composed of xylose monomer and glucose [13]. The higher the hemicellulose composition, the more xylose available. On the contrary, Unpad hybrid corn cobs possess cellulose content of 20.89 (% DB) which is smaller in value when compared to cellulose content of corn cobs in general, i.e. 34.40 (% db). Cellulose is a polymer composed of glucose monomers [14], so that after the hydrolysis process occurs, the resulted glucose concentration was expected to be lower. The high glucose concentration in hydrolysate will inhibit the fermentation process of xylose into xylitol [5]. Unpad hybrid corn cobs have a lignin content of 16.26 (% DB) where the number is less compared to the lignin content of corn cobs in general, i.e. 18.80 (% db). Lignin is a three-dimensional polymer consisting of phenylpropane units tied by the ether linkage (C-O-C) and carbon bond (C-C). Lignin is resistant to hydrolysis because of arylalkyl bond and ether bond, and is insoluble in water, acid, and a hydrocarbon solution. Meanwhile, the other fiber components such as hemicellulose which has xylose monomer in the substrate will stick together with each other. Therefore, the lignin content that is too high on the substrate will inhibit hydrolysis process to produce xylose [9]. Based on the test results of the three components of the fiber, Unpad hybrid corn cobs have a higher potential to be used as the substrate of hydrolysis which produces xylose in xylitol when compared to corn cobs in general because it has higher hemicellulose content.

1.5. Effects of Hydrolysis Types on The Composition of Hydrolysate

As expected from the corn cob composition analysis, the hydrolysis results showed that a sample of corn cobs produced a primary product in the form of xylose and byproducts in the form of glucose and acetic acid (Table 2).

Table 2. Components of corn cob hydrolysate.

Components (g/L)	Type of Hydrolysis	
	Enzymatic Hydrolysis	Acid Hydrolysis
Xylose	2.070 ± 0,060	5.530 ± 0,140
Glucose	3.590 ± 0.120	4.070 ± 0,080
Acetic acid	1.150 ± 0,020	0.280 ± 0,000

Xylose was produced from the hydrolysis of hemicellulose, while glucose was the hydrolysis result of cellulose. Even though the solid loading applied in the acid hydrolysis was lower than enzymatic hydrolysis, which was 12.5% in the acid hydrolysis and 15% in the enzymatic hydrolysis, the concentration of xylose produced by acid hydrolysis was higher than those produced by the enzymatic hydrolysis. Besides that, both acid and enzymatic hydrolysis produced glucose, which glucose of enzymatic hydrolysis was higher than acid hydrolysis. This might be caused by the high temperature used in the hydrolysis process, 121°C, that could facilitate the breaking of xylose and cellulose from the substrate [6]. The higher concentration of acetic acid in the enzymatic hydrolysate when compared to those in acid hydrolysate was related to the addition of acetate buffer in the enzymatic hydrolysis process.

1.6. Effect of Enzymatic and Acid Hydrolysate on Microorganism's Growth During Fermentation

The fermentation using acid hydrolysate went through a phase of adaptation for 2 hours and the logarithmic phase from the 2nd until the 24th hour (Figure 1). The fermentation lasted for 96 hours. Fermentation using enzymes hydrolysate went through a phase of adaptation for 2 hours and the phase of logarithmic from the 2nd hour till the 24th hour. Based on previous studies [15] about logarithmic phase on fermentation using Oil Palm Empty Fruit Bunches (OPEFB) enzymatic hydrolysates at similar fermentation conditions, it was obtained that microbes during fermentation went through the logarithmic phase until the 46th hours. Faster logarithmic phase was thus obtained in this study.

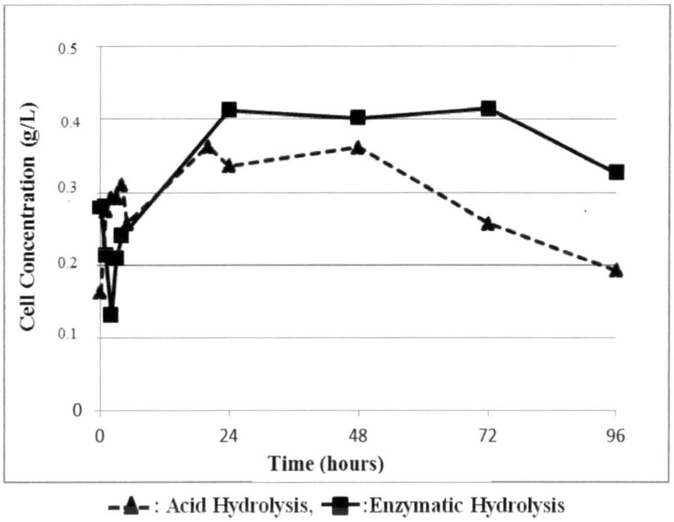

Figure 1. Profile of the growth of *D. hansenii* in different hydrolysate during fermentation.

Fermentation using acid hydrolysate went through a phase of adaptation for 2 h and the logarithmic phase from the 2 - 24 h fermentation time. The fermentation lasted for 96 hours. Fermentation using enzymes hydrolysate went through a phase of adaptation for 2 hours and the phase of logarithmic from untill the 24th hour. Based on previous studies [15] about logarithmic phase on fermentation using Oil Palm Empty Fruit Bunches (OPEFB) enzymatic hydrolysates at similar fermentation conditions, it was obtained that microbes during fermentation went through the logarithmic phase at 46 h fermentation time. Faster logarithmic phase was thus obtained in this study.

Fermentation using an enzymatic hydrolysate had the specific growth rate (μ) of 0.039 h^{-1}, whereas fermentation using acid hydrolysate had a μ of 0.0056 h^{-1}. From the data obtained, the fermentation using enzymes hydrolysate had a specific growth rate value that was greater than the fermentation using acid hydrolysate.

μ value resulted was small or nearly zero. It indicates that there is no significant growth of the cells. Slow cell growth caused by the tendency of microbial cells utilize sugars to form a metabolite product under conditions of high initial cell concentration. Limitations of available oxygen can also be one of the reasons for the absence of the cell growth in the fermentation with the condition of a high initial concentration of cells [7].

1.7. Effects of Types of Hydrolysis on The Production of Metabolic Products

Fermentation using enzymatic hydrolysate as well as acid hydrolysate showed that the utilization of glucose tended to be higher compared to the xylose utilization (Table 3).

Table 3. Fermentation product of enzyme hydrolysate and acid hydrolysate.

Products	Type of Hydrolysis	
	Enzymatic Hydrolysis	Acid Hydrolysis
Initial Xylose Concentration (g/L)	2.070 ± 0.060	5.530 ± 0.140
Final xylose concentration (g/L)	1.040 ± 0.070	3.550 ± 0.020
Xylose Utilization (%)	49.750 ± 2.290	35.800 ± 2.660
Initial Glucose Concentration (g/L)	3.590 ± 0.120	4.070 ± 0.080
Final Glucose Concentration (g/L)	0.380 ± 0.010	1.880 ± 0.040
Glucose utilization(%)	89.420 ± 0.090	53.810 ± 2.530
Initial Xylitol Concentration (g/L)	0.000 ± 0.000	0.000 ± 0.000
Final Xylitol Concentration (g/L)	0.090 ± 0.020	0.200 ± 0.010
Xylitol Yield from xylose t = 72 h (g/g) ($Y_{P/S}$)*	0.216 ± 0.000	0.000 ± 0.000
Xylitol Yield from xylose t = 96 h (g/g) ($Y_{P/S}$)*	0.100 ± 0.050	0.100 ± 0.490
Initial Cells Concentration (g cell/L)	0.281 ± 0.900	0.163 ± 0.015
Final Cells Concentration (g cell/L)	0.326 ± 0.001	0.193 ± 0.012
Biomass Yield of Substrate (g/g) ($Y_{X/S}$)*	0.031 ± 0.001	0.013 ± 0.015
Xylitol Yield of Concentration cells (g/g) ($Y_{P/X}$)*	1.500 ± 0.140	1.000 ± 0.089
Specific Growth Rate (h^{-1}) (μ)	0.039 ± 0.001	0.0056 ± 0.001

The higher glucose utilization was due to the fact that glucose was used dominantly as the food source by yeasts during the fermentation process. Something similar happened to the research Parajo (1998) [16] where the fermentation occurred using wood hydrolysate that had been concentrated as a substrate.

Based on Table 3, xylitol began produce in the fermentation using enzymatic hydrolysate at the 72 hour fermentation time, with the xylitol yield of 0.216 g xylitol/g xylose (xylitol concentration of 0.16 g/L), while for the acid hydrolysate, xylitol began to form at the 96 hour or in the late time of fermentation with yield 0.100 g xylitol/g xylose (xylitol concentration of 0.2 g/L). At the 96th hour, the xylitol produced by the enzyme hydrolysis decreased to 0,100 g of xylitol/g xylose. The decrease of xylitol yield might be caused by the consumption of xylitol by microbes as a source of food due to low levels of xylose and glucose at the end of fermentation. In another study, the yield of xylitol using the enzymatic hydrolysate with fermentation conditions similar to this study equaled to 0.098 g xylitol/g xylose [15]. Meanwhile, for the xylitol yield using acid hydrolysate for the same fermentation condition was 0.084 g of xylitol/g xylose. The yield of biomass on fermentation using enzymatic hydrolysate possessed a higher value when compared to the fermentation using acid hydrolysate [8]. The higher value of this yield can be caused by not accounting for acetic acid content as a substrate in the fermentation of hydrolysate (indicated by the yield of acetic acid to the substrate which is negative).

Xylose and glucose from both types of hydrolysates continued to decline as it continued to be used by microbes to produce xylitol and other metabolite products (Figure 2). The concentration of glucose in the enzymatic hydrolysate decreased faster than the xylose concentration in the acid hydrolysate. This resulted in higher values of glucose utilization in the enzyme hydrolysate. The decline in xylose concentration, both acid and enzyme hydrolysate, was not as fast as the decrease in the concentration of glucose. Based on the decrease of concentration, the substrate produced greater utilization value than the xylose utilization, both for the acid hydrolysate and enzymes hydrolysate. In this case, the enzymatic hydrolysis had a higher utilization value, both glucose and xylose utilization. The high utilization of glucose indicated that the yeast that conducts the fermentation process tended to consume glucose so that the yeast will be more likely to produce other metabolite products like ethanol than xylitol.

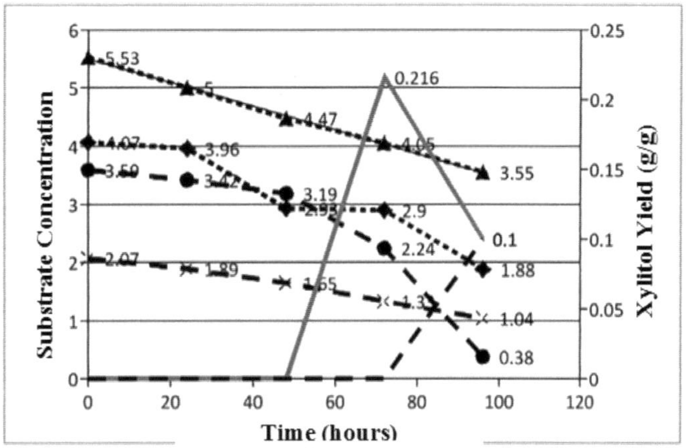

Xylose Concentration (acid), Xylose Concentration (enzymatic) : Glucose Concentration (acid), Glucose Concentration (enzymatic), Xylitol (enzyme)

Xylitol (acid)

Figure 2. The concentration of xylose and glucose, and yield of xylitol per 24 hours.

Fermentation using enzymatic hydrolysate provides greater and faster xylitol yield and when compared to the acid hydrolysate. The higher yield of xylitol on fermentation using enzyme hydrolysate can be caused by the greater amount of xylose consumed compared to acid hydrolysate. In addition, the higher yield of xylitol on enzyme hydrolysate fermentation indicated the absence of inhibitor compounds that can interfere with the activity of *D. hansenii* [17]

1.8. Effectsof Enzyme and Acid Hydrolysate on other Metabolites Products

The production of metabolite products such as ethanol and acetic acid continued to increase, even the acquisition is indicated to continue to increase along with the increase of duration of fermentation (Table 4). A tendency to produce ethanol than xylitol and is indicated by the greater yield of ethanol when compared to the yield of xylitol.

Because the fermentation process takes place under anaerobic conditions, the oxygen dissolved in the broth of low fermentation. The low dissolved oxygen causes microbes to tend to convert glucose into ethanol and a small portion xylose into ethanol [18]. It also indicates that the ratio of glucose to xylose in the substrate is too big.

IOP Conf. Series: Earth and Environmental Science **141** (2018) 012019 doi:10.1088/1755-1315/141/1/012019

Table 4. Yield of other metabolites product.

Information	Fermentation Medium	
	Enzymatic Hydrolysis	Acid Hydrolysis
Ethanol Yield of Substrate t = 0 h (g/g) ($Y_{P/S}$)*	0.000 ± 0.000	0.000 ± 0.000
Ethanol Yield of Substrate t = 24 h (g/g) ($Y_{P/S}$)*	0.104 ± 0.000	0.000 ± 0.000
Ethanol Yield of Substrate t = 48 h (g/g) ($Y_{P/S}$)*	0.130 ± 0.000	0.000 ± 0.000
Ethanol Yield of Substrate t = 72 h (g/g) ($Y_{P/S}$)*	0.151 ± 0.009	0.000 ± 0.000
Ethanol Yield of Substrate t = 96 h (g/g) ($Y_{P/S}$)*	0.177 ± 0.140	0.000 ± 0.000
Acetic Acid of Substrate t = 0 h (g/g) ($Y_{P/S}$)*	0.271 ± 0.012	0.067 ± 0.001
Acetic Acid of Substrate t = 24 h (g/g) ($Y_{P/S}$)*	0.344 ± 0000	0.094 ± 0.000
Acetic Acid of Substrate t = 48 h (g/g) ($Y_{P/S}$)*	0.366 ± 0.000	0.098 ± 0.000
Acetic Acid of Substrate t = 72 h (g/g) ($Y_{P/S}$)*	0.375 ± 0.000	0.108 ± 0.000
Acetic Acid of Substrate t = 96 h (g/g) ($Y_{P/S}$)*	0.394 ± 0.007	0.137 ± 0.004
Glycerol Yield Of Substrate (g/g)($Y_{P/S}$)*	0.000 ± 0.000	0.000 ± 0.000

Acid hydrolysate fermentation has a smaller glucose conversion value than the enzymes hydrolysate fermentation. This small value of the glucose conversion causes the absence of formation of ethanol and glycerol which are metabolite products produced by microbes when consuming glucose as substrate.

1.9. Scanning Electron Microscope (SEM) Analysis

Material test by means of SEM (Scanning Electron Microscope) serves to quantitatively and qualitatively determine the morphology of the surface of the material. SEM (Scanning Electron Microscope) is one type of electron microscopes capable of producing a high-resolution picture of a sample's surface. Therefore, the image generated by SEM has qualitative characteristics in two dimensions because it uses electrons instead of light waves and is useful for determining the structure of the sample's surface. SEM characterizes the material in the form of a thin layer having a thickness of 20 μm from the surface, images of surface topography in the form of protrusions, indentations and the thickness of a thin layer of a cross-sectional.

SEM or scanning electron microscope focuses the electron beam at the surface of the object and takes a picture by detecting electrons that appear on the surface of the object. Various types of SEM allow for different usages, such as morphological study, high-speed composition analysis, surface roughness, porosity, particle size distribution, homogeneity of materials or environmental studies on the subject of the sensitivity of the material. The test result of corn cob substrate by means of SEM can be seen in Figure 3-8.

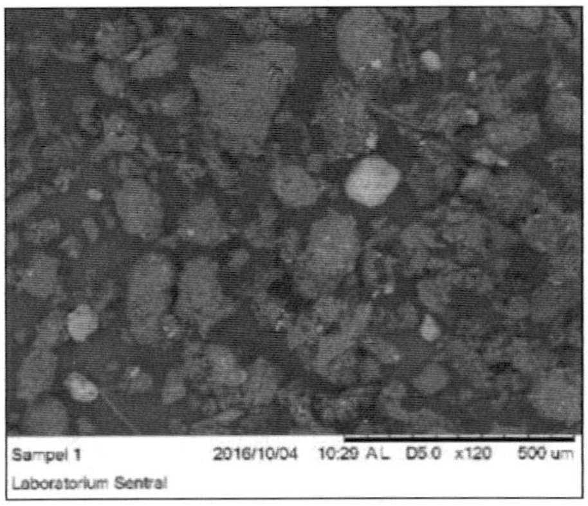

Figure 3. Particle shape (120x magnification) of corn cob substrate before hydrolysis.

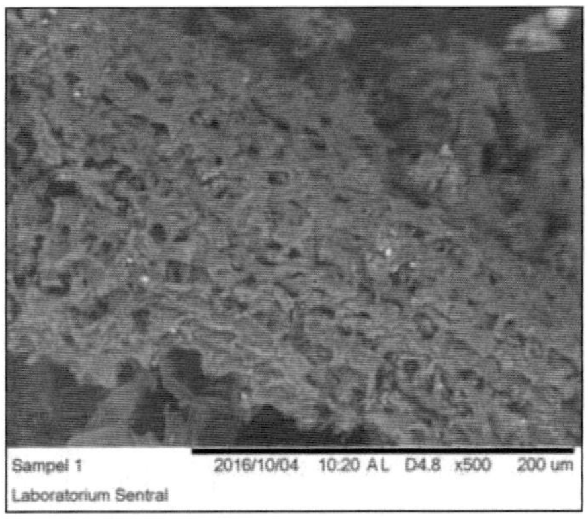

Figure 4. Particle pores (500x magnification) of corn cob substrate before hydrolysis.

Figure 5. Particle shape (40x magnification) of corn cob substrate after acid hydrolysis.

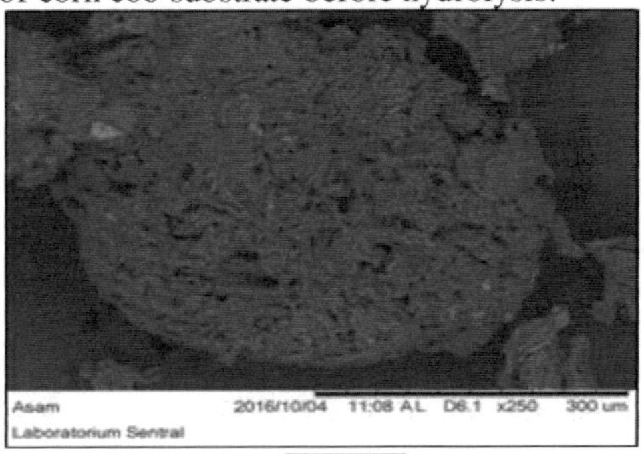

Figure 6. Particle pores (250x magnification) of corn cob substrate after acid hydrolysis.

Figure 7. Particle shape (40x magnification) of corn cob substrate after enzymatic hydrolysis.

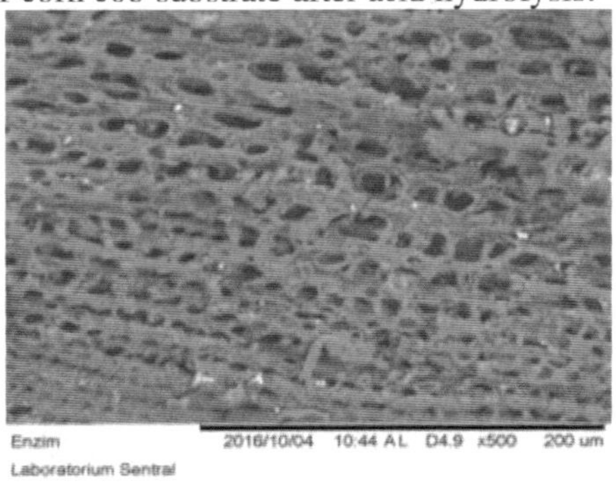

Figure 8. Particle pores (500x magnification) of corn cob substrate after enzymatic hydrolysis.

Enzymatically-hydrolyzed corncobs had a much larger average area of the particle (4.49 x 10^4 μm^2) when compared to corn cobs before being hydrolyzed (19.35 x 10^2 μm^2) and after being hydrolyzed with acid (4.20 x $10^4 \mu m^2$). The average size of the corncob's pores after hydrolyzed by the enzyme (33.17 μm^2) was also greater when compared to corn cobs before hydrolyzed (11.22 μm^2) and after acid hydrolysis (27.54 μm^2) (Table 5).

Table 5. Morphological test on corn cob substrate before and after hydrolysis

Components (g/L)	Average Particle Size (μm^2)	Average Pore Size (μm^2)
Before Hydrolysis	19.35×10^2	11.22
After Acid Hydrolysis	4.20×10^4	27.54
After Enzymatic Hydrolysis	4.49×10^4	33.17

The result of enzymatic hydrolysis enlarged the pore size to be more than the original corncob substrate and more than the result of acid hydrolysis. It was caused by incubation time in the hydrolysis process, namely 72 hours. Meanwhile, the particle size of enzymatic hydrolysis result was enlarged due to the clotting of the substrate by buffer acetate used during hydrolysis process. The enlarged particle size of corn cob substrate on acid hydrolysates is because acid can destruct some crude fibers by lowering the pH value. At low pH conditions, crude fiber contained in a material will clot. The clotting causes the particles to grow in size. Meanwhile, the enlargement of the pore of corn cob fiber's particle occurs due to the disconnection of fiber chains into sugar monomer that is dissolved in the hydrolysate liquid [8].

Conclusions

This study was explored the effect of the enzymatic and acid hydrolysis on xylitol fermentation product. The type of hydrolysis applied affected the growth of *D. hansenii* and correspondingly, the xylitol production by these microbes. Fermentation using enzymatic hydrolysate gave higher xylitol yield ($Y_{P/S}$) and specific growth rate (μ) than the fermentation using acid hydrolysate. These phenomena were very likely to be caused by the different concentration of glucose and fermentation inhibitors in the hydrolysate.

References

1. Brunzell JD. Use Of Fructose, Xylitol, Or Sorbitol As A Sweetener In Diabetes Mellitus. *Diabetes Care*. 1978;1(4).
2. Parajó JC, Dominguez H, Dominguez JM. Biotechnological Production Of Xylitol Part I: Interest Of Xylitol And Fundamentals Of Its Biosynthesis. *Bioresour Technol*. 1998;65:191-211.
3. Nigam P, Singh D. Process for Fermentative Production Of Xylitol – A Sugar Substitute. *Pro Biochem*. 1995;30:117-124.
4. Granström TB, Izumori K, Leisolla M. A Rare Sugar Xylitol Part II: Biotechnological Production And Future Applications Of Xylitol. *Appl Microbiol Biotechnol*. 2007;74:273-276.
5. Kresnowati M, Mardawati E, Setiadi T. Production of Xylitol from Oil Palm Empty Friuts Bunch : A Case Study on Bioefinery Concept. *Mod Appl Sci*. 2015;9(7):206-213. doi:10.5539/mas.v9n7p206.
6. Rahman SHA, Choudhury JP, Ahmad AL. Production of xylose from oil palm empty fruit bunch fiber using sulfuric acid. 2006;30:97-103. doi:10.1016/j.bej.2006.02.009.
7. Mardawati E, Wira DW, Kresnowati M, Purwadi R, Setiadi T. Microbial Production of Xylitol from Oil Palm Empty Fruit Bunches Hydrolysate : The Effect of Glucose Concentration. *J Japan Inst Energy*. 2015;94:769-774.
8. Mardawati E, Werner A, Bley T, Kresnowati M, Setiadi T. The Enzymatic Hydrolysis of Oil Palm Empty Fruit Bunches to Xylose. *J Japan Inst Energy*. 2014;93:973-978.
9. Mardawati E, Purwadi R, Kresnowati M, Setiadi T. Evaluation of The Enzymatic Hydrolysis Process of Oil Palm Empty Fruit Bunch Using Crude Fungal Xylanase. *ARPN Eng Appl Sci*. 2017;12(18):5286-5292.
10. Ruswandi D, Suryadi E, Marta H. Multi Environmental Test High Efficiency Maize Hybrids and High Nutrition Content in West Java. In: *Superior Research of Higher Education*. Bandung:

Padjadjaran University; 2014.
11. Datta R. Acidogenic Fermentation of Lignocellulose-Acid Yield and Conversion of Components. *Biotechnol Bioeng*. 1981;XXIII:2167-2170.
12. Mardawati E, Trirakhmadi A, Kresnowati M, Setiadi T. Kinetic study on Fermentation of xylose for The Xylitol Production. *J Ind Inf Technol Agric*. 2017;1(1):1-6.
13. Sjöström E. *Chemical Wood, Fundamentals and Usage*. 2nd ed. (Press U, ed.). Yogyakarta; 1995.
14. Fengel D, Wegener. *Wood: Chemistry, Ultrastructure Reactions*. Yogyakarta: UGM Pess; 1995.
15. Tantra TM, David. *Microbial Xylitol Production from Oi Palm Empty Fruit Bunches Hydrolysate*. Bandung; 2013.
16. Parajó JC, Dominguez H, Dominguez JM. Biotechnological Production Of Xylitol Part 3: Operation In Culture Media Made From Lignocelluloses Hydrolysates. *Bioresour Technol*. 1988;66:25-40.
17. Winkelhausen E, Kuzmanova S. Microbial Conversion of D-Xylose to Xylitol. *J Ferment Bioeng*. 1998;86(1).
18. Oh DK, Kim SY. Increase Of Xylitol Yield By Feeding Xylose And Glucose In Candida Tropicalis. *Appl Microbiol Biotechnol*. 1998;50:419-425.

IOP Conf. Series: Earth and Environmental Science **141** (2018) 012020 doi:10.1088/1755-1315/141/1/012020

Socioeconomic factors of smallholder farmers' behavior in biomass burning around palm oil plantation in Indonesia

Maswadi [1,4], Arifudin [2], Nurmelati Septiana [3] and Maulidi [1]

[1] Universitas Tanjungpura, Universitas Riau[2], dan Universitas Lambung Mangkurat[3]
[4] Email Corresponding : maswadi@faperta.untan.ac.id

Abstract. Indonesian peatland fires has been revealed as the cause of haze disaster in Indonesia, while oil palm plantation's concesion owned both by companies and smallholder farmers are accused as the main cause of this problem, especially in practice of land clearing. It is very important to conduct research on socioeconomic factors of farmers' behavior in burning the peatland, while peatland one of the megabiomass storage in nature. The research was conducted in Kalimantan barat, where in province has been choosen two villages as the sample. Observation, interview with quiestionare, and focus group discussion were used in collecting data. In term of analysing the data, regression analysis (ordinary least square) was performed using SPSS Program. The result show that: (1). The socio economics factor that are affecting the burning behavior, were extension's activities, degree of knowledge, consideration to burn, degree of participation on organisation and degree of cosmopolite. On the other hand, degree of burning frequent, was affected by land productivity, extension activities, and degree of participation in organisation, and finally the size of land' burning is affected by, the kind of burning's activities, the mutual aid (social capital), consideration of land burning, degree of awarness, and degree participation on organization.

1. Introduction

The community is one of the most important elements in the implementation of fire prevention efforts. Society is all unity of human life that interact according to custom system which is bound by a sense of common identity. The concept of society according to Hughes et al. (2002) referred by Main (2010) [15], society refers to a group of people living within the same territory and sharing a particular culture. The culture makes people have the same values and norms and usually have a common language.

Based on the Regulation of the State Minister for the Environment, Land is a terrain of terrestrial ecosystems used for business or field activities or gardens for the community, whereas according to West Kalimantan Regulation [10] , land is an area outside the forest area, Vegetation (alang-alang, bush, cultivated and other crops) or non-vegetation intended for development in agriculture, plantation, forestry, transmigration, mining and others. Based on some of these definitions, the definition of land in this study is the area outside the forest area used for the cultivation activities in the form of peat soil.

The results of a study conducted by [9] suggest the reasons for the danger of fire in peatlands because if fires occur below the surface, there is no tool capable of extinguishing them. Very perfect peat land hold fire, fire can be extinguished only with the rain that fell with a swift. If rain falls relatively small (not to inundate peatland), peat fires will cause greater smoke, other than that rapid fire creeps on the inside of dry peatlands, which often leads to new fire spots in some places.

Because of the incident, many human activities are disturbed [2]. Fires cause huge losses and problems in various aspects, ranging from health, social and economic aspects. Disadvantages in the health aspects of the smoke caused can interfere with breathing and cause various diseases. Losses on social aspects affect the political relations between neighboring countries, because the smoke that spreads beyond the borders of the country is considered environmental pollution that can disturb visibility. Losses on the economic aspects that can be estimated only include losses in timber value, but many other non-timber losses cannot be accurately estimated, such as germplasm, ecotourism, water resources and water regulator, erosion control and soil conservation and nutrient cycles . Forest and land fires also have negative impacts on the vegetation, wildlife, soil, water and air that can be felt by the community not only in the fire location, but also into the area even to neighboring countries.

Due to the magnitude of losses caused by frequent fires, it is necessary to make efforts to prevent fire. Fire prevention is better as the first action than doing much more difficult and costly blackouts and rehabilitation. Precautions in the management of forest and land fires have the objective of preventing fires, minimizing fires, minimizing the impact of fires and maintaining and preserving forest resources from the dangers of land fires [1]. This study aims to determine the factors that influence the behavior of burning the peat land, especially around community-based oil palm plantation areas.

2. Methods

2.1. Research Design

This study uses an explanatory study is research method to retrieve data from the sample and questionnaire as a data collection tool. Once the data is obtained, the results are explained in the explanatory, and the data were analyzed to test the hypothesis proposed at the beginning if study [4].

2.2. Location and Time of Study

Determination of research area is based on criteria that have a close relationship with the issues raised in the study. These criteria include the West Kalimantan province which has the location of the second most frequent fires in Indonesia after Riau Province. Bengkayang Regency is a fire-prone district located around the area of oil palm plantations.

2.3 Population and Sample

Populations in this research are farmer owner and peat land farmer whose land has been burnt and its land is in the surrounding land which had burned around palm oil plantation. The sampling technique was done by using census method or total sampling. Therefore, this study includes population studies or census studies, in line with those expressed by [3] hich refers [6] that the appropriate sample size for census studies is 100-200. Based on the informants (in this case the village head of each research village), the number of farmers who owned and cultivated the land that had burned and been around the land had been burned is as follows:

Tabel 1. Number of Population and Sample

District	Village	Number of Population/unit	Number of Sample
Sui raya Kepulauan	Sui Raya	60	60
Sui Raya Kepulauan	Sui Keran	60	60
Total		120	120

The total sample in this research is 120 respondent of farmer owner and cultivator of land that has been burnt and farmer owner and cultivator of land which is around the land had burned and land around palm oil plantation area.

IOP Conf. Series: Earth and Environmental Science **141** (2018) 012020 doi:10.1088/1755-1315/141/1/012020

3. Data Analysis

This study used multiple regressions with OLS (ordinary least square) method with prediction of factors influencing burning behavior seen from three aspects, that is the opportunity of burning behavior of the land (Y_1), the frequency of burning the land (Y_2) and the area of burned land (Y_3). Therefore, the model for the analysis of factors affecting burn behavior consists of three models. The model used is as follows [8]:

$$Y_{1n} = \alpha+\beta_1X_1+\beta_2X_2+\beta_3X_3+\beta_4X_4+\beta_5X_5+\beta_6X_6+\beta_7X_7+\beta_8X_8+\beta_9X_9+\beta_{10}X_{10}+\beta_{11}X_{11}+\beta_{12}X_{12}+\beta_{13}X_{13}+\beta_{14}X_{14}+\beta_{15}X_{15}+\varepsilon$$

$$Y_{2n} = \alpha+\beta_1X_1+\beta_2X_2+\beta_3X_3+\beta_6X_6+\beta_7X_7+\beta_8X_8+\beta_{10}X_{10}+\beta_{11}X_{11}+\beta_{12}X_{12}+\beta_{13}X_{13}+\beta_{14}X_{14}+\beta_{15}X_{15}+\varepsilon$$

$$Y_{3n} = \alpha+\beta_1X_1+\beta_2X_2+\beta_3X_3+\beta_6X_6+\beta_7X_7+\beta_8X_8+\beta_{10}X_{10}+\beta_{11}X_{11}+\beta_{12}X_{12}+\beta_{13}X_{13}+\beta_{14}X_{14}+\beta_{15}X_{15}+\varepsilon$$

4. Results and Discussions

4.1. Analysis of Burning Behavior of West Kalimantan Province

Analysis on the behavior of burning land in West Kalimantan Province shows there are severa nine variables that influence the burning behavior significantly, as follows:

Table 2. Result of Behavioral Burning Analysis in West Kalimantan Province

Variabel		Y1 (Land Burning Behavior)			Y2 (Burning Frequency)		Y3 (Burning Area)	
		Coef.	Sig.	Exp(B)	Coef.	Sig.	Coef.	Sig.
X1	Area of farmland	**.104**	**.009**	**1.110**	.002	.975	**-.171**	**.010**
X2	Age of respondent	.000	.476	1.000	.006	.920	.010	.861
X3	Income	-.055	.110	.946	.080	.214	.029	.554
X4	Land productivity	**.195**	**.000**	**1.216**	-.018	.768	**-.131**	**.019**
X5	Extension activities	**.120**	**.000**	**1.128**	.073	.202	-.057	.282
X6	Number of plant species	-.160	.091	.853	-.020	.735	.008	.889
X7	Number of land plots	.002	.256	.998	.042	.441	-.079	.122
X8	Type of activity to burn	-.007	.464	.993	.072	.243	**.127**	**.028**
X9	Number of types of gotong royong	.000	.856	1.000	-.069	.210	-.030	.559
X10	Level of education	.017	.396	1.018	-.089	.119	.069	.192
X11	Knowledge level	**.104**	**.009**	1.009	.005	.928	.068	.199
X12	Rating burns	.031	.961	1.031	.060	.380	**.116**	**.071**
X13	Level of concern for the burning of land	.256	.649	1.291	**-.210**	**.044**	**-.121**	**.014**
X14	Activity of organization	.002	.907	1.002	**.166**	**.003**	**.104**	**.043**
X15	Cosmopolitan level	**-.002**	**.000**	**.998**	**.142**	**.031**	-.079	.122
Constant		-13.572	.002	.000		.357		.054
R^2		**.716**			**.437**		**.575**	
N		120			120		120	

Source: Primary data analysis results (2017)
*Remark: * = Significant at 95% confidence level*

The results of the analysis show that there are severa nine factors influencing the burning behavior of the land, namely:

Table 3. Significant Variables that Affect Behavior of Land Burning

Independent Variable	Y1	Y2	Y3
Farmland Area X1)	Landowners tend not to burn land		Land owners who are still burning tend to burn land with a large area
Land Productivity X4)	Respondents with low productivity levels, land burning behavior tend to be high		Respondents with low productivity levels tend to burn large areas of land
Extention Activity X5)	Respondents with high level of education tend not to burn land		
Type of activity to burn (X8)			Respondents with type of activity burning a lot, tend to burn land with large area
Level of Knowledge (X11)	Respondents with a high level of knowledge tend not to burn land		
Rating burns (X12)			Respondents with good burn ratings tend to burn with a smaller area
Level of concern for the burning of land (X13)		Respondents' level of concern for high land burning tends to rarely burn	Respondents' level of awareness of high land burning, tend to burn with a smaller area
Activity of organization (X14)		Respondents with high organizational activity, tend to rarely burn land	Respondents with high organizational activeness, tend to burn land with small area
Cosmopolitan level (X15)	Respondents with high cosmopolitan levels, tended not to burn land	Respondents with high cosmopolitan rates tend to rarely burn	

Source: Primary Data Analysis Result (2017)

Factors that influence the behavioral opportunities of burning land are the area of farming land, land productivity, extension activities, level of knowledge, and cosmopolitan level. Thus, to reduce the chances of burning behavior can be done through the approach of the five variables. Village peat land fire prevention activities around the palm oil company is a social engineering in the direction of social change by paying attention to local wisdom (Indigenous knowledge). Local wisdom can be interpreted as local knowledge of the community in addressing the existing environmental conditions so as to be in harmony with nature. Physical changes of peat that have wet nature, causing some local wisdom are no longer suitable to be practiced, for example clearing land by burning. The habit of burning land is a local wisdom for people who used to do shifting cultivation, but now this practice is very dangerous to do, because changes in drained peat conditions and extreme weather changes cause vulnerable peat

IOP Conf. Series: Earth and Environmental Science 141 (2018) 012020 doi:10.1088/1755-1315/141/1/012020

land to occur small fires can spread into large fires. Nevertheless, many communities still have local wisdom relevant to practice, such as customary rules, such as forbidden forests, which prohibit the conversion of forest functions into non-forest areas. Another example is the local wisdom of the people who plant sago plants that are very compatible with the nature of peat in Sungai Tohor Village, Meranti Regency, Kepulauan Riau, [14]. Local community wisdom in every village that is relevant and in harmony with nature is very well explored and developed in the activities of fire prevention of land.

Several studies on local wisdom have been conducted by several researchers, in various types of society in Indonesia, such as the wisdom of the Osing community, Banyuwangi, East Java. Their local wisdom can be seen from their knowledge, values, morals and ethics, and the norms applied in the form of suggestions, rules and sanctions, and words of wisdom as a guideline for them to behave and act in safeguarding, preserving and preserving the eyes Water, especially in keeping a constant stream of water discharge, they protect the trees and plants around. In maintaining this local culture, the older generation pass on values, morals, ethics, and norms including Islamic norms as guidance on how to behave and act in the tradition and instinct to respect the environment for their families, neighbors, relatives and children -grandchildren [12]. Other local wisdom is in Sundanese society, where research conducted by [7] shows that directly or indirectly nature is actually "the earth of residence as well as the life of the living" for the Sundanese. Local wisdom is on its development to be 'custom and culture. Sundanese people who are still bound by tatali paranti karuhun) have a role in maintaining the sustainability and balance of nature. The Sea Tribe Society, Condong Indragiri Hilir Riau, also has local wisdom in preserving nature [17]. While local community wisdom research on peat swamp land has not been done yet, but there are some studies that indicate that there is local wisdom possessed by society such as research conducted by [16], who found that in cultivating, Dayaks have a particular way of managing agriculture and production technology, especially manifested in the local concept of malacak, manatak and maimbul, which is a manifest in hoeing, bantangan and agricultural cycles. Thus, basically local people in each region have their own local wisdom, in accordance with natural conditions and their history travel to adapt to nature. Utilization and excavation of local wisdom will be faster with a social engineering that encourages social change towards a better, environmentally friendly and sustainable.

The burning behavior of land viewed from burning frequency is also influenced by several factors, ie income and activeness of organization, while the area of burned land is affected by the area of farming land, the productivity of the land, the type of burning activity, the burning assessment, the level of awareness of the burning of the land and the activity of the organization. The main factors causing fires in sub optimal areas including on peat lands are burning behavior when clearing land, throwing cigarette butts indiscriminately, and land conflicts. The village community's control over the learner behavior is still low [5]. Therefore, it needs strengthening to reduce the human factor as the cause of fire or fire source through the assistance of fire prone villages. Company firefighting teams and community counseling concerning fires that have only been busy extinguishing fires during the dry season and are often overwhelmed, especially if the fires occur in peat areas, can be optimized by strengthening their capacity. In addition, the application of fire control regulations and sanctions to reduce the human factor as a source of fire is in dire need of strict supervision in the field (village) in order to sustain the impact. [5] indicated that weak land governance, highly dependent land-based economic activities, and poor community institutions have contributed to the lack of social control over land fires.

5. Conclusions

The burning behavior of community land around the oil palm plantation is influenced by the area of farming land, income, land productivity, extension activity, burnt activity type, level of knowledge, burning assessment, the level of awareness of the burning of land, the activity of organizing and the cosmopolitan level. Sustainability in reducing burning behavior is social dimension (cosmopolitan level), economic dimension (farmer exchange rate) and ecology dimension (fertilizer use).

Acknowledgments
We would like to thank the Palm Oil Plantation Fund Management Agency (BPDPKS) for research funding to researchers and the Palm Research Committee on suggestions and inputs to researchers. We also thank the local government of Bengkayang Regency, West Kalimantan Province for their direct assistance and cooperation at the local and field level. We also greatly appreciate the participation of various parties including the Head of Sui Raya Village and Sui Keran Village where we did the research and members of the Coordinating Team for the Prevention of Peat land Fire prevention in the vicinity of the Oil Palm Plantation Area

References
[1] Akbar A. 2011. Studi kearifan lokal penggunaan api persiapan lahan: studi kasus di hutan Mawas. *Jurnal Penelitian Sosial dan Ekonomi Kehutanan*. **8**(3):211-230.
[2] Arifudin, Nasrul B, Maswadi. 2013. Program of community prevents forest fires in Indonesia peat land. *Procedia EnvironmentalSciences*. **17**:129-134.
[3] Arikunto S. 2002. *Prosedur Penelitian suatu Pendekatan Praktek*. Jakarta (ID):PT.Rineka Cipta.
[4] Cresswell, J, W. 1992. Research Design Qualitative and Quantitative Approaches. Thousand Oaks London New Delhi. Pp 117-142
[5] Departemen Kehutanan. 2014. Laporan akuntabilitas Kinerja Instansi Pemerintah Dtijen PHKA.
[6] Hair, J. F., Jr., R. E. Anderson, R. L. tatham & W. C. Black (1995). Multivariate Data Analysis with Readings, Englewood Cliffs, NJ: Prentic Hall.
[7] Indrawardana,I. 2012. Kearifan lokal adat masyarakat sunda dalam hubungan dengan lingkungan alam **4** (1) (2012): 1-8.
[8] Maswadi. 2014. Tipologi Sebaran Perilaku Pembakaran Lahan Gambut di Kabupaten Kuburaya dan Bengkayang. *Jurnal Social economic Agriculture,* **Vol. 3 No.1** april 2014, hal 4.
[9] Pasaribu SM, Friyatno S. 2008. Memahami penyebab kebakaran hutan dan lahan serta upaya penanggulangannya: kasus di Provinsi Kalimantan Barat. Bogor. Pusat Analisis Sosial Ekonomi dan Kebijakan Pertanian Litbang Pertanian Bogor.
[10] [PERDA KALBAR] Peraturan Daerah Kalimantan Barat. 1998. *Peraturan Daerah Kalimantan Barat Nomor 6. 1998 Tentang Pencegahan dan Penanggulangan Kebakaran Hutan dan Lahan*. Kalimantan Barat (ID).
[11] Sammy, B C. 2013. *Social Engineering Theory: A Model for the Appropriation of Innovations with a Case Study of the Health MDGs*. University of Yaounde 1 and Cameroon Academy of Sciences. Cameroon.
[12] Sumarmi, 2015. *Local Wisdom Of Osing People In Conserving Water Resources*. Jurnal Jurnal Komunitasresearch & Learning In Sociology And Anthropolog 43-51. Unesa. Semarang.
[13] Sztompka. P, 2008. Sosiologi perubahan sosial. Jakarta Premada Media.
[14] Thorburn, CG., Kull, CA.2015. *Peatlands And Plantations In Sumatra, Indonesia: Complex Realities For Resource Governance, Rural Development And Climate Change Mitigation. Asia Pacific Viewpoint,* vol. 56, pp153–168 issn 1360-7456
[15] Utama S. 2010. Pemberdayaan masyarakat hutan dengan pendekatan kelompok. Kasus pengelolaan hutan bersama masyarakat pada areal hutan produksi perum perhutani unit I Provinsi Jawa Tengah [disertasi]. Bogor (ID): Institut Pertanian Bogor.
[16] Wahyu& nasrullah, 2012. *Malacak, Manatak, Maimbul*: kearifan lokal petani dayak bakumpai dalam pengelolaan padi di lahan rawa pasang surut. *jurnal komunitasresearch & learning in sociology and anthropolog*, **4** (1) (2012) : 36-45

[17] Zulkarnain., Agustar, A., Febriamansyah A, 2008. Kearifan lokal dalam pemanfaatan dan pelestarian sumberdaya pesisir. Studi Kasus di Desa Panglima Raja Kecamatan Concong Kabupaten Indragiri Hilir Propinsi Riau. *70 Jurnal Agribisnis Kerakyatan*, Volume 1, No.1. Juli 2008:69-84

Design of transportation and distribution Oil Palm Trunk of (OPT) in Indonesia

Defi Norita[1], Yandra Arkeman[2]
(1) Postgraduate Student of Bogor Agricultural University, defi.norita@gmail.com
(2) Department of Agroindustrial Technology, Faculty of Agricultural Technology, Bogor Agricultural University
yandra_ipb@yahoo.com

Abstract. This research initiated from the area of oil palm plantations in Indonesia 13 million hectares, triggering consternation of abundance of oil palm trunk when garden regeneration is done. If 4 percent of the area is rehabilitated every year, almost 100 million cubic feet of oil palm will be trash. Biomass in the form of pellets can be processed from oil palm trunk. It is then disseminated back to the palm oil processing area into biomass. The amount of transportation cost of the used ships and trucks was defined as parameters. So the objective function determined the type and number of ship and truck trips that provide the minimum transportation cost. To optimize logistics transportation network in regional port cluster, combining hub-and-spoke transportation system among regional port with consolidation and dispersing transportation systems between ports and their own hinterlands, a nonlinear optimization model for two-stage logistics system in regional port cluster was introduced to simultaneously determine the following factors: the hinterlands serviced by individual ports and transportation capacity operated between each port and its hinterland, cargo transportation volume and corresponding transportation capacity allocated via a hub port from an original port to a destination port, cargo transportation volume and corresponding transportation capacity allocated directly from an original port to a destination port. Finally, a numerical example is given to demonstrate the application of the proposed model. It can be shown that the solution to the proposed non-linear model can be obtained by transforming it into linear programming models.

1. Introduction

Literature on port logistics has only developed over the last two decades or so, for example,
UNCTAD defines 'third generation' ports as those offering value-added services (e.g. warehousing, packaging) in addition to cargo handling, and 'fourth generation' ports as those that are separated geographically but with common operators or administration, such as by global multi-port companies [1]. Distinguished between general logistics services (GLS) and value-added activities or logistics (VAL) in an effort to assess the logistics potential of ports [2], analyzed the components, functions, and optimization principles of port logistics [3], port logistics development strategies [4],[5], advocated the application of 'agility' to the port environment, proposing that ports should be proactive rather than reactive
along supply chains in a modern globalized world economy [6]. Performed a simulation study for the logistics planning of a container terminal in view of supply chain management [7]. Furthermore, much of the other literature advocates the future of ports as logistics centers and highlights their nodal role in the changing patterns of maritime and intermodal transport [8-11]. Most of abovementionedstudies related to port logistics focus on qualitative analysis and are lacking in systematic theoretic study. In addition to that, these studies only consider internal logistics in a port terminal or a port's overall logistics instead of systematically considering overall logistics network in regional port cluster.

Research positions.

NO	Researchers	Years	Objection Function	Methode
1	Adler	2001 [12]	Develop, evaluate and select network hub and spoke on the airlines	Non linier Programming
2	Suprayogi	2002 [13]	Forming clusters of a set of ports.Determine the number of container container flows, ship size and number of ships and containers used	Mathematic Model
3	Sasaki	2003 [14]	Presents a formulation of a hub and spoke model with a capacity involving limited arc capacity as a hub capacity constraint	Branch and bound dan relaksasi lagrang
4	Chaug-Ink	2007 [15]	Using hub and spoke networks in formulating two objective models to determine optimal ship routes, ship size and display frequency by minimizing shipping costs	Pareto
5	Wang	2008 [16]	Optimizing the transport network on the logistics network within the internal port	
6	Defi	2011 [17]	Determine the number of ships and trucks in a certain time unit, and determine the number of travel modes transportnya. The combination of Wang (2008) and Suprayogi (2002)	Mathematic Model
7	Defi	2017	Design of Transportation and Distribution Oil Palm Trunk	MILP

The system discussed is the delivery of goods from hinterland to certain ports. The hinterland consists of eight hinterland and five ports. At each hinterland, trucks are used as a means of transportation. Trucks of a certain type take goods from areas in their hinterlands. Then the goods are brought to the port. From the port the goods delivered to the destination port by using aa ship with a certain type as well. The ABCD region is defined as hinterland 1. IJLK is defined as hinterland 2. The CDFE region is defined as hinterland 3, and the KLNM region is defined as hinterland 4. As the initial hinterland, hinterland 1,2,3,4 can be seen in Figure 1.

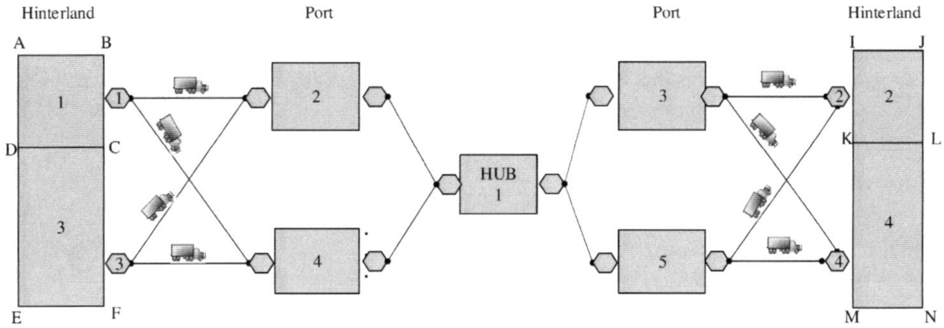

Figure 1. Pure Hinterland

Trucks are required to collect items from IJLK. When the truck's capacity has not been met the truck is required to collect goods from the PO area until it reaches its trucking capacity, so that the IJLPOK area is defined as hinterland 6. While the OPNM area is defined as hinterland 8.
Furthermore, if the truck picking up the goods from the CDFE area has also not reached the capacity, then the truck can take additional goods from the GH area, so the GHDFEC area is defined as hinterland 7. While the ABHG area is defined as hinterland 5. For that hinterland 5,6,7, 8 is called an alternative hinterland as shown in Figure 2.

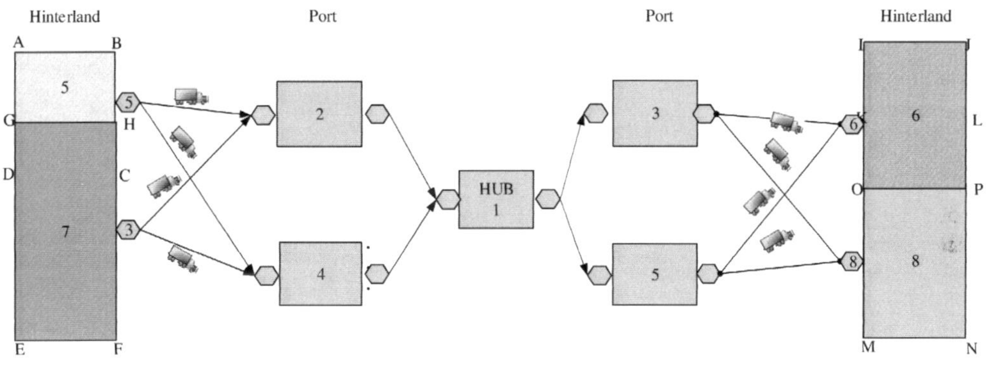

Figure 2. Alternatif Hinterland

2. Materials and Methods
Problem Formulation

The conceptual model can describe the whole system represented in the model. All parameters, decision variables are determined to produce a specific purpose function. In this study, the parameters are the number of goods taken from all hinterland with a certain type of truck with a certain capacity also to be brought to the nearest port. From that port the item is redistributed to the destination port by using a particular type of vessel and a certain capacity as well. The amount of transportation cost of used ships and trucks is defined as parameters, so it can be determined the type and number of vessel and truck travel that gives minimum value to total transportation cost. The conceptual model can be seen in Figure 3.

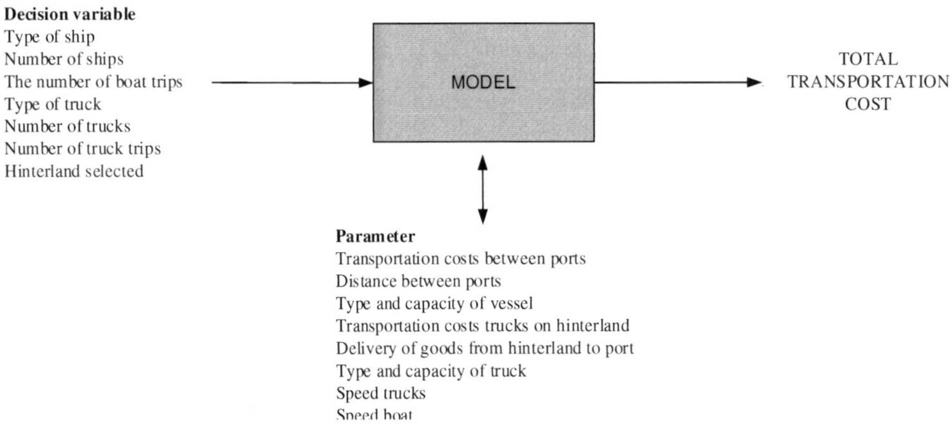

Figure 3. Conceptual Model

Figure 3. shows clearly that the objective function to be achieved is the total transportation cost with the decision variable is the selected hinterland, the number and type of vessel, the number of vessel trips, the number and type of truck, and the number of truck trips on selected hinterland later.

Identification of Model Variables

After making the conceptual model above, then in this research identification of notation and related variables to minimize total transportation cost incurred.

Notation of purpose function:

B : Total Transportation Cost [Rp/week]

Index notation:

I, j : Index for ports

(I, j) : The relationship index between ports i and j

K : Index of ship type

L : Index of truck type

Set Notation

M : The set of ports

N : The set of hinterland

A_p : Number of alternative hinterland for p ε I

E_p : Number of hinterland except region p, p ε I

I : Number of all hinterland allowed

I_i : Number of hinterland owned by port i

Parameter
CT_{lp} : Transportation cost of l type truck at hinterland p [Rp / travel]
CS_{kij} : Ship cost type from port i to j [Rp / travel]
CV_{kij} : Cost of ship type k from port i to j [Rp / ton]
CM_{kij} : Maintenance costs of ship type k from port i to j [Rp / ship]
D_{pj} : Number of goods distributed from hinterland p to port j [ton / week]
JH_{pi} : The distance between hinterland p and port i [km]
J_{ij} : Distance to port i with j (i.j) [mile]
JMK_{ijk} : Minimum number of ship type k on link i to j

$$JMK_{ij}^{k} = \emptyset\left(\frac{\frac{J_{ij}}{24v_k} + \xi}{365}\right), \forall i,j \in M; k \in K$$

$\emptyset$: Number of weeks in a year [52 weeks / yr]
S^k : Ship k capacity [ton / ship]
ST_p^l : Truck l capacity on*hinterlandp* [ton/truck]
v_k : Speed of speed k [knot or mile/hour]
ξ : Operating ports per week [7 day]
VT^l : Truck l type speed [km/hr]
Decision Variable :
If the selected hinterland is zp = 1 instead zp = 0
R_{ij}^k : Frequency of ship k from*porti*to*j* [times/weeks]
JMT_p^l : Number of type l trucks required for each hinterland p [truck]
T_p^l : Number of trips of type l trucks operating at hinterland p [times/week]
x_{ij} : The number of items distributed from port*i* to*portj* [ton/week]
Y_k : Number of ship type k required [ship]
z_p : The variable that is conditioned as binary.
If the selected *hinterland*is $z_p = 1$ instead $z_p = 0$

Final Model
After identification of the index notation, parameter notation and decision variable notation, then compiled the model to be built.

Function Purpose:
Minimize total transportation costs between two ports and transportation costs at each hinterland.

Min B = Ship Cost + Truck Cost

$$\text{MIN } B = \sum_{i=1}^{M}\sum_{\substack{j=2 \\ j\neq i}}^{M}\sum_{k=1}^{K} CS_{ij}^{k} R_{ij}^{k} + \sum_{i=1}^{M}\sum_{\substack{j=2 \\ j\neq i}}^{M}\sum_{k=1}^{K} CV_{ij}^{k} x_{ij} + \sum_{i=1}^{M}\sum_{\substack{j=2 \\ j\neq i}}^{M}\sum_{k=1}^{K} CM_{ij}^{k} Y^{k}$$
$$+ \sum_{p=1}^{I}\sum_{l=1}^{L} CT_p^l T_p^l z_p + \sum_{p=1}^{I}\sum_{l=1}^{L} CT_p^l JMT_p^l Z_p$$

Subject to:

$$x_{i1} + \sum_{\substack{j=1 \\ j\neq i}}^{M} x_{ij} = \sum_{\substack{j=1 \\ j\neq 1}}^{M}\sum_{p=1}^{li} D_{pi} z_p, i,j = 1,2,\ldots,M, i \neq j \qquad (1)$$

$$x_{1i} + \sum_{\substack{j=1 \\ j \neq i}}^{M} x_{ji} = \sum_{\substack{j=1 \\ j \neq 1}}^{M} \sum_{p=1}^{Ii} D_{pj} z_p, i, j = 1,2, \dots, M, i \neq j \quad (2)$$

$$x_{ij} \leq \sum_{p=1}^{Ii} D_{pj} z_p, i, j = 0,1,2, \dots, M \; i \neq j \quad (3)$$

$$x_{ij} \leq \sum_{k=1}^{K} JMK_{ij}^{k} R_{ij}^{k} S^{k} \;\; i, j = 1,2 \dots, M \quad \forall i, j \in M \quad (4)$$

$$Y^{k} = \sum_{i} \sum_{j} JMK_{ij}^{k} R_{ij}^{k}, \quad \forall k \in K; \forall i, j \in M \quad (5)$$

$$\left(\sum_{\{j|(i,j) \in M|\}} R_{ij}^{k} = \sum_{\{j|(i,j) \in M|\}} R_{ji}^{k} \; \forall k \in K; \forall i, j \in M \right) \quad (6)$$

$$\sum_{i=1}^{M} D_{pi} z_p \leq \sum_{k=1}^{K_p} T_p^{l} ST_p^{l}, p = 1,2, \dots I \quad (7)$$

$$JMT_p^{l} \geq \frac{T_p^{l} \;\; \sum_i JH_{pi}}{168 \, VT_l} \quad , \forall l, \forall p \quad (8)$$

$$\sum_{p=1}^{A_p} z_p = 1, p = 1,2, \dots, I \quad (9)$$

$$z_p \sum_{p=1}^{E_p} z_p = 0, p = 1,2, \dots, I \quad (10)$$

$$\sum_{p=l}^{I} z_p = N \quad (11)$$

$$x_{ij} \geq 0, i, j = 0,1,2, \dots, M \; i \neq j \quad (12)$$

$$R_{ij}^{k}, Y_{k,} T_p^{l} JMT_p^{l} > 0, integers, i, j = 0,1,2, \dots, M \; p = 1,2, \dots, I \quad (13)$$

z_p : If hinterland is selected $z_p = 1$ otherwise $z_p = 0$

Constrains 1. is Ensure that the amount of goods distributed from port i to port hub plus total goods from each port i to port j equals the total number of items from the selected hinterland p to port i.

3. Results and Discussions

Data

The data used in this numerical example are data from Norita (2011) model and hypothetical data. Data from Norita's model (2011).

The data from Norita (2011) model used are:

A. Data on the quantity of goods distributed (Dpj)

Within a week there is a total amount of goods that must be distributed from hinterland to port. In this case the port used is port 2, port 3, port 4, while port 1 is used as a hub port, as shown in Table 1 and Figure 1.

Table 1. Number of items distributed from hinterland to all ports [ton / week].

H	Area	*Port*[ton/week]			
		2	3	4	5
1	ABCD	0	1230	1290	1770
2	IJLK	1100	0	850	3180
3	CDFE	1160	1020	0	4290
4	KLNM	580	1550	2440	0
5	ABGH	0	560	670	910
6	IJLPOK	1400	0	2180	3650
7	GHDFEC	1160	1650	0	5200
8	OPNM	310	780	1250	0

a. Type and capacity of ship (*k* dan *V*)

For delivery between ports used two types of ships with different capacity and speed. Can be shown with Table 2.

Table 2. Type, capacity and speed

Ship	Capacity S^k(ton)	Speed V^k(knot)
1	5000	8.01
2	8000	10.25

b. Ship Cost

The cost of the vessel is the cost incurred when the vessel travels from the original port to the destination port, as shown in Table 3.

Table 3.Operating cost of ship from port of origin to destination port (000000IDR/week).

Port	2		3		4		5		1	
	Ship		Ship		Ship		Ship		Ship	
	1	2	1	2	1	2	1	2	1	2
2	0	0	1,5	2,0	2,0	2,5	1,8	2,5	0,75	0,75
3	1,2	1,5	0	0	0,9	1,5	2,5	3,0	0,60	0,70
4	1,5	2,0	1,3	1,8	0	0	3,1	3,5	0,80	1,00
5	0,8	1,2	1,6	2,0	2,0	2,5	0	0	0,90	1,20
1	0,75	0,90	0,60	0,70	0,80	1,00	0,90	1,20	0	0

c. Type, capacity and speed of truck (ST^l dan VT^l)

Type, capcity, and speed of truck can be seen in table 4.

Table 4. Type, capacity and speed of truck

Jenis Truk	Kapasitas truk V^l(ton)	Kecepatan truk VT^l (km/jam)
1	3	30
2	5	35

d. Truck cost (CT)

The cost of the truck, is the cost incurred for truck operations starting from picking up the goods from the areas of a particular hinterland and delivering it to the nearest port within a week. The cost of the truck can be seen in Table 5.

Table 5. The cost of trucks at each hinterland (000000Rp/week)

Truk	Hinterland							
	1	2	3	4	5	6	7	8
1	0,30	0,40	0,35	0,10	0,10	0,35	0,40	0,10
2	0,40	0,50	0,40	0,15	0,15	0,50	0,45	0,15

Hypothetical Data

There are some hypothetical data used are:

a. Distance between ports (J_{ij})

Distance between ports is the distance between the origin port, port hub and destination port in a cluster port. Distance between ports can be seen in table 6.

Table 6.Distance between ports (mil)

Port	1	2	3	4	5
1	0	236	392	794	521
2	236	0	189	591	677
3	392	189	0	458	823
4	794	591	458	0	1118
5	521	677	823	1118	0

b. The distance between hinterland to port

The distance between the hinterland to the port is the distance between the circumference of the hinterland through which the boundary moves toward the port can be seen in Table 7.

Table 7.The distance between hinterland to port(km)

H	Daerah	Port			
		2	3	4	5
1	ABCD	40	100	80	150
2	IJLK	100	40	150	80
3	CDFE	80	100	40	150
4	KLNM	150	80	100	40
5	ABGH	45	105	75	145
6	IJLPOK	105	45	145	75
7	GHDFEC	75	105	45	145
8	OPNM	145	75	105	45

Model Solutions

After the data obtained from research data Norita (2011) and hypothetical data, which is required to complete the model. Then the data input process into the model. The search for an optimal solution for this example is done using the V10 Lingo program where lingo programming is based on a research operational method of Branch and Bound Algorithm (ABB). The solution obtained is nonlinear integer. This model can be made in linear form in a way, the model is made 4 cases, where set combination of 8 hinterland which support certain part:

Table 8.Cases of possible hinterland selected

Kasus	Hinterland							
	1	2	3	4	5	6	7	8
1	0	0	0	0	1	1	1	1
2	0	1	0	1	1	0	1	0
3	1	1	1	1	0	0	0	0
4	1	0	1	0	0	1	0	1

So we get a solution that is global optimal.

Table 9. Comparison of calculation results from 4 cases

Notation	Decision Variables	Hinterland selected			
		1234(11110000)	1368(10100101)	2457(01011010)	5678(00001111)
B	Total cost	1,515,150	1,659,000	1,440,500	1,583,900
Y	Ship type 1	3	4	5	6
	Ship type 2	2	2	1	1
R_{ijk}	Frequency of links (1,2)	Ship type 1 with 1 trip	Ship type 1 with 1 trip	Ship type 1 with 1 trip	Ship type 1 with 2 trip
	Frequency of links (1,3)	Ship type 2 with 1 trip	Ship type 2 with 1 trip	Ship type 2 with 1 trip	Ship type 2 with 1 trip
	Frequency of links (1,4)	Ship type 2 with 1 trip	Ship type 2 with 1 trip	Ship type 1 with 2 trip	Ship type 1 with 2 trip
	Frequency of	Ship type 1 with	Ship type 1 with	Ship type 1 with	Ship type 1 with 2

	links (1,5)	2 trip	2 trip	2 trip	trip
	Frequency of links (2,1)	Ship type 1 with 1 trip	Ship type 1 with 1 trip	Ship type 1 with 1 trip	Ship type 1 with 2 trip
	Frequency of links (3,1)	Ship type 2 with 1 trip	Ship type 2 with 1 trip	Ship type 2 with 1 trip	Ship type 2 with 1 trip
	Frequency of links (4,1)	Ship type 2 with 1 trip	Ship type 2 with 1 trip	Ship type 1 with 2 trip	Ship type 1 with 2 trip
	Frequency of links (5,1)	Ship type 1 with 2 trip	Ship type 1 with 2 trip	Ship type 1 with 2 trip	Ship type 1 with 2 trip
T_2	Type truck 2	54	54	65	27
	Type truck 2	65	82	58	91
	Type truck 2	82	91	27	1001
	Type truck 2	58	30	101	30
JMT	Truck Frequency 2	858	858	1026	428
	Truck Frequency 2	1026	1294	914	1446
	Truck Frequency 2	1294	1446	428	1602
	Truck Frequency 2	914	468	1602	468

4. Conclusions

Of the four cases above which provide the minimum value is case 2, with the results are as follows: The solution is optimally global with transport costs being 1440,500. For the number of ships type one needed is 10 vessels and 2 units of ship type 2. Table 10. shows the frequency of ship trips per week in sending goods by ship of choice from the port of origin to the destination port with details as follows: Ship type 1 as much as 1 unit sending goods as much as 2840 tons from port 1 to the port hub that sent 1 trip. Ship type 2 as much as 1 unit sending goods as much as 3760 tons from port 3 to port hub that sent 1 trip. Ship type 1 as much as 2 units sending goods as much as 3960 tons from port 4 to port hub that sent 2 trips. Ship type 1 as much as 2 times the journey sending goods as much as 9290 tons from port 5 to port hub.

Conversely from the port hub to port 2 items shipped as much as 2140 tons with 1 boat trip type 1. From the port hub to port 3 items shipped as much as 5130 tons with 1 trip with ship type 2. From the port hub to port 4 items shipped as much 8010 tons with 2 trips by ship type 1. From port hub to port 5 items shipped as much as 4570 tons with 2 trips by ship type1, as shown in Figure 2.

Table 10. Number of ship travel and ship type used for case 2

Ship Type	Port of origin	Port of distination	The amount of goods	The number of boat trips / week
1	1	2	2840	1
2	1	3	3760	1
1	1	4	3960	2
1	1	5	9260	2
1	2	1	2140	1
2	3	1	5130	1
1	4	1	8010	2
1	5	1	4570	2

Next to the hinterland section can be seen in Table 10. where between hinterland 1 and hinterland 5 selected as supplier for port 2 is hinterland 5. For hinterland 2 and hinterland 6 selected as supplier for port 3 is hinterland 6. Next to hinterland 3 and hinterland 7 selected as a supplier for port 4 is hinterland 7. While hinterland 4 and hinterland 8 are selected as suppliers for port 5 is hinterland 8.
There are similarities between the types of trucks used in transporting goods on selected hinterland ie truck type 2, with varying amounts. At hinterland 5 the number of trucks used is 27 trucks with 428 trips. At hinterland 6 the number of trucks used is 91 trucks with 1446 trips. At hinterland 7 the number of trucks used is 101 trucks with 1602 trips. At hinterland 8 the number of trucks used is 30 with 468 trips.

Table 11. Hinterland was selected and the number of trucks and truck trips for case 2

NO	Hinterland of origin	Port of distination	Type truck	Number of trips	Number of trucks
1	2	3	2	1026	65
2	4	5	2	914	58
3	5	2	2	428	27
4	7	4	2	1602	101

Table 11. explains that on hinterland 2 used type 2 truck of 65 units with the number of travel 1026 times. At hinterland 4 used type 2 truck as many as 58 units with the number of trips 914 times. At hinterland 5 used type 2 truck of 27 units with the number of trips 428 times. At hinterland 2 used type 2 truck counted 101 units with the number of trips 1602 times.

References

[1] Groom S, Romano P, Giannakis M. Supply chain management:*an analytical framework for critical literature review.*European Journal of Purchasing and Supply management, 2000, 6: 67–83.
[2] Harding A, d Juhel M H, 1997 *Ports, cities and the challenge of global logistics.6th International Conference of Cities and Ports.* Montevideo, Uruguay, November, 1997.
[3] Zhen H, 1999*Optimization of port logistics*. Chinese Port.
[4] Mo B M, Zhang S Z, 2000 *Guideline for developing port logistics industry*. Chinese Port.
[5] Hou X D, 2002. *Study on port logistics development*. Port Economy
[6] Paixao A C , Marlow P B, 2003*Fourth Generation Ports-A Question of Agility* International Journal of Physical Distribution & Logistics Management.
[7] Lee T W, Park N K, Lee D W, 2003 *A simulation study for the logistics planning of a container terminal in view of SCM.* Maritime Policy and Management.
[8] Yang C X. 1999 *Development of Port logistics center*. LogisticsTechnology, : 6–8.
[9] Lu R F. 2000: 37–38.*On the Construction of MIS within port logistics center.*Chinese Port,
[10] Lu W.2001: 2–4.*Implement of port logistics center*. Shipping Management
[11] Bichou K, Gray R. A. , 2004, 31:47–67*logistics and supply chain management approach to port performance measurement.* Maritime Policy& Management..
[12] Adler, Nicole and Joseph Berecham, 2001. *Evaluating Optimal Multi-Hub Networks ina Deregulate Aviation Market With An Application to Western Europe*. Transportation Research Part A, Vol. 35, pp. 373-390
[13] Suprayogi. (2002), *A study on the Transportation System Design Based on Mathematical Models*, Doctoral thesis. Department of Enviromental and Ocean Engineering, Graduate School of Engineering, The University of Tokyo
[14] Sasaki, Mihiro and Masao Fukushima. (2003). *On the Hub-and-Spoke Model With Arc Capacity Constraints.* Journal of The Operation Research Society of Japan, vol. 46, pp. 409 – 428.
[15] Hsu, Chaug-Ing and Yu-Ping Hsieh,2007. *Routing, Ship Size, and Sailing Frequency Decision-Making for a Maririme Hub-and-Spoke Container Network.* Mathematical and Computer Modelling, vol. 45, pp. 899 - 916
[16] Lin, Cheng-chang, Yu-Jen Lin, and Dung-Ying Lin. (2003). *The Economic Effects of Center-to-Center Directs on Hub-and-Spoke NetworksFor Air Express Common Carriers.* Journal of Air Transport Management, vol. 9, pp. 255 – 265.
[17] Mingjun JI, Maoying HE (2010). *Optimization of Two-Stage Port Logistics Network of Dynamic Hinterland Based on Bi-level Programming Model* Transportation Management College, Dalian Maritime University, China
[18] Rodrigue JP.(2005) *The Geography Transport System.*Dept of Economics & Geography Hofstra University
[19] Wang, Chun-xu. (2008). *Optimization of Hub-and-Spoke Two-stage :Logistics in Regional Port Cluster*. System Engineering – Theory and Practice vol. 28 (9).

Biomass and carbon stock potential of *Gliricidia Sepium* as an alternative energy at Timor Tengah Utara Regency, East Nusa Tenggara Province, Indonesia

F H Prima [a], Hariyadi[b], A Hartono[c]

[a] *Post Graduate Student, Bogor Agriculture University, Dramaga, Bogor 16680 —fidelharmanda@gmail.com*
[b] *Department of Agronomy and Horticulture, Faculty of Agriculture, Bogor Agriculture University, IPB Dramaga, Bogor 16680*
[c] *Department Soil Science and Land Resource Management, Faculty of Agriculture, Bogor Agriculture University, IPB Dramaga, Bogor 16680*

Abstract. The utilization of biomass from plants is one efforts for the fulfillment an availability of alternative energy in indonesia. *Gliricidia sepium* is a tolerant species that can grow in dry land. However its utilization as renewable energy source is non-optimized. This study aims to analyze the potential carbon stocks and biomass from *Gliricidia sepium* as a raw material for alternative energy in East Nusa Tenggara. This study was conducted in November 2015 and located in Humusu Sainiup, Timor Tengah Utara Regency, East Nusa Tenggara Province. The method used in collecting data was applied in three different land-use, namely monoculture *Gliricidia sepium*, polyculture between *Gliricidia sepium* and *Leucaena leucocephala*, and polyculture between *Gliricidia sepium* and *Zea mays*. We used the allometric equation from Ketterings namely $B = 0.11\rho D^{2+0,62}$ and $C = 0.5 \times B$. The results showed that the different land-use will give different value of carbon stocks which is in this study the biggest value of carbon stocks was found in monoculture of Gliricidia sp (35.35 tC ha^{-1}) compared with Gliricidia sp + Leucaena sp (18.83 tC ha^{-1}), and Gliricidia sp + Zea mays (13.79 tC ha^{-1}). The value of biomass and carbon stocks was influenced by wood density, trees density, and diameter at breast height (dbh).

Keywords: *biomass, carbon stocks, renewable energy, Gliricidia sepium*

1. Introduction

Population growth encourage increased energy needs for the improvement of human well-being from year to year. Increased energy consumption caused by two factors: population growth, which reached more than 20% and the improvement of living standards that drive energy consumption per capita increased by almost 40%.

The need for energy is a crucial issue for national development requires the availability of sufficient energy, safe, quality, and diverse and spread evenly throughout Indonesia affordable. The high growth of industry and population resulted in a gap between demand and supply of energy. This can lead to an energy crisis. Indonesia has energy resources which consist of non-renewable energy sources and renewable energy sources. The energy source has an important role in development as a crucial factor for other sectors such as industry, transport, households, commercial, and others [1].

Law No. 30 of 2007 on Energy [2] and Government Regulation No. 78 of 2014 on National Energy Policy [3] has been set up management and utilization of energy resources. National energy policy mandates a target of achieving an optimal mix of primary energy in 2025 that the role of renewable energy at least 23% and by 2050 at least 31%. Utilization of Indonesia's energy consumption has increased with an average growth of 1.9% per year which is still dominated by fossil fuels, the increase was due to higher activity in the transportation, industrial, and household [4]. The energy problem faced by Indonesia, which are: energy subsidies increased to Rp 255.6 trillion in 2011, the number of people who do not have access to electricity was 87.69 million population, limitations of domestic infrastructure challenges and problems in meeting domestic energy needs. Dependence on oil is still dominant reached 49.7%, while the utilization of renewable energy was approximately 6% [5]. Indonesia has the landscape and geographical location on the equator have some type of alternative energy sources that can be developed such as solar energy, wind energy, geothermal energy, ocean thermal energy and biomass energy. Among the sources of alternative energy, biomass energy is an alternative energy source that needs to be given priority in development compared with other energy sources.

Efforts to comply with the availability of alternative energy is the utilization of biomass plants. Biomass-based energy sources have the advantage in ensuring the availability of alternative energy sources as energy needs. Gliricidia plants is one of the renewable energy sources available in Indonesia, particularly in the province of East Nusa Tenggara. *Gliricidia* have a place to grow tolerant characteristics in upland and utilization of renewable energy sources is not optimal. This study aims to analyze the potential of carbon stocks and biomass in *G. sepium*.

2. Materials and Methods

The study was conducted in October 2015 in the secondary forest in Humusu Sainiup, Timor Tengah Utara Regency, East Nusa Tenggara Province. Location of the study there are three types of land cover, namely on *G. sepium* monoculture, polyculture between *G. sepium* and *L. leucochepala*, and polyculture between *G. sepium* and *Z. mays*.

The materials used for the observation and measurement of the vegetation is in a location that includes seedlings, saplings, poles, trees. The equipment used for to conduct of research include GPS, Compass, Roll-meter, Meter Sewing, Hagameters, Tally Sheet,, Cameras, and Stationery.

The type of experimental plot is square with 20x20 cm, then made 3 subplot each sized 10x10 cm, 5x5 cm, and 2x2 cm. Plot with sized 20x20 cm be destined for tree, 10x10 cm for poles, 5x5 cm for stakes, and 2x2 cm for seedlings. Each type of land cover has made 5 experimental plots. After determining and make the plot as the sample plot, then do inventary to obtain data and information carefully according to circumstances on the ground in each plot were made, the variety of data and information is the location plots, plant species, and the diameter breast height (1.3 m). Biomass data analysis done using the allometric equation $B = 0.11\ \rho\ D^{2+0.62}$ [6] and $C = 0.5 \times B$ [7].

IOP Conf. Series: Earth and Environmental Science **141** (2018) 012022 doi:10.1088/1755-1315/141/1/012022

3. Results and Discussions

3.1 Inventory of biomass and carbon storage on monoculture G. sepium

Inventory of biomass and carbon storage on monoculture *G. sepium* is located at coordinates S 09^0 29' 16" E 124^0 56' 05" and at an altitude of 333 meters above sea level. Inventory made on 5 experimental plots, inventory data for biomass and carbon storage of data obtained by the above ground biomass that include categories stakes, poles, and trees. Table will present the results of the inventory of biomass and carbon storage contained in monoculture *G. Sepium*.

Table 1. Inventory of Biomass and Carbon Storage in Monoculture *G. sepium*

Category	Biomass (kg/ha)	Carbon Storage (kg/ha)
Stakes	17,007.10	8,503.55
Poles	33,511.38	16,755.69
Trees	20,223.60	10,111.80

According to the table 1, the results of the calculation of the largest biomass and carbon storage are in the category pole of 33,511.38 kg / ha, equivalent to 33.51 tonnes / ha, while for carbon storages 16,755.69 kg / ha, equivalent to 16.75 tonnes / ha. This is because the sample plots commonly found plants that fall into the category pole density and very high frequency, which resulted in having the potential of biomass and carbon storage are greatest when compared to a stake or a tree well. Biomass and carbon storage value depends heavily on the density and frequency contained in each respective category in field trials. The higher level of density and frequency in a region, the greater the value of the biomass and carbon storage in the region. Based on study conducted by Fuwape and Akindele [8] obtained the results of research on Gamal plant biomass of 37.4 tons / ha. The value of biomass and carbon stock depends greatly on the density and frequency levels found in each category on the experimental field.

3.2 Inventory of biomass and carbon storage on intercropping G. sepium and L. leucochepala

Inventory of biomass and carbon storage on intercropping *G. sepium* and *L. leucochepala* is located at coordinates S 09^0 27' 51" E 124^0 57' 65" and at an altitude of 148 meters above sea level. Inventory made on 5 experimental plots, inventory data for biomass and carbon storage of data obtained by the above ground biomass and carbon storages that include categories stakes, poles, and trees. Table will present the results of the inventory of biomass and carbon storage found on imtercropping *G. sepium* and *L. Leucochepala*.

Table 2. Inventory of Biomass and Carbon Storage on Intercropping *G. sepium* and *L. leucochepala*

Category	Biomass (kg/ha)	Carbon Storage (kg/ha)
Stakes	9,516.38	4,758.19
Poles	7,700.95	3,850.47
Trees	20,437.94	10,218.97

According to the table 2, the results of the calculation of the largest biomass and carbon storage are in the category tree of 20,437.94 kg / ha, equivalent to 20:44 tons / ha, while for carbon deposits generates 10,218.97 kg / ha, equivalent to 10:22 tons / ha. The results of a quantitative analysis of vegetation found indicates that stands found in the category of samplings have higher numbers than other categories. Density and high-frequency plant on intercropping *G. sepium* and *L. leucochepala* no contribution of biomass and carbon storage means. Biomass and carbon storage value is also influenced by the size of the diameter of the plant. Study from Rahayu *et al.* [9] in Nunukan stated that the diversity of the size of the diameter, the presence of trees with a diameter >30 cm in a land use systems, contributing to significant biomass to the total carbon storage. This causes that the biomass content at the level of the tree has a value greater than the rate of saplings and poles.

3.3 Inventory of biomass and carbon storage on intercropping G.sepium and Zea mays

Inventory of biomass and carbon storage on intercropping *G.sepium* and *Zea mays* are on the coordinates S 09^0 29' 43" E 124^0 56' 86" and at an altitude of 281 meters above sea level. Inventory made on 5 experimental plots, inventory data for biomass and carbon storage of data obtained by the Biomass stand up (Above Ground Biomass) and carbon deposits that include categories stakes, poles, and trees. Table will present the results of the inventory of biomass and carbon storage contained in Gamal and corn fields.

Table 3. Inventory of biomass and carbon storage on intercropping *G.sepium* and *Zea mays*

Category	Biomass (kg/ha)	Carbon Storages (kg/ha)
Stakes	25,426.81	12,713.41
Poles	1,053.19	526.595
Trees	1,105.16	552.58

According to the table 3, the results of the calculation of the largest biomass and carbon storage is at stake category of 25,426.81 kg / ha, equivalent to 25.42 tonnes / ha, while for carbon deposits generates 12,713.41 kg / ha, equivalent to 12.71 tonnes / ha. This result is obtained because the sample plots *G. sepium* and *Z. mays* crops is very much a crop plant regeneration. The sample plots are also found *Z. mays*, because when the research was ongoing planted corn has been harvested by the local community. The difference in the acquisition of biomass is influenced by the density of vegetation, the diversity of its diameter size and density distribution of vegetation [9]. Data on the ground suggest that the category of the stake has a number of stands more than the category resulting poles and tree biomass and carbon storage values obtained at different stake levels much to the value of biomass and carbon storage in small trees and tree level.

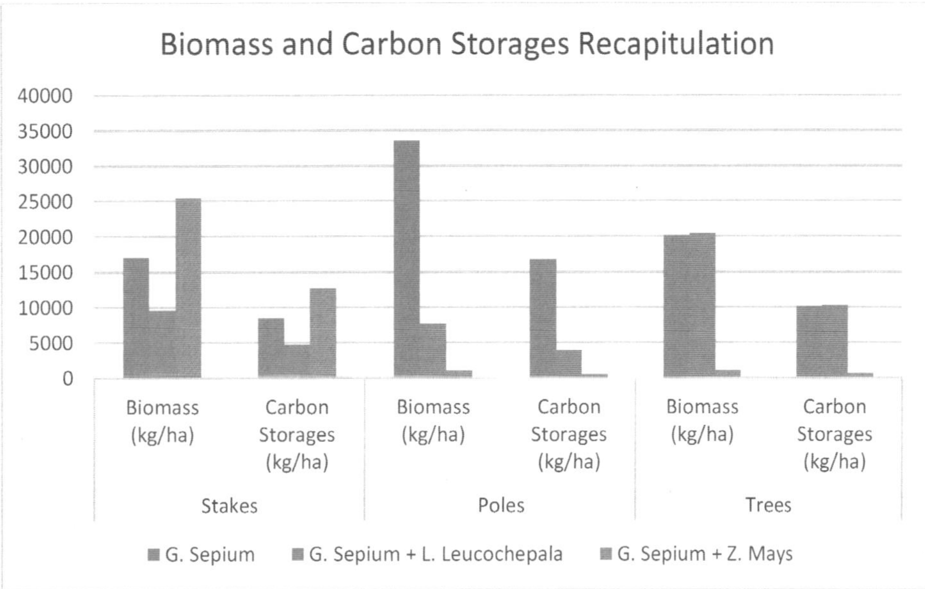

Figure 1. Biomass and Carbon Storages Recapitulation

According to the figure 1, the category of stakes has a value of biomass and carbon storage site that is on the sample plot *G. sepium* and *Z. mays* amounting to 25,426.81 kg / ha and 12,713.41 kg / ha. Category poles at monoculture sample plots *G. sepium* has the highest value than other sample plots that have a biomass value of 33,511.38 kg / ha with a total carbon storage of 16,755.69 kg / ha. Category trees that has value is the largest biomass plant sample plots *G. sepium* and *L. leucochepala* with a value of 20,437.94 kg / ha and the total carbon storage of 10,218.97 kg / ha. Based on previous research conducted by Yuniati [10] in Kupang, the highest carbon stocks amounted to 148.48 tons / ha using allometric equations Ketterings. Value to its lowest carbon stocks worth 106.59 tons / ha. This value was very different from what obtained in this study. The biomass value of 70 tons / ha with a total carbon storage of 35 tonnes / ha. This is caused by the frequency and the relative abundance of low crop, the size of the diameter of a small plant, density of plants, nutrients in the soil that affect biomass and carbon storage value obtained from this study.

Tree growth is determined by the interaction between three factors namely genetic factors, environmental factors and cultivation techniques or silviculture techniques applied [11]. Genetic factors and cultivation techniques can be manipulated through tree breeding and the selection of appropriate cultivation or silviculture techniques that can increase tree productivity that is closely related to increased tree biomass and carbon storage.

All this time, *Gliricidia* are used as garden barriers, as firewood, as ruminant animal feed, and some studies have explained that *Gliricidia* have high carbon stores that can reduce air pollution, as well as large biomass and calorific values that can be developed as materials raw alternative energy. Based on study conducted by Amirta *et al* [12] showed that calorific value of *Gliricidia* higher than another fast growing species. *Gliricidia* has calorific value amounted to 4026.77 kCal/kg. These result explained that *Gliricidia* has a high calorific value compared to other fast growing species. In addition to, with the advantages of tolerant and one of the fast growing species, *Gliricidia* is one of raw materials that can be cultivateed in dry land and also has a low electrification ratio as an alternative energy.

4. Conclusions

Based on the results of the research potential of biomass and carbon storage in Humusu Sainiup, Timor Tengah Utara Regency, East Nusa Tenggara Province obtained the largest potential of biomass and carbon storage in land monoculture *G. sepium.* This is a preliminary study of the potential biomass and carbon stocks *G. sepium* as an alternative energy. Assessing the theoretical potential of biomass energy provides a large energy estimation because the energy contained in each biomass is taken into account in a region. But not all theoretical potential can be used as an energy source. Barriers such as climate, policy, local wisdoms can hinder access to theoretical potential. Therefore, it is necessary further research for the development *G. sepium* as an alternative energy source in Humusu Sainiup, East Nusa Tenggara Province.

References

[1] Triatmojo, F., 2013. Dinamika kebijakan diversifikasi energi di Indonesia: Analisis kebijakan pengembangan energi terbarukan di Indonesia [Dynamics of energy diversification policy in Indonesia: Analysis of renewable energy development policy in Indonesia]. *Jurnal Ilmiah Administrasi Publik dan Pembangunan*, **4**(2), pp.2087-0825.

[2] Nomor, U.U.R.I., 30. tahun 2007 tentang Energi.

[3] Nomor, P.P.RI., 78. tahun 2014 tentang Kebijakan Energi Nasional.

[4] Nasional, D.E., 2014. Outlook Energi Indonesia 2014. *Jakarta: Dewan Energi Nasional.*

[5] Energi, K., 2015. Rencana Strategis Kementrian Energi dan Sumber Daya Mineral 2015-2019. *Jakarta: Kementrian Energi dan Sumber Daya Mineral.*

[6] Ketterings, Q.M., Coe, R., van Noordwijk, M. and Palm, C.A., 2001. Reducing uncertainty in the use of allometric biomass equations for predicting above-ground tree biomass in mixed secondary forests. *Forest Ecology and management*, **146**(1-3), pp.199-209.

[7] Brown, S., 1997. *Estimating biomass and biomass change of tropical forests: a primer* (Vol. 134). Food & Agriculture Org.

[8] Fuwape, J.A. and Akindele, S.O., 1997. Biomass yield and energy value of some fast-growing multipurpose trees in Nigeria. *Biomass and Bioenergy*, **12**(2), pp.101-106.

[9] Rahayu, S., Lusiana, B. and Van Noordwijk, M., 2007. Pendugaan cadangan karbon di atas permukaan tanah pada berbagai sistem penggunaan lahan di Kabupaten Nunukan, Kalimantan Timur. *Bogor: World Agroforestry Centre.*

[10] Yuniati, D. and Kurniawan, H., 2011. Potensi Simpanan Karbon Hutan Tanaman Jati () Studi Kasus Di Kabupaten Kupang Dan Belu Provinsi Nusa Tenggara Timur Tectona Grandis. *Jurnal Penelitian Sosial dan Ekonomi Kehutanan*, **8**(2).

[11] Kramer, P.J. and Kozlowski, T.T., 1960. *Physiology of trees.* Physiology of trees.

[12] Amirta, R., Yuliansyah, A.E., Ananto, B.R., Setiyono, B., Haqiqi, M.T., Septiana, H.A., Lodong, M. and Oktavianto, R.N., 2016. Plant diversity and energy potency of community forest in East Kalimantan, Indonesia: Searching for fast growing wood species for energy production. *Nusantara Biosci*, **8**(1), pp.22-31.

IOP Conf. Series: Earth and Environmental Science **141** (2018) 012023 doi:10.1088/1755-1315/141/1/012023

Absorption of CO_2 from modified flue gases of power generation Tarahan chemically using NaOH and Na_2CO_3 and biologically using microalgae

Elida Purba[1*], Dewi Agustina[1], Finka Putri Pertama[1], and Fita Senja[1]
[1]The Department of Chemical Engineering, Faculty of Engineering, University of Lampung, Lampung, Indonesia
*Author to whom any correspondence should be addressed
elida.purba@eng.unila.ac.id

Abstract. This research was carried out on the absorption of CO_2 from the modified flue gases of power generation Tarahan using NaOH (sodium hydroxide) and Na_2CO_3 (sodium carbonate). The operation was conducted in a packed column absorber and then the output gases from the packed column was fed into photo-bioreactor for biological absorption. In the photo-bioreactor, two species of microalgae, *N. occulata* and *T. chuii*, were cultivated to both absorb CO_2 gas and to produce biomass for algal oil. The aims of this research were, first, to determine the effect of absorbent flow rate on the reduction of CO_2 and on the decrease of output gas temperature, second, to determine the characteristics of methyl ester obtained from biological absorption process. Flow rates of the absorbent were varied as 1, 2, and 3 l/min. The concentrations of NaOH and Na_2CO_3 were 1 M at a constant gas flow rate of 6 l/min. The output concentrations of CO_2 from the absorber was analyzed using Gas Chromatography 2014-AT SHIMADZU Corp 08128. The results show that both of the absorbents give different trends. From the absorption using NaOH, it can be concluded that the higher the flow rate, the higher the absorption rate obtained. The highest flow rate achieved maximum absorption of 100%. On the other hand, absorption with Na_2CO_3 revealed the opposite trend where the higher the flow rates the lower the absorption rate. The highest absorption using Na_2CO_3 was obtained with the lowest flow rate, 1 l/min, that was 45,5%. As the effect of flow rate on output gas temperature, the temperature decreased with increasing flow rates for both absorbents. The output gas temperature for NaOH and Na_2CO_3 were consecutively 35 °C and 31 °C with inlet gas temperature of 50°C. Absorption of CO_2 biologically resulted a reduction of CO_2 up to 60% from the input gas concentration. Algal oil was extracted with mixed hexane and chloroform to obtain algal oil. Extracted oil was transesterified to methyl ester using sodium hydroxide as a catalyst. The results of in-situ transesterification method cannot be identified. Both microalgae achieved maximum yield at 2% catalyst concentration. *Nannochloropsis occulata* achieved the highest yield of algal oil that is 88.5%. The highest content of methyl ester from *Nannochloropsis occulata* was undecanoic acid methyl ester by 55.42% and the result from *Tetraselmis chuii* was palmitic acid methyl ester by 81.58%.

Keywords: CO_2, chemical absorption, biological absorption, microalgae

1. Introduction

Pollution has caused a lot of damage to the environment. One of waste that can be a threat to the environment is CO_2, which is increasingly increasing and can degrade air quality. Most of the CO_2 that is released into the environment is a result of human activities, for example on steam power generation Tarahan in South Lampung. The amount per hour is not massive but, the buildup can occur if CO_2 is continuously released without further treatment.

Action can be taken by absorption of CO_2 chemically and biologically. The biological absorption is carried out by using microalgae, where CO_2 will be absorbed by microalgae in the photosynthesis process. Besides reducing CO_2, this process can produce biodiesel using algal oil from microalgae. Microalgae have the lipid content of more than 30% and productivity of microalgae 200 times more than other vegetable sources, so it can be used as an alternative source of biodiesel [1]. On the other hand, the chemical absorption is applied using absorbents where absorbents are contacted with CO_2 in the absorber.

Research on CO_2 absorption has been published in literature. Previous study [2], show that the initial absorption decreased at a flow rate of 0.2 l / min to 0.4 l / min, and then absorption increased at a flow rate of 0.4 l / min to 0, 6 l / min. The initial condition was due to unsteady process. He found that the greater the flow rate of fluid the lower the k_Ga due to the unsteadiness of flow. When the flow rates is steady then the value of k_Ga is greater. The k_La decrease with increasing the flow rate but then it is increase with higher flow rates. An optimum concentration in the CO_2 reduction process was obtained by previous study [3] that is 25% by weight of Na_2CO_3. The effect of the addition of boric acid (H_3BO_3) to the solution in has been determined where the best boric acid concentration was at 3% by weight with the CO_2 gas absorption of 67.81%. However it increased the absorption rate up to 2.4 times.

Theoretically, chemical gas absorption is a process of mass transfer between two phases, gas and liquid, in which mass transfer occur when there is a driving force from one phase to another. The force of motion is a collision between molecules during mixing of fluid due flow rate velocity or forced mixing. Therefore, the greater the collision between molecules, the higher the absorption rate. The operating temperature affects the size of an absorber in the absorption operation. High temperatures affect the absorption capacity where the temperature enhances the reaction mechanism in the absorber [4].

Many research has been done by utilizing microalgae as a CO_2 absorbent with different types of microalgae. The commercial production of biodiesel derived from microalgae is still in the research and development stage, mainly due to the current prohibitive high costs associated with the biomass production and fuel conversion process. The schemes of research and development are optimizing parameters of operation such as the reaction time, the ratio of alcohol and microalgae lipid, catalyst concentration, and reaction temperature. In conventional trans-esterification, the extraction of lipid from microalgae was carried out by percolation using mixture of polar and non-polar solvents, such as methanol and chloroform. The oil was then trans-esterified into methyl ester in batch. Many research have been carried out to determine the best operating condition for the extraction of microalgal oil. However, this method takes a long time and therefore it requires high cost. Therefore, the more attractive alternative is being developed, namely the in-situ method or trans-esterification of biodiesel without the extraction step. In this research, the experiment was carried out not only about the ability of alkaline solutions and microalga in reducing CO_2, but also the potential of microalgae species in producing algal oil.

Previously, the studies were carried out separately either chemically or biologically. In this study, the absorption is carried out both chemically and biologically. The flow rates of the absorbent were 1, 2, dan 3 l /min. This is higher than the flow rate used by Hasnan et al. [2]. This is because the height of absorber in this study is larger than that applied by Hasnan et al. [2]. Similarly, as explained in previous research [5,6] high absorber and flow rates have an effect on overall mass displacement.

2. Materials and Methods

2.1 Chemical Absorption

Materials

The main equipment is packed column made from acrylic with a diameter of 7.5 cm and height of 127 cm). The packed column was filled with packing (based aluminum pipe mixture with a diameter of 8 mm, length of 2 cm and height of stuffing packing on columns 70 cm). Supporting equipment are a heater, air pump, gas flowmeter, liquid flowmeter, thermometer, and gas chromatography for analysis.

The absorbent was 1 M NaOH and Na_2CO_3. They were made from 40 grams of NaOH crystal dissolved in 10 liters of water and 1.06 kg of Na_2CO_3 dissolved in 10 liters of water. Na_2CO_3 was mixed with the H_3BO_3 as a catalyst of 3% weight. The CO_2 from Aneka Gas Industri in Natar-Lampung Selatan was made with concentration of 15% in air. The variables was the flow rates of the absorbent of 1, 2, and 3 l/min for both absorbents.

Experimental set-up and Procedure

Figure 1 shows the experimental schematic diagram for chemical absorption.

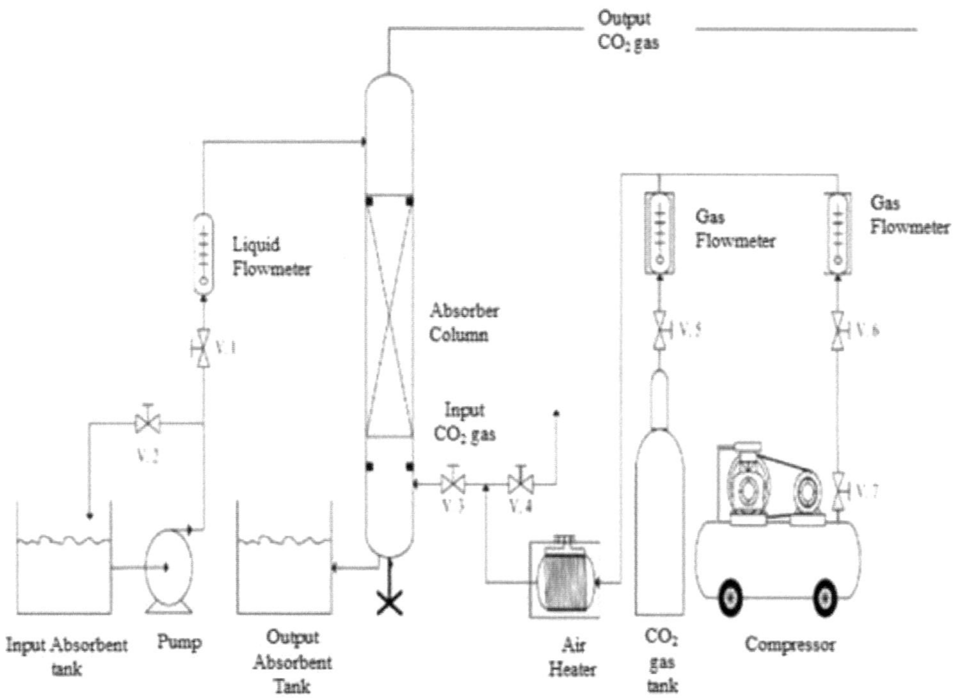

Figure 1. Schematic of CO_2 absorption tool setting

The absorbent was pumped into to the top of the packed column as setting volumetric flow rates (1,2,3 l/min). In the same time, the mixture of 15% CO_2-air was fed at constant flow rate of 1 l/min and contacted counter-currently with the absorbent. Input gas mixture temperature was set to 50 °C. The operation was applied until steady state condition was obtained. The output gas of the packed column was fed into the foto-bioreactor for biological absorption using microalgae. Gas sampling was applied using sampling bag and injected it into CO2 analyser (2014-AT Gas Chromatography Picture /SIMADZU Corp 08128).

2.2 Biological Absorption

Foto-bioreactor was filled with 2 liter culture of microalgae which was cultured for 6 days. The microalgae was then cultivated after 6 days. In the same time, the concentration of the output gas of the packed column was analyzed and the temperature was also measured. Nannochloropsis oculata and Tetraselmis chuii were obtained from The Laboratory of Marinculture Development, Lampung, Indonesia. They were cultured as explained above. After the cultivation, the microalgae was extracted in a soxhlet using chloroform: methanol and n-hexane as solvent [7]. As a result, algal oil was obtained. The alga oil was converted into methyl ester using sodium hydroxide in transesterification process with sulfuric acid as a catalyst.

IOP Conf. Series: Earth and Environmental Science **141** (2018) 012023 doi:10.1088/1755-1315/141/1/012023

3. Results and Discussions

3.1 Chemical Absorption

3.1.1 The Effect of Absorbent Flow Rates on CO_2 Absorption

The effect of NaOH Flow Rate on CO2 Absorption is shown in Figure 2.

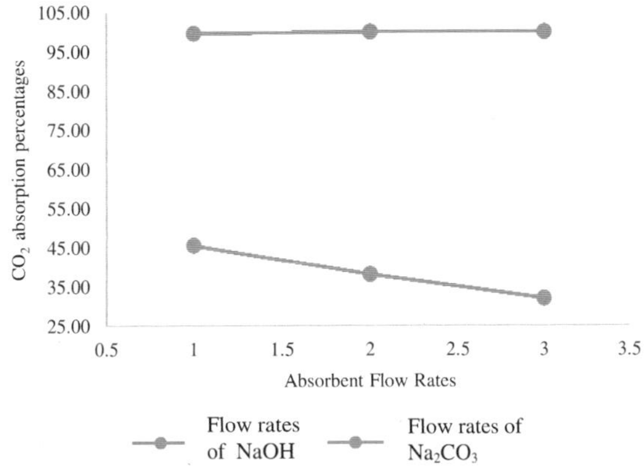

Figure 2. The effect of flow rates on the percentage of CO_2 absorption

As can be seen from Figure 2, the absorption of CO_2 by NaOH is much greater than that of by Na_2CO_3. However, both absorbents gives different trend. As for NaOH, the higher the flow rates the higher the percentage of absorption. In contrast, as for Na_2CO_3 increasing the flow rates of absorbent decreasing the percentage of absorption.

As for NaOH, the absorption of CO_2 using NaOH increases with increasing flow rate is due to the increasing the flow rate means increasing the availability of absorbent for CO_2 to dissolve as well as to react. This is in accordance with research [8] about the effect of flow rate of CO_2 gas absorption. In the study, it was described that an increase in flow rate can increase the rate of absorption in the gas because the increase of gas flow rate will increase the value of gas mass transfer coefficient. The higher the value of mass transfer coefficient of the gas the higher the liquid absorption capacity of the gas. On the other hand, the absorption of CO_2 using Na_2CO_3 shows that increasing the flow rates of absorbent decreasing the percentage of absorption.

3.1.2 The Effect of Absorbent Flow Rate on output gas temperature

The effect of flow rates on output gas temperature is shown in Figure 3. As can be seen form the figure 3, the higher the flow rates the lower the percentage of output gas temperature. This means that the higher the flow rates the gas output temperature is increasing. Higher flow rates was expected to lower the gas output temperature, however the results show the opposite. This is because the higher flow rates means shorter contact time with the gas. Therefore, the gas outlet temperature is higher with higher flow rates than that of lower flow rates.

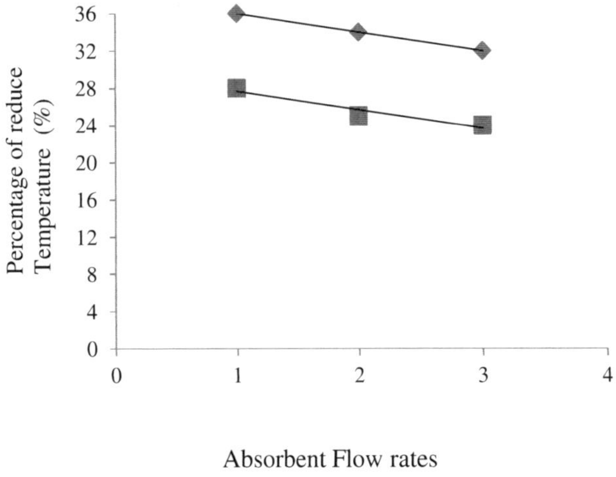

Figure 3. The Effect of Absorbent Flow Rate on the percentage of reduce output temperature

3.1.3 Extraction of microalgal oil

The lipids content or microalgal oil was extracted from microalgae in a soxhlet and show the results in Tabel 1.

Table 1 Lipid Content extracted from Microalgaae

Run	*Nannochloropsis oculata*	*Tetraselmis chuii*
I	11.0%	10.3%
II	13.0%	10.0%
III	12.5%	10.0%

Cultivation and harvesting processes affect the difference in the lipid. The culture was evaporated under vacuum to release the solvent using rotary evaporator. Then, the oil was obtained. It can be seen from the table that the oil extracted from *N. oculata* is much more that the oil from *T. chuii*.

The algal oil was converted into methyl ester by means of transesterification. The reaction occurs in two stages, they are esterification and transesterification. Then, the oil produced from each algal species was mixed with a mixture of catalyst and methanol with stirring. Esterification must be carried out due to the high content of fatty acid, that is more than 5%. A Certain amount of algal biomass of each species was reacted methyl ester.

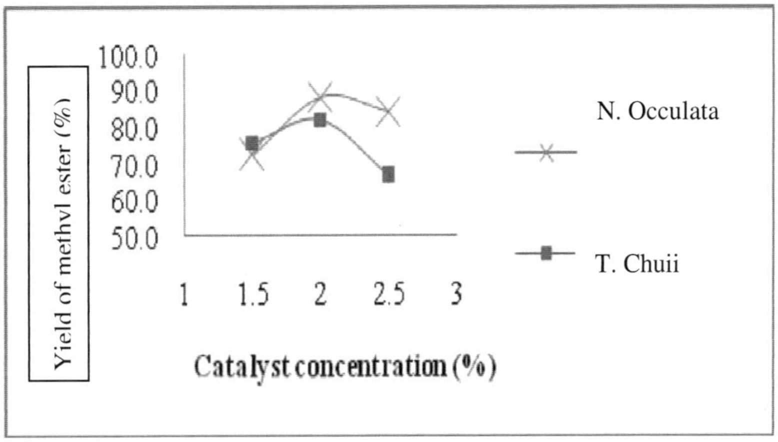

Figure 4. The effect of catalyst concentration on yield methyl ester of *N. oculata* and *T. chuii*

Figure 4 showed that the yield of methyl ester increase with increasing of catalyst concentration up to 2,0%. The concentration of catalyst in this state is able to optimally break the bond on lipid and exchange with methanol, thus forming FAME and glycerol. The mechanism of reaction can be seen from the Figure 5 below.

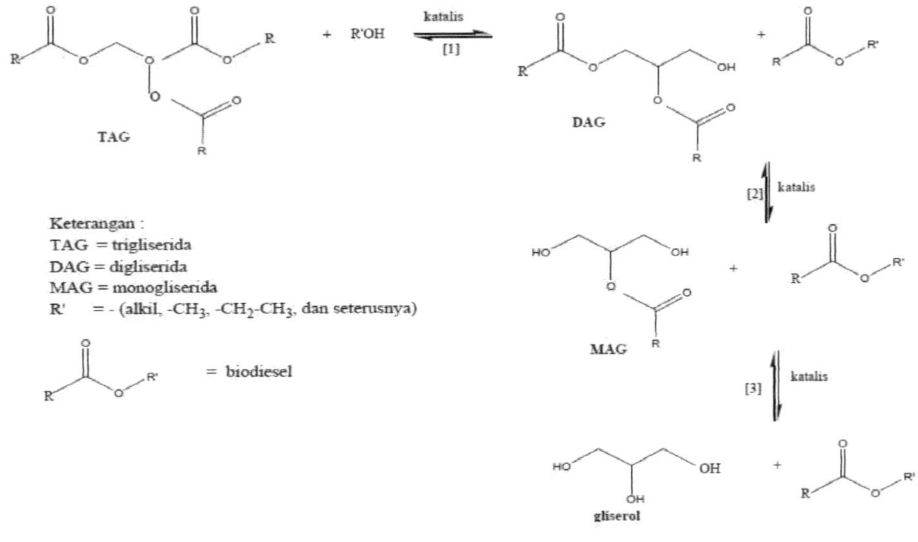

Figure 5. The mechanism of transesterification reaction with base catalyst [9]

Figure 5 showed that the mechanism of the transesterification reaction takes 2 steps . The first step, NaOH catalyst will bind to alcohol and wait for the contact with triglycerides [10]. After contacting between alcohol and triglycerides, the Na+ ions help break the bond contained triglycerides. The disconnected ties react with alcohol and Na+ ions back to form a compound NaOH.

It also occurs in the second step, in order to obtain the compound of alkyl esters (biodiesel). In the third Step, H + ions produced from the break up of alcohol will bind to O=, thus forming glycerol. At a concentration of 1.5% catalyst, methyl ester produced is minimum. When insufficient amount of catalyst is available, the catalyst is not able to optimally promote the lipid break the bond, so the reaction is slow. It takes a little longer to achieve optimum yield. The highest yield in both microalgae is obtained in the catalyst concentration of 2%, that is 88.5% in microalgae *N. oculata* and 82.3% in microalgae *T. chuii*. Maximum yield of FAME obtained from *N. oculata* is slightly higher than that of *T. chuii*.

3.1.4 Characterisation of biodiesel by GC-MS

The biodiesel produced from *N. oculata* and *T. chuii* were analyzed and compared with standards of fatty acids and methyl ester by gas chromatography analyzer. Based on Figure 6 and 7, there are two peaks that indicate the presence of methyl ester component on both microalgae. Figure 6 shows the result of GC-MS analysis from *N. oculata*.

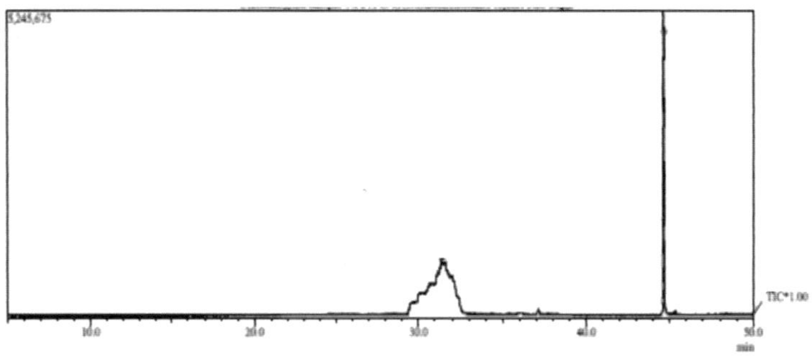

Figure 6. GC-MS result for *Nannochloropsis oculata* FAME

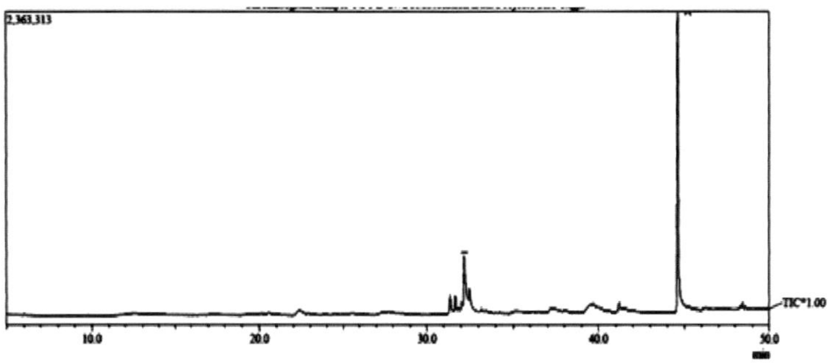

Figure 7. GC-MS result for *Tetraselmis chuii* FAME

The results indicated that the peak at R. time 31.390 corresponding to the presence of undecanoic acid methyl ester by 55.42% and at R. time 44.620 corresponding to the presence of glyceryl - 1,2 – isopropylidene – 3 - laurin by 44.58%. The GC-MS analysis result of *Tetraselmis chuii* methyl ester is shown in Figure 7. The first peak appeared at R. time 32.165 minutes. Based on the data bank of methyl ester, it shows that the peak is dodecanoic acid methyl ester by 18.42%. The second peak that appeared at the time R. 44.750 minutes corresponding to the presence of palmitic acid methyl ester by 81.58%.

4. Conclusions

The absorption of CO_2 by NaOH is greater when compared with that of Na_2CO_3. This is because the solubility and reaction rate of CO_2 in or with each NaOH and Na_2CO_3 is different. The absorption of CO_2 with NaOH increases with increasing flow rate and the opposite is for Na_2CO_3. Extracted oil was transesterified to biodiesel using sodium hydroxide as a catalyst. Both of microalgae obtained maximum yield at 2% catalyst concentration. *Nannochloropsis oculata* gives the highest yield that is 88.5%. The highest content of methyl ester from *Nannochloropsis oculata* is undecanoic acid methyl ester by 55.42% and the result from *Tetraselmis chuii* is palmitic acid methyl ester by 81.58%.

Acknowledgement

Elida Purba is in the Chemical Engineering Departement, Universitas Lampung, Bandar Lampung 35145, Indonesia. Elida Purba acknowledges government funding as research grant via The Ministry of Research, Technology, and Higher Education Republic of Indonesia

References

[1] Christi, "Biodiesel from Microalgae," *Biotechnology Advances 25*, 2007, pp. 294-306.
[2] Hasnan A, M., Najib, Prima, P., Kumaeti, N., dan A. Aji Hapsoro. 2012. *Studi Pengaruh Variabel Laju Alir NaOH dalam Proses Absorpsi Gas CO_2.* Jurusan Teknik Kimia, Fakultas Teknik, Universitas Dipenegoro. Semarang
[3] Cundari, L.; Selpiana; Redian, Bobby; dan Zaidan, Achmad. 2014. *Pengaruh Penambahan (H_3BO_3) pada Larutan Na_2CO_3 Terhadap Absopsi CO_2 dalam Biogas Menggunakan Spray Column.* JurusanTeknik Kimia, Fakultas Teknik, Universitas Sriwijaya. Palembang
[4] Ndĩritũ, H.M.; Kĩbicho, K; dan Gathitũ. 2013. *Effect of Heating on the absorption of CO_2 as Greenhouse Gas in Structured Packed Scrubber.* Department of Mechanical Engineering. Jomo Kenyatta University of Agriculture and Technology (JKUAT)
[5] E. Purba, D. Saragih, K. Ranti, "Ekstraksi Minyak Alga dari Alga Kering Nannochloropsis oculata dan Tetraselmis chuii Menggunakan Etanol dan N-Heksana, 2010, " *Proceedings of the 3rd National Seminar on Science and Technology*, Lampung Universitas, Bandar Lampung, Indonesia.
[6] Javed, K., H., Mahmud, T., and Purba, E., 2010." *The CO_2 Capture on a High-Intensity Vortex Spray Scrubber*". Chemical Engineering Journal, Vol. 162, pp. 448-456.
[7] Purba, E., Mahmud, T., and Javed, K.H., *Enhancement of Mass Transfer in a Spray Tower Using Swirling Gas Flow.* Jurnal International *Chemical Engineering Research and Desaign*, Volume 84, Issue 6, Hal 448-446. Juni 2006
[8] Said, Imam Noor; Saputri. Irma; Pamungkas M. Dawam. 2014. *Absorbsi Gas Karbondioksida Dengan Larutan NaOH.* Jurusan Teknik Kimia, Fakultas Teknik, Universitas Diponegoro
[9] S. P. Singh, D. Singh, "Biodiesel Production Through The Use of Different Sources and Characterization of Oils and Their Esther and Substitute of Diesel, 2010, " *Renewable and Sustainable Energy Reviews 14*, pp. 200-216.
[10] X. Ma, "Biodiesel Production from Algae through In Situ Transesterifikasi Technology,"2012, M.Sc. Thesis, Minnesota Univ., Minnesota, USA.

Biogas production from oil palm empty fruit bunches of post mushroom cultivation media

Agus Purnomo[1], Suprihatin[2], M Romli[2], Udin Hasanudin[3]

[1]*PhD student of Natural Resources and Environmental Management Studi Program, Bogor Agricultural University*
[1] *Departement of Health Analyst, Health Polytechnic of Tanjungkarang*
[2]*Department of Agro-Industrial Technology, Bogor Agricultural University, Bogor-16680, Indonesia*
[3]*Department of Agro-Industrial Technology, Faculty of Agriculture, University of Lampung, Bandar Lampung-35145, Indonesia*

Abstract. The Empty fruit bunches are one of the palm oil industry wastes, which can be used for mushroom cultivation. Post-cultivation of mushroom from former EFB-mushroom media (EFBMM) has the potential to be processed into biogas. The purpose of this research was to examine optimum co-digestion conditions for biogas production of EFBMM.The research was carried out in an anaerobic digester with three different conditions - dry fermentation (Water content (WC)/Total Solid (TS) ratio 1.5 - 3.5), semi-wet fermentation (WC/TS ratio = 4.0 - 5.7) and wet fermentation (WC/TS ratio> 9.0) conditions. Digester of capacity 50L was used. Fermentation was done using 20% cow feces as inoculum which then added with circulation system for 70 days. The results showed that optimum biogas production were produced in semi-wet fermentation conditions (WC/TS ratio = 4). It was produced 37.462 liters (2.420 liters CH_4/Kg Volatile Solid (VS)) of biogas with methane contain about 26.231%. Total volume of inoculum during process was 19.6 liters (1: 4 w/v) with absorbed TS inoculum ratio, TS/I = 0.4 (1:2.5 w/v). The result of research also showed that biogas which was produced from control about2.865 liters (0.041 liters CH_4/KgVS) , with TS absorbed inoculum ratio, TS/I = 0.5 (1: 5w/v).

Keywords: *Biogas,EFBMM, fermentation, WC/TS ratio, anaerobic digester*

1. Introduction

Several efforts to manage the solid waste of palm oil (EFB) have been investigated by utilizing them as raw materials for bioethanol production [1], raw material for composting mixed with mud from Palm Oil Mill biogas production [2], raw material of Biomass power plant [3], and materials for Raw production of biogas [4-7].Biogas is one type of energy that can be generated from the anaerobic fermentation process of organic waste such as EFB, produced every day by the palm oil industry in a very abundant amount. Currently EFB waste is generally used as mulch and compost fertilizer for oil palm plantations. EFB utilization process is less profitable because the decomposition process runs very slow if its used as mulch or compost.

EFB has a high potential for organic material, which is carbon as much as 480-490 kg/ton dry weight of EFB and nitrogen 7.4-9.8 kg/ton dry weight of EFB. The value of C/N ratio ofEFB is 50-65, causing the decomposition process to run slowly when used as mulch. The high content of organic materials in EFB has the potential to be used as biogas raw materials. Currently EFB has begun to be utilized as biogas raw material. The formation of biogas takes place through the decomposition process of the organic compounds in the anaerobic atmosphere by anaerobic bacteria by adding activators such as palm oil mill effluent (POME) and livestock manure (dissolved cow manure).EFB dried at 95°C for 48 hours can be used as biogas feedstock using POME activator. The results showed that the mixture of EFB and POME (F/I) ratio (2:1) incubated at 37°C for 45 days resulted in the highest methane gas with production of 55 m^3 CH_4 / ton [4]. EFB treated with NaOH and drying first then mixed with POME (F/I) ratio (6.8: 1) can produce maximum methane gas as much as 82.7 m^3 CH_4/ton [8].

Sawdust and sawdust fungus media (Baglog) can also be used as raw material for making biogas with cow dung activator. It turned out that the volume of biogas produced from baglog raw material (C/N ratio = 57.79) was 28% higher than the volume of biogas from sawdust raw material (C/N ratio = 88.30). The fermentation process lasts for 32 days [9].In the dry anaerobic digester process, the parameters of; inoculum concentration, pH, moisture, temperature, total solids and C/N ratio are the main parameters affecting biogas production. In addition to the C/N ratio, the choice of raw materials used affects the volume of biogas produced [10].

Currently EFB has also begun to be utilized as a medium of planting mushroom with 2 monthly cycles. Post-utilization of EFB as baglog mushroom (EFBMM) has been used as a material of organic fertilizer (compost) but need to be left first for ± 2 months. Based on some of the previous reference, this research was done with utilization of raw materials as follows, 20% cow dung inoculum using dry anaerobic digestion.

The purpose of this study was to obtain an exact inoculum ratio, according to Brummeler and Koster (1989) the addition of an appropriate inoculum ratio would be advantageous during the start-up process [11]. In this study the exact inoculum determination was done by calculating the volume of inoculum that will be circulated in the fermentation process based on the total ratio of EFBMM + inoculum mixed water content (WC) with total solid (TS) of EFBMM + inoculum mixture (WC/TS ratio). To obtain the optimum condition, the fermentation was done with 3 variations on treatment, dry fermentation with the content of TS 22-40% and water content with 60-78% (WC/TS ratio 1.5 – 3.5), semi wet fermentation With TS content of 15-20% and water content 80-85% (ratio WC/TS = 4.0 - 5.7) and wet fermentation with TS <10% and moisture content> 90% (WC/TS ratio> 9.0). This research also conducts fermentation with the same condition for fresh EFB as control.

2. Materials and Methods

2.1. Materials Research

The materials used in this research are: Bunches of Blended Palm Oil former media of mushroom (EFBMM) as substrate obtained from strawmushroom farmers and cow dung which then added with water proportions of 1 kg: 5 liters as inoculum.

2.2 Research Utilities

The production of biogas was conducted using a digester. The fermented material (EFBMM) was stored in each tank of digester with a capacity of 50 liters. On the inside of the digester there was a screen that serves as a support for the raw materials to be fermented and to prevent clogging on the bottom of the faucet. The inside/top of the digester was equipped with a hose that was perforated and forms a spiral so that the addition/circulation of the inoculum can spread evenly on the EFBMM. While on the outer lid of the digester was equipped with 2 systems, namely (1) a hose with a flange connected with a gas flowmeter serves to measure the volume of biogas produced and (2) circulation pipe system used to circulate the inoculum. At the bottom of the digester was equipped with a stop faucet and gallon that serve to accommodate stock and inoculum to be circulated. In detail the digester images used in the study are presented in Figure 1.

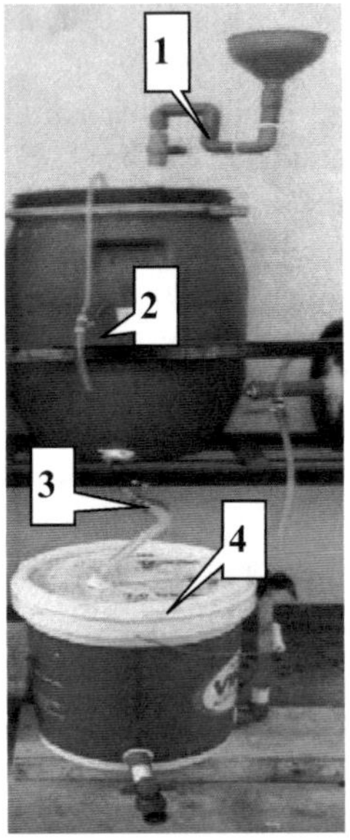

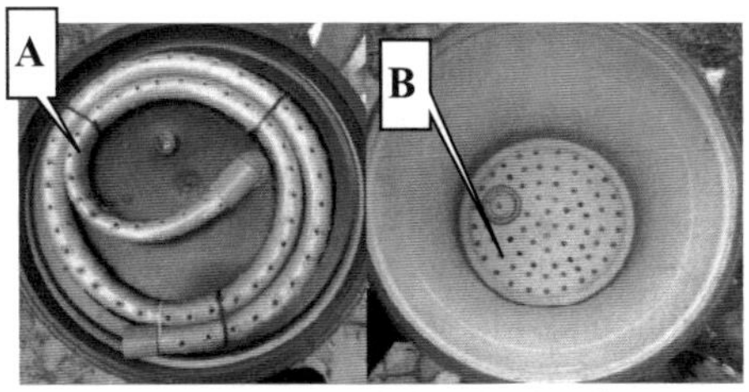

A = a hose perforated lid inside the digester
B = Screen bottom of the digester

Description:

1. Funnel **U** and circulation pipe equipped inoculum stop the tap
2. Place of sampling andbiogas volume measurement
3. The bottom valve of digester for the flow of inoculum
4. Circulating inoculum storage vessel

Figure 1. A series of biogas making tools with circulation system

2.3 Implementation of Research

The parameters studied were to determine the optimum condition of biogas production (fermentation process). EFBMM with inoculum of cow dung on each digester with 3 treatments, i.e. wet fermentation (TS material <10%), semi-wet fermentation (TS material 15-20%) and dry fermentation (TS material 22-40%). The fermentation conditions on each digester can be determined by adjusting the volume of inoculum which is added/circulated by using the comparison equation between water content (WC) (Substrate + Inoculum) with total solid (TS) content (Substrate + Inoculum) using the following equation :

$$Ratio\frac{WC}{TS} = \frac{(WC\ EFBMMx\,Weight) + (WC\ InoculumxVolume)}{(TS\ EFBMMx\ Weight) + (TSInoculumxVolume)}$$

From the above equation, the volume of inoculum that was circulated daily on each digester is presented in Table 1.

Table 1. Characteristic material and volume of inoculum that is circulated daily at each digester

Code	WC (%)	TS (%)	VS (%)	VI (L/day)	Ratio (WC/TS)	Fermentation
R-1	27.2	72.8	65.4	**6**	2.3	dry
R-2	30.7	69.3	60.8	**10.0**	3.0	dry
R-3	32.7	67.3	59.0	**13.0**	4.0	semi-wet
R-4	45.7	54.3	45.1	**14.5**	5.7	semi-wet
R-5	50.4	49.6	42.6	**22.0**	9.0	wet
R-6	58.0	42.0	33.7	**23.5**	11.5	wet
control-R7	48.3	51.7	48.3	**8.7**	4.0	semi-wet
control-R8	38.0	62.0	59.1	**5.7**	2.3	dry
control-R9	73.1	26.9	24.6	**9.4**	9.0	wet

Description : WC = water content; TS = total solid, VI = volume of inoculum that is circulated daily
VS = Volatile Solid; Weight EFBMM each digester = 5 Kg
Water content of inoculum = 90.3% and TS inoculum = 9.7%

The experiment was carried out by inserting the 5 kg EFBMM into each digester with screened plaster on the bottom, then on the inside of the lid silicone glue was added and then closed tightly and ensured to be

airtight/no leakage. Each EFBMM-filled digester was added inoculum via a circulation pipe with volumes as presented in Table-1 and anaerobic fermentation process was then started. Process of fermentation takes place every day, during the process the measurement of biogas volume was calculated, also measured was the volume of inoculum which was channeled to the containment gallon. The volume-measured inoculum was then recycled into the same volume as the initial circulation volume. If there is a reduction/volume depreciation, then conduct the addition of inoculum onthe stock gallon and if there was an excess volume, the excess volume of inoculum was returned to the stock gallon. Measurement of biogas volume and circulation process was carried out continuously during the anaerobic fermentation process on the digester.
Biogas composition of each digester was measured once every week (7 days) . The biogas samples were taken directly on the biogas flow line of the digester using gas syringes and directly analyzed for the biogas composition (CH_4 , CO_2 and N_2) using Gas Chromatography (Shimadzu GC 2014) with thermal conductivity detectors at 200 °C and 100 °C injection temperature. GC analysis used a Shincarbon column of 4.0 ms longs and 3 mm inner diameter with Helium gas as a gas carrier with a flow rate of 40 ml/min .

3. Results and Discussions

3.1. Material Characteristics In Digester

The raw materials used in the dry anaerobic digestion process are EFBMM which has been described, fresh EFB which had already pressed from the factory, then chopped as raw material for control and 20% cow feces as inoculum. The result showed that EFBMM used in research had average water content 41.10%, TS content 59.90%, C content 25.44%, N rate equal to 1.20%, and Ratio C/N amounted to 16.639. The presence of C and N content shows that EFBMM still has high organic and mineral element content, while low C/N ratio and high TS at EFBMM show that the raw material are susceptible to degradation, thus has the potential to be used as raw material for biogas production by dry Anaerobic digestions. Various parameters such as inoculum concentration, pH, moisture, temperature, total solids and C/N ratios are the main parameters affecting biogas production. In addition to the C/N ratio, the choice of raw materials used affects the volume of biogas produced [10]. The characteristics of EFBMM and inoculum in each digester are presented in Table 1. Raw materials (EFB-Fresh, EFBMM, and Cow dung) characteristics used in dry anaerobic digestions are presented in Table 2.

Table 2. Characteristics of raw materials

Parameters	Unit	Average levels		
		EFB-Fresh	EFBMM	Cow dung 20%
C content	%	34.97	25.44	24.92
N content	%	0.73	1.20	1.50
C/N Ratio	-	47.83	16.63	16.61
Water Content	%	53.13	41.10	90.30
TS content	%	46.87	59.90	9.70
VS content	%	44.00	52.01	8.29

3.2. Biogas production resulted from fermentation process variations

Production of organic solid waste biogas with cow feces fluid (inoculum) using a circulatory system are a dry anaerobic digestion process. This process is one of the innovations in handling organic biological solids by an anaerobic microbial consortium contained in the inoculum to restore the potential of renewable energy sources and nutrient-rich fertilizers to manage sustainable solid waste. This process generally takes place in solids with concentrations greater than 10% (TS> 10%) and allows for higher loading rates of organic matter, minimal material handling, low energy requirements for heating, limited environmental consequences and effective performance . As for retention time, poor initial performance, poor homogeneous mixing and simple fatty acid accumulation (VFAs) were considered as weaknesses in the dry anaerobic digestion process [11].

IOP Conf. Series: Earth and Environmental Science **141** (2018) 012024 doi:10.1088/1755-1315/141/1/012024

Measurement of biogas volume of dry, semi-wet and wet fermentation process were done by using gas flow meter. The biogas production accumulation curve in each digester treatment was presented in Figure 2. Figure 2 shows that the highest biogas production was produced by R-3 with a total volume of 37.47 liters (7.49 liters/kg EFBMM) with solid residence time (SRT) 70 days with WC/TS ratio of raw materials of EFBMM 4.0 . In the same condition the ratio of WC/TS and SRT produced in the control digester (R-7) using fresh raw material EFB is 2.86 liters (0.57 liters/kg EFB). While the result of measurement and calculation of cumulative volumes of biogas compared with the amount of volatile solid (VS) in each digester of the study is presented in drawing 3. Figure 2 shows that the accumulative volume of biogas in dry fermentation group , control digester (R-8) produces biogas Higher than the R-1 and R-2 digesters, but in Fig. 3 it was found that the volume of methane gas (CH_4) produced was still the lowest in the treatment group, ie R-8 0.408 liters/KgVS , R-1 0.571 liters/KgVS and R-2 1.189 liters/KgVS . Physically EFB fiber has a strong fiber, so it takes a long time to be decomposed. The presence of biogas produced in the dry anaerobic digesters process was derived from organic materials such as oil residues from the pressing process of the plant and fibers attached to the EFB [12] .

Figure 2 showed thatthe average of all the digester treatments showed an increase in biogas production after 4 weeks, the same as reported by Lu et al. that during the dry anaerobic process, the pH decreased from the initial conditions of 6.7 to 4.2 under mesophilic conditions and pH 5.8 under thermophilic conditions for the first 2 weeks. Furthermore the pH gradually increased to 7.6 after 4 weeks with a digester temperature of 35°C and pH 8 after 3 weeks with digester temperature of 55°C [11].

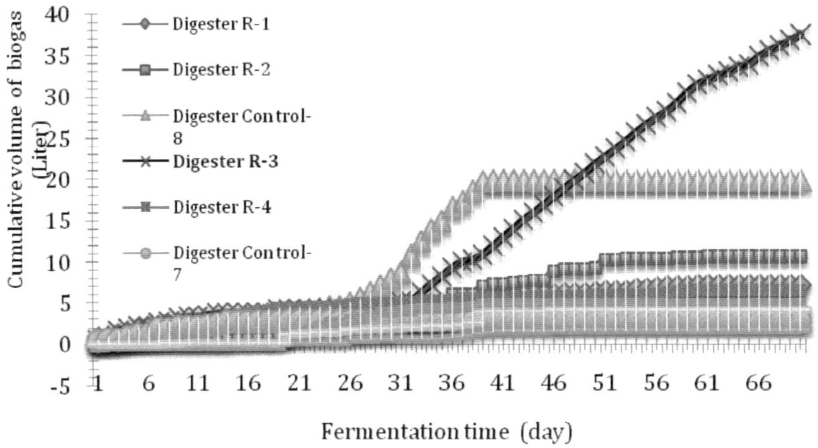

Figure 2. Biogas Production on each digester

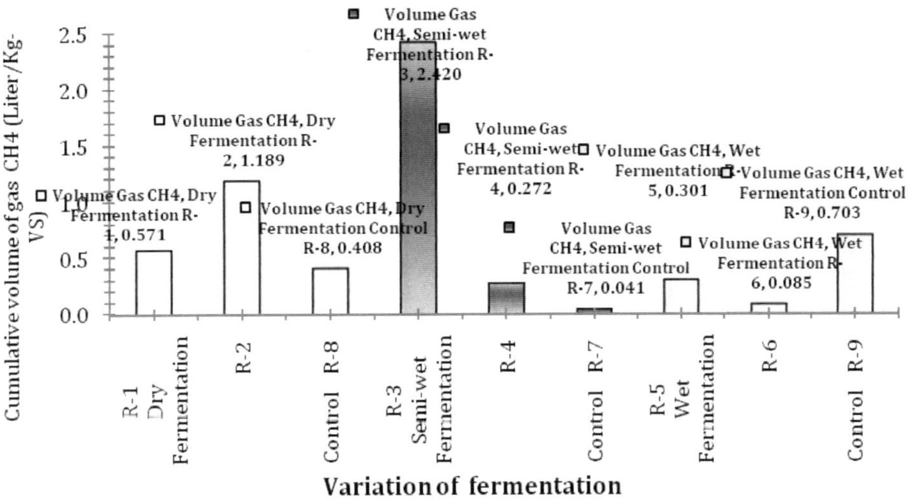

Figure 3. Cumulative volume of gases CH_4based on levels VS

3.3. Optimization of fermentation process

In the determination of fermentation optimization EFBMM uses 20% cow dung inoculum which is circulated with variation on treatment of R-1 (WC/TS 2,3), R-2 (WC/TS 3.0), R-3 (WC/TS 4.0), R-4 (WC/TS 5.7), R-5 (WC/TS 9.0) and R-6 (WC/TS 11.5) results of said biogas produced were 1.42 L/Kg; 2.12 L/kg; 7.49 L/kg; 1.31 L/kg; 0.98 L/kg and 0.34 L/kg EFBMM. From the data, the optimum condition of the fermentation process can be clearly seen in the curve of Figure 4. From Figure 4, it shows that EFBMM fermentation using 20% cows dung inoculum with circulation system in semi-wet condition (ratio of WC/TS = 4) can produce more biogas compared to the conditions of dry fermentation and wet fermentation. The drier or wetter the fermentation condition, can decrease biogas production. If calculated on the basis of the VS content of each EFBMM raw material, the CH_4 gas volume produced under the optimum condition of the semi-wet fermentation process (R-3) was 2.42 L KgVS.When compared with the results reported by Hasanudin et al. 2015, the cumulative volume of biogas generated in this research turned out to be only 10 percent of biogas production from anaerobic composting process of fresh EFB and POME which expected [12].

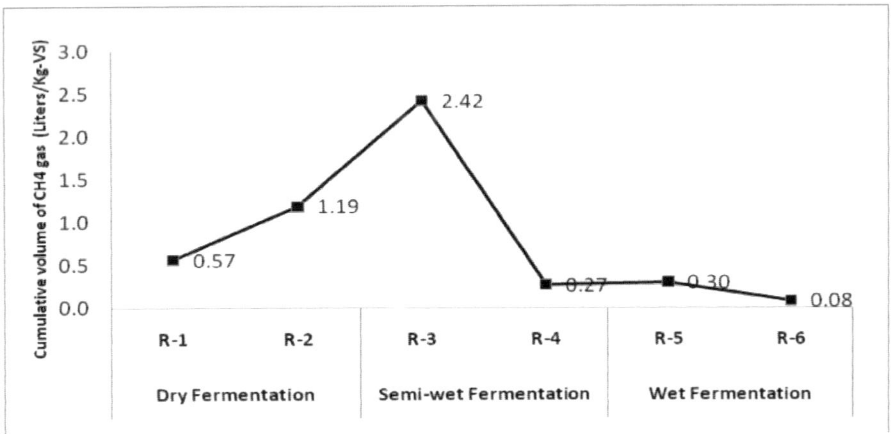

Figure4. Fermentation optimization curve

3.4. The composition of biogas produced on the variety of the fermentation process

In the process of determining the optimum conditions of fermentation in this study, no observations were made on other factors such as temperature and pH digester. In the determination of optimum conditions the researchers only wanted to know the best WC/TS ratio of theusage of inoculum circulation process. Based on the data presented in Table 3, the CH_4 gases content generated from each digester ranged from 3.218% to 24.63%. CH_4 gas content of this research was lower when compared with previous studies. Biogas formed from into co-composting process of EFB and POME can produce biogas with CH_4 gas content of 40.1% [12]. Other research results reported by Amelia JR et al. (2017) stated that CH_4 gases content in the anaerobic process of composting fresh fruit bunch (FFB) with 20 litres of POME per day, can produce 51.08% CH_4. The CH_4 gas content exceeding 45% will be highly flammable and can be a proper fuel because the high calorific fuel value [13]

Table 3. Gas composition CH_4each digester

Parameter	Treatment ratio (WC/TS)								
	R-1	R-2	Control-R8	R-3	R-4	Control-R7	R-5	R-6	Control-R9
CH_4 (%)	17.977	18.236	24.630	16.356	12.498	18.269	11.686	3.218	18.027
CO_2 (%)	31.712	32.668	41.816	30.710	27.505	35.987	25.771	20.587	25.975
N_2 (%)	50.277	48.759	32.515	52.570	59.997	46.752	62.509	76.116	55.982

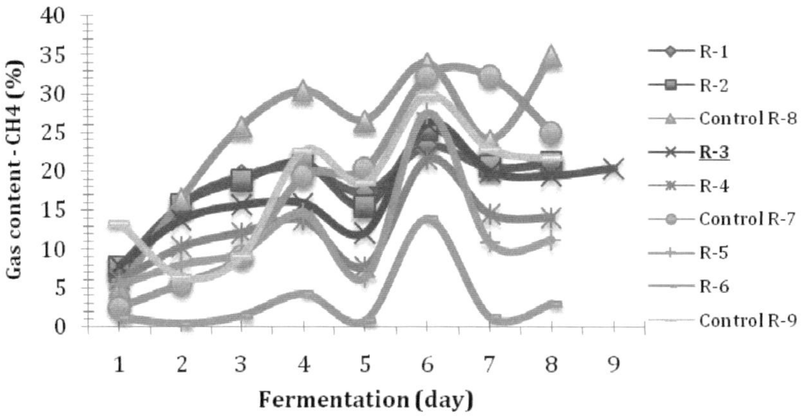

Figure5. CH_4 gas trend curveon fermentation process variation

Figure 5 also showed that during the fermentation process, with increasing time, the CH_4 gas content generated from each of the average digester tends to increase, but fluctuates (up and down). As stated earlier that in this fermentation process, researchers did not observe changes in temperature and pH in the digester. Jha AK et al. 2011 stated that the bacteria in the reactor were very sensitive to changes in temperature and the rise in temperature of about 2°C per day can lead to the termination of the fermentation process. In addition, the more important thing was the pressure in the fermentation reactor. High gas pressure within the reactor was reactive to methane bacteria and slows the methane formation process. When the inoculum circulation was done, the fermentation process just continuously goes again [14]. Kunatsa T et al. 2013 also states that the anaerobic decomposition process occurs in the mesophilic range and has a 15 day retention time. In tropical countries, the unheated digesters tend to be at an average temperature between 20-30°C. In addition to temperature effects, biogas production levels also depends on the pH and the dryness of the feedstock [15].

3.5. The volume of inoculum that was circulated on the variety of fermentation processes

In the anaerobic fermentation process, the solid organic matter in the digester requires the inoculum to speed up the reaction process. Inoculation can improve digestion efficiency of the digester and may reduce the residence time of solids [11].

Table 4. Volume of inoculum in each digester

Description	R-1	R-2	Control R-8	R-3	R-4	Control R-7	R-5	R-6	Control R-9
Vol. of circulation-Inoculum (L/day	6	10	5.7	13	14.5	8.7	22	23.5	9
Weight TS EFBMM (Kg)	3.64	3.465	3.365	2.715	2.48	2.1	2.585	3.1	1.345
Absorbed Inoculum Vol (L)	3.55	5.35	0.15	6.6	10.15	4.05	10.5	6.35	0.35
The Ratio TS/I adsorbed	1.0	0.6	22.4	0.4	0.2	0.5	0.2	0.5	3.8

Table 4 showed that variations in circulating volume of circulating inoculum based on calculation of WC/TS ratio in each digester being absorbed differently. Overall absorption capacity of EFBMM raw material was greater than absorption of control raw material (EFB). This is because the physical structure of EFBMM was relatively soft/not hard and easily decomposed, so it has larger surface area and pores than the control material that has denser and harder physical texture.The calculation of TS ratio of inoculum (TS/I) absorbed in each digester ranged from 0.2 to 1.0. The highest biogas production of this research was obtained in absorbed inoculum with a TS/I ratio of 0.4. Brummeler et al. (1992) reported that the optimum ratio of 'solid inoculum form' to solids at start-up was 0.5 to 0.6 with 30-day solid residue time. The time of solids retention will be longer at a lower inoculum ratio, since the leachate circulation was not optimum. When the inoculum ratio falls to 0.4 or lower, the residence time of solids will increase to 50 days or more, since other conditions were not optimal, such as low pH and high organic acid concentrations [11]

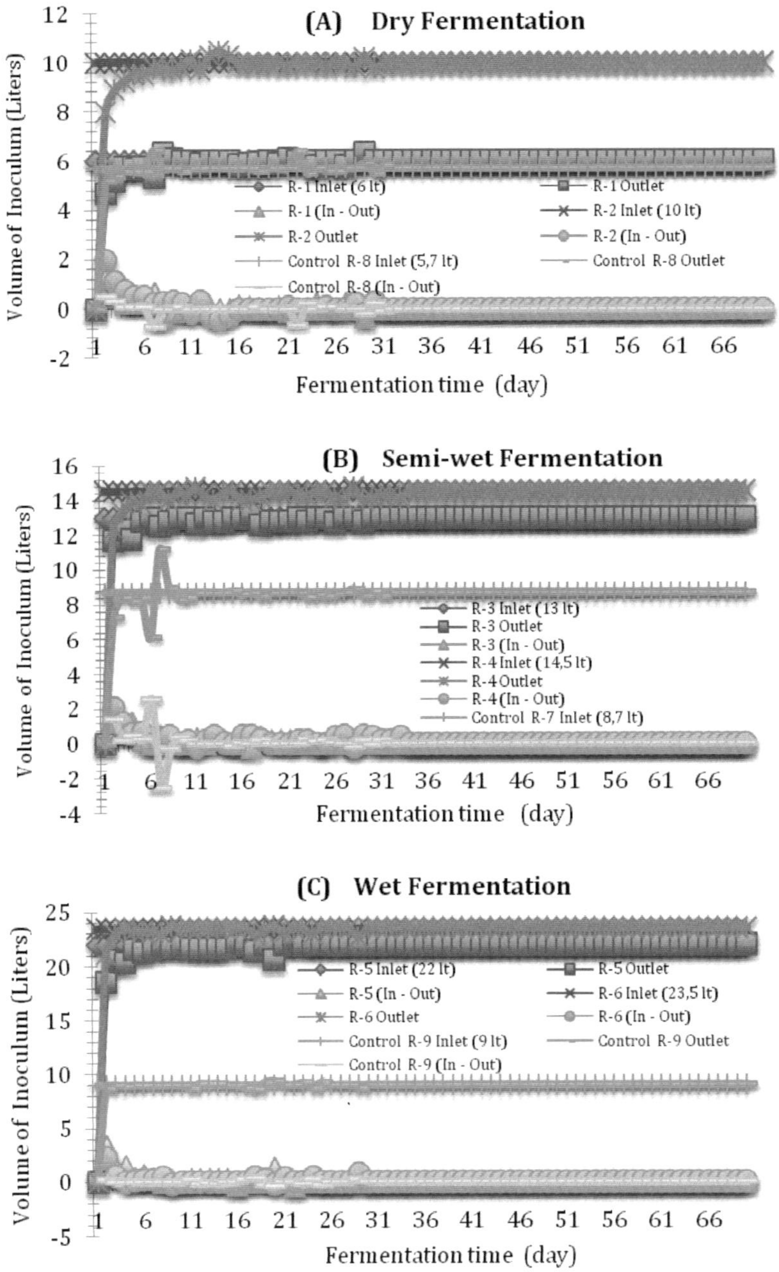

Figure6. The inoculums circulation volume curva in dry (A), semi-wet (B) and wet (C) fermentation

Figure 6 shows that the average daily circulation volume of all new digester treatments was stable after week 4th. Entering the 5th week onwards, the volume of inoculum that was circulated/introduced into the digester equals the volume of inoculum output. This was in line with the biogas production presented in Figure 2 which shows that biogas production begins to increase after 4th weeks.

4. Conclusions

The optimum condition of biogas production using EFBMM raw material takes place in semi-wet fermentation with WC/TS ratio was 4, with biogas volume produced 37.462 liters (2.420 liters CH_4/KgVS). While control samples biogas yields was 2.865 liters (0.041 liters CH_4/KgVS). Under optimum condition, the average concentration of CH_4 gas produced were 16.356% with the lowest CH_4 concentration being 7.872% at startup and the highest concentration was 26.231% at maximum production. The total volume of inoculum used at optimum conditions during the circulation process were 19.6 liters with a TS/I ratio of 0.4 or (1:2.5 w/v).) and the total volume of inoculum in the control were 12.75 Liters with a ratio of TS/I = 0.5 or (1:5 w/v)

Acknowledgements
The authors would like to thank the Center for Human Resources Development (PPSDM) Health Ministry of Health of the Republic of Indonesia which has provided financial support (Cost of Education), thank you also to the Study Program of Natural Resources Management and Environment IPB, Faculty of Agriculture of UNILA, and Analytical Laboratories of Department Agro-Industrial Technology UNILA, for the provision of data, information and discussion.

References

[1] Sudiyani Y, Styarini D, Triwahyuni E, Sudiyarmanto, Sembiring KC, Aristiawan Y, Abimanyu H, Han MH 2013 *Utilization of biomass waste empty fruit bunch fiber of palm oil for bioethanol production using pilot-scale unit.* Journal Energy Procedia **32**31-38.
[2] Nutongkaew T, Duangsuwan W, Prasertsan S, Prasertsan P 2014 *Effect of inoculum size on production of compost and enzymesfrom palm oil mill biogas sludge mixed with shredded palm empty fruit bunches and decanter cake*Journal of Science and Technology**36** 275-281.
[3] Wijono A 2014*PLTU biomasa tandan kosong kelapa sawit studi kelayakan dan dampak lingkungan.* Simposium nasional RAPI XIII-2014 *FT UMS*, ISSN 1412-9612.
[4] Chaikitkaew S, Kongjan P, O-Thong S 2015*Biogas production from biomass residues of palm oil mill 3by solid state anaerobic digestion*. Journal Energy Procedia **79**838 – 844.
[5] Nieves DC, Karimi K, Horváth IS. 2011. *Improvement of biogas production from oil palm empty fruit bunches (OPEFB)*. Journal Industrial Crops and Products **34** 1097–1101.
[6] Nurliyana MY, H'ng PS, Rasmina H, Kalsom MSU, Chin KL, Lee SH, Lum WC, Khoo GD 2015*Effect of C/N ratio in methane productivity and biodegradability during facultative co-digestion of palm oil mill effluent and empty fruit bunch* Industrial Crops and Products **76** 409-415.
[7] Suksong W, Kongjan P, Prasertsan P, Imai T, O-Thong S 2016 *Optimization and microbial community analysis for production of biogas from solid waste residues of palm oil mill industry by solid-state anaerobic digestion*Journal Bioresource Technology **214**166-174.
[8] O-Thong S, Boe K, Angelidak I 2012*Thermophilic anaerobic co-digestion of oil palm empty fruit bunches with palm oil mill effluent for efficient biogas production*Journal Applied Energy **93** 648-654.
[9] Mulyadi D, Mukmilah L, Kusumawati D 2016 *Efektifitas pemanfaatan serbuk gergaji dan limbah media tanam jamur (baglog) sebagai bahan baku pembuatan biogas* Jurnal Kimia Valensi **2** 11-16.
[10] Dioha IJ, Ikema CH, Nafi'u T, Soba NI, Yusuf MBS 2013 *Effect of carbon to nitrogenratio on biogas production* International Research Journal of Natural Sciences **1** 1-10.
[11] Jha AK, Li J, Nies L, Zhang L 2011 *Research advances in dry anaerobic digestion process of solid organik wastes* African Journal of Biotechnology **10** 14242-14253
[12] Hasanudin U, Sugiharto R, Hariyanto A, Setiadi T, Fujie K 2015 *Palm oil mill effluent treatment and utilization to ensure the sustainability of palm oil industries* Journal Water Science & Technology.
[13] Amelia JR, Suprihatin S, Indrasti NS, Hasanudin U, Fujie K 2017 *Performance evaluation of integrated solid-liquid wastes treatment technology in palm oil industry*IOP Conf.Serisc: Earth and Environmental Science **65** 012036.

[14] Pilat P, Patsch M, Jandacka J 2012*Analysis of problems with dry fermentation process for biogas production* EPJ Web Conferences **25** 01075
[15] Kunatsa T, Madiye L, Chikuku T, Shonhiwa C, Musademba D 2013 *Feasibility study of biogas production from water hyacinth A case of lake Chivero – Harare, Zimbabwe*International Journal of Engineering and Technology **3**119-1128.

IOP Conf. Series: Earth and Environmental Science **141** (2018) 012025 doi:10.1088/1755-1315/141/1/012025

Enzymatic saccharification of liquid hot water and dilute sulfuric acid pretreated oil palm empty fruit bunch and sugarcane bagasse

L Risanto[1], Fitria[1], T Fajriutami[1] and E Hermiati [1,2]

[1] Research Center for Biomaterials, Indonesian Institute of Sciences (LIPI), Jl. Raya Bogor Km 46, Cibinong, Bogor 16911, Indonesia
[2] Author to whom any correspondence should be addressed
euis.hermiati@lipi.go.id

Abstract. Oil palm empty fruit bunch (OPEFB) and sugarcane bagasse (SB) are potential feedstocks for the production of bioethanol. In this study OPEFB and SB were pretreated by liquid hot water and dilute sulfuric acid (3% H_2SO_4), and continued with enzymatic saccharification. Heating treatment for both methods was conducted in an autoclave at 121 °C for 1 hr. The saccharification was performed up to 72 hours with cellulase enzyme loading of 10, 20, and 30 FPU per g biomass. Results showed that OPEFB and SB pretreated with H_2SO_4produced higher reducing sugars than those pretreated by liquid hot water. Higher enzyme loading also resulted in higher reducing sugars. Reducing sugars obtained from enzymatic saccharification of OPEFB were higher than those obtained from SB. The highest total reducing sugars (50.48 g/100 g biomass) was obtained from OPEFB pretreated with 3% H_2SO_4 at enzyme loading of 30 FPU per g biomass.

1. Introduction

Oil palm empty fruit bunch (OPEFB) and sugarcane bagasse (SB) are among potential feedstocks abundantly available in Indonesia for producing bioethanol[1,2]. The conversion of the lignocellulosic biomass to ethanol involves the pretreatment, enzymatic hydrolysis of cellulose to simple sugars, and fermentation of the sugars to ethanol.

Since pretreatment of biomass is an important step in the conversion, there have been numerous efforts to pretreat the lignocellulosic biomass, including OPEFB and SB[3-8]. Acidic pretreatment is one of pretreatment methods that can effectively modify the chemical structures of lignocellulose. In this case, acids serve as catalysts to hydrolyze carbohydrates, especially hemicellulose, so that they loosen the lignin-hemicellulose barrier, which can facilitate the penetration of cellulase enzyme to cellulose matrices[9]. Hot water pretreatment is another method of pretreatment that mainly aimed at removing of hemicellulose from lignocellulosic biomass. Compared to dilute acid pretreatment, hot water pretreatment is considered environmentally friendier and has some advantages, for example lower risk of equipment corrosion, and less xylose degradation, which means lower amount of inhibitors in the hydrolysate [10].In the enzymatic hydrolysis process, enzyme loading is one of factors that affect amount of sugars obtained. The amount of enzyme added to the pretreated biomass should be appropriate, so that the saccharification process is efficient.

The present study was condcuted to compare the hot water and dilute acid pretreatment in the production of reducing sugars, both in the hydrolysate of preteatment and in the hydrolysate of enzymatic saccharification. Besides that, the effects of enzyme loading on the production of the reducing sugars during enzymatic saccharification were also observed.

2. Materials and Methods

Raw materials used in this study are OPEFB and SB, which were obtained from Sukabumi and PG Rajawali, Subang Indonesia, respectively. The materials were dried and ground into particle sizes 40-60 mesh, then they were kept in sealed plastic bags and stored in a container. All other chemicals used were analytical grade without further purification.

2.1. Chemical component analysis of OPEFB and SB

Moisture content of biomass was determined according to TAPPI Test Method T 264 cm-97[11], while ash and extractive content were determined according to TAPPI Test Method T 211 om-02 12] and TAPPI Test Method T 204 cm-97[13], respectively. NREL LAP 003 [14] was used as a reference for the determination of acid insoluble ligninand Wise method [15]was used for the determination of holocellulose. The α-cellulose content was determined according to Rowell (2005)[16]. Hemicellulose content was obtained from subtraction of holocellulose and α-cellulose.

2.2. Pretreatment of OPEFB and SB

The materials were put in an Erlenmeyer flask. Water or 3% sulfuric acid were added to the flask, so that the solid loading of the suspension was 6%. The Erlenmeyer flask was tightly closed, and put in an autoclave for heat treatment. Heating in the autoclave was performed at 121 °C for 1 hour. After the process completed, the samples were cooled down, and continued with separation between soluble and insoluble fraction by filtration. The insoluble fraction, also called pulp, was neutralized using 1% NAOH and was washed with distilled water. The pulp recovery of the pretreated samples were determined by measuring the weight of the pulp and determination of moisture content of the pulp. Both fractions were kept in a refrigerator for further processes or analyses.

2.3. Enzymatic saccharification of pretreated OPEFB and SB

Enzymatic saccharification of the pretreated OPEFB and SB were conducted according to NREL. As much as 0.3 g of pretreated samples were put in centrifuge tubes. About 15 mL of 0.1 M citrate buffer pH 4.8 was added to the sample, continued with the addition of 300 μL of 2% sodium azide and 10, 20 and 30 FPU of cellulase enzyme (Meicellase from Meiji Sheika, enzyme activity 200 FPU/mL). The citrate buffer was then added to get the total volume of 30 g. The samples were incubated in a waterbath shaker at 150 rpm, 50 °C for 72 hours. Sampling of the hydrolysates were conducted every 24 hours. After the saccharification was completed, the samples were heated in boiled water for 5 minutes to inactivate the enzyme. The samples were kept in refrigerator for further analyses.

2.4. Analysis and determnation of reducing sugars

Determinations of reducing sugars in the soluble fraction and in the hydrolisates of enzymatic saccharified samples were performed using dinitro salycilic acid (DNS) method[17]. The absorbance of the DNS preparations were observed using UV-Vis Spectrophotometer (Hitachi-U-2001) at 520 nm. The yields of the reducing sugars from the pretreated samples were determined based on the weight of the initial biomass.

3. Results and Discussions

3.1. Chemical compositions of OPEFB and SB

OPEFB contained lower ash, acid insoluble lignin and hemicellulose than did SB, but higher extractives and α-cellulose (table 1). Based on these compositions, it can be predicted that OPEFB is more potential in producing sugars than is SB. All the chemical contents, except hemicellulose content of OPEFB and SB used in this study were slightly lower than those used in our previous study [18,19].

Table 1. Chemical compositions of oil palm empty fruit bunch (OPEFB) and sugarcane bagasse (SB).

Components	OPEFB	SB
Ash (%)	2.91 ± 0.03	4.57 ± 0.09
Extractives (%)	2.69 ± 0.11	2.36 ± 0.06
Acid insoluble lignin (%)	23.62 ± 0.08	25.42 ± 0.15
α-cellulose (%)	41.23 ± 0.77	36.53 ± 0.18
Hemicellulose (%)	26.55 ± 0.57	27.04 ± 0.89

3.2. Pulp Recovery

The effects of each pretreatment on the pulp recovery (figure 1) show that pretreatment using sulfuric acid resulted in lower pulp recovery than that using only water. These were observed either in pretreated OPEFB or in pretreated SB, and suggested that sulfuric acid degraded more chemical components in the material into lower molecular weight compounds that dissolved in the acid solution and became soluble compounds. These phenomena were confirmed by the results of reducing sugars in the soluble fractions (figure 2). The pulp recovery from pretreated OPEFB was slightly higher than that from pretreated SB, either in that obtained from LHW or from sulfuric acid pretreatment. These indicated that OPEFB was slightly harder to degrade than was SB. The amount of pulp recovered from sulfuric acid pretreatment of OPEFB was comparable with those from hydrogen peroxide enhanced oxalic acid pretreatment process with alkaline impregnation, amounted 60.7-65.2% [3], but much lower than those obtained from fungal pretreatment (85-90%) [18]

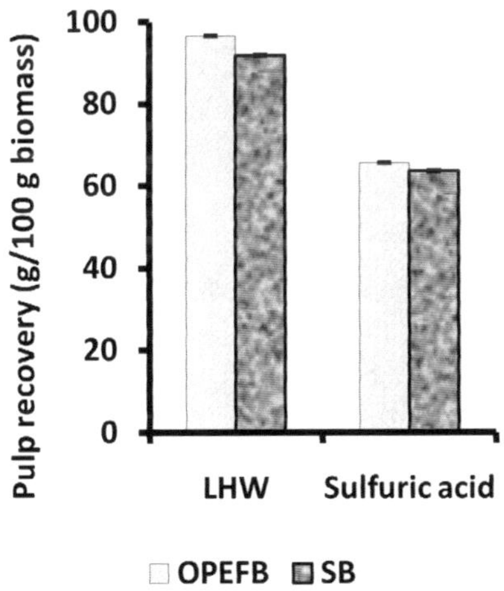

Figure 1. Pulp recovery from oil palm empty fruit bunch (OPEFB) and sugarcane bagasse (SB) after pretreatment using liquid hot water (LHW) and 3% sulfuric acid at 121 °C for 1 hour.

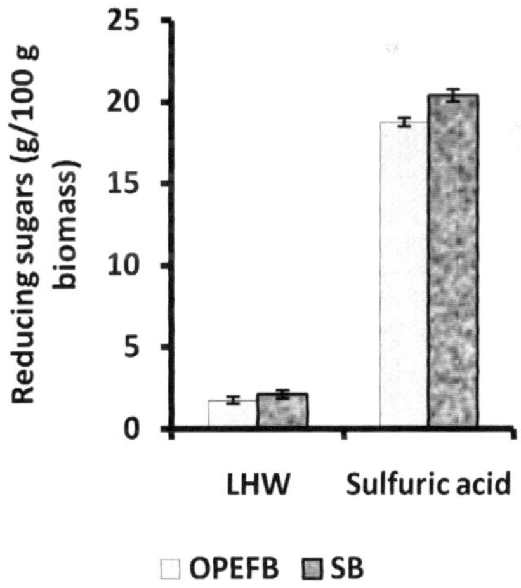

Figure 2. Reducing sugars in soluble fraction after pretreatment of oil palm empty fruit bunch (OPEFB) and sugarcane bagasse (SB) using liquid hot water (LHW) and 3% sulfuric acid at 121 °C for 1 hour.

3.3. Reducing sugars from soluble fractions

Hot water and sulfuric acid could hydrolyze polysaccharides in the pretreated biomass as shown in figure 2. However, the acid could better degrade the polysaccharides, which was shown by the much higher reducing sugars in the soluble fractions of sulfuric acid pretreatment than those in the soluble fractions of LHW pretreatment. The catalysis of water in the LHW pretreatment was due to the hydronium ions (H_3O^+) generated in situ by auto-ionization of water, and acetic acid generated by hydrolysis of hemicellulose[20,21]. This was not as strong as the catalysis of sulfuric acid that could produce hydrogen ions (H^+) with low pKa. Figure 2 also shows that there were slightly higher reducing sugars from SB than those from OPEFB in the soluble fractions. This was probably due to the slightly higher hemicellulose in SB than that in OPEFB (table 1).

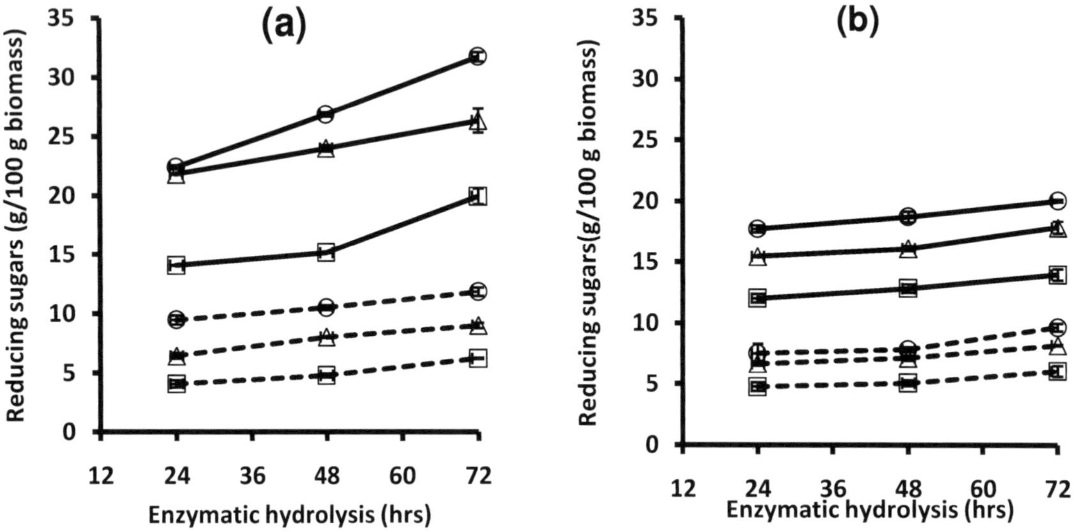

Figure 3. Reducing sugars release after enzymatic saccharification of (a) oil palm empty fruit bunch (OPEFB) and (b) sugarcane bagasse (SB) pretreated by liquid hot water (LHW) (dashed lines) and 3% sulfuric acid (solid lines) at 121 °C for 1 hour with enzyme loading of 10 (□), 20 (Δ), and 30 (○) FPU/ g substrate.

3.4. Reducing sugars from enzymatic saccharification

Data in figure 3a and 3b show that reducing sugars obtained from enzymatic saccharification were increased with increase of enzyme loading and with longer hydrolysis. The data also show that LHW pretreatment produced lower reducing sugars than did sulfuric acid pretreatment. The highest reducing sugars obtained from the pretreatment OPEFB and SB using LHW were 11.86 and 9.67 g/100 g biomass, respectively, while those from sulfuric acid pretreatment were 31.74 and 20.0 g/100 g biomass, respectively. Much higher differences of reducing sugars obtained from LHW and sulfuric acid pretreatment were observed when the total reducing sugars from the soluble fractions were added to those obtained from enzymatic saccharification. As much as 50.48 and 40.41 g/100 g biomass of reducing sugars could be obtained from sulfuric acid pretreated OPEFB and SB, respectively, while there were only 13.59 and 11.73 g/100 g biomass of reducing sugars obtained from LHW pretreated OPEFB and SB, respectively (figure 4a and 4b).Of the two biomass feedstocks used in this study, SB produce higher reducing sugars in the soluble (liquid) fraction (figure 2). However, OPEFB could produce much higher reducing sugars from enzymatic saccharification of the insoluble fraction (figure 3). Therefore, the total amount of reducing sugars from OPEFB was higher than those from SB (figure 4). Compared to our previous study on pretreatment of OPEFB using fungal pretreatment, the reducing sugars obtained in this study were much higher than those using fungal pretreatment 8.98-13.08% [18]. On the other hand, reducing sugars obtained from acid pretreatment of SB were lower than those obtained from lime pretreatment (17.51-35.69%) and sodium hydroxide pretreatment (29.31-45.69%) [19]. Nevertheless, it would be much easier to recover sugars from soluble fractions of acid pretreated biomass than that of alkaline pretreated biomass.

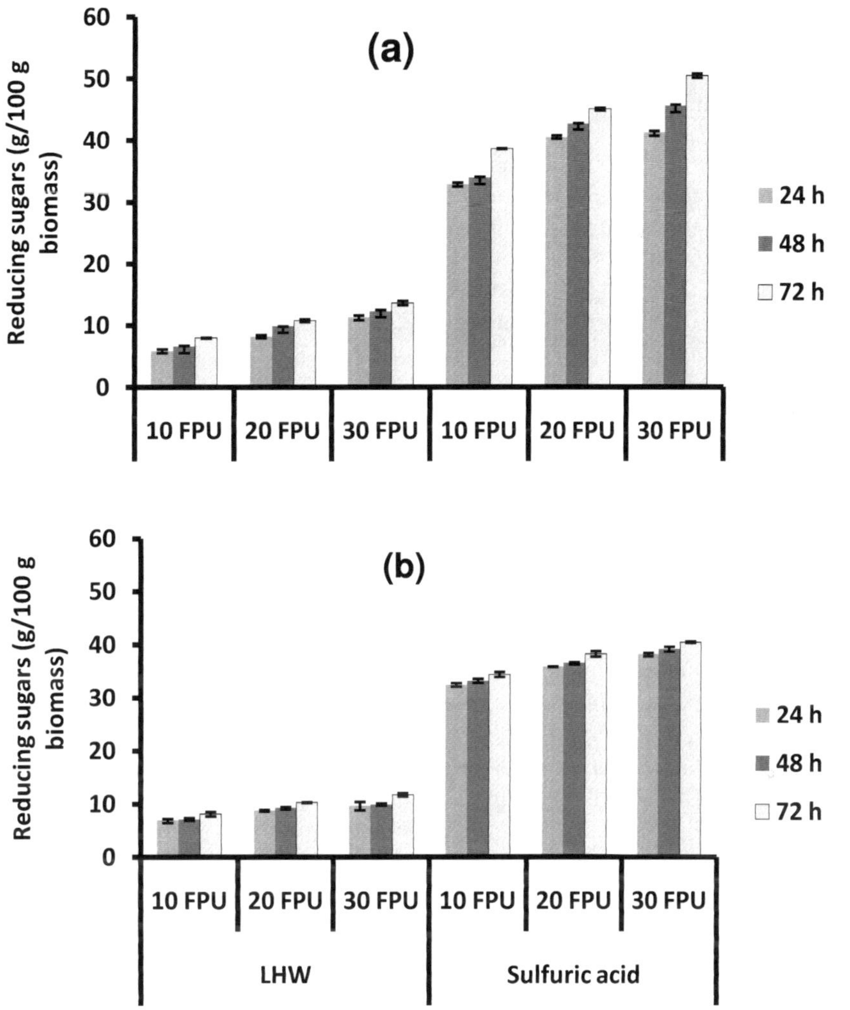

Figure 4. Total reducing sugars obtained from soluble fractions and from enzymatic saccharification of (a) oil palm empty fruit bunch (OPEFB) and (b) sugarcane bagasse (SB) pretreated by liquid hot water (LHW) and 3% sulfuric acid at 121 °C for 1 hour with enzyme loading of 10, 20, and 30 FPU/ g substrate.

4. Conclusions

Biomass pretreated with H_2SO_4produced higher reducing sugars than those pretreated by liquid hot water. Higher enzyme loading also resulted in higher reducing sugars. Reducing sugars obtained from enzymatic saccharification of OPEFB were higher than those obtained from SB. The highest total reducing sugars (50.48 g/100 g biomass) was obtained from OPEFB pretreated with 3% H_2SO_4 at enzyme loading of 30 FPU per g biomass.

Acknowledgments
This study was supported by DIPA Project of the Research Center for Biomaterials, Indonesian Institute of Sciences in the fiscal year 2014 and also part of the Japan Scince and Technology Agency (JST) – Japan International Collaboration Agency (JICA) for Science and Technology Research Partnership for Sustainable Development (SATREPS) Project Innovative Biorproduction Indonesia (Ibiol): Integrated Bio-refinery Strategy to Promote Biomass Utilization using Super Microbes for Fuels and Chemicals Production (2013-2018).

References

[1] SudiyaniY and HermiatiE 2010 *Indo. J. Chem.* **10**261-267.
[2] Hermiati E, Mangunwidjaja D, Sunarti TC, Suparno O and Prasetya B 2010*J.Litbang Pertanian***29** 121-130.
[3] Tan L, Yo Y, Li X, Zhao J, Qu Y, Choo YM and Loh SK2013*Bioresour. Technol.* **135** 275-282.
[4] Nomanbhay SM, Hussain R and Palanisamy K 2013 *J. Sust. Ener. Syst.***3** 7-17.
[5] Akhtar J, Teo CL, Lai LW, Hassan N, Idris A and Aziz RA 2015 *Bioresources* **10** 588-596.
[6] Isroi, Ishola MM, Millati R, Syamsiah S, Cahyanto MN, Niklasson C and Taherzadeh MZ 2012 *Molecules***17** 14995-15012.
[7] Sasaki C, Takada R, Watanabe T, Honda Y, Karita S, Nakamura Y and Watanabe T 2011 *Bioresour. Technol.***102** 9942-9946.
[8] Rezende CA, de Lima MA, Maziero P, de Azevodo ER, Garcia W and Polikarpov I 2011 *Biotechnol. Biofuels***4** 54.
[9] Jung YH and Kim KH *2015Pretreatment of Biomass: Processes and Technologies* (Amsterdam: Elsevier) p 27.
[10] Vallejos ME, Dibzambon M, Area MC, da Silva Curvelo AA 2012 *Green Chem.* **14** 1982-1989.
[11] TAPPI 1997*TAPPI T 264 cm-07 Preparation of Wood for Chemical Analysis.*
[12] TAPPI 2002*TAPPI Test Method T 211 om-02 Ash in Wood, Pulp, Paper, and Paperboard: Combustion at 525 °C.*
[13] TAPPI 1997*TAPPI Test Method T 204 cm-97 Solvent Extractives of Wood and Pulp.*
[14] Templeton D and Ehrman T 1995*Determination of Acid Insoluble Lignin in Biomass*, NREL LAP 003.
[15] Wise LE, MurphyM and Addieco1946 *Pap. Trade J.***122**35-43.
[16] Rowell RM 2005 *Handbook Wood Chemistry and Wood Composites*, 1st ed. (Florida: CRC Press) pp. 71-72.
[17] Adney B and Baker J 2008 *Measurement of Cellulase Activities* NRL Tech. Report NREL/TP-510-42628.
[18] Hermiati E, Risanto L, Anita SH, Aristiawan Y, Sudiyani Y, Hanafi A and Abimanyu H 2014 *J. Penelitian Hasil Hutan***32** 111-122.
[19] Fajriutami T, Fatriasari W and Hermiati E 2016 *J. Riset Industri***10** 147-161.
[20] Girio FM, Fonseca C, Carvalheiro F, Duarte C, Marques S and Bogel-Lukasik R 2010 *Bioresour. Technol.***101** 4775-4800.
[21] Mosier N, Wyman C, Dale B, Elander R, Lee YY, Holtzapple M and Ladisch M 2005 *Bioresour. Technol.***96** 673-686.

IOP Conf. Series: Earth and Environmental Science **141** (2018) 012026 doi:10.1088/1755-1315/141/1/012026

Formulation and performance test of palm-based foaming agent concentrate for fire extinguisher application

M Rivai[1*], E Hambali[1,2], A Suryani[1,2], R Fitria[1], S Firmansyah[1], G Pramuhadi[3]

1*Surfactant and Bioenergy Research Center, Bogor Agricultural University*, Kampus IPB Baranangsiang, Jl. Raya Pajajaran No. 1 Bogor Indonesia

2*Department Agroindustrial Technology, Bogor Agricultural University*, Kampus IPB Dramaga Bogor, Indonesia

3 *Department of Mechanical and Biosystem Engineering, Bogor Agricultural University*, Kampus IPB Dramaga Bogor, *Indonesia*

* mira.rivai2705@gmai.com

Abstract. The utilization of foaming agent for fire extinguisher application improves the efficiency of water as a fire extinguishing agent, lowers surface tension, and acts as a foaming agent. The formed foam cools the fire down and covers the burned material to avoid it from further contact with oxygen which may reignite the fire. This study aimed to produce and assess the performance of foaming agent concentrate from palm oil as a fire extinguisher agent. In the performance test, measurements were taken on foam stability, foaming ability, surface tension, interfacial tension, viscosity, contact angle, density, and specific gravity. The formulation was conducted by using the best produced potassium palmitate, potassium methyl ester, and sodium lauric combined with diluents, chelating agent, and other additives at various composition comparisons. The produced foaming agent concentrate was found to be in a rather paste and liquid form with viscosity of 2.34 – 253 cP. It was also found that the resulted foaming agent concentrate dissolved in water at the concentration rate of 1% had a foam stability level of 30-91%, foaming ability of 288 – 503%, surface tension of 19.68 – 25.05 dyne/cm, interfacial tension of 0.54 – 4,20 dyne/cm, viscosity of 1.00 – 1.05 cP, contact angles of 53.75 – 63.79^{o} at 0 minute and 11.84 – 22.42 ^{o} at minute 10, density of 0.99586 – 0.99612 g/cm^3, and a specific gravity of 1.00021 – 1.00046. Based on foam stability, foaming ability, and surface tension parameters, it was concluded that NF5 and NF17 were the best formulas.Compared to the other formulas, NF5 formula had the best droplet diameter (minimum 0.14 mm) and droplet density (maximum 3056 droplets/cm^2).

Keywords: palm oil, potassium palmitate, potassium methyl ester, sodium lauric, foaming agent concentrate, fire extinguisher

1. Introduction

Fire-fighting agent or fire-fightingfoam consists of surfactant as the main component which increases the efficiency of water utilization, reduces surface tension of a liquid, and improves the stability of foam formation. Imported surfactant which is relatively more expensive has long been used for fire extinguishment. Therefore, it is important to seek for an alternative surfactant from potential agricultural products such as oil palm in Indonesia in order to cut off the country's high dependency on imported surfactant. Palm oil-based surfactant is environmentally friendly as it is highly degradable. In addition, it has low toxicity level and sustainable availability as palm oil feedstock in Indonesia is plenty. It is expected that the utilization of palm oil as a foaming agent increase the value added of palm oil, diversification of palm oil processed products, and reduce dependency on imported surfactant.

Foam is defined as a dispersion system consisting of gas bubbles covered by liquid film. Significant differences in density levels of bubbles and liquid medium make the system separate quickly to form two layers with the bubbles being the top layer. When gas bubbles are formed under the liquid surface, the bubbles will immediately break up as a result of drainage created by a gravitational or downward force. Therefore, pure liquid will have no bubbles unless a surfactant is added in as surfactant reduces gas/liquid interfacial tension allowing the dispersion of gas in liquid to occur so that foam is formed [1, 2].

Fire is an event in which flame in uncontrollable. In the Regulation of Manpower and Transmigration Ministry Number 04/MEN/1980 Chapter I Article 2, Paragraph 2, there 4 categories of fires, namely A, B, C, and D. Meanwhile, National Fire Protection Association (NFPA) classifies fires into 5 categories including A, B, C, D, and K. Some countries have even determined an extra fire class (E). Each type of fires can be handled with different type of foam. In Indonesia, A fire class is the most common fire type found. This fire types occurs on solid materials wood, paper, fabric, plastic and other materials except steel. This fire can be handled effectively by using sand, soil/mud, extinguisher powder, foam, and water. The suitable foam to handle this type of fire is class A foam.

Class A foam consisted of a mixture of water, foam concentrate, and water. This foam has different characteristics according to the production method used, concentration of foam solution, type of concentrate used, length of hose, type of nozzle, and aeration. The concentration of class A foam used for class A fire is considerably low, namely 0.1 to 1% (U.S. Department of Homeland Security 1996). Class A foam concentrate has physicochemical properties which can be used as the standard of concentrate production. These include 35.57 cP viscosity at 25°C, 1.0240 specific gravity at 15.6°C, 26.51 dyne/cm (concentrate) and 30.69 dyne/cm (0.3% of concentrate) surface tension, and 1.025 g/ml density at 20°C [3]. The use of 0.3% concentration is selected based on results of several studies which showed that in class A foam solution, higher concentration did not result in any changes in surface tension.

Class A foam has several advantages. Compared to water, it extinguishes fire more rapidly providing more efficient use and conservation of water. Hose containing this foam is lighter than that containing water making fire fighters easier to maneuver in using it. Class A foam forms a blanket covering the surface of fuel so that it can lower the temperature of the burnt materials, avoiding contact between fuel and oxygen, cover wider surface, and lasts longer making it suitable for fighting fire in areas with limited water supply. This type of foam can reduce surface tension allowing water to go deeper into the center of the flame in the burnt materials.

Studies have been conducted to assess the utilization of plant oil derivatives as environmentally friendly firefighting agents. A firefighting agent was made from sodium oleic (10% w/w), potassium lauric (6.87% w/w), potassium palmitic (0.1% w/w), GLDA (40% w/w), and diluents (43.03% w/w) {4]. Ethyl estersas the result of palm oil trans esterification also were used as material for a firefighting agent [5]. Abiodegradable firefighting agent was made by using potassium lauric and potassium oleic at the ratio of 4:6, and potassium lauric and sodium oleic at the ratio of 4:6, and a *biodegradable chelating agent*(MGDA), with an expansion ratio of 9.6 times that of commercial firefighting foam for structural fire [6].

Based on the above considerations, this study was aimed at producing a foaming concentrate from palm oil and assessing the performance of it as firefighting agent. The foaming agent concentrate was developed for class A fire.

2. Materials and Methods

The main materials used in this study included potassium palmitic, potassium methyl ester, sodium ethyl ester, sodium lauric, sodium oleic, sodium methyl ester, diluents (propylene glycol, hexylene glycol), EDTA chelating agent, emulsifier, and water. Homogenizerwas used to make the formula mixed. Foaming stability and ability of solution were measured by diluting 1% sample in 99% distilled water. The dilution was stirred for 15 minutes by using a homogenizer before it was transferred into 100 ml capped graduated cylinders, shaken up 20 times in constant speed. The volume of foam was recorded on day 0, 1, 2, and 3. AnAnton Paar DMA 4500M densitometer was used to measure the density and specific gravity of foaming agent concentrate formula. A Brookfield DV-III Ultra viscometer was used to measure the viscosity of foaming agent concentrate and foaming agent concentrate solution. Contact angle of foaming agent concentrate solution was measured by using a Phoenix 300 contact angle analyzer. Interfacial tension of foaming agent concentrate solution was measured by using a TX500C spinning drop tensiometer. TASCO MD-150 mist blower was used to measure droplet diameter and density. Droplet size was measured by using Sharp Develop application. The pixel unit produced from this application was converted to micrometer unit. Fire extinguishment performance of foaming agent concentrate formula was assessed by using a portable foam educator. The foaming agent concentrate test was conducted at 1% concentration.

2.1. Formulation of Foaming AgentConcentrate

The resulted foaming agent concentrate formulation was modified from [4] and [6], by using combinations of six types of foaming agent products including potassium palmitic acid, sodium palm ethyl ester, sodium lauric acid, sodium oleic acid,sodium palm ethyl ester and potassium palm methyl ester produced by [7]. The concentration variations included potassium palmitic acid 2-3%, sodium palm ethyl ester 0-4%, sodium lauric acid 2.5-5%, sodium oleic acid 0-3%, sodium palm methyl ester 0-2,5%, and potassium palm methyl ester 0-2.5%, with other additional and additive materials consisting of diluents 2-6%, emulsifier 0-15%, chelating agent 1-2%, and water 71-80%. The mixing process was done at room temperature by using a homogenizer with a stirring speed of 2000 rpm for an hour. Analyzes were done on foam stability, foaming ability, surface tension, interfacial tension, viscosity, contact angle, density, and specific gravity.

2.2. Performance Test of Foaming Agent Concentrate for Fire Extinguishment

The performance of foaming agent concentrate for fire extinguishmentwas tested by measuring foaming effectiveness and using the foaming agent concentrate to extinguish fire. Foaming effectiveness was determined based on droplet diameter and droplet density measured by using a mist blower TASCO MD-150. A portable foam eductor was used in fire extinguishment test.

3. Results and Discussions

3.1. Formulation of Foaming Agent Concentrate

The process of foaming agent concentrate formulation and homogenation at 25 liter scale is depicted in Figure 1. This formulation stage resulted in many formulas of foaming agent concentrate with rather sticky and liquid consistency. Seven formulas including NFA, NF3, NF5, NF7, NF15, NF17, and NF18 were selected as they had relatively better foam stability as their foam could stay intact until day 3. In addition to these 7 formulas, parameter testing was also done to AFFF, a commercial fire extinguishment chemical available in the market which was used as control. In the test, foaming agent concentrate was dissolved in water at the concentration of 1% and results of this test are listed in Table 1.

The value range of 30-91% was obtained in the measurement of foam stability of the seven formulas on day 3. The formulas of foaming agent concentrate had significantly higher foam stability than commercial AFFF formula did (0%). Foam stability of commercial AFFF dropped drastically to 12.8% on day 1, 5.3% on day 2, and 0% on day 3. NF5 formula was found to have the highest foam stability (91%) followed by NF17 (88%).

Figure 1. Formulation and homogenation process of foaming agent concentrate at 25 liter scale

Table 1. Results of parameter analysis of foaming agent concentrate dissolved in 1% water and commercial product

Parameter	Formula of Foaming Agent Concentrate							AFFF
	NFA	NF3	NF5	NF7	NF15	NF17	NF18	
Foam stability (%, day 3)	58	83	91	61	77	88	30	0
Foaming capability(%)	479	314	288	318	503	372	310	204
Surface tension (dyne/cm)	22.13	19.89	21.19	20.53	19.68	25.05	19.74	19.81
Interfacial tension (dyne/cm)	4.2	0.54	0.86	2.23	1.12	1.67	1.61	0.55
Viscosity (cP):								
- 1%	1.04	1.03	1.00	1.05	1.02	1.02	1.04	0.99
- 100%	2.64	3.64	2.34	224.33	253.27	7.15	52.79	2.71
Contact angle (°):								
- Minute 0	63.79	58.16	63.49	61.46	55.09	54.77	53.75	60.41
- Minute 10	22.42	19.52	11.84	17.47	15.27	19.70	16.99	25.51
Density (g/cm3)	0.99586	0.99600	0.99608	0.99597	0.99612	0.99601	0.99601	0.99577
Specific gravity	1.00021	1.00034	1.00043	1.00032	1.00046	1.00036	1.00036	1.00007

Foaming ability of the seven formulas of foaming agent concentrate was 288-503%. These figures were higher than that of commercial AFFF (204%). Surface tension of formulas of foaming agent concentrate was 19.68–25.05 dyne/cm. These were relatively similar to that of commercial AFFF (19.81 dyne/cm). The lowest surface tension (19.68 dyne/cm) was found in NF15 formula and the highest (25.05 dyne/cm) was found in NF17 formula. This highest figure was still within the requirements of class A foam [3]. At 1% concentration, the interfacial tension of AFFF foam and AFFF/PEO foam was about 52-63 dyne/cm, and increment of the concentration of foaming agent concentrate gave significant effects on the reduction of interfacial tension [8].

Interfacial tension of formulas of foaming agent concentrate was 0.54–4.20 dyne/cm. These figures were almost similar and close to that of AFFF product (0.55 dyne/cm). The seven formulas of foaming agent concentrate were rather thick and liquid with viscosity levels of 2.34-2.53 cP. Meanwhile, AFFF product was liquid with a viscosity level of 2.71 cP. Only 4 formulas including NFA, NF3, NF5, and NF17 were liquid similar to AFFF product. At a solution concentration of 1%, the viscosity levels of formulas of foaming agent concentrate were about 1.00-1.05 cP and that of AFFF product was 0.99 cP. NF5 formula was found to have a viscosity level closest to that of AFFF product.

Contact angles of formulas of foaming agent concentrate were 53.75-63.78° at minute 0 and 11.84 – 22.42° at minute 10. For AFFF product, the contact angles were 60.41° at minute 0 and 25.51° at minute 10. Higher reduction of contact angle of foaming agent concentration solution at minute 10 indicated that the concentrate solution had stronger ability to penetrate into the media than AFFF product. Density levels (0.99586 – 0.99612 g/cm^3) of foaming agent concentrate formulas were slightly higher than that of AFFF product (0.99577 g/cm^3). However, these figures were considerably lower than that of class A foam concentrate (1.025 g/ml) [3].

Specific gravity of foaming agent concentrate ranged from 1.00021 to 1.00046. These figures were slightly higher than that of AFFF commercial product (1.00007). However, these figures were considerably lower than that of class A foam concentrate (1.0240) [3].

3.2. Performance Test of Foaming Agent Concentrate for Fire Extinguishment

Determination of parameters of foaming effectiveness of foaming agents is important to do before the agents are applied in fire extinguishment. In principle, fire extinguishment is covering or filling a surface and porous space with a foaming agent liquid to avoid contact between oxygen and fire or coals of fire so that the fire or coals of fire can be extinguished. Fire extinguishment by using a foaming agent will be more effective with finer droplet size and higher droplet density. Results of foaming effectiveness analysis showed that at constantly maintained pressure of 7.0 kgf/cm^2 and effective spraying height of 20-43 cm, NF5 foaming agent formula produced the smallest droplet diameter of 0.14 mm and the highest droplet density of 3056 droplets/cm^2. The graphs of droplet diameter and droplet density of various concentration of foaming agent solution are depicted in Figures 2 and 3, respectively.

The application test of foaming agent as fire extinguisher was conducted by using a portable foam eductor. The eductor was connected by a hose to water source and a foaming agent concentrate collecting tank. The flow rate of foaming agent concentrate was adjusted to about 1 L/7 seconds and water flow rate to about 5 L/second. Spraying was then done as shown in Figure 4. Results of fire extinguishment test showed that fire was put out rapidly after foam solution was sprayed into the source of fire.

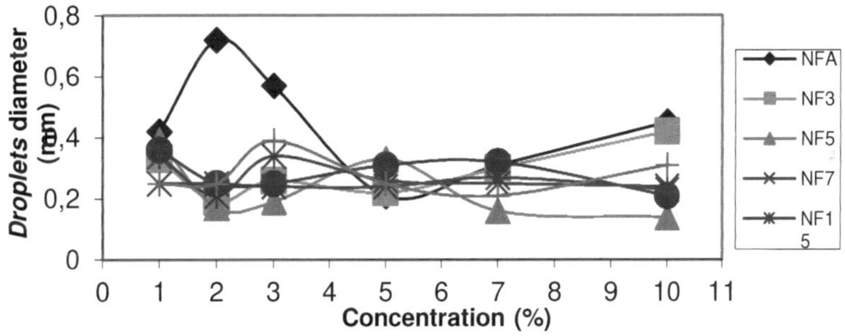

Figure 2. Graph of droplet diameter in various concentrations of foaming agent solution

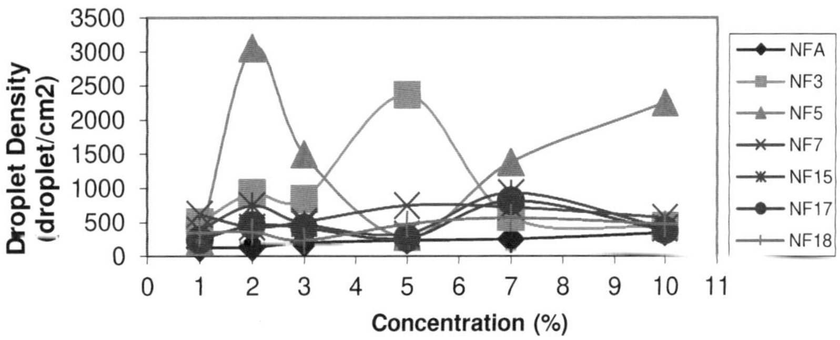

Figure 3. Graph of droplet densityin various concentrations of foaming agent solution

Figure 4. Fire extinguishment trial by using resulted formulas of foaming agent concentrates

4. Conclusions

The foaming agent concentrate products were produced in a rather paste and liquid forms with viscosity levels of about 2.34–253 cP. Results of formula analysis of foaming agent concentrates dissolved in 1% water had a range of foam stability levels of 30-91%, foaming ability of 288-503%, surface tension of 19.68-25.05 dyne/cm, interfacial tension of 0.54–4.20 dyne/cm, viscosity level of 1.00-1.05 cP, contact angle of 53.75-63.79° at minute 0 and 11.84-22.42° at minute 10, density of 0.99586–0.99612 g/cm^3, and specific gravity of 1.00021-1.00046. Based on foam stability, foaming ability, and surface tension parameters, the best foaming agent concentrate products were NF5 and NF17 formulas which also had better performance than commercial product did. NF5 formula had the smallest droplet diameter of 0.14 mm and the highest droplet density of 3056 droplets/cm^2.

Acknowledgements

Research funding assistance from The Indonesian Oil Palm Estate Fund (BPDPKS) and laboratory facility assistance from Surfactant and Bioenergy Research Center IPB were acknowledged.

5. References

[1] Tadros, TF. 2005. Applied Surfactant : Principles and Applications. 259-263. Wiley-VCH Verlag GmbH & Co. KgaA, Weinheim.
[2] Exerowa D dan Kruglyakov PM. 1998. Foam and Foam Films: Theory, Experiment, Application, 1-3, 494, Elsevier, Netherlands.
[3] Carey, W M. 1994. Structural Fire Fighting – Room Burn Test Phase II. Nasional Class A Foam Research Project Technical Report. Quinci, Massachusetts, USA. : National Fire Protection Research Foundation.

IOP Conf. Series: Earth and Environmental Science **141** (2018) 012026 doi:10.1088/1755-1315/141/1/012026

[4] Mizuki H, K Ueza, T Kawano, T Kadono, M Kobayashi, S Hatae, Y Oba, S Iwamoto, S Mitumune, Y Nagatomo, MOwari, H Umeki and KYamaga. 2007. Novel environmental friendly soap-based fire-fighting agent. *J.Environ. Eng. Manage.* 17(6) : 403-408.
[5] Oguike RS. 2013. Study of Fire Fighting Foam Agent from Palm Oil for Extinguishing of Petrol Fires. Science Postprint 1(1): e00007.doc10.14340/spp.2013,12A0002.
[6] Kawahara T, S Hatae, T Kanyama, Y Ishizaki, and K Uezu. 2016. Development of eco-friendly soap-based firefighting foam for forest fire. Environ. Control Biol., 54 (1) : 75-78.
[7] Rivai M, E Hambali, A Suryani, R Fitria, S Firmansyah, and J Pradesi. 2017. Synthesis of palm oil fatty acid as foaming agent for firefighting application. IOP Conf. Series: Earth and Environmental Science 65 (2017) 012047, doi :10.1088/1755-1315/65/1/012047.
[8] Figueredo RCR and E Sabadini. 2003. Firefighting foam stability: the effect of the drag reducer poly(ethylene) oxide. Colloids and Surfaces A: Physicochem. Eng. Aspects 215: 77-86.

IOP Conf. Series: Earth and Environmental Science **141** (2018) 012027 doi:10.1088/1755-1315/141/1/012027

Cellulose nanofiber isolation from palm oil Empty Fruit Bunches (EFB) through strong acid hydrolysis

Dwi Setyaningsih[1] , Uju[1], Neli Muna[1],Isroi[2], Nyoman Budi Suryawan[3], Ami Azid Nurfauzi[3]

[1]Surfactan and Bioenergy Research Center, Bogor Agricultural University , Pajajaran Road No . 1, Bogor- Indonesia , 16144
dwisetya.sbrc@gmail.com
[2]Indonesian Research Institute for Biotechnology and Bioindustry, No.1, Taman Kencana Road , Bogor- Indonesia 16151
[3]Departement of Agroindustrial Technology, Bogor Agricultural University

Abstract. The palm oil industry produces about 25-26% of palm oil empty fruit bunches. The empty fruit bunch of palm oil contains cellulose up to 36.67%. This is a good opportunity for the synthesis of cellulose nanofiber (CNF). Cellulose nanofiber is a nano-sized cellulose material that has unique physical and mechanical properties. The synthesis was performed using a strong acid method with sulfuric acid. Sulfuric acid removes the amorphous region of cellulose so that the crystalline part can be isolated. CNF yield measurement showed that temperature, time, acid concentration, and interaction between each factor were affecting significantly to CNF yield. The result showed that yield of 14.98 grams, was obtained by hydrolysis at 35°C for 6 hours and 55% acid concentration. The crystallinity measurement showed that the temperature, time, acid concentration, and interaction between each factor during hydrolysis were not affected significantly to percent value of CNF crystallinity. The result showed that 31.1% of crystallinity, was obtained by hydrolysis at 45°C for 3 hours and 55% of acid concentration. The size measurement showed that the temperature, time, acid concentration and interaction between each factor were affected significantly. The result showed 894.25 nm as the best result, obtained by hydrolysis with 35°C and 60% acid concentration for 6 hours. CNF color was white with the best dispersion of hydrolysis at 35°C of 55% for 6 hours.

1. Introduction

Isolation of cellulose nanocrystals from palm oil empty fruit bunches becomes an excellent research and industrial opportunity because the palm oil processing industry produces empty fruit bunches of palm oil tree around 25-26% [1]. Cellulose is a carbon-rich material (Cellulose 36.67%, hemicellulose 13.5%, and lignin 31.16%) [2]. The high content of cellulose in palm oil empty fruit bunches that reaches 36.67% making palm oil empty fruit bunches as a potential source of cellulose nanofiber (CNF). The nano size of the CNF provides excellent properties such as high crystalline content, good optical properties, and excellent mechanical properties [3]. CNF can be utilized in various products such as strengthening the tensile strength of plastics, as the filler of nanocomposites to fix the mechanical and barrier properties, aerogel, lithium battery production, as a mechanical strengthening agent for lowthickness polymer electrolytes, improving the physical properties of mechanical adhesive materials, and many other applications [4]. If you don't wish to use the Word template provided, please use the following page setup measurements.

Cellulose is composed of amorphous and crystalline parts. The cellulose nanocrystals can be isolated by hydrolyzing the amorphous portion. Hydrolysis is carried out on pure cellulose with strong acids under controlled temperature and time. However, the optimum conditions to isolate the crystalline part of cellulose has not been defined. Therefore, this study was conducted to determine the effect of acid concentration, temperature, and hydrolysis time on cellulose nanocrystal (cellulose

IOP Conf. Series: Earth and Environmental Science **141** (2018) 012027 doi:10.1088/1755-1315/141/1/012027

nanofiber) isolation, and to characterize the CNF produced. This study was expected to provide information on the conversion of cellulose of palm oil empty fruit bunches into cellulose nanofiber (CNF) using acid hydrolysis.

2. Materials and Methods

The materials used in this study were α-cellulose from empty fruit bunches of palm oil, aquadest, and H_2SO_4 (55%, 60%, 64%). The equipment used in this research was hotplate, magnetic stirrer, glass beaker, measuring cup, erlenmeyer, centrifuge, waterbath, sonicator, oven, refrigerator, tea strainer, vacuum filter, and blender.

Research Procedures

2.1 Preparation of α-cellulose from Palm Oil Empty Fruit Bunches

The dry α-cellulose material was immersed in aquades at a ratio of 1: 5 for 24 hours until the cellulose expanded and the texture was smoother. After that, the α-cellulose was crushed so that the size was smaller and smoother and then filtered to remove excess water.

2.2 Isolation of Cellulose Nanofiber from α-cellulose Pulp

Acid hydrolysis was performed by using acid-biomass ratio of 1:10, by adding 30 ml acid into 3 g α-cellulose. A total of 12.77 grams of pulp with a moisture content of 76.5% (3 grams of dry α-cellulose) were dissolved in sulfuric acid with three different concentrations of 55%, 60%, and 64%. Acid hydrolysis was performed with two different time treatments, i.e. 3 hours and 6 hours, and different temperatures used were 35°C and 45°C. Hydrolysis was performed on a hotplate with a stirring speed of 85 rpm. After the hydrolysis process was completed, aquadest was added doubled of the amount of acid added. The hydrolysis suspension was washed by centrifugation at 10,000 rpm for 40 minutes, and the washing was performed repeatedly until reaching neutral pH or acid-free. The suspension was then sonicated by an ultra-sonication process for 20 minutes to obtain well dispersed cellulose nanocrystalline and increase the pulp weight after hydrolysis process of CNF.

2.3 Characterization of CNF

The isolated cellulose nanofiber (CNF) was characterized by the crystallinity index (User Manual Standard Operational Procedure Measurement Center XRD D8 Advance Brucker), particle size (User Manual Standard Operational Procedure MALVERN Zetasizer), morphology (Standard Operational Procedure Scanning Electron Microscopy Hitachi S-4800), and FTIR (User Manual Standard Operational Procedure Scanning Electron Microscopy Hitachi S-4800).

3. Results and Discussions

3.1 Pulp Weight After Hydrolysis

Hydrolysis at 35°C produced higher pulp weight after hydrolysis compared to hydrolysis occurred at 45°C. The highest weight of pulp after hydrolysis was obtained at the treatment or sample A1B2C1 which was hydrolysis condition at 35°C for 6 hours with 55% acid and A1B1C1 which was hydrolysis condition at 35°C for 3 hours with 55% acid. From the result of variance test (ANOVA α = 0.05) the interaction between temperature factor (A) and time (B) was not significantly different, so that the selected process was treatment A1B1C1 with shorter processing time. Temperature, time, and acid concentration during acid hydrolysis process certainly affected the obtained results. Temperature can affect the energy produced, so the higher the temperature, the more energy that can be used to achieve the activation energy [5]. The temperature of 35°C produced higher pulp weight after hydrolysis, meaning that CNF can be produced at lower temperatures. Higher temperatures will lead to more reactive reactions, so that when the process of destruction of the amorphous portion of cellulose, the crystalline part will also be damaged. Damage of the crystalline portion of cellulose will facilitate the breakdown of cellulose and convert it into glucose and other compounds [6]. Similar to the time factor, the longer hydrolysis time was used, the more hydrolysis of the material will occur. However, if the time taken was too long, the hydrolyzed material will be converted into other products such as glucose.

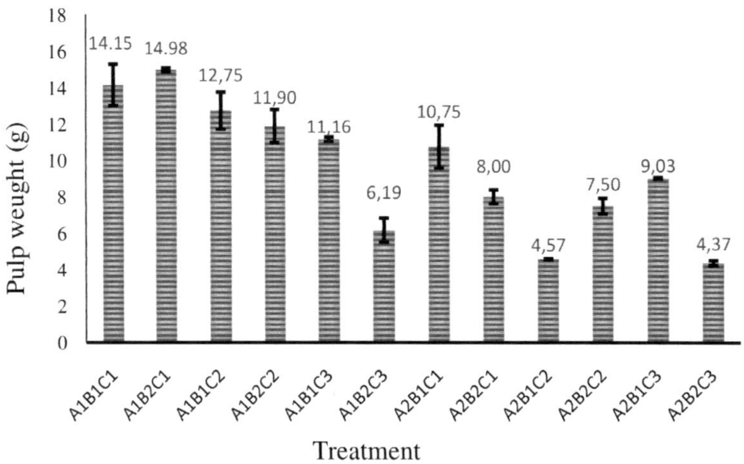

Note: A: Temperature (A1 = 35°C and A2 = 45°C)
B: Time (B1 = 3 hours and B2 = 6 hours)
C: Acid concentration (C1 = 55%, C2 = 60%, C3 = 64%)

Figure 1. Pulp weight after hydrolysis of CNF at various treatments.

3.2 Crystallinity

The X-ray diffraction pattern (XRD) can be used as a method of identifying and characterizing a substance. In this way, the pattern of crystallinity and amorphous levels contained in cellulose can be determined [7]. The crystallinity index was calculated for SNF obtained from various hydrolysis experiments conducted, so that it the crystallinity index for each treatment can be compared. The crystallinity index of the starting material used should be known first, given that the hydrolysis treatment of the cellulose was to increase the value of the crystallinity index. The α-cellulose used has a crystallinity index value of 25%. Figure 2 showed no significant difference in the crystallinity index values of the various treatments

Figure 3 showed that the treatment of A1B1C1 or hydrolysis at temperature 35°C for 3 hours using 55% acid concentration achieved the highest crystalline percentage of 31.1%. Very high acid concentration for a long time can make the crystalline part of cellulose damaged and cellulose will be converted into glucose and other compounds. The crystallinity of the obtained product was still lower when compared to the results of other studies which reported the crystallinity index for CNF was 54-88% [8]. Crystallinity index of cellulose nanocrystalline was 72.5%. The research used bagasse as a material for cellulose nanocrystalline production, and achieved the crystallinity index of bagasse 35.5%, whereas the crystallinity of bagasse waste separated from lignin and its hemiselulose was 63.5% [9]. Another research uses cotton fiber with crystallinity index of 66.9% with cellulose nanocrystal product of crystallinity index 88% [7]. The previous studies indicated that the removal of lignin and hemicellulose will increase the crystallinity index of the cellulose. The crystallinity index obtained in this study was still very low. This may be because the α-cellulose used still contains lignin and hemicellulose. The presence of high lignin and hemicellulose content in hydrolyzed cellulose will inhibit the hydrolysis process.

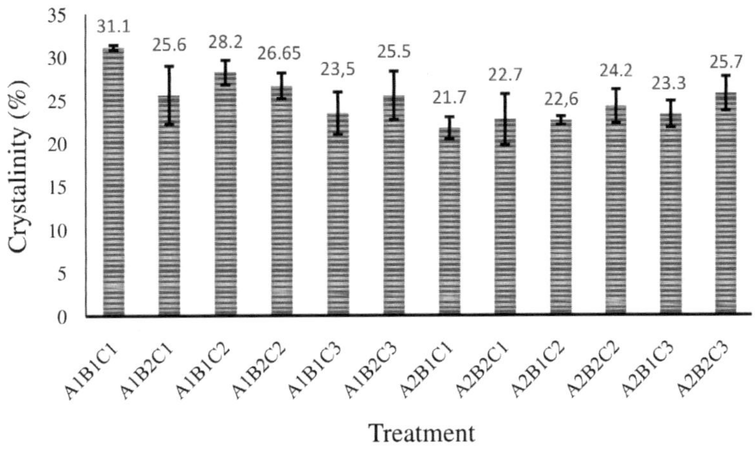

Note: A: Temperature (A1 = 35°C and A2 = 45°C)
B: Time (B1 = 3 hours and B2 = 6 hours)
C: Acid concentration (C1 = 55%, C2 = 60%, C3 = 64%)

Figure 2. Crystallinity of CNF at various treatments.

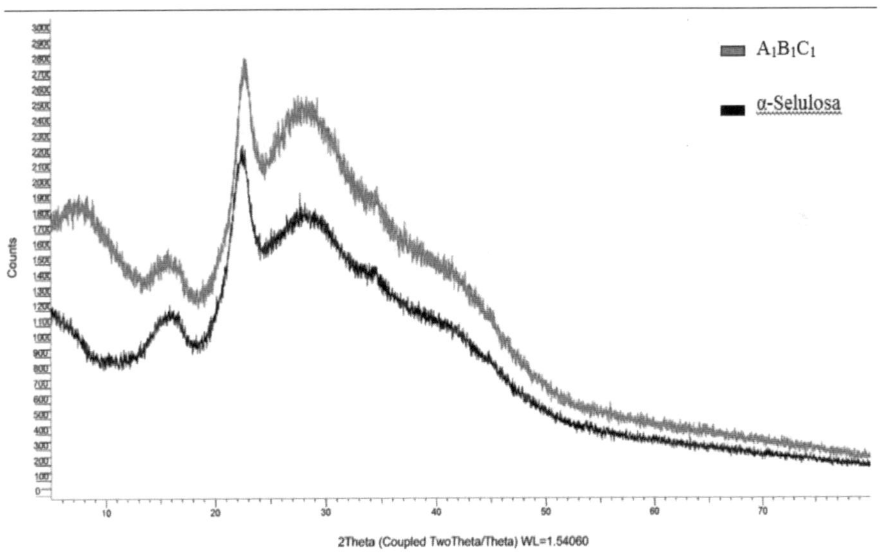

Figure 3. XRD pattern of α-cellulose and A1B1C1 treatment (hydrolysis at 35°C for 3 hours with 55% sulfuric acid).

3.3 Particle Size

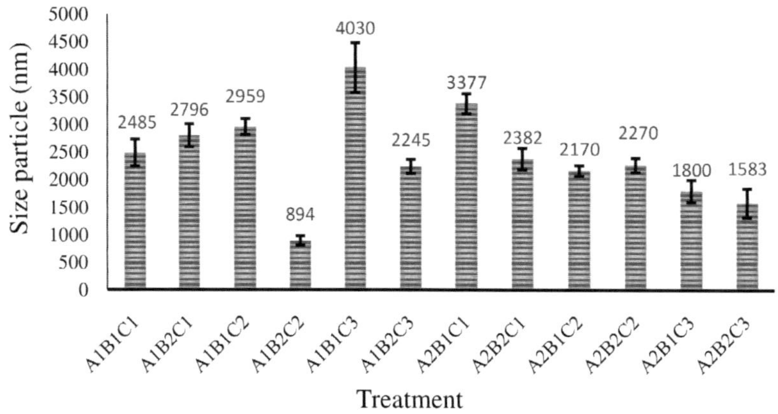

Note:
A: Temperature (A1 = 35°C and A2 = 45°C)
B: Time (B1 = 3 hours and B2 = 6 hours)
C: Acid concentration (C1 = 55%, C2 = 60%, C3 = 64%)

Figure 4. Mean size of CNF at various treatments.

Figure 4 showed that at temperature treatment of 45°C, the higher the acid concentration used and the longer hydrolysis time will lead to smaller particle size produced. The hydrolysis time had a significant effect on the size reduction of cellulose, in which with longer hydrolysis time, the cutting of the cellulose structure by the acid will be occurred more or the cellulose will be more hydrolysed. Acid concentration had no significant effect on the reduction of size of cellulose, but rather had a significant effect on the crystallinity index [10]. The best result obtained based on the average size of the particles was in the treatment of A1B2C2, i.e. it was optimum at 35°C, using 60% acid concentration for 6 hours.

3.4 Functional Groups with FTIR (Fourier Transform Infra-Red)

The α-cellulose and CNF (A1B2C1) showed similar FTIR spectra results (Figure 5). This indicated that the chemical composition of both materials was similar. In general, the sample generated two transmittance percentage groups, i.e. low wavelengths in the range 1000-1633 cm^{-1} and high wavelengths in the range 2000-3302 cm^{-1}. Absorption peaks at 3302 cm^{-1} and 1633 cm^{-1} showed the vibration of the O-H group [11]. The presence of an O-H group was influenced by the presence of water in the tested sample [3]. The absorption peak of 2915 cm^{-1} indicated the presence of stretching vibration of the C-H group of cellulose, hemicellulose, and lignin [12]. The presence of peak at 1429 cm^{-1} was due to the presence of various lignin components contained in the cellulose [12]. Peak at 1158-1034 cm^{-1} indicated that the CNF obtained was formed from cellulose Iβ [13]. Cellulose Iβ is a cellulose easily soluble in NaOH that has a concentration of 17.5% at 20°C and settled in acid solution [14].

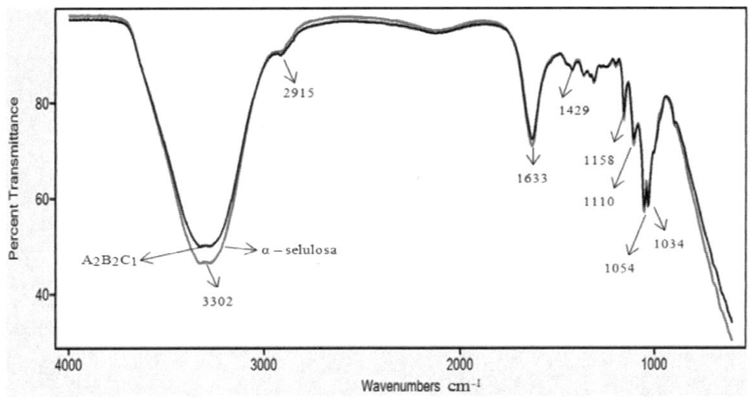

Figure 5. FTIR spectra of α-cellulose and CNF A1B2C1 (hydrolysis at 35°C for 6 hours with 55% sulfuric acid).

3.5 Physical Characteristics of CNF

The best dispersion was observed in the treatment A1B2C1 (hydrolysis with temperature 35°C for 6 hours with 55% acid concentration) both after cooled for 24 hours and 48 hours, in which the resulting dispersion was stable and not settled. The best dispersion treatment was observed with SEM and compared with α-cellulose. From Figure 7, the structure of cellulose nanocrystal was not observed. The diameter of α-cellulose was 6721 nm and the size of treatment diameter A1B2C1 was 2708 nm. The shape of the stem of CNF had not been seen because the crystallinity index obtained was still low.

(a)

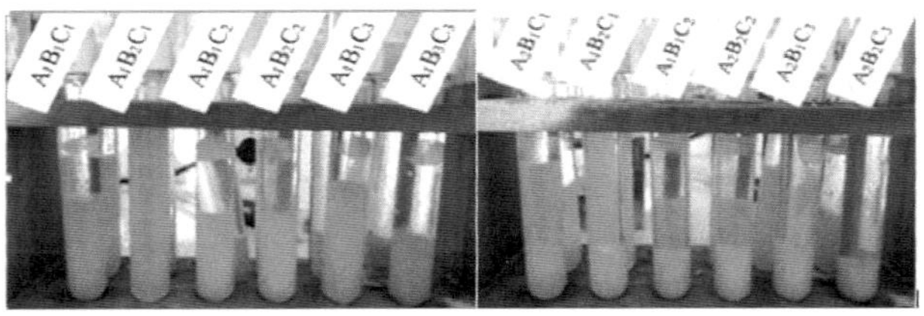

(b)

Note: A: Temperature (A1 = 35°C and A2 = 45°C)
B: Time (B1 = 3 hours and B2 = 6 hours)
C: Acid concentration (C1 = 55%, C2 = 60%, C3 = 64%)

Figure 6. Visualization of sample color and dispersion of various treatments (a) 24 hours (b) 48 hours.

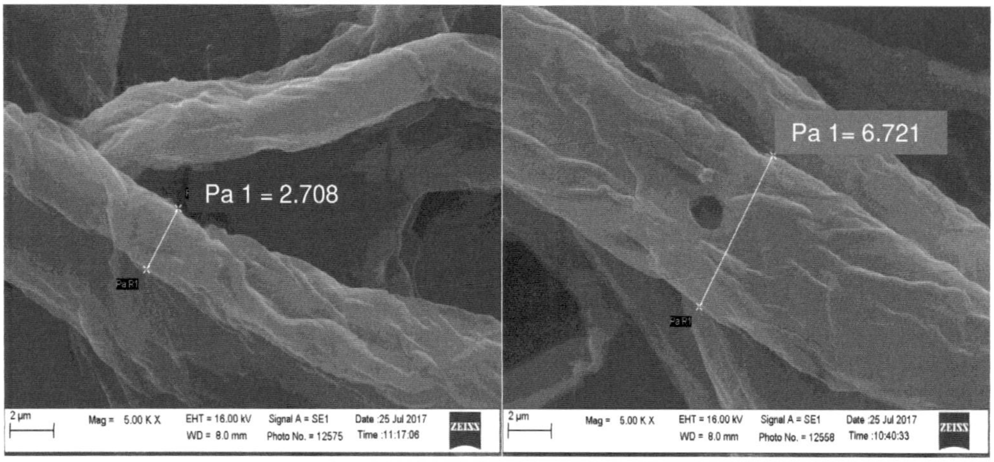

Figure 7. Measurement of SEM (A) treatment A1B2C1 and (B) α-cellulose.

4. Conclusions

The process of synthesis of cellulose nanocrystal by acid hydrolysis method was influenced by several factors, i.e. temperature, time, and acid concentration during hydrolysis process. The results of pulp weight after hydrolysis of CNF showed that temperature, time, acid concentration, interaction between temperature and acid concentration, and interaction between time and acid concentration were significantly affected pulp weight after the hydrolysis of CNF. The highest pulp weight after the optimum hydrolysis was 14.98 g, obtained by hydrolysis at 35°C for 6 hours using 55% acid concentration. The crystallinity test showed that the temperature, time, acid concentration, and interaction between factors during hydrolysis were not significantly affected the crystallinity percentage of CNF. Crystallinity percentage was 31.1%, obtained by hydrolysis at 45°C for 3 hours with acid concentration of 55%. The results of mean size showed that the temperature, time, acid concentration, and interaction between factors were significantly different with the best particle size of 894.25 nm, obtained by hydrolysis at 35°C with 60% acid concentration for 6 hours. The resulting CNF color was white with the best dispersion generated by hydrolysis at 35°C and acid concentration of 55% for 6 hours.

References

[1] Herawan T and Rivani M. Pemanfaatan limbah padat kelapa sawit untuk produksi geen product. Prosiding Pertemuan Teknis Kelapa Sawit 2013. JCC Jakarta. 2013 7-9 Me; ISBN 978-602-7539-16-7: 181-190.

[2] Isroi; Cifriadi A;Panji T.Bioplastic production from oil palm empty fruit bunch. International Conference on Biomass 2016: 10-11th October 2016, Bogor, Indonesia.

[3] Abraham E; Deepa B; Pothan L A; Jhon M, Narine S S; Thomas S; Anandjiwala R. Physicomechanical properties of nanocomposites based on cellulose nanofibre and natural rubber latex. Cellulose. 2012; 20: 417-427.

[4] Brinchi L; Cotana F; Fortunati E; Kenny J M. Production of nanocrystal cellulose from lignocellulosic biomass: Technology and applications. Carbohydrate Polymer. 2013; 94: 154-169.

[5] Evon P; Vandenbossche; Pontalier P Y; Rigal L. Direct extraction of oil from sunflowering seeds by twin-screw extruder according to an aqueous extraction process: feasibility study and influence of operating conditions. Industrial Crops and Products. 2007; 26: 351-359.
[6] Ozvaldo z; Putra P; and Faizal M. Pengaruh konsentrasi asam dan waktu pada proses hidrolisis dan fermentasi pembuatan bioetanol dari alang-alang. Teknik Kimia. 2012; 2(18): 52-62.
[7] Ozvaldo z; Putra P; and Faizal M. Pengaruh konsentrasi asam dan waktu pada proses hidrolisis dan fermentasi pembuatan bioetanol dari alang-alang. Teknik Kimia. 2012; 2(18): 52-62.
[8] Moon RJ; Martini A; Nairn J; Simonsen J;Younblood J. Cellulose nanomaterials review: structure properties and nanocomposites. Chemical Society Reviews. 2011; 40: 3941-3994.
[9] Kumar A, Barrett D M, Michael J, Delwiche, Stroeve P. 2014. Characterization of cellulose nanocrystals produced by acid-hydrolysis from sugarcane bagasse.
[10] Wulandari W T; Rochliadi A; Arcana I M. Nanocellulose prepared by acid hydrolysis of isolated cellulose from sugarcane bagase. IOP. 2016; 107: 1-17.
[11] Moran JI; Alvarez V A; Cyras V P; Vzquez A. Extraction of cellulose and preparation of nanocellulose fromsisal fibers. Cellulose. 2008; 15:149-159.
[12] Cherian BM; Pothan L A; Chung T N; Mennig G; Kottaisamy M; Thomas S. A novel method for the synthesis of cellulose nanofibril whiskers frombanana fibers and characterization. J Agi Food Chem. 2008; 56: 5617-5627.
[13] Leung AC; Hrapovic S; Lam E; Liu Y; Male K B; Khaled A; Luong J H T. Characteristics and properties of carboxylated cellulose nanocrystals prepared from a novel one-step procedure. Small. 2011; 7(3): 302-305.
[14] Nuringtyas TR. 2010. Karbohidrat. Yogyakarta (ID): Gajah Mada University.

IOP Conf. Series: Earth and Environmental Science **141** (2018) 012028 doi:10.1088/1755-1315/141/1/012028

Antibacterial activities of leave extracts as bactericides for soaking of skin or hide

Ono Suparno*, Tania Panandita, Amalia Afifah, Marimin, Rini Purnawati

Department of Agroindustrial Technology, Faculty of Agricultural Engineering and Technology, Bogor Agricultural University (IPB), Darmaga Campus, Bogor 16680, Indonesia

*ono.suparno@apps.ipb.ac.id

Abstract. Antibacteria, a subtance inhibiting the growth of bacteria, can be obtained from tropical-almond (*Terminalia catappa*), morinda (*Morinda citrifolia*), and white leadtree (*Leucaena leucocephala*) plants, since the plants have phytochemical content functioning as antibacterial agent. Commonly, part of plant that contains higher antibacterial substances is its leaf. The objectives of this study were to determine antibacterial activity of tropical-almond, morinda, and white leadtree leaves extracts, and to analyse the potency of the three extracts as natural bactericide for soaking of skin or hide. The responses measured in this study were phytochemical contents, total flavonoid, tannin content, the inhibition zone, the minimum inhibitory concentration (MIC), and the minimum bactericidal concentration (MBC). Phytochemical contents containing the three leaves extracts were alkaloid, flavonoid, tannin, saponin, phenolic, and glycoside. Total flavonoid and tannin contents of the three extracts were tropical-almond extract of 1.14 % and 1.51 %, respectively; morinda extract of 0.61 % and 0.36 %, respectively; and white leadtree extract of 0.60 % and 4.82 %, respectively. White leadtree leaf extract gave the highest inhibition zone against *B. subtilis*, *S. aureus* and *E. coli*, i.e. 1.50, 1.3, and 1.65 cm, respectively; and the lowest MIC and MBC against *B. subtilis*, *S. aureus* and *E. coli*, i.e. 1500, 3000, and 1500 µg/ml, respectively. Therefore, the white leadtree leave extract had more potential as bactericide for soaking of skin or hide compared to those of the tropical-almond and morinda leaves extracts.

1. Introduction

Natural antibacterial sources can be obtained from some parts of plants. The part of plant containing more antibacterial substances is leaf [1]. The antibacterial agents can be obtained from plant leaves by extraction.

Tropical-almond (*Terminalia catappa*) leaf extract has good antibacterial activity against bacteria *Klebsiella pneumoniae*, *Staphylococcus aureus*, *Escherichia coli*, *Streptococcus pneumoniae*, and *Salmonella typhi* [2]. White leadtree (*Leucaena leucocephala*) leaf extract can inhibit the growth of bacteria commonly found in skin or hide preservation, such as *Bacillus subtilis*, *Staphylococcus aureus*, and *Escherichia coli* [3]. Phytochemical contents such as flavonoids, tannins, alkaloids, and saponins contained in leaf extract act as an antibacterial [4][5].

Leather manufacture commences with soaking of cured skin or hide in water to rehydrate and to cleanse it [6]. The soaking is one of the processes in the leather industry that serves to rehydrate the skin [7]. Bactericide is needed in the soaking process to prevent collagen damages by microorganisms [8].

Utilization of plant leaf extract should be developed as a natural bactericide. The use of natural bactericides from plant extracts may be potentially used in eco-friendly soaking.

This research was conducted to determine antibacterial activity of tropical-almond, morinda, and white leadtree leaves extracts, and to analyse the potency of the three leaves extracts as natural bactericide for soaking of skin or hide in leather manufacturing.

2. Materials and Methods

2.1. Materials

The materials used in this study were tropical-almond (*Terminalia catappa*) and morinda (*Morinda citrifolia*) leaves obtained from Darmaga Campus, Bogor Agricultural University (IPB), Bogor, Indonesia. Isolates of *Bacillus subtilis*, *Stapylococcus aureus*, and *Escherichia coli* were obtained from the Department of Biology, IPB, Bogor, Indonesia. Other chemicals used were Muller Hinton agar (MHA), ethanol 70%, chloramphenicol, amoxicillin, ciprofloxacin, nurient broth (NB), and dimethyl sulfoxide (DMSO).

2.2. Chemical Components Tests

The chemical components tests of the leaves were performed by method reported by Harborne [9] and Hasan et al. [10]. The tests consisted of alkaloids, flavonoids, saponins, tannins, phenolics, triterpenoids, steroids and glycosides.

2.3. Preparation of Extracts

The fresh leaves were dried under the sunlight for 2 - 3 days until the constant weight. The dried leaves were grinded using a blender and stored in an airtight container.

The extraction technique used in the extraction was maceration. The powder was soaked using 70% ethanol at 1:10 w/v for 24 hours and then stirred with a shaker. Use of a 1:10 w / v ratio between simplicia and solvent based on a study conducted by Margeretha et al. [11]. The 70% mixture of simplisia and ethanol was filtered through fabric and filter paper. The filtrate was evaporated with a vacuum rotary evaporator, water bath heating at 60 °C for an hour, then the water bath heater temperature was increased to 70 °C, thus leaf condensed extract was obtained [12]. The condensed extract was heated in a water bath to obtain an extract with a solid total of about 50%-60%. The total solid measurement was performed using a moisture analyser.

2.4. Total Flavonoid Test [13]

Test of total flavonoid extract was initiated by making reagent solution. The reagents used were 0.5% w/v HMT (hexamethyleneetrine) solution, 25% HCl solution, 5% glacial acetic acid solution in methanol, and 2% $AlCl_3$ solution in glacial acetic acid solution. The next step was making the mother liquor used as the sample solution. The procedure for making a mother liquor was an extract equivalent to 200 mg of simplicia poured into a round bottom flask and then added 1 ml of HMT solution, 20 ml of acetone, and 2 ml of HCl solution. The mixed solution mixed with the sample is then hydrolysed by reflux for 30 minutes. The mixture was then filtered using cotton, filtrate poured into a 100 ml measuring flask. The next step was the resulting residue refluxed with 20 ml of acetone for 30 minutes, filtered, and mixed in a 100 ml measuring flask. The filtrate mixture in the measuring flask was supplemented with acetone to 100 ml. The filtrate taken 20 ml was poured into the separating funnel, added with 20 ml of water, and extracted three times with 15 ml of ethyl acetate. Ethyl acetate fraction was collected and added with ethyl acetate to 50 ml in a measuring flask. The sample solution to be tested was taken from the mother liquor. The sample solution came from 10 ml of the mother liquor and was added with 1 ml of $AlCl_3$ solution and acetic acid solution up to 25 ml in a measuring flask. In addition to the preparation of the sample solution, the blank solution was made as a comparison. The blank solution was taken 10 ml from the mother liquor, then added with a solution of glacial acetic acid to 25 ml in a measuring flask. Measurements were conducted 30 minutes after the addition of $AlCl_3$ using a spectrophotometer at 425 nm wavelength with a quercetin standard.

Calculation:
% = Cp (As - Abs) / (Ap - Abp) x 1.25 x 100 / Sample weight
where:
Cp = standard concentration
As = absorption of sample
Abs = absorption of blank sample
Ap = absorption of standard
Abp = absorption of blank standard

2.5. Test for Tannin Content [14]
2 grams samples were weighed and then heated with 500 ml boiling water for 30 minutes. Then stuck for a few minutes and poured slowly through a lump of cotton into a 250 ml flask. The filtration was repeated several times until the solution when reacted with iron (III) ammonium sulfate did not show any tannin. Furthermore, the liquid was cooled and added with water up to 250 ml. The 25 ml solution was poured into a 1000 ml flask and added with 750 ml of water and 25 ml of LP indigo sulfonic acid, titrated with 0.1 N potassium permanganate to a golden yellow solution. 1 ml of potassium permanganate 0.1 N was equivalent to 0.004157 grams of tannin. Titration was carried out also for a blank.
The indigo sulfonic acid LP was made with 1 gram of indigo of carmin P dissolved in sulfuric acid P, then added with 25 ml of P sulfuric acid again and diluted with water to 1000 ml.

2.6. Inhibition Zone Test
The inhibition zone test carried out in this study used the well method reported by Mukti [15] and Suparno et al. [3].

2.7. Minimum Inhibitory Concentration and Minimum Bactericidal Concentration Tests
Minimum inhibitory concentration (MIC) and minimum bactericidal concentration (MBC) tests of the leaves extracts against *E. coli, B. subtilis*, and *S. aureus* were carried out using liquid dilution method repoted by Nuraina [16] and Suparno et al. [3].

3. Results and Discussions

3.1. Phytochemical Contents
Phytochemical test in this study was conducted to show the presence of substances or phytochemicals in tropical-almond, morinda, and white leadtree leaves extracts. Phytochemicals play a role in inhibiting microbial activity. The results of qualitative tests of the three leaves extracts were presented in Table 1. Phytochemical contents of some types of leaves were in accordance with previous studies. According to Kadam et al. [17], the tropical-almond leaves have a blackish-green leaf colour and a distinctive odour. The leaves of this plant contain phytochemicals such as carbohydrates, amino acids, glycosides, flavonoids, essential oils, alkaloids, tannins, and steroids. Krishnaiah et al. [18] reported that the main components of morinda are terpenoids, alkaloids, glucoside flavonoids, and some amino acids. The contents of tannins, saponins, flavonoids, phlobtanins, steroids, and glycosides in white leadtree leaves show good antimicrobial and antioxidant properties [19] [3].
Table 1 shows that morinda leaf contains alcaloids, saponins, tannins, phenolics, flavonoids, triterpenoids, steroids, and glycosides, whereas the tropical-almond and white leadtree leaves did not contain triterpenoids. Widiyati [20] states that triterpenoids are secondary metabolite compounds whose skeletons derived from acyclic C30 hydrocarbon derivatives and have alcohol groups, aldehydes or carboxylic acids. Most of the morinda plant components contain triterpenoids that are used as medicines. Ghalib [21] reported that active compounds such as alkaloids, saponins, and flavonoids in the leaves act as antifungals, thereby preventing mould activity. Quantitatively, the total contents of phytochemicals such as total flavonoids and tannin contents are presented in Table 2.

Table 1. Phytochemical test of the leaves extracts.

Phytochemicals	Tropical-almond	Morinda	White leadtree[a]
Alcaloids	+	+	+
Saponins	+	+	+
Tannins	+	+	+
Fenolics	+	+	+
Flavonoids	+	+	+
Triterpenoids	-	+	-
Steroids	+	+	+
Glycosides	+	+	+

[a]Suparno et al. [3]

Table 2. Total flavonoids and tannin content.

Leaves Extracts	Total Flavonoids (% w/w)	Tannin Content (%)
Tropical-almond	1.14	1.51
Morinda	0.61	0.36
White-leadtree	0.60	4.82

[a]Suparno et al. [3]

White leadtree leaf extract contained the highest tannin content (4.82%). According to Kumar et al. [22], the morinda fruit extract contains total tannins of about 3.5 mg/g. The tannin content in tropical-almond leaves can be an anti-quorum sensing (QS) in *Chromobacterium violaceum* and *Pseudomonas aeruginosa* [23]. High levels of tannin affect the antibacterial inhibition. Tannins can precipitate proteins and inhibit the enzyme reverse transcriptase and DNA topoisomerase, so that no bacterial cell formation [24].
The highest total flavonoid was obtained from tropical-almond leaf extract (1.14%). According to Muryati et al. [25], tropical-almond leaves ethanol extract at concentration of 1% contain total flavonoid of 0.29%. Flavonoids can reduce the thickness of the organism because it has a broad spectrum on antimicrobial activity [26].

3.2. Yields
The yield of the extracted simplicia was measured after the ethanol extract was evaporated using a rotary evaporator. The yield was influenced by the amount of compound which can be extracted by the solvent and had the same polarity level [27]. The yields of the leaves extracts are shown in Table 3.

Table 3. Yields of the leaves extracts.

Leaves	Yields (%)
Tropical-almond	22±0.70
Morinda	29±0.70
White leadtree[a]	24±0.79

[a]Suparno et al. [3]

Table 3 shows that morinda leaf extract gave the highest yield (29%). It was influenced by the number of polar compounds that can be extracted by 70% ethanol [27]. The yield was a yield of leaf extract with a total solid content of 50%. The morinda leaf contained more polar compounds than the other leaves so that the extracted compounds increased, so that increased the extract yield (Table 3). The compounds that can be extracted with 70% ethanol were alkaloids, flavonoids, saponins, terpenoids, and tannins [28]. Several extraction yields with 70% ethanol maseration process are 22.6% for purple cabbage powder and 22.53% for the henna tree leaf [29]. The yields obtained from our research were similar with those of some other extracts with the same solvent and type of extraction.

3.3. Inhibition Zones

The antimicrobial activity of the leaf extract was measured by inhibition zone, minimal inhibitory concentrations (MIC), and minimum bactericidal concentrations (MBC) tests. Inhibition zone test is performed to determine the activity of extracts or substances in inhibiting bacterial growth. The inhibitory zone is characterized by a clear zone around the well or sample hole indicating the absence of bacterial activity [30]. Measurement of inhibition zone using concentrated extracts was carried out to show the potential of antimicrobials in the extract. The inhibition zone are presented in Table 4.

Table 4. Inhibition zones (cm) of the leaves extracts against *B. subtilis, S. aureus,* and *E. coli.*

Extracts	*Bacillus subtilis*	*Staphylococcus aureus*	*Escherichia coli*
Tropical-almond	0.91±0.11	0.82±0.12	1.14±0.25
Morinda	0.23±0.14	1.04±0.25	0.57±0.01
White leadtree[a]	1.50±0.00	1.30±0.10	1.65±0.06
Chloramfenicol[a]	2.03±0.06	-	-
Ciprofloxacin[a]	-	3.26±0.11	-
Amoxicillin[a]	-	-	2.63±0.06

[a]Suparno et al. [3]

Table 4 shows that the inhibition zones of tropical-almond leaf extract against *B. subtilis, S. aureus,* and *E. coli* were 0.91 cm, 0.82 cm and 1.14 cm, respectively. The study conducted by Chanda et al. [31] showed that tropical-almond methanol extract is able to inhibit the growth of *B. subtilis*, *S. Aureus*, and *E. coli* with the inhibition zones of 0.9 cm, 1.3 cm, and 1.2 cm, respectively. Muhammad and Mudi [2] stated that the difference in results is influenced by the type of solvent used during the extraction of the leaves. The tropical-almond leaf extract contains alkaloids, tannins, and saponins that act as antimicrobials. Riskitavani and Purawani [32] reported that the development of almond-tree plant extracts can also be used to inhibit weeds grass (*Cyperus rotundus*) because it contains alkaloid, tannin, and saponin metabolite compounds.

The inhibition zones of morinda leaf extracts were 0.23 cm for *B. subtilis*, 1.04 cm for *S. aureus*, and 0.57 cm for *E. coli*. Pandey et al. [33] reported that inhibition zones are resulted in by the presence of flavonoid, alkaloid, and saponin metabolite compounds that inhibit the bacterial activity.

The inhibition zone of white leadtree leaf extract was the widest compared to those of moringa, cucumber tree, and cherry leaves extracts [3]. In this research, the white leadtree leaf extract showed also the widest inhibition zone compared to those of tropical-almond and morinda leaves extracts. White leadtree leaf extract can inhibit *B. subtilis*, *S. aureus,* and *E. coli* with inhibition zones of 1.5 cm, 1.3 cm, and 1.7 cm, respectively [3]. Research conducted by Aderibigbe et al. [34], white leadtree leaf extracted with n-hexane resulted in inhibition zone of 1.4 cm on 100% extract concentration against *B. subtilis*, 1.2 cm against *S. aureus*, and 1.8 cm against *E. coli*. The inhibition zones formed showed antimicrobial activities of the extracts because of their flavonoids, alkaloids, saponins, tannins, and triterpenoids contents.

The white leadtree leaf contains lupeol, an antibacterial active compound [35]. According to Soedarjo and Borthakur [36], white leadtree plant parts including its leaf contain mimosine that harmfull when used as food. Mimosine compound is able to inhibit DNA replication of yeasts [37]. These compounds support the high inhibition zone of white leadtree leaf extract among the other two leaf extracts.

3.4. Minimum Inhibitory Concentration and Minimum Bactericidal Concentration

Tests of minimum inhibitory concentration and minimal bactericidal concentration were performed to measure antibacterial activity of the leaves extracts. According to Nuraina [16], the minimum inhibitory concentration is the lowest concentration of a substance that has the power to inhibit the growth of microorganisms (indicated by the absence of turbidity in the test tube) which has been incubated for 24 hours at 37 °C. The results of minimum inhibitory concentrations are presented in Table 5.

Table 5. Minimum inhibitory concentration (μg/ml) of the extracts against *B. subtilis*, *S. aureus*, dan *E. coli*.

Extracts	*Bacillus subtilis*	*Staphylococcus aureus*	*Escherichia coli*
Tropical-almond	3000	> 6000	3000
Morinda	> 6000	> 6000	6000
White leadtree[a]	1500	3000	1500

[a]Suparno et al. [3]

Minimum bactericidal concentration (MBC) is the lowest concentration of a substance that can kill the bacteria characterized by the absence of bacterial growth on the solid medium of 1 *ose* suspension of the minimum inhibitory concentration tested [16]. The lower the minimum bactericidal concentration, the stronger the antimicrobial activity. Table 6 shows the results of MBC of the leaves extracts. Quantitative results from minimum inhibitory concentrations were performed through the minimum bactericidal concentration test.

Table 6. Minimum bactericidal concentration (μg/ml) of the extracts against *B. subtilis*, *S. aureus*, dan *E. coli*.

Extracts	*Bacillus subtilis*	*Staphylococcus aureus*	*Escherichia coli*
Tropical-almond	3000	> 6000	3000
Morinda	> 6000	> 6000	6000
White leadtree[a]	1500	3000	1500

[a]Suparno et al. [3]

Table 5 and Table 6 show that morinda leaf extract was only capable of inhibiting and killing *B. subtilis* at concentration of more than 6000 μg/ml and for *E. coli*. at 6000 μg ml. MIC and MBC of tropical-almond leaf extract were at 3000 μg/ml for both bacteria. The antibacterial activity of white leadtree leaf extract against *B. subtilis* and *E. coli* on MIC and MBC of 1500 μg/ml [3]. Low concentrations of MIC and MBC of white leadtree leaf extract compared to the other extracts were supported by its inhibition zone.

MIC and MBC of tropical-almond and morinda leaves extracts against *S. aureus* were more than 6000 μg/ml. The MIC and MBC of white leadtree leaf extract against *S. aureus* were 3000 μg/ml [3]. The presence of tannin in the white leadtree leaf has potential as antioxidant and antimicrobial [38].

Compounds that play a role in the MIC test are metabolite compounds such as alkaloids, flavonoids, steroids, tannins, and quinones [16]. Saponin and flavonoid compounds are capable to alter the permeability of bacterial cell membranes, thus denature proteins in bacterium cell wall [26]. In addition, metabolite compounds that play a role in inhibiting the bacterial activity is tannins.

4. Conclusions

Tropical-almond, morinda, and white leadtree leaves extracts had antibacterial activity. White leadtree (*Leucaena leucocephala*) leaf extract gave the highest antibacterial activity compared to those of tropical-almond and morinda leaves extracts, as indicated by the inhibition zone, minimal inhibitory concentrations (MIC), and minimum bactericidal concentrations (MBC) tests. Therefore, white leadtree leaf extract was more potential as a bactericide for the soaking process in leather manufacturing.

Acknowledgments
The authors acknowledge *Direktorat Riset dan Pengabdian Masyarakat* (DRPM), Indonesian Ministry of Research Technology and Higher Education for its financial support and Bogor Agricultural University for its facility for conducting the research.

5. References

[1] Joshi D, Patel R, Patel N, Patel D and Pandya C. Antimicrobial evaluation of leaf and stem extract of *Cordia macleodii*. *Pharmaceutical Science Journal*. 1:1-3

[2] Muhammad A and Mudi SY. 2011. Phytochemical screening and antimicrobial activities of *Terminalia catappa* leaf extract. *Biokemistri*. 23(1):35-39.

[3] Suparno O, Afifah A, Panandita T, Marimin, Purnawati R. 2017. The potency of plant extract as antimicrobial for the skin or hide preservation. *AIP Conference Proceedings 1823, 020098 (2017),* The American Institute of Physics, doi: 10.1063/1.4978171.

[4] Sibi G, Naveen R, Dhananjaya K, Ravikumar KR and Mallesha H. 2012. Potential use of *Muntingia calabura* L. extracts against human and plant pathogens. *Phcog. J.* 4(34):44-47.

[5] Nepolean P, Anitha J and Renitta RE. 2009. Isolation, analysis, and identification of phytochemicals of antimicrobial activity of *Moringa oleifera* Lam. *Current Biotica*. 3(1):33-39.

[6] Suparno O. 2005. Phenolic reactions for leather tanning and dyeing. PhD Thesis. Leicester (UK): University of Leicester.

[7] Covington AD. 2009. Tanning chemistry: the science of leather. Northampton (UK): The Royal Society of Chemistry.

[8] Beghetto V, Zancanaro A, Scrivanti A, Matteoli U and Pozza G. 2013. The leather industry: a chemistry insight, part i: an overview of the industrial process. *Science At Ca' Foscari*. 1:13-22.

[9] Harborne JB. 1987. Phytochemical Methods. London (UK): Chapman and Hall.

[10] Hasan AEZ, Mangunwidjaja D, Sunarti TC, Suparno O and Setiyono A. 2014. Investigating the antioxidant and anticyctotoxic activities of propolis collected from five regions of Indonesia and their abilities to induce apoptosis. *Emirates of Journal of Food and Agriculture*. 26(5): 390-398.

[11] Margeretha I, Suniarti DF, Herda E and Mas'ud ZA. 2012. Optimization and comparative study of different extraction methods of biologically active components of Indonesian porpolis *Trigona spp*. *Journal of Natural Product*. 5:233-242.

[12] Isnarianti R, Wahyudi IA and Puspita RM. 2013. *Muntingia calabura* L. leaves extract inhibits glucosyltrausferase activity of *Streptococcus mutans*. *Journal of Dentistry Indonesia*. 2013 20(3):59-63.

[13] [BPOM] Badan Pengawas Obat dan Makanan. 2004. Monography of Indonesian medicinal plants Extracts. Volume 1. Jakarta (ID): Badan Pengawas Obat dan Makanan Republik Indonesia.

[14] Departemen Kesehatan. 1980. Materia Medika Indonesia. Jakarta (ID): Departemen Kesehatan.

[15] Mukti D. 2012. Antimicrobial effectivity test of *Mormodica charantia* L. ethanol extract on *Streptococcus mutans* causing dental caries. Bogor (ID): Universitas Pakuan.

[16] Nuraina. 2015. Antimicrobial activity test of *Garcinia benthami* Pierre leaf extract using dilution method. Jakarta (ID): UIN Syarif Hidayatullah Jakarta.

[17] Kadam PV, Yadav KN, Narappanawar NS, Shivatare RS, Bhusnar HU and Patli MJ. 2011. Development of quality standards of *Terminalia catappa* leaves. *Pharmacognosy Journal*. 3 (26): 19-24.

[18] Krishnaiah D, Nityanandam R and Sarbatly R. 2012. Phytochemical constituents and activities of *Morinda citrifolia* L. Sabah (MY): Universitas Malaysia Sabah.

[19] Mohammed RS, El Sauda SS, Taie HAA, Moharram ME and Shaker KH. 2015. Antioxidant, antimicrobial activities of flavonoid glycoside from *Leucaena leucocephala* leaves. *Journal of Applied Pharmaceutical Science*. 5(6):138-147.

[20] Widiyati E. 2006. Determination of triterpenoid compound existence and biological activity test at some species traditional medicinal plants of Bengkulu village. *Jurnal Gradien*. 2(1):116-122.

[21] Ghalib D. 2009. Inhibition test of senggani (*Malestoma malabathricum* L.) leaf on *Trichoplyton mentagrophytees* and*Candida albicans*. *Jurnal Ilmu-ilmu Hayati*. 9(5):523-527.
[22] Kumar SNK, Suresh M, Kumar SA and Kalaiselvi P. 2014. Bioactive compounds, radical scavenging, antioxidant properties and FTIR spectroscopy study of *Morinda citrifolia* fruit extracts. *International Journal of Current Microbiology and Applied Sciences*. 3(2):28-42.
[23] Taganna JC, Quanico JP, Perono RMG, Amor EC and Rivera WL. 2011. Tannin-rich fractio from *Terminalia catappa* inhibits quorum sensing (QS) in *Chromobacterium violaceum* and the QS-controlled biofilm maturation and LasA staphylolytic activity in *Pseudomonas aeruginosa*. *Journal of Ethnopharmacology*. 134:865-871.
[24] Nuria MC, Faizatun A and Sumantri. 2009. Antibacterial activity test of ethanol extract of *Jatropha curcas* L. leaf on *Staphylococcus aureus* ATCC 25923, *Escherichia coli* ATCC 25922, and *Salmonella typhi* ATCC 1408. *Jurnal Ilmu-ilmu Pertanian*. 5(2), 26 – 37.
[25] Muryati S, Yuniawati DR and Mutiara EV. 2009. Abtioxidant activity test using linoleic-thiocyanate method and determination of phenolic and total flavonoid of tropical-almond (*Terminalia catappa* L.) leaf ethanol extract. Semarang: STIFAR - Sekolah Tinggi Ilmu Farmasi.
[26] Naidu AS and Clemens RA. 2000. Natural Food Antimicrobial System: Probiotics. CRC Press. New York, 431-462.
[27] Sani RN, Nisa FC, Andriani RD and Maligan JM. 2014. Yield analysis and phytochemical screening of sea microalgae *Tetraselmis chuii* ethanol extract. *Jurnal Pangan dan Agroindustri*. 2(2):121-126.
[28] Rusdi NK, Sediarso and Fadila SH. 2010. Antibacterial activity test of 70% ethanol fraction of leaf extract of *Phaleria macrocarpa* (Schreff) Boerl.) on *Streptococcus mutans* bacteria. *Farmasains*. 1(2):89-94.
[29] Senja RY, Issusilaningtyas E, Nugroho AK and Setyowati EP. 2014. Comparison of extraction method and solven variation on yield and antioxidant activity of purple cabbage extract. *Traditional Medicine Journal*. 19(1):43-48.
[30] Dewi FK. 2010. Antibacterial activity of morinda fruit ethanol extract (*Morinda citrifolia* L.) on fresh meat rotten bacteria. Surakarta (ID): Universitas Sebelas Maret.
[31] Chanda S, Rakholiya K and Nair R. 2011. Antimicrobial activity of *Terminalia catappa* L. leaf extract against some clinically important pathogenic microbial strains. *Journal Scientific Research*. 2:171-177.
[32] Riskitavani DV and Purwani KI. 2013. Study of bioherbiside potency of tropical almond leaf (*Terminalia catappa*) on weeds grass (*Cyperus rotundus*). *Jurnal Sains dan Seni*. 2(2):59-63.
[33] Pandey A, Pandey RD, Tripathi P, Gupta PP, Haiden J, Bhatt S and Singh AV. 2011. *Moringa oleifera* Lam. (sahijan) – a plant with a plethora diverse therapeutic benefits: an updated retrospection. *Medical & Aromatic Plants*. 1(1):1-8.
[34] Aderibigbe SA, Adetunji OA and Odeniyi MA. 2011. Antimicrobial and pharmaceutical properties of the seed oil of *Leucaena leucocephala* (Lam.) de wit (Leguminosae). *African Journal of Biomedical Research*. 14:63-68.
[35] Sartinah A, Astuti P and Wahyuono A. 2010. Isolation and identification of antibacterial compound from white leadtree (*Leucauena leucocephala* (Lam.) De Wit.). *Majalah Obat Tradisional*. 15(3):146-152.
[36] Soedarjo M and Borthakur D. 1996. Simple procedure to remove mimosine from young leaves, pods and seeds of *Leucaena leucocephala*. *International Journal of Food Science and Technology*. 31: 9-103
[37] Levenson V and Hamlin JL. 1993. A general protocol for evaluating the spesific effect of DNA replication inhibitors. *Nucleic Acid Research*. 21(17):3997-4004.
[38] Zarin MA, Wan HY, Isha A and Armania N. 2016. Antioxidant, antimicrobial and cytotoxic potential of cendensed tanins from *Leucaena leucecophala* hybrid-*rendang*. *Food Science and Human Wellness*. 5:65-75.

Effects of gamma irradiation on the performance of Jatropha *(Jatropha curcas L.)* accessions

M Surahman[12]*, E Santosa[1], H Agusta[12], S I Aisyah[1], F N Nisya[2]

[1]Departemen Agronomi dan Hortikultura, Fakultas Pertanian, Jl. Meranti, Dramaga, Bogor.16680.Jawa Barat. Indonesia
[2]Surfactant and BioenergyResearch Center.LPPM IPB. Jl. Pajajaran No.1 Bogor.Jawa Barat. Indonesia. 16144. E-mail:memensurahman@yahoo.com/08128231160

Abstract. This study aimed to assess the effects of mutation by using gamma ray on the performance of jatropha plants. The study was conducted at PAIR BATAN. Jatropha seeds obtained from the collection farm of *SBRC LPPM IPB* and PT Indocement Tunggal Prakarsa Tbk in Gunung Putri, Bogor, were irradiated. The irradiated seeds were grown in Jonggol Trial Farm of IPB. Gamma irradiation was conducted by using a GCM 4000A device. Treatments consisted of irradiation doses, irradiation methods, and accessions. Irradiation doses given were 175, 200, 225 Gy, and no irradiation (control). Irradiation methods consisted of acute, intermittent, and split-dose. Accessions used in this study were Dompu, Medan, Bima, Lombok, ITP II, IP2P, and Thailand. Results of the study were analysed until 5 months after planting showed that gamma ray mutation gave stimulating and inhibiting effects on similar traits. Irradiation dose of 225 Gy was good to be given in acute, intermittent, and split-dose methods. Irradiation effects were found to be significant in jatropha accessions. Effects of irradiation on production will be published soon.

Key words: Jatropha, mutation, gamma ray, performance

1. Introduction

Jatropha produce fruits whose seeds contain oil that can be used as fuel for cooking, lighting, and biodiesel feedstock (1). Depending on their places of origins, jatropha seeds contain oil of up to 40.96% with an average figure of 36.43% (2). Jatropha fruits are found in inflorescents formed from pollinated female flowers. Most fruits contain three seeds but some may have up to four seeds containing oil. The amount of oil that can be produced was found to be positively correlated with the number of inflorescents, number of harvested fruits per plant, and number of total fruits per plant, dry seed weight, and seed diameter (3).

Dry seed weight is the idiotype commonly used as an indicator of superior variety of jatropha plant as it is correlated with oil production. Dry seed production was significantly correlated with plant height, branch diameter, number of secondary branches per plant, number of productive branches per plant, number of inflorescents per plant, number of fruits per plant, number of seeds per plant, and dry weight of 100 seeds (4). Another research also shown that jatropha production was affected by number of inflorescents per plant and number of seeds per plant (5). Therefore, these characters can be used as important characters for the selection of jatropha plant.

However, not all characters which were correlated with high production of jatropha had high variability and heritability (3). Therefore, not all of those characters can be used as the selection

criteria for high producing jatropha plants. Other research showed that only number of inflorescents per plant and number of fruits per plant could be used selection criteria for high productivity of jatropha as they had wide genetic variability, high heritability, and high genetic advance (3,4,5,6).
Narrow genetic variability from the observed characters indicated that the accessions of jatropha from various places in Indonesia were relatively uniformed although they came from geographically different origins. By using molecular markers to detect genetic variability of jatropha in India, it was found high level of uniformity among different genotypes. This meant that the accessions of jatropha currently available had narrow variability. In other words, current population was relatively uniformed, introduction from other countries, crossbreeding, or mutation was needed (7,8).
Mutation is an effective method to improve genetic variability. It has been proven to improve the genetic variability of jatropha (9, 10, 11). Gamma irradiation is very important in mutation breeding. Mutagenesis can be used to improve genetic variation (12). Therefore, this study was aimed at analyzing the performance of jatropha produced by gamma irradiation.

2. Materials and Methods

The study was conducted at the Center for Isotopes and Radiation Application (PAIR), National Nuclear Energy Agency (BATAN). Jatropha seeds obtained from the collection farm of *SBRC LPPM IPB* and PT Indocement Tunggal Prakarsa Tbk in Gunung Putri, Bogor. The irradiated seeds were grown in Jonggol Trial Farm of IPB. As the jatropha seed availability was limited, the irradiation of it was done in two stages in March and April 2014.
Gamma irradiation was conducted by using a GCM 4000A device. Treatments consisted of irradiation doses, irradiation methods, and accessions. Irradiation doses given were 175, 200, 225 Gy, and no irradiation (control). Irradiation methods consisted of acute, intermittent, and split-dose. Accessions used in this study were Dompu, Medan, Bima, Lombok, ITP II, IP2P, and Thailand. In acute irradiation method, irradiation was done only once in the proposed doses. In intermittent method, irradiation was done twice in full doses. The irradiation was repeated within 1 week interval. In split-dose method, irradiation was done twice. The dose in the second irradiation was half of it in the first irradiation.
Irradiated seeds were grown on a nursery bed for 2 months before they were moved into the field. Planting was done in a planting distance of 2 m x 2 m. Planting holes were prepared in 3 weeks prior to planting time. Five kilograms of manure (organic fertilizer) was put into each planting hole. One seed was planted in each planting hole.
Morphological observation was done to assess the effects of irradiation on jatropha plants resulted from irradiated seeds (M1). The observation was done on seed germination, plant height, number of leaves, number of branches, number of inflorescents, and number of fruits per inflorescent. Means of plant growth rate in nursery bed were measured to assess the effects of irradiation and accessions used. Meanwhile, observation data from the field were obtained from individual plant. Data were subjected to an analysis of variance by using SAS 9 application.

3. Results and Discussions

3.1 Effects of irradiation on mutant seed germinability (M1)

The germination rate (%) of M1 irradiated seeds is shown in Table 1. Theoretically, irradiated seeds undergo genetic or physiological changes. Yet, when irradiated seeds can still germinate and grow into intact plants, then this is an indication that the irradiation done creates changes which are naturally acceptable (13). From the germination rate indicated that the irradiation dose and method applied were naturally acceptable for jatropha seeds.
Among all irradiated jatropha accessions, ITP II accession had the highest germinability (80.8%) although it was not different from those of Lombok (80.7%), Bima (60.6%), Medan (59.3%), and Dompu (79.25%) accessions. It was shown that some irradiated seeds had higher germination rates than control and lower doses groups. These included seeds of Lombok accession treated with intermittent dose of 175 Gy, Dompu accession treated with intermittent dose of 200 Gy, and Lombok, Bima, and Medan accessions treated with split dose of 225 Gy. Another research also found that gamma irradiation at the doses of up to 200 Gy did not cause a seed damage although it lowered seed germination rate especially in jatropha of IP-3A variety (14). The germination rates found in this

study also showed a stimulation effect of irradiation on seed germination. Higher doses of irradiation can effect on higher germination rate of jatropha seeds (14,15). Responses to gamma irradiation were also found to be different in each accession.

Table 1. Effects of gamma irradiation on jatropha seed germinability (%)

Accession	*Intermittent (%)*			*Split dose (%)*		*Acute (%)*			Control
	175+175	200+200	225+225	175+87.5	225+112.5	175	200	225	
ITPII	90	90	80	70	90	80	80	67	90
Lombok	92	67	69	90	100	92	67	69	90
Bima	38	83	46	80	90	38	79	31	80
Medan	58	67	21	90	100	50	67	21	90
Dompu	83	92	67	-	-	83	92	58	90
Mean	72	80	57	83	95	69	77	49	88

Effects of irradiation on mutant plant (M1) growth in nursery bed

In the nursery bed, mutant plants had various growth rates as indicated by their plant height, number of leaves, and number of branches. In stage 1 irradiation, in 2 months after planting (MAP), ITP II accession had the highest growth rate.

Jatropha plants of ITP II accession started to have branches in nursery bed. Similar development was shown by Lombok accession irradiated acutely with 225 Gy. Like germination rate, response of each accession on plant growth to gamma irradiation was varied. Gamma irradiation gave either stimulating or inhibiting effects on plant height. For Dompu, Bima and Lombok accessions, acute irradiation doses of 175 and 200 Gy inhibited plant height and number of leaves increment in nursery bed. In contrast, acute irradiation at 225 Gy was found to result in higher plant height and number of leaves. However, in nursery bed, two jatropha plants of Dompu accession in control group were found to have, on average, six branches. In stage 2 irradiation, increased growth rate and number of leaves were found in jatropha plants of Medan and ITP II accessions treated with 225 Gy acute irradiation. Growth rate of mutant plants in 2 MAP is shown in Table 2.

At second stage irradiation, in 1 month after planting (MAP), jatropha plants of Medan accession tended to have better growth as shown by their higher number of leaves (11.10) and plant height (24.91 cm). For plants of ITP II accession, repeated irradiation was found to inhibit their growth in nursery bed as indicated from their plant height and number of leaves. Inhibiting effect of gamma irradiation on plant height was also observed in Medan accession.

Gamma irradiations at 225 Gy (acute) and 175 Gy (split dose) increased plant height and number of leaves of Lombok accession in nursery bed. Similar effect on plant height and number of leaves was found in Bima accession treated with split dose irradiation at 225 Gy. On the contrary, inhibiting effects on plant height and number of leaves of Bima accession in 1 MAP were shown by acute irradiation at 225 Gy and split dose irradiation at 175 Gy. These findings indicated that both stimulating and inhibiting effects of irradiation were already experienced by plants since 1 MAP. Plant height and number of leaves in jatropha plants in second stage irradiation are given in Table 3.

Table 2. First stage mutant plant growth in 2 months after planting

Jatropha accession	Irradiation doses (Gy)	Plant height (cm)	Number of leaves	Number of branches
Dompu	175	9.2	6.1	0
Dompu	200	12.1	7.2	0
Dompu	225	25.1	20.4	0
Dompu	Control	22.2	13.8	2
Bima	175	6.0	4.3	0
Bima	200	10.8	5.6	0
Bima	225	23.9	20.1	0
Bima	Control	15.6	4.4	0
Lombok	175	9.8	5.4	0
Lombok	200	12.9	7.1	0
Lombok	225	32.3	26.2	3.5
Lombok	Control	16.1	5.0	0
Medan	175	7.5	4.8	0
Medan	200	11.6	6.4	0
Medan	225	12.7	13.0	0
ITP II	200	38.7	36.2	3.8
ITP II	225	43.0	32.0	2.75

Table 3. Second stage mutant plant growth in 1 month after planting

Treatment	Mutation treatment (Gy)	Height (cm)	Number of leaves
ITP II	control	32.60	12.88
ITP II	175+175	26.43	9.28
ITP II	200+200	14.13	7.88
ITP II	225+225	19.60	9.56
ITP II	225+112.5	20.67	9.76
ITP II	175+87.5	22.57	12.21
Lombok	control	15.00	5.33
Lombok	175+87.5	20.40	10.78
Lombok	225	17.33	5.14
Lombok	225+112.5	29.57	12.40
Medan	control	36.00	21.50
Medan	175+87.5	14.33	4.11
Medan	225	30.14	16.00
Medan	225+112.5	30.25	13.20
Bima	control	30.33	16.00
Bima	175+87.5	11.00	4.50
Bima	225	21.00	13.00
Bima	225+112.5	34.33	13.33

Effects of gamma irradiation in the field

Based on field observation within 5 months after planting, it was found that the treatments given to the seeds resulted in various plant growths. Gamma irradiation gave significant effects on the growth of jatropha plants of different accessions. The growth of treated jatropha plants is shown in Table 4.

Like on seed germination, gamma irradiation gave stimulating and inhibiting effects on jatropha plant growth. These were indicated by plant height in 5 MAP, number of inflorescents and number of fruits per inflorescent in 4 MAP. The average plant height in 5 MAP was 84.5 cm with average number of branches of 5.3. In 4 MAP, the average number of inflorescents per plant was 2.7.

As shown in Table 4, irradiation doses and methods were found to be effective in improving plant growth. In plant breeding, induced mutation is an effective way to improve the variability of the existing germplasms and it can be used to improve plant varieties (16). The distribution frequency of plant height, number of branches, number of inflorescent per plant, and number of fruits per inflorescent of all M1 mutant plants is presented in Figure 1. It was known from this figure that M1 mutant plants had various number of branches, number of inflorescent per plant, and number of fruits per inflorescent.

It was revealed that plants resulted from irradiated seeds were either shorter or taller than those in control group. Irradiation dose of 225 Gy given in acute, intermittent, and split dose methods gave positive effects on plant height, number of branches, number of inflorescent per plant, and number of fruits per inflorescent in all plant accessions except ITP II accession. ITP II accession have negative impact of split dose irradiation at 225 Gy at plant growth and production. Meanwhile, irradiation doses of 175 and 200 Gy gave both positive and negative effects on plant height, number of branches, number of inflorescent per plant, and seed weight per harvested plant. Other report shown that mutation gave positive or negative effects on plant height (15,17). This phenomenon was also found in gladiolus plant, where gamma irradiation, at certain dose, was found to not only stimulate but also inhibit plant height growth (18). This might be caused by the fact that irradiation not only stimulates but also inhibits the division and elongation of meristematic cells including plant bud meristematic cells. Inhibition results in plant dwarfism.

For Bima accession, irradiation doses of 200 and 175 Gy which was repeated twice at a full dose resulted in plants which were shorter than those in control group. Meanwhile, irradiation of 225 Gy given in acute, intermittent, and split dose methods and irradiation of 175 Gy given in a split dose method resulted in plants which were higher than those in control group. The opposite results were found in Dompu accession for which irradiation doses of 200 and 175 Gy given twice in full doses resulted in plants which were taller than those in the control group. This result was similar to that given by 225 Gy irradiation.

For ITP II accession, acute and intermittent 225 Gy irradiation resulted taller plant than control group. Split dose 225 Gy irradiation, like acute 200 Gy irradiation, however, gave a negative effect on plant height. For ITP II accession, intermittent 200 Gy irradiation, acute and split dose of 175 Gy irradiation resulted in taller plant than control group plants. For Lombok accession, all irradiation treatments, except intermittent 175 Gy irradiation, resulted taller plant than control group plants. For Medan accession, acute, intermittent, and split dose 225 Gy irradiations resulted taller plant than control plants while 200 and 175 Gy irradiation in all methods produced plants shorter than control plants.

Table 4. Effects of mutation on plant height, number of branches, number of inflorescents, and seed weight of jatropha plant

Accession	Irradiation (Gy)	Plant height (cm)	Number of branches	number of inflorescents	Seed weight (g)
Bima	control	73.46 f-j	3.82 e-i	2.36 c-f	17.5
Bima	225	126.50 a	6.75 bcd	7.50 a	45.02
Bima	225+225	82.29 c-i	5.00 b-i	3.29 cde	29.77
Bima	200+200	67.00 ij	5.18 b-h	0.24 f	-
Bima	175+175	71.50 f-j	4.00 d-i	0.33 f	-
Bima	225+112.5	99.67 bcd	4.89 b-i	5.89 b	71.48
Bima	175+87.5	77.88 e-j	4.00 d-i	3.00 c-f	-
	Mean	**85.47**	**4.80**	**3.23**	
Dompu	control	73.67 f-j	4.89 b-i	1.67 c-f	13.56
Dompu	225+225	91.2 b-e	10.6 a-b	3.2 cde	20.65
Dompu	200+200	84.14 b-i	6.43 b-e	3.71 cd	13.65
Dompu	175+175	74.82 f-j	3.64 f-i	0.09 f	-
	Mean	**80.96**	**6.39**	**2.17**	
ITP	control	84.57 b-h	4.48 c-i	2.19 c-f	58.08
ITP	225	89.06 b-h	5.31 b-h	1.50 c-f	3.72
ITP	200	84.19 b-i	4.88 b-i	1.38 efd	1.28
ITP	175	87.06 b-h	4.22 d-i	1.44 c-f	11.57
ITP	225+225	109.25 bc	6.50 b-e	2.33 c-f	15.31
ITP	200+200	103.56 bc	6.11 b-f	7.56 a	17.12
ITP	225+112.5	83.71 b-i	4.41 d-i	0.94 efd	27.42
ITP	175+87.5	92.43 b-e	5.50 b-h	2.79 c-f	30.72
	Mean	**91.73**	**5.18**	**2.52**	
Lombok	control	60.22 j	2.67 g-j	0.89 ef	-
Lombok	225	67.50 hij	- j	0.33 f	-
Lombok	225+225	-	-	7.11 a	40.66
Lombok	200+200	68.75 g-j	2.50 hij	0.12 f	4.75
Lombok	175+175	58.40 J	2.00 ij	0.40 ef	13.47
Lombok	225+112.5	92.00 b-e	4.80 b-i	1.60 c-f	53.28
Lombok	175+87.5	79.67 d-j	3.33 ghi	2.22 c-f	13.47
	Mean	**71.09**	**3.06**	**1.81**	
Medan	control	91.50 b-e	5.67 b-g	2.67 c-f	30.68
Medan	225	102.29 bc	7.43 bc	7.36 a	132.68
Medan	225+225	98.00 b-e	12.00 a	4.00 bc	-
Medan	200+200	81.25 c-j	5.25 b-h	1.50 c-f	-
Medan	175+175	69.00 f-j	6.40 b-f	0.40 ef	-
Medan	225+112,5	106.05 B	7.55 b	8.25 a	179.52
Medan	175+87.5	72.65 f-j	2.88 ghi	0.82 ef	-
	Mean	**88.68**	**6.74**	**3.57**	

Plant height is one of the most important characters in the selection of superior jatropha plant. Farmers prefer shorter plants so it easier to reach during harvest time. However, plant height is not the only one to consider. Other production characters including number of inflorescents per plant, number of fruits per inflorescent, and dry seed production are also important. Tall plants with higher number of inflorescents and fruits are preferred to shorter plants with fewer inflorescents and fruits.

In branch growth character, mutagen treatment resulted in plants with higher or fewer number of branches than control plants. However, the results were not significant in Bima, Dompu, ITP II, and Lombok accessions. For Medan accession, intermittent 225 Gy irradiation produced plants having 12 branches. Branch in jatropha plant is an important character as it relates to fruit production characters. Flower production in jatropha plant is always related to plant branch growth. Flowers or inflorescents can be formed in plant axial or terminal positions. If inflorescents are formed in stem terminal, branches could be formed after the plant forms flowers. Meanwhile, if flowers are formed in axial position, the plant would have indeterminate growth and few branches tended to be formed. Pprevious studies showed that branches in jatropha plants could be formed in early growth after the plants had 5-10 leaves (Nisya, 2010) or 60-70 leaves or within 4-5 MAP (19). In order to increase fruit production in jatropha plant, pruning is always done to increase number of branches. More inflorescents are expected with higher number of branches.

Number of inflorescents per plant was affected by irradiation. The effects of irradiation on number of inflorescents were both positive and negative. The highest number of inflorescents (8.25) was found in Medan accessions treated with split dose 225 Gy irradiation. However, it was not different with Medan accessions treated with acute 225 Gy irradiation (7.36), Lombok accessions treated with intermittent 225 Gy irradiation (7.11), ITP II accessions treated with intermittent 200 Gy irradiation (7.56), and of Bima accessions treated with acute irradiation (7.50).

Inflorescent in jatropha plants is a cluster of flowers consisting of female and male flowers. However, in some accessions in Indonesia inflorescents were found to consist of male and hermaphrodite flowers (3). In this study, within 1 MAP, some plants had already had flowers. For all accessions, intermittent irradiation treatment of 225 Gy was found to result in plants which produced flowers early within 1 MAP. It was indicated by the finding that more than 50% of all plant accessions treated with intermittent 225 Gy irradiation had flowers. Mutant plants having early flowering were also found in Bima and Medan accessions treated with 225 Gy split dose irradiation, ITP II accessions treated with 200 Gy intermittent irradiation, Lombok accessions treated with 175 Gy split dose irradiation, Medan accessions treated with 225 Gy acute irradiation, and Medan accessions treated with no irradiation.

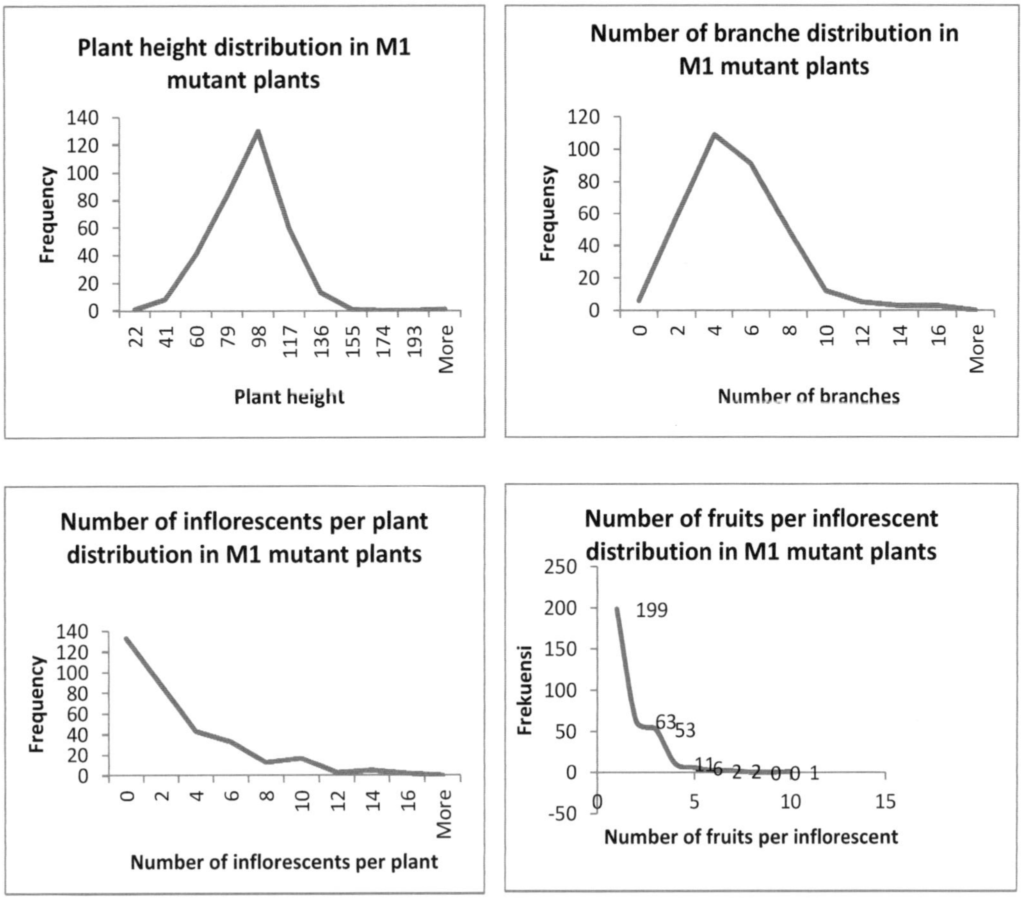

Figure 1. Distribution frequency of morphological and agronomical characters of mutant plants at 4 weeks after planting (WAP)

This study shown that for different accession, the effects of irradiation were different. Radio sensitivity of the irradiation given varied among plant species (20). The difference in radio sensitivity among cultivars or breeding lines is directly related to their genetic. This might occur as each accession has different base sequence. The frequency of mutant occurrence is affected by different base sequences in various plant genetic(21).

Conclusions

It was concluded that gamma irradiation gave both stimulating and inhibiting effects on the morphological and agronomical characters observed. Acute, intermittent, and split dose 225 Gy irradiation tended to give stimulating effects in jatropha. Irradiation effects were different among accessions.

Acknowledgement

Funding support from Ministry of Research and Technology through the 2014 Sinas Incentive Program was acknowledged.

References

[1] Jongschaap REE, Coree WJ, Bindraban PS, Bradenburg WA. 2007. Claim and Facts on *Jatropha curcas* L. Global *Jatropha curcas* evaluation, breeding and propagation programme. Plant Research International B.V., Wageningen. Stichting Het Groene Woudt, Laren. p.158

[2] Surahman M, Santosa E, Nisya FN. 2009. Karakterisasi dan Analisis Gerombol Plasma Nutfah Jarak Pagar Indonesia dan Beberapa Negara Lain Menggunakan Marka Morfologi dan Molekuler. J. Agron. Indonesia 37 (3) : 256 – 264.

[3] Nisya FN. 2010. Analisis keragaman genetik jarak pagar (*Jatropha curcas* L.) berdasarkan karakter morfologi, agronomi dan marka RAPD [Tesis]. Bogor: Sekolah Pascasarjana. IPB.

[4] Das S, Misra RC, Mahapatra AK, Gantayat BP, Pattnaik RK. 2010. Genetic variability, character association, and path analysis in Jatropha curcas. *World Applied Sciences Journal* 8 (11): 1304-1308.

[5] Arisanti Y. 2010. Analisis karakter agronomi dan pola pita isozim jarak pagar (*Jatropha curcas* L.) di daerah beriklim basah [Tesis].Bogor: Sekolah Pascasarjana. IPB.

[6] Mohapatra S dan Panda PK. 2010. Genetic Variability on Growth, Phenological and Seed Characteristics of *Jatropha curcas* L. *Not Sci Biol* 2 (2) : 127-132.

[7] Basha SD, Sujatha M. 2007. Inter and intra-population variability of *Jatropha curcas* (L.) characterized by RAPD and ISSR markers and development of population specific SCAR markers. *Euphytica* 156: 375-386

[8] Jubera, M.A., Janagoudar, B.S., Biradar, D.P. 2009 Genetic diversity analysis of elite Jatropha curcas (L.) genotypes using randomly amplified polymorphic DNA markers. Karnataka Journal of Agricultural Sciences 22 (2), 293-295

[9] Dhillon RS, Saharan RP, Jattan M, Rani T, Sheokand RN, Dalal V, GeorgeVon Wuehlisch.2014. Molecularcharacterization ofinduced mutagenesis through gamma radiation using RAPDmarkers in*Jatropha curcas*L. AfricanJournalofBiotechnology. 13(7), pp. 806-813

[10] Zainudin A, Maftuchah, Fitriani H. 2014. Analysis of Genetic Diversity on Mutants JatrophacurcasUsing RAPD.Energy Procedia 47. hlm 1 – 6.

[11] Chiangmai P Na, Pootaeng-on Y, Meetum P, Jankomon N, Muangnoi D, Kitthip D. 2014. Mutation Induction in Physic Nut (Jatropha curcasL.) by Colchicine Treatments. Silpakorn U Science & Tech J 8(2): 28-39.

[12] Kiong, A., A., Ling Pick, S.H. Grace Lai and A.R. Harun. 2008. Physiological responses of Orthosiphon stamineus plantlets to gamma irradiation. Am-Eurasian J. Sustain. Agric., 2(2): 135-149

[13] Yudiwanti, Darwati I, Rosita,Pulungan MY,Roostika I. 2013. Keragaan Mutan Putatif Purwoceng (Pimpinella pruatjanMolk.) dari Benih Diiradiasi Sinar Gamma pada Tiga Ketinggian Tempat. J. Agron. Indonesia 41 (1) : 77 – 82.

[14] Purwati, RD, Hariyono B. 2010. Aplikasi teknologi radiasi dalam pemuliaan jarak pagar di Balittas. Prosiding Simposium dan Pameran Teknologi Aplikasi Isotop dan Radiasi. Batan

[15] Dhakshanamoorthy D, Selvaraj R, Chidambaram a. 2010. Physical and chemical mutagenesis in *jatropha curcas* L. To induce variability in seed germination, growth and yield traits. Rom. J. Biol. – Plant Biol., Volume 55, No. 2, P. 113–125, Bucharest.

[16] Micke A, Donini B, Maluszyonski M. 1990. Induced Mutations for Crop Improvement.Mutation Breeding Revm. 7: 1–41

[17] Okuno K, Kawait T. 1978. Genetic analysis of induced longculm mutants in rice. Jap.J.Breed. 28. pp. 336-342.

[18] Sukartini T. 1992. Pengaruh Radiasi Corm dengan Sinar Gamma terhadap Keragaman Pertumbuhan dan Bunga Gladiol (Gladiolus hybridus cv. Queen Occer). Skripsi. Departemen Budidaya Pertanian, Fakultas Pertanian, IPB. Bogor.

[19] Hasnam. 2006. Karakteristik Pembungaan Jatropha curcas L. *Info Tek Jarak Pagar* (*Jatropha curcas* L.). 1:5.

[20] Broertjes C, van Harten AM. 1988. Applied Mutation Breeding for Vegetatively Propagated Crops. Elsevier, Netherland. p. 345.

[21] Barshile JD, Apparao BJ, 2009. Genetic improvement of chickpea (*Cicerarietinum L*) using induced mutations. In Q.Y. Shu (ed). Induced Plant Mutations in the genomicsera. FAO of the UN. Rome. pp 91-94.

Preliminary plant design of *Escherichia coli* BPPTCC-EgRK2 cell culture for recombinant cellulase production using Oil Palm Empty Fruit Bunch (OPEFB) as substrate

E A Surya, S F Rahman, S Zulamraini, M Gozan*

Chemical Engineering Department, Faculty of Engineering, Universitas Indonesia, Kampus UI, Depok 16424 – Indonesia
*corresponding author: misrigozan@gmail.com

Abstract. An economic analysis of recombinant cellulase production from *E. coli* BPPTCC Eg-RK2 was conducted to support the fulfilling of Indonesia's energy roadmap for ethanol production. The plant use oil palm empty fruit bunch (OPEFB) as primary substrate in cellulase production, with the expected lifetime of 12 years. The plant is assumed to be built in Indonesia and will fulfill 1% of total market demand. The effect of different pretreatment process (alkaline, steam explosion, and sequential acid-alkaline) on the economic value was also studied. A simulation using SuperPro Designer was used to calculate the mass and energy balance based on the kinetic parameter of *E. coli* BPPTCC-EgRK2. Technology evaluation show that alkaline pretreatment gave the highest yield with no known inhibitors formed. The steam explosion show the lowest lignin and hemicellulose removal and known to form known fermentation inhibitors. The net present value of alkaline, steam explosion, and sequential acid-alkaline pretreatment were USD 7,118,000; - USD 73,411,000 and USD -114,013,000 respectively, which mean alkaline pretreatment is the only economically feasible pretreatment method for recombinant cellulase production.

1. Introduction

Cellulase is enzyme that hydrolyzed cellulose, the most abundant compound in plant cell walls, and widely use in different industry [1]. It is used in ethanol production, denim processing, animal feed additives, and detergents [2, 3]. However, according to Juniarto, as quoted by BPPT in 2013, 99% of the enzymes used for industries in Indonesia are imported from other countries. The import value of enzyme is estimated to reach 121.85 billion rupiah and will continue to increase with a rate of increase of 6.67% per annum. This is unfortunate, as there is enormous potential in Indonesia's natural resources that should be able to meet those needs.

One example of Indonesia's natural resource potential is oil palm [4]. The processing of palm leads to the formation of several by-products and residues that have economic potential [5, 6]. Empty fruit bunches (EFBs) are solid residue and they have a high cellulose content, at 45.80% of EFB dry weight. In this case, the high cellulose content of EFB is used as a substrate for bacteria cultivation to produce cellulase. By using EFB as an alternative substrate, the production cost of cellulase on an industrial scale can be suppressed.

Currently, most cellulase enzymes are produced from saprophytic microorganisms [3], with fungi from *T. reseei* being one of the most used in industrial applications [7]. However, industrial applications often require a cost-effective production of enzymes on a large scale, which is a common bottleneck. An overexpression of individual cellulose enzymes using recombinant DNA is one of the common solutions [3]. *Escherichia coli* and *Bacillus* are two most common microbes used to express recombinant proteins [3, 7]. The feasibility of recombinant cellulase production from *E. coli* will be evaluated in this study using the SuperPro Designer Simulator.

In prior to simulate a cellulase plant design, kinetic studies of bacteria used in cultivation is needed to create an accurate simulation. These studies was done in prior research and used as primary data in this research.

2. Materials and Methods

2.1. Process description

The production of recombinant endo-β-1,4-glucanase is based on kinetic study from this and previous study [8] and the process scheme was shown in Figure 1. Oil palm empty fruit bunch (OPEFB) was used as raw materials for cellulase enzyme production in this study. OPEFB was collected from local palm oil mill and its characteristics were shown in Table 1. The OPEFB was pretreated with three different procedure, alkaline [9], steam explosion [10], and sequential acid-alkaline [11]. The pretreated OPEFB was fermented using *Escherichia coli* BPPTCC-EgRK2. The microbes was obtained from Indonesia's Agency for the Assesment and Application of Technology.

The fermentation was conducted in two seed fermenter and one production fermenter operating in stagered mode. All of the fermenter was regarded as a completely mixed reactor. The seeding fermenter to fermenter ratio was 1:40, with fermentation time of 4-7 h. The optimum production conditions were obtained from previous study at pH 7, agitation at 150 rpm and temperature of 37°C and known to produce 3.5 IU/mL cellulase [8,12]. The known cellulase concentration was used to estimate the scale up enzyme production yield. The fermentation reaction was:

$$C_6H_{12}O_6 + 0.69\ C_{10}H_{14}N_2O_5 + 1.56\ NH_3 + 0.085\ O_2 \rightarrow 12.25\ CH_{1.77}O_{0.49}N_{0.24} + 2.33\ H_2O + 0.64\ CO_2$$

The following assumption is used for the mass balance: 1) nitrogen sources come from fish powder which represent as thymidine, 2) cellulase was intracellular product represent by *E. coli,* 3) glucose, cellulase and cell concentration was calculated based on experimental data.

The resulting cell was separated from its medium using centrifugation to concentrate the cell before disrupted. Cell was disrupted to release its cellulase using high pressure homogenizer at 800 bar with 3 passes. Cellulase was then separated from cell debris using centrifugation. The crude cellulase was then stabilized using maltodextrin [13] and spray dried to produce crude enzyme powder. The spray dried product was expected to have activity loss up to 37%.

2.2. Economic analysis

The mass and energy balance was calculated from SuperPro Designer 9.0 based on the described process. The same simulation was also used to estimate various economic parameters using data such as equipment purchase cost, direct cost and indirect cost. There were two costs used to evaluate the economic parameters of the plant, total capital investment (TCI) and annual operating cost. TCI was calculated based on main equipment purchase cost. The equipment cost was estimated using equation [14] and graph [15]. The additional direct/indirect costs was estimated from the equipment purchase cost [15]. Annual operating cost was estimated from raw materials cost, labor dependent cost, facility dependent cost, utilities and depreciation cost.

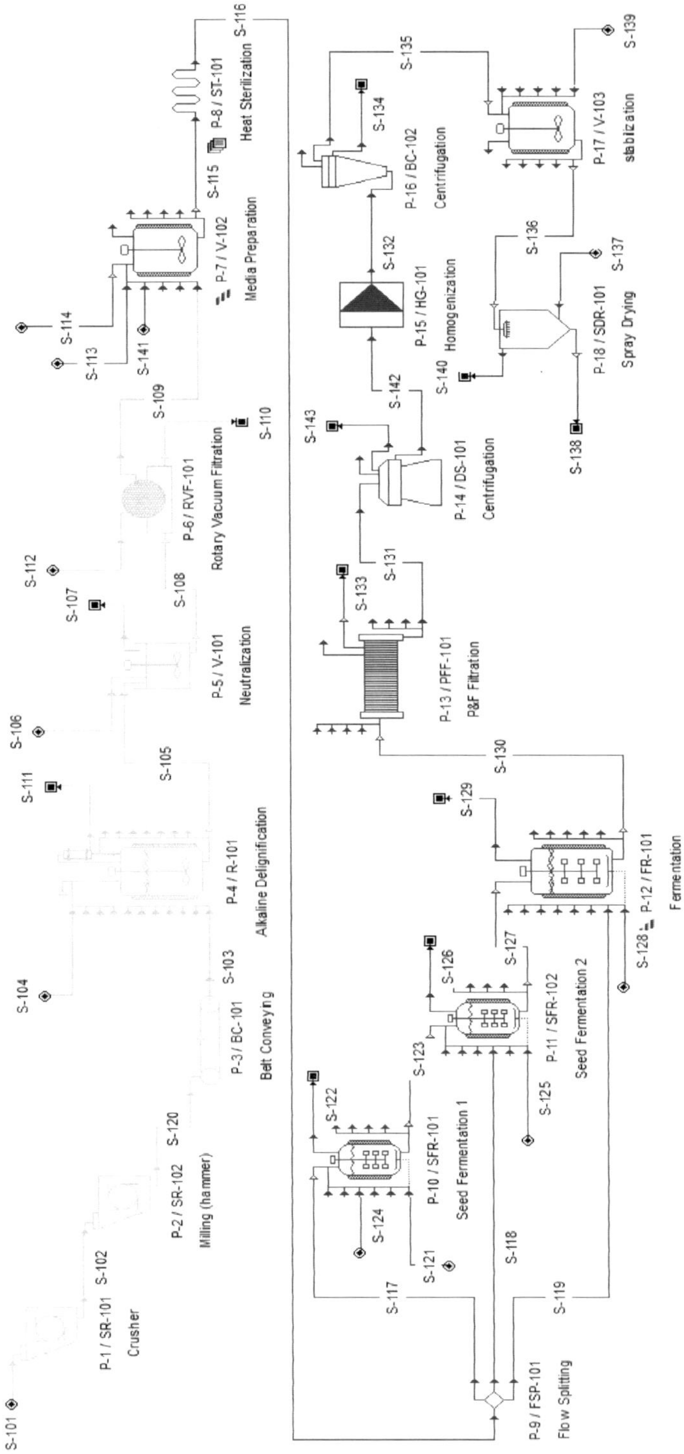
S-101
P-1 / SR-101
Crusher
S-102
P-2 / SR-102
Milling (hammer)
S-120
P-3 / BC-101
Belt Conveying
S-103
S-104
P-4 / R-101
Alkaline Delignification
S-111
S-105
S-106
P-5 / V-101
Neutralization
S-107
S-112
S-108
P-6 / RVF-101
Rotary Vacuum Filtration
S-110
S-109
S-141
S-113
S-114
P-7 / V-102
Media Preparation
S-115
P-8 / ST-101
Heat Sterilization
S-116
P-9 / FSP-101
Flow Splitting
S-117
S-118
S-119
S-121
S-124
P-10 / SFR-101
Seed Fermentation 1
S-122
S-123
S-125
P-11 / SFR-102
Seed Fermentation 2
S-126
S-127
S-128
P-12 / FR-101
Fermentation
S-129
S-130
P-13 / PFF-101
P&F Filtration
S-133
S-131
P-14 / DS-101
Centrifugation
S-143
S-142
P-15 / HG-101
Homogenization
S-132
P-16 / BC-102
Centrifugation
S-134
S-135
P-17 / V-103
stabilization
S-139
S-136
S-137
P-18 / SDR-101
Spray Drying
S-140
S-138

(A)

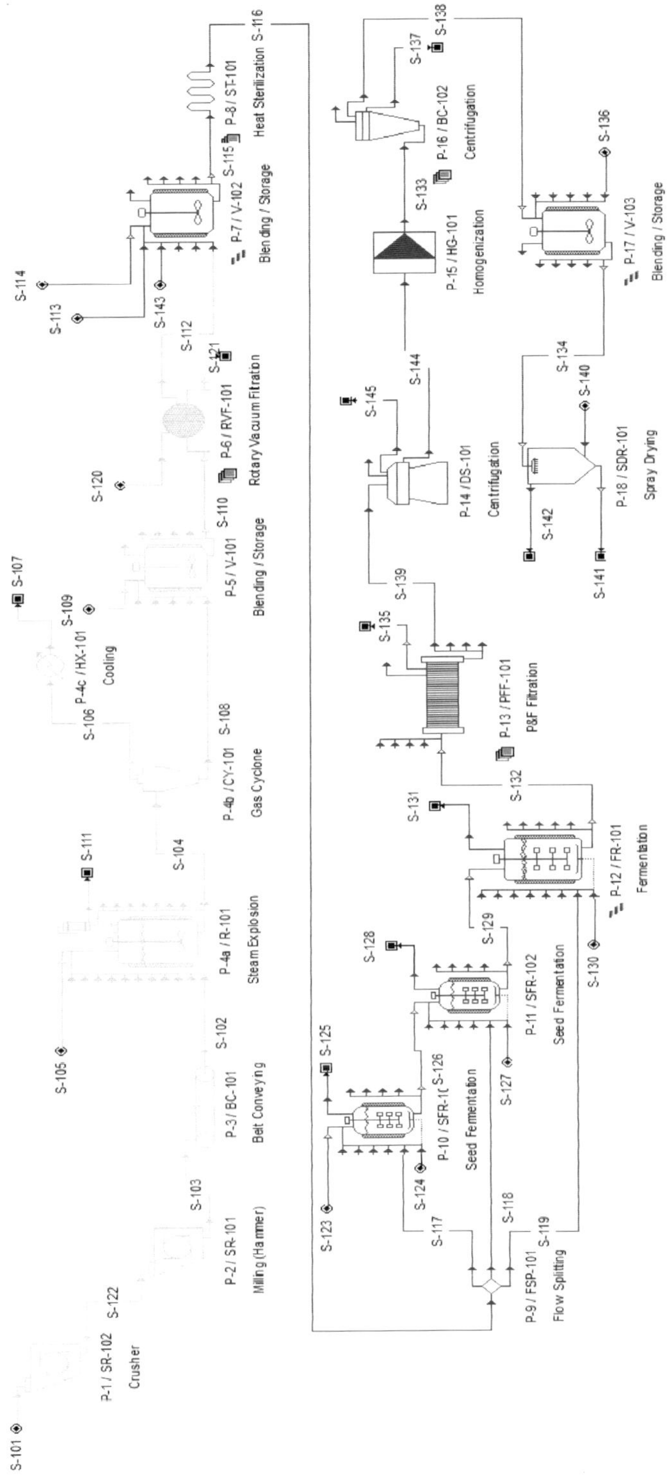
S-101
P-1 / SR-102
Crusher
S-122
P-2 / SR-101
Milling (Hammer)
S-103
P-3 / BC-101
Belt Conveying
S-102
S-105
P-4a / R-101
Steam Explosion
S-111
S-104
P-4b / CY-101
Gas Cyclone
S-108
S-106
P-4c / HX-101
Cooling
S-109
S-107
P-5 / V-101
Blending / Storage
S-110
S-120
P-6 / RVF-101
Rotary Vacuum Filtration
S-121
S-112
S-113
S-114
S-143
P-7 / V-102
Blending / Storage
S-115
P-8 / ST-101
Heat Sterilization
S-116
S-117
S-118
S-119
P-9 / FSP-101
Flow Splitting
S-123
S-124
S-125
P-10 / SFR-1(
Seed Fermentation
S-126
S-127
S-128
P-11 / SFR-102
Seed Fermentation
S-129
S-130
S-131
P-12 / FR-101
Fermentation
S-132
P-13 / PFF-101
P&F Filtration
S-135
S-139
P-14 / DS-101
Centrifugation
S-145
S-144
P-15 / HG-101
Homogenization
S-133
P-16 / BC-102
Centrifugation
S-137
S-138
S-136
P-17 / V-103
Blending / Storage
S-134
S-140
P-18 / SDR-101
Spray Drying
S-142
S-141

(B)

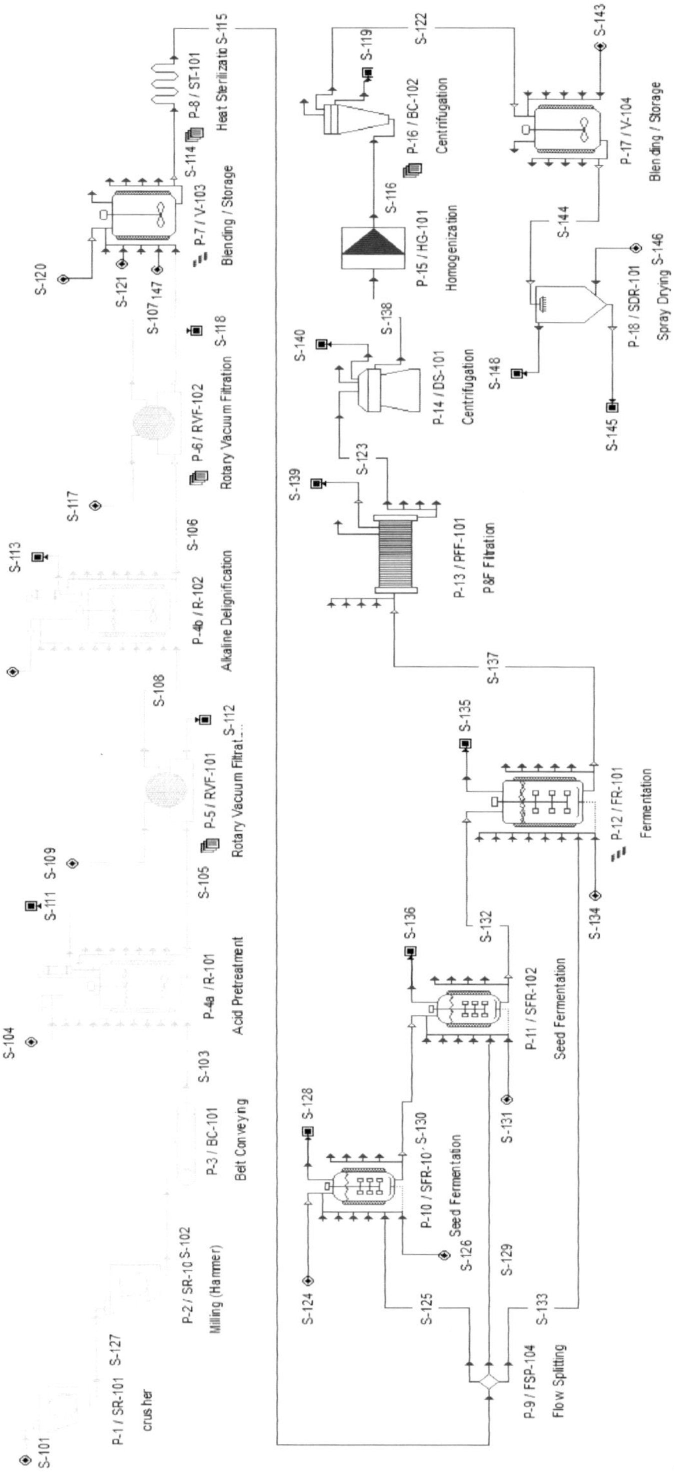

(C)

Figure 1. Cellulase Production Process scheme with A) Alkaline pretreatment, B) steam explosion pretreatment, and C) sequential acid-alkaline pretreatment.

IOP Conf. Series: Earth and Environmental Science **141** (2018) 012030 doi:10.1088/1755-1315/141/1/012030

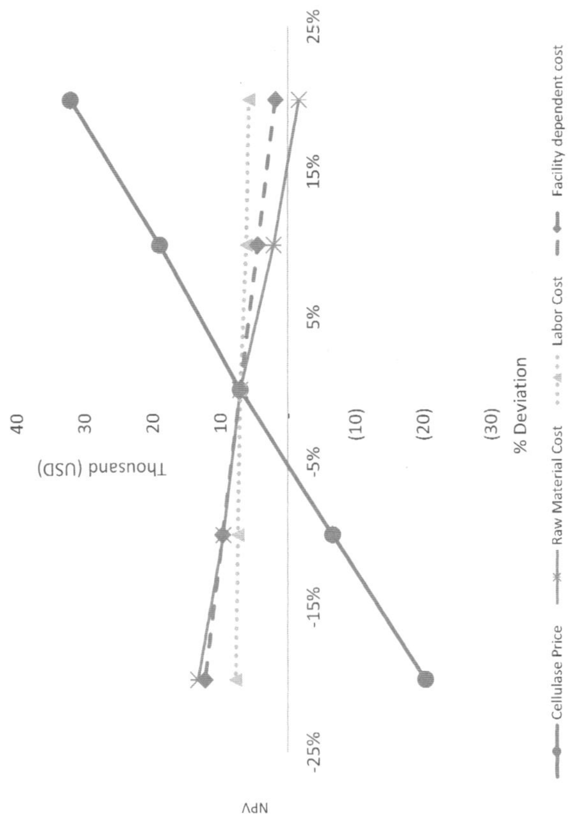

Figure 2. Sensitivity analysis of recombinant cellulase production from OPEFB

3. Results and Discussions

3.1. Techno-Economy analysis of cellulase production

In order to evaluate the most suitable cellulase production process, the yield, batch duration, number of unit, pretreatment performance and economic parameter were compared. The alkaline pretreatment process was able to fulfill 1% of Indonesia's market share using 8.5 ton of OPEFB as substrate, smaller then the acid-alkaline and steam explosion pretreatment which required 19 ton and 29 ton OPEFB, respectively. The alkaline pretreatment also give the highest product yield, 0.03, compared with acid-alkaline and steam explosion, which gives yield of 0.021 and 0.008.
As for delignification efficiency, the sequential acid-alkaline gives the highes lignin removal, by removing 90% lignin from OPEFB [11]. Steam explosion did not remove any lignin from OPEFB but gives high hemicellulose removal, which increase its permeability and increase enzimatic digestibility [10]. However, this process also produce fermentation inhibitors such as levulinic acid (0.11 g/L), formic acid (0.21 g/L), hidroxymethyl furfural (0.71 g/L) and furfural (0.23 g/L) [10]. Similar result also achieved by another study [16]. The performance of each simulation is shown in table 1.

Table 1. Simulation summary of different pretreatment method effect on project profitablility

Parameter	Pretreatment Method		
	Alkaline (CHEMEX) [1]	***Steam Explosion* [2]**	**Acid-Alkaline [3]**
OPEFB Feed/batch (Ton)	8.5	29	19
Yield p/s	0.03	0.008	0.021
Number of units	18	20	21
Fermentation inhibitor	-	Present	-
Removal efficiency			
-Cellulose	19%	30%	41%
-Hemicellulose	67%	82%	86%
-Lignin	80%	0%	90%
Total Capital Investment (USD)	35,195,000	53,227,000	83,383,000
Operating Cost (USD)	23,094,000	34.948,000	41,692,000
Internal Rate of Return	14.61	N/A	N/A
Net Present Value (9.6%) (USD)	7,118,000	-73,411,000	-114,013,000
Pay back period	4,84	N/A	N/A

3.2. Economic analysis

The TCI was calculated based on purchase and instalation cost of each equipment, piping, instrumentation, facility dependent cost, engineering and construction fee. Table 1 show the total capital investment and annual operating cost of each simulation. The TCI was 35,195,000 USD, 53,227,000 USD and 83,383,000 USD for alkaline pretreatment, steam explosion pretreatment and acid-alkaline pretreatment. High capital investment on steam explosion and sequential acid-alkaline pretreatment was due to its number of unit. Steam pretreatment have 20 different unit and sequential acid-alkaline pretreatment used 21 units with multiple rotary vacuum filtration unit, resulting in higher capital investment on equipment and installation cost. The annual operating cost for recombinant cellulase production were highly influence by raw materials cost, labor-dependent cost, and waste treatment cost. Acid-alkaline pretreatment had the highest operating cost at 41,692,000 USD due to energy intensive pretreatment process. Steam explosion also give high operational cost due to low conversion rate, resulting in high raw materials cost and high waste treatment cost.

The plant revenue came solely from cellulase sales. The annual cellulase production was 154 ton (1% market share), with revenue of 30,800,000 USD. NPV was the key to determine the feasibility of a project. Simulation show that alkaline pretreatment gives the highest NPV, 7,118,000 USD. Both steam explosion pretreatment and sequential acid-alkaline pretreatment process gives negative NPV value due to higher operational cost.

3.3. Sensitivity Analysis

Project feasibility is also evaluated from its sensitivity to key variables, such as selling price, raw materials cost and labor cost. A project that highly affected with external factors are considered riskier. Figure 2 show the effect of changes in labor cost, raw materials cost, selling price, and facility dependent cost to NPV value. It was found that cellulase selling price have the highest influence on the NPV. However, as this project was intended to fulfill Indonesia's government renewable energy target, a steady demand is expected in near future. Both labor cost, raw materials cost, and facility dependent cost do not affect the project as much as the selling price, showing less risk on these parameters.

4. Conclusions

In this study, the techno-economic analysis of recombinant cellulase production from OPEFB was investigated. A comparison of three different pretreatment process were evaluate to find the most feasible process. Recombinant cellulase production using alkaline pretreated OPEFB was found to give highest NPV, internal rate of return, and payback period with value of 7,118,000 USD, 14.61% and 4.84 years, respectively. This project also found to be sensitive to cellulase selling price, as it gives highest impact on the net present value.

Acknowledgments

The authors would like to thank PITTA Ristekdikti funds, Universitas Indonesia and BPDPKS, for providing the funding needed for this research.

References

[1] Kuhad R C, Deswal D, Sharma S, Bhattacharya A, Jain K K, Kaur A, Pletschke B I, Singh A, and Karp M 2016 Revisiting cellulase production and redefining current strategies based on major challenges *Renew. Sust. Energ. Rev.* **55** 249

[2] Zhang X-Z and Zhang Y-H P 2013 *Bioprocessing Technologies in Biorefinery for Sustainable Production of Fuels, Chemicals, and Polymers* (USA: John Wiley & Sons) p 131-146

[3] Juturu V and Wu J C 2014 Microbial Cellulases: Engineering, production and applications. *Renew Sust. Energ. Rev.* **33** 188

[4] Badan Pusat Statistik, 2014. *Statistik Kelapa Sawit Indonesia,* Jakarta: Badan Pusat Statistik

[5] Piarpuzan D, Quintero J A and Cardona C A 2011 Empty Fruit Bunches from Oil Palm as Potential Raw Material for Fuel Ethanol Production. *Biomass Bioenergy* **35(3)** 1130

[6] Chiew Y L and Shimada S 2013 Current state and environmental impact assessment for utilizing oil palm empty fruit bunches for fuel, fiber and fertilizer - A case study of Malaysia. *Biomass* Bioenergy **51** 109

[7] Hasunuma T, Okazaki F, Okai N, Hara K Y, Ishii J and Kondo A 2013 A review of enzymes and microbes for lignocellulosic biorefinery and the possibility of their application to consolidated bioprocessing technology. *Bioresour Technol.* **135** 513

[8] Ariyani L P 2015 *Produksi Selulase dengan Bacillus amyloliquefaciens BPPTCC RK2 dan Rekombinan EgRK2 Berbasis Tandan Kosong Kelapa Sawit: Pengaruh pH dan Suhu serta Uji Aktivitas* (Depok: Universitas Indonesia)

[9] Sudiyani Y, Styarini D, Triwahyuni E, Sudiyarmanto, Sembiring K C, Aristiawan Y, Abimanyu H and Han M H 2013. Utilization of biomass waste empty fruit bunch fiber of palm oil for bioethanol production using pilot - scale unit. *Energy Procedia* **32** 31

[10] Medina J D C, Woiciechowski A, Filho A Z, Nigam P S, Ramos L P and Soccol C R 2016 Steam explosion pretreatment of oil palm empty fruit bunches (EFB) using autocatalytic hydrolysis: A biorefinery approach. *Bioresour. Technol.* **199** 173

[11] Kim S, Park J M, Seo J-W and Kim C H 2012 Sequential acid/alkali-pretreatment of empty palm fruit bunch fiber. *Bioresour. Technol.* **109** 229

[12] Setiawan R 2014 *Kloning dan Ekspresi Endo-B-1,4-Glukanase dari Isolat Lokal BPPTCC-RK2,* (Depok: Universitas Indonesia)

[13] Belghith H, Chaabouni S E and Gargouri A 2001 Stabilization of Penicilium occitanis cellulases by spray drying in presence of Maltodextrin *Enzyme Microb. Technol.* **28** 253

[14] Seider W D, Seader J D and Lewin D R 2003 *Product and Process Design Principles: Synthesis, Analysis, and Evaluation.* (New York: John Wiley and Sons, Inc)

[15] Petrides D 2003 *Bioprocess Design and Economics* (Oxford: Oxford University Press) p. 441-510

[16] Duangwang S, Ruengpeerakul T, Cheirsilp B, Yamsaengsung R and Sangwichien C 2016. Pilot-scale steam explosion for xylose production from oil palm empty fruit bunches and the use of xylose for ethanol production *Bioresour Technol.* **203** 252

Study of oil palm root architecture with variation of crop stage and soil type vulnerable to drought

Lisma Safitri[1], Sri Suryanti[1], Valensi Kautsar[1], Agung Kurniawan[2] and Fajar Santiabudi[2]
[1] Instiper Yogyakarta, Jl. Nangka II, Sleman, Yogyakarta, Indonesia
[2] Research Centre PT Bumitama Gunajaya Agro, Pundu, Kalimantan Tengah, HO : Jl. Melawai Raya No. 10 Kebayoran Baru, Jakarta 12160, Indonesia
Author to whom any correspondence should be addressed
Email: lisma.safitri86@gmail.com

Abstract. Root arhitecture is affected by watertable level, characteristic of soil, organic matter and also the crop stages. Root architecture spread horizontally and vertically which each consist of primary, secondary, tertiary and quaternary downward root. The oil palm root observation with variation of crop stage and soil type showed that the root of oil palm plant year 2008 on spodosols soil spread along 650 cm horizontally from the trunk and penetrate downward in range of 9-28 cm vertically. Planted in the same type of soil, the root of oil palm plant year 2004 spread along 650 cm horizontally and reached to downward in a larger range from 3 to 57 cm vertically. As a comparison, the root architecture of oil palm on inceptisols soil established the range much greater vertically than the previous. The root of oil palm plant year 2008 spread along 640 cm horizontally and penetrate downward in range of 52-90 cm vertically. With the variation of crop age, the root of oil palm plant year 2003 spread along 650 cm horizontally and reached to downward in a larger range from 150 to 200 cm vertically. Based on this study, root architecture of oil palm was varied and need to be detailed. The precise root architecture of oil palm allows a better understanding on hydrological properties of oil palm root particularly which is cultivated on soil type vulnerable to drought. Referring to this root architecture, it was enable to develop the study on early drought detection of oil palm to optimise production and towards oil palm sustainability.

Keywords: Root Architecture, Oil Palm, Drought, Spodosols Soil, Inceptisol Soil

1. Introduction

Root architecture describes root topology and distribution [8]. Root architecture varies among plant species, it even varies among individual plants within the same species. Root architecture is an important factor of root function such as nutrient and water intake [12]. The roots of oil palm can grow only to a depth of 1 m if the plants get enough water and nutrients. However, on condition of limited water availability, it can reach more than 5 m. In mature palm oil plants, total dry weight of roots increased after a radius of 2 m to 3.5 - 4.5 m from the stem [2].

According to Jourdan et al. [3], the oil palm (*Elaeis guineensis* Jacq.) has a root system consisting of primary (or order 1) roots, which are either orthogravitropic (R1 VD, with positive gravitropism) or diagravitropic (R1 H). Order 2 roots may have vertical upward (R2 VU) or downward growth (R2 VD) or even horizontal growth (R2 H). Order 3 roots are usually agravitropic. When they grow

IOP Conf. Series: Earth and Environmental Science 141 (2018) 012031 doi:10.1088/1755-1315/141/1/012031

upwards, R3 VU, their amyloplasts are located near the proximal wall. Likewise, the growth direction of R4 varies, but they have little or no statolith Sedimentation.

Jourdan and Rey [4] described the growth dynamics and architecture of the oil-palm root system consisted of a transitional juvenile phase and eight different morphological types of roots have been distinguished according to their development pattern and state of differentiation: primary vertical and horizontal roots, secondary horizontal roots, upward growing secondary vertical roots and downward growing secondary vertical roots, superficial and deep tertiary roots and quaternary roots. The relative position of these types of roots determines a morphological and functional unit ofthe root system called "root architectural unit" of the oil palm. The quantitative parameters of different root types of oil palms in the juvenile phase. RI1_5-primary root emitted for a 1 to 5-month-old oil palm, RI5_ 12 primary root emitted for a 5 to 12 month-old oil palm.

The development of oil palm is currently directed to marginal land, such as spodosols and inceptisols. There are limiting factors on spodosols such as the depth of spodic layer and sandy soil texture and acidic for tropical area. The depth of spodic layer is the main factor for root growth because it is related to the ability of roots in penetrating soil, whereas the sandy soil texture will reduce the soil ability to retain water as well as greater chance of nutrients leaching. Other limiting factors that could potentially inhibit plant growth are poor drainage and soil acidity.

According to Kasno and Subarja [5], the depth of spodic layer on spodosols ranges from 30 to 70 cm below soil surface. Oil palm requires solum depth equal to or greater than 80 cm without layers of rock for optimal growth and development [7] and at least the layer is more than 75 cm depth [13].

Inseptisols is one of the acid mineral soils with low nutrient availability. The productivity of oil palm planted on the soil is low, and there are symptoms of decreased productivity in certain months of the year. The use of insepticols for agricultural land has faced many problems in physical, biological and chemical properties of the soil. The problem with its physical properties is that the topsoil has a coarse texture, but less coarse in the lower texture. Therefore, the permeability is rapid on the top surface but low in the lower layer. The topsoil structure is granular or crumb and the lower layer is unstructured, with lower density on the surface and increased with the increasing depth. The problems related to biological properties such as species, population, and biota activities. And the problems related to chemical properties are high C-organic content (5,06 – 5,39%) while N-total is relatively low (0.15 – 0.42%) resulted in moderate C/N ratio which is about 12 – 35%. Cation Exchange Capacity is relatively moderate at about14.1-17.3 me/100 g, while base saturation is low 24-29%. This low base saturation results in low nutrient and CEC availability [10].

Empty fruit bunches are soil and nutrients enhancers for plants. Empty fruit bunches contain organic matter and have relatively high nutrient content. The utilization of empty fruit bunches can be done by direct application as mulch or composting. It has high C/N rate which is about 45-55.

The results from research in Pontianak, West Kalimantan using electrical resistivity method showed that on the peat land the oil palm plant has a horizontal root distribution reaching 3.5 m with a depth of 1.74 m [9]. The results of oil palm resistivity measurements at PTPN 13 showed that tertiary root distribution and the dominant quarter were in the horizontal direction between 2.5 - 4.0 m from the trunk with maximum depth of 0.3 m. While the secondary roots were generally located between 0.3 m - 0.5 m depth and a small part appear to the surface. The primary roots are at a depth of 0.5 - 1.08 m and tend to move downward along vertical direction.

The distribution of roots in spodosols is limited only to the surface of soil layer. On sandy soil, around 90% lateral root is found from the surface to 30 cm depth [1]. Root distribution has an important role in water and nutrient intake. It is also closely related to groundwater level on a field. The studies regarding empty fruit bunches application on spodosols and inceptisols were conducted to observe the influence of this application to the changes of root architecture on oil palm. The observation included distribution/range of primary, secondary, and tertiary roots both in vertical and horizontal direction.

Root arhitecture was affected by water table level, soil characteristics, soil organic matter, and as well as crop stage. The ability to grow at any time in response to soil conditions might be an essential

IOP Conf. Series: Earth and Environmental Science **141** (2018) 012031 doi:10.1088/1755-1315/141/1/012031

prerequisite for phreatophytes (deep-rooted plant) if they are to survive fluctuating water table conditions in seasonally water-limited environments. Based on this study, root architecture of oil palm was varied and need to be detailed. The precise root architecture of oil palm allows to better understanding on hydrological properties of oil palm root particularly which is cultivated on soil type vulnerable to drought. Referring to this root architecture, it enables to develop the study on early drought detection of oil palm to optimize production and towards oil palm sustainability.

This research was aimed to study the root distribution of different soil type: spodosols and inceptisols, root distribution of different crop age: plant year 2008 and 2004, and root distribution in response to water level and empty fruit bunches (EFB) application.

2. Materials and Methods

The research was conducted in Pundu Central Kalimantan on Avril to May 2017, where the humidtropical climate with marked seasons is characterized by (i) average annual rainfall was 3002 mm/year(ii) average annual temperatures varied between21.4 – 33.8 °C and, (iii) the average sunshine hour was around5.9 hours of sunshine per year (based on the climate data series 2008-2015 from climate station of Pundu Plantation). The observation was on oil palm field planted in 2004 and 2008 on spodosols and inceptisols covering the area of 24.457,7 hectare.

The observation of root distribution deep phase based on growth position (upright, sideways and falldown) used destructive method by unloading the root zone using excavator,digging a trench along the side of the plant at a certain depth and of a convenient width.The observation of root distribution using two plants on spodosols and inceptisols soil type.

The measuring water table using automatic water level sensor HOBO U20-001-04-Ti (13-Foot Depth Titanium Water Level Data Logger). The water level HOBO was installed 9 separate points, with details of 4 tools at the location of the oil palm plant, 4 tools on the surface of the nearest river and one device mounted for reference. The equipment was installed by drowning it in groundwater as deep as ± 20 cm or in a river as deep as ± 1 meter. It is intended that the water fluctuations that occur will still drown the water level HOBO. In the meantime as a reference, HOBO should be installed in a condition not submerged in water, therefore mounted above ground level. Each HOBO is recorded depth of mounting or tool height, as a basis for determining the depth of the groundwater level. Pressure data retrieval was done automatically every 30 minutes during the observation. The readable pressure on each device will be adjusted to the reference HOBO, so that the actual pressure difference is known. Furthermore, from the pressure value recorded by the HOBO will be converted to the depth of groundwater at each observation point.

The effect of Empty FruitBunches on root distributionin spodosols and inceptisols observed with the variation of oil palm age (plant year 2008 and 2004). The Empty Fruit Bunches application to the oil palm field near the mill for nutrient cycling.

3. Results and Discussions

The roots of oil palm can only grow to 1 m depth when the plants receive adequate water and nutrients, but on condition where there is limited water availability, the roots can reach more than 5 m. In mature palm oil plants, total root dry weight increased after a radius of 2 m to 3.5 - 4.5 m from the stem [2].

The observation of root distribution was conducted in marginal land (spodosols and inceptisols) in Central Kalimantan. Spodosols with the soil texture consist of 89.29% sands, 3.44% silt and 7.28% loam and spodic layer has root distribution spread on the soil surface and shallow water table. While inceptisols which has soil texture 52.38% sands, 16.24% silt and 31.38% loam and deep water table,the roots were spread similarly shown in Figure 1. Root distribution of oil palm plant year 2004 Root distribution of oil palm plant year 2004 and 2008 could be seen in details in Table 1. Visually, the root architecture on spodosols is illustrated in Figure 2.

Figure 1. The observation of oil palm root distribution on spodosols (left) and inceptisols (right)

Table 1. Oil palm root distribution on spodosols plant year 2008 and 2004

Observation Parameter	Spodosols 2008 No EFB	Spodosols 2004 EFB
Water table	22 cm	14 cm
Spodic layer	56 cm	56 cm
Primary horizontal	28 cm	57 cm
Primary vertical downward	28 cm	57 cm
Secondary vertical downward	28 cm	57 cm
Tertiary vertical downward	28 cm	3 cm
Quaternary vertical downward	9 cm	3 cm

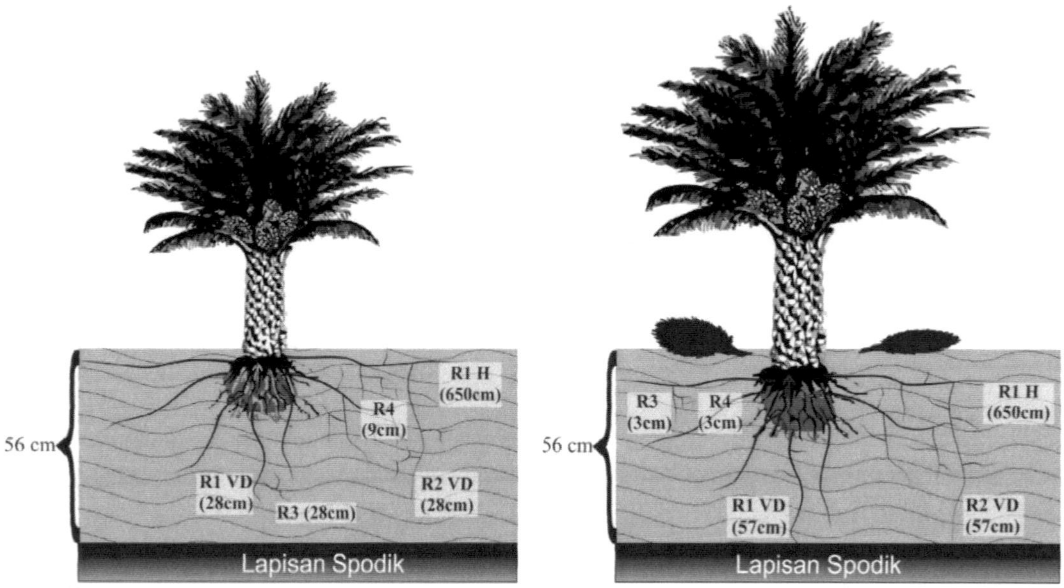

Figure 2. The architecture of oil palm root on spodosols plant year 2008 (left) and 2004 with EFB application (right)

The oil palm root observation with variation of crop stage and soil type showed that the root of oil palm plant year 2008 on spodosols spread along 650 cm horizontally from the trunk and penetrate downward in range of 9-28 cm vertically. Planted in the same type of soil, the root of oil palm plant year 2004 spread along 650 cm horizontally and reach downward in a larger range from 3 to 57 cm vertically.

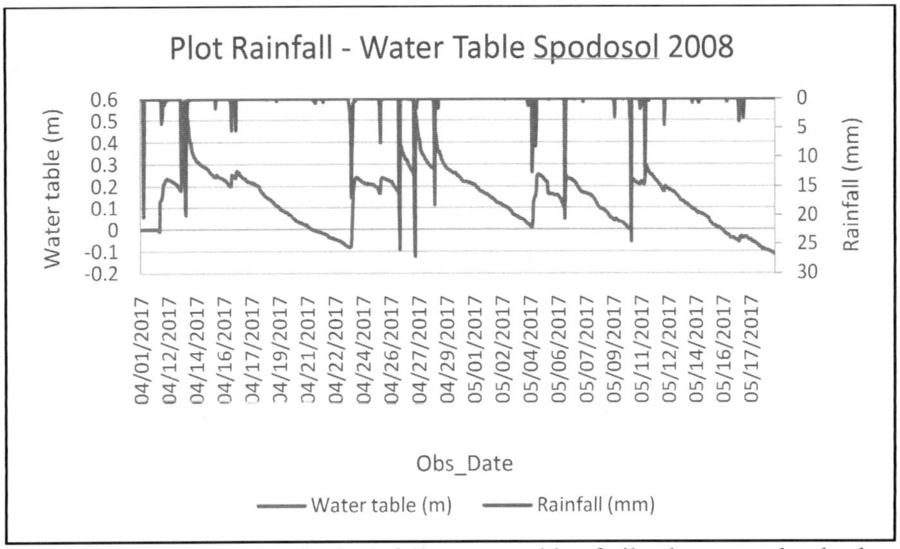

Figure 3. Observation graph of rainfall – water table of oil palm on spodosols plant year 2008

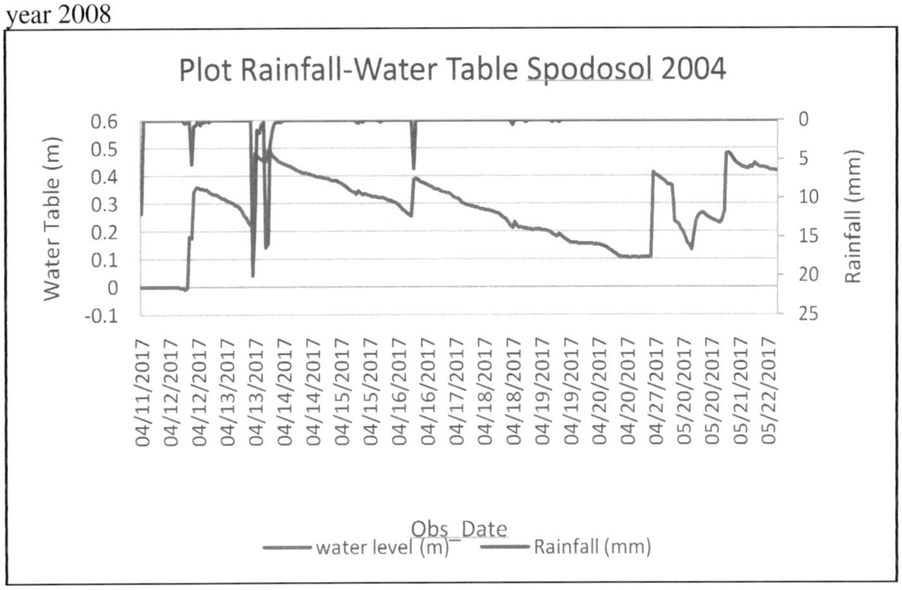

Figure 4. Observation graph of rainfall– water table of oil palm on spodosols plant year 2004

The observation of root distribution on spodosols in Pundu Central Kalimantan indicated that the depth of the spodic layer in both sites was 56 cm depth. Primary root coverage in horizontal direction with and without the application of empty fruit bunches showed no difference. The range of secondary root with the application of empty fruit bunches was further than without the application. The application caused the tertiary and quarter roots to appear in soil surface within 3 cm, while without the application the tertiary and quarter roots reached 28 and 9 cm, respectively. The use of empty bunches can improve the ability of soil to store water, improve physical and chemical properties of soil that the growth of tertiary and secondary roots appeared much in the soil surface.

The addition of empty fruit bunches in spodosols is very beneficial for plants due to more roots spread in the surface. Vertical roots distribution of tertiary and quarter roots range of 28 and 9 cm, respectively could not penetrate the spodic layer.

IOP Conf. Series: Earth and Environmental Science **141** (2018) 012031 doi:10.1088/1755-1315/141/1/012031

Table 2. Oil palm root distribution on inseptisols plant year 2008 and 2004

Observation Parameter	**Inceptisols; 2008 EFB**	**Inseptisols; 2004 No EFB**
Water table	600 m	1000 cm
Primary horizontal	640 cm	650 cm
Primary vertical downward	60 cm	200 cm
Secondary vertical downward	90 cm	200 cm
Tertiary vertical downward	52 cm	150 cm
Quaternary vertical downward	52 cm	150 cm

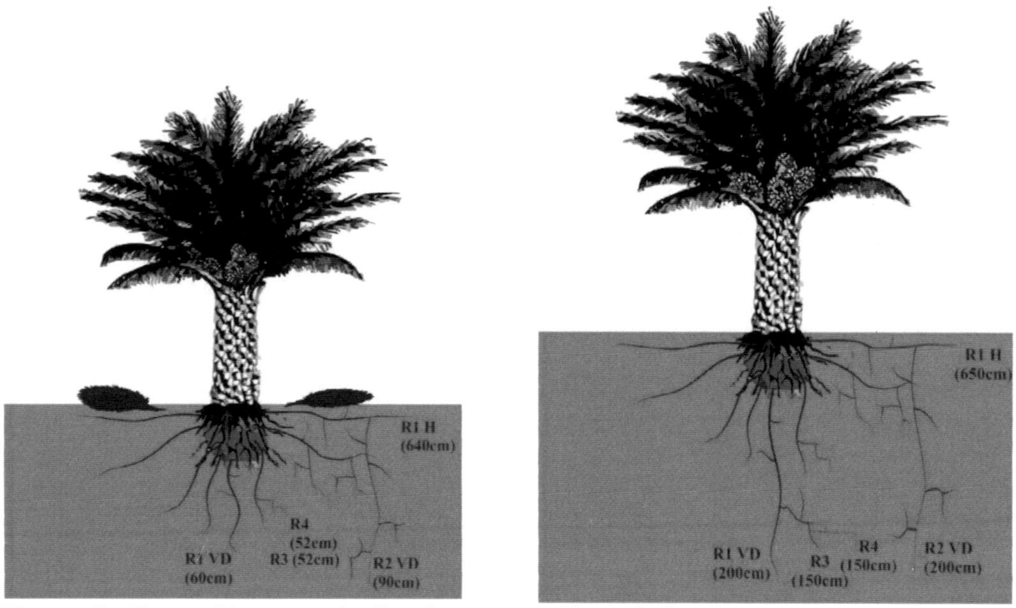

Figure 5. The architecture of oil palm root on inceptisols plant year 2008 with EFB application (left) and 2004 (right)

As a comparison to different type of soil, the root architecture of oil palm on inceptisols established greater range vertically than the previous. The root of oil palm plant year 2008 spread along 640 cm horizontally and penetrate downward in range of 52-90 cm vertically. With the variation of crop age, the root of oil palm plant year 2003 spread along 650 cm horizontally and reach to downward in a larger range from 150 to 200 cm vertically.

Palm oil roots in inceptisols without the application of empty fruit bunches have a deeper vertical root range compared to with the application of empty fruit bunches (Table 2). Inceptisols is infertile so that it has low nutrient availability. Empty fruit bunches acted as soil enhancer with the capability of binding groundwater. The application of empty fruit bunches resulted in shorter primary, secondary, tertiary and quarter roots to spread vertically than without the application. This result was similar to the research conducted by Kheong, *et al.*, 2010, which stated that the application of empty fruit bunches caused the quarter root of oil palm at 0 - 45 cm depth more than without the application of empty fruit bunches.

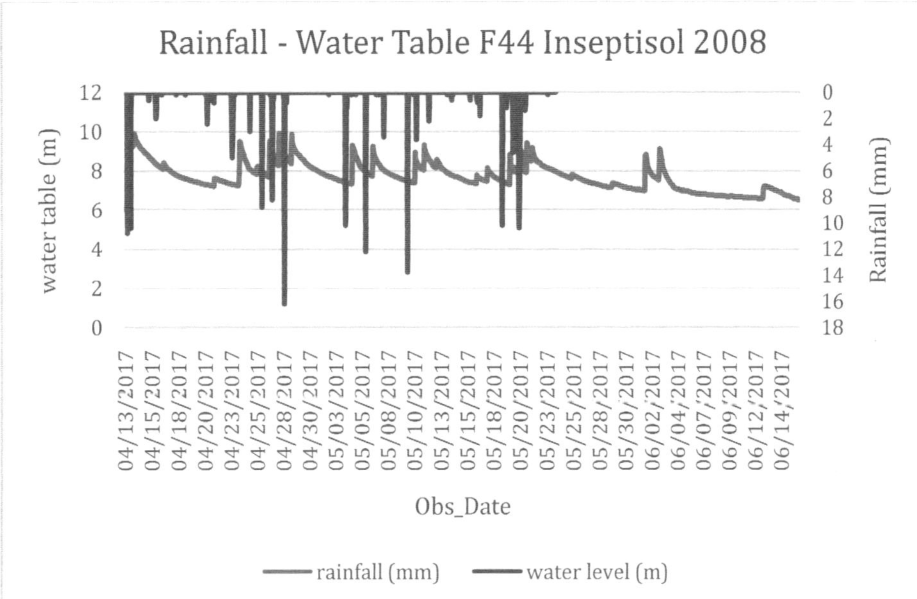

Figure 6. Observation graph of rainfall– water table of oil palm on inceptisols plant year 2008

Soil physical properties such as texture, structure, effective depth of soil, groundwater level, soil thickness, soil permeability are the factors which influence the development of plant root to support water and nutrient availability as well as supporting plant upright. If the texture is dominated by clay, the drainage will be hampered. On the contrary, if it is dominated by sand, soil will be drained quickly resulting in disturbed root development. Shallow effective depth of soil or high groundwater level (shallow) means that root distributions are limited [11].

Table 3. Root zone levelling for oil palm on inceptisols and spodosols plant year 2008 and 2004

Root Zone	Depth (cm)			
	Inceptisols		Spodosols	
	2008	2004	2008	2004
Zone 1	0-30	0-50	0-9	0-5
Zone 2	30-60	50-150	9-18	5-25
Zone 3	60-90	150-200	18-28	22-57

In inceptisols, high groundwater level affected the distribution of primary, secondary, tertiary and quarter roots vertically. The roots spread deeper by the increasing of groundwater level. The root zone levelling for oil palm on inceptisols deeper than on spodosols (Table 3). The application of empty fruit bunches could increase the groundwater level from 1000 cm to 600 cm, and reduce the coverage of primary, secondary, tertiary, and quarter roots vertically. The application of empty fruit bunches did not change much the distribution of horizontal primary roots (Table 2).

4. Conclusions

Based on the research observation about oil palm root distribution on two type of soils, i.e. spodosol and inceptisol, it can be concluded that :

- The primary root spread almost similar along horizontal side.
- The vertical downward root reached much greater area in inceptisols than spodosols.

- The deeper water table, the longer vertical downward root which penetrated the soil.
- The application of EFB on top soils led to the tertiary and quaternary vertical downward root distribution spread upward.

5. Suggestions

Based on the result of root distribution on spodosol and inceptisol with the variation of oil palm age (plant year 2008 and 2004), it could be suggested the root zone levelling as showed in Table 3. The root zone levelling could be the reference for the observation of soil moisture distribution in oil palm root zone.

Acknowledgement

This research was fully funded by Oil Palm Plantation Fund Management Agency, Indonesia. We also thank our colleagues from PT Bumitama Gunajaya Agro, Indonesia who provided the location, accommodation and the labours that greatly assisted this research.

References

[1] Brockway DG, Outcalt KW and Boyer WD. 2006 Longleaf pine regeneration ecology and methods. In : The longleaf Pine Ecosystem, Ecology, Silviculture and Restoration. *Springer Science + Business media LLC.*
[2] Corley and Tinker. 2003. *The Oil Palm (fourth edition)* (USA: Blackwell Science Ltd)
[3] Jourdan C, Ferrière NM, Perbal G. 2000. Root System Architecture and Gravitropism in the Oil Palm. Annals of Botany, Volume 85, Issue 6, 1 June 2000, Pages 861–868.
[4] Jourdan C and Rey H. 1997. Architecture and development of the oil-palm (*Elaeis guineensis* Jacq.) Root System. *Plant and Soil.* **189**: 33–48.
[5] Kasno A and Subardja D. 2010 Soil fertility and nutrient management on spodosol for oil palm. *Agrivita.* **32**(3): 285 - 292
[6] Kheong L V, Rahman Z A, Musa H N and Hussein A . 2010. Empty fruit bunch application and oil palm root proliferation. *Journal of Oil Palm Research.* **22** : 750 - 757
[7] Lubis AR. 2008. KelapaSawit di Indonesia. Pusat Penelitian Bandar Kuala Marihat Pematang Siantar : Sumatera Utara
[8] Lynch J.1995. Root architecture and plant productivity. *Plant Physiology.* **109**: 7-13.
[9] Miftahuddin. 2016. Pendugaan sebaran akar kelapa Sawit pada lahan gambut dengan menggunakan metode geolistrik resistivitas. *Prisma Fisika.* **4** (3) : 114 - 120
[10] Nasrul B, Hamzah A and Anom E. 2002. Klasifikasi tanah dan evaluasi kesesuaian lahan Kebun Percobaan Fakultas Pertanian Universitas Riau. *Jurnal Sagu.* **2**: 16- 26
[11] Syakir M, Zulkarnai.P, Syafaruddin., and Rumini W. 2010. *Budidaya Kelapa Sawit.* (Bogor: Aska Media)
[12] Wang H, Siopongco J, Wade L and Yamaguchi A. 2009. Fractal analysis on root systems of Plants in response to drought stress. *Environmental and Experimental Botany Journal.* **65** (2-3) : 338 – 344.
[13] Wiratmoko, Darlan N H, Winarna, and Purba AR.2015.Teknologi Pengelolaan Lahan Sub Optimal untuk Optimalisasi Produksi Kelapa Sawit. Disampaikan pada Seminar Optimalisasi Pemanfaatan Lahan Marginal untuk Usaha Perkebunan. Surabaya.

Evaluation of sweet sorghum (*Sorghum bicolor* L. [Moench]) on several population density for bioethanol production

Suwarti[1], R Efendi[1], R Massinai[2] and M B Pabendon[1]

[1.] Indonesia Cereals Research Institute, Jl. Dr. Ratulangi no. 274 Maros, 90514, South Sulawesi
[2.] Indonesia Center for Estate Crops Research and Development, Bogor
Email: warti.smile@gmail.com

Abstract
Sweet sorghum (*Sorghum bicolor* L. [Moench]) crop management that is use for raw source of bioethanol for industrial purpose in Indonesia is less developed. The aim of this research was to evaluated sweet sorghum variety at several population to determine optimum density for juice production. Experiment design was set on split-plot design with three replications, conducted on August to December 2016 at the Indonesian Cereals Research Institute Research Station, Maros South Sulawesi. Main plot were six variation of plant row, and sub plot were three sweet sorghum varieties. Result of the study showed that plant population was high significanty affect to stalk weight, total biomass yield, leaf weight, and also significantly affect bagass weight and juice volume. Varieties were high significantly different in plant height, juice volume, and number of nodes. Super 1 variety on population at 166,667 plants/ha (P1) was obtained the highest juice volume (19,445 lHa^{-1}), meanwhile the highest brix value obtained from Numbu at the same plants population. Furthermore juice volume had significant correlation with biomass weight at the r=0.73. Based on ethanol production, Super 2 and Numbu had the highest volume at 83.333 plants/ha density (P3) and Super 1 at 166.667 plants/ha density with the ethanol volume were 827.68 l Ha^{-1}, 1116.50 l/ha and 993.62 l Ha^{-1} respectively.

Keywords: sweet sorghum, population, biomass, bioethanol.

1. Introduction

Sweet sorghum has many advantages compared to other cereals food crops, regarding to its funcionally to serve as a source of food, feed and also bioenergy sources (1). Sorghum is main staple for more than 500 million people in 30 semi arid countries (2). This sweet sorghum stalk contains approximately equal soluble (glucosa and sucrosa) and insoluble (celullosa and hemicelullose) carbohydrat and has been considered as promising alternative feedstock from by product of bioethanol production in simple and easy way (3) (4). Sweet sorghum is efficient C-4 photosintetic pathway, requireless fertilizer and water intake, possibillity of mutiple cropping per year, it is resilient to able to grow on marginal area especially drought prone area, has ratoon capability, wide adaptability geographic, and huge potential breeding (5) (6) (7). Sorghum flour fractions from its grain (whole and decorticated flours and bran) were evaluated by (8) for diet induced obese rats for oxidative stress and obesity parameters. Sorghum flour fractions reduced liver fat and whole sorghum flour decreased fasting glucose, improved glucose tolerance, insulin resistance and reduced insulin secretion compared to high fat-fructose groups. Sorghum flour showed a slight improvement in antioxidant status. This result might be aplicated for human being.

Beside grain production, sweet sorghum juice is a potential source for bioethanol, which can directly mixed with gasoline, or react with isobutylene to form of ethyl *tert*-butyl ether (ETBE), an octane additive for gasoline (9). Another research, by Barcelos *et al.* (2016) reported that under optimum hydrolysate with *Saccharomyces cerevisiae* strain, the result of ethanol was 87 gL^{-1}. from stalk juice, concentration of ethanol produced from bagasse was 72 g L^{-1}, after acid pre treatment to remove hemicelullose and hydrolysis and a flocculant of strain *Scheffersomyces stipitis*was used for fermeability of the hemicellulosic hydrolysate for the 23 hours pre treatment, yielded ethanol at 30 gL^{-1} concentration.

Sweet sorghum cultivation for bioenergy production is an attractive option to cope challenges of climate changes (11). Appropriate cropping technique option will increasing yield per hectare. Commonly, sorghum planting spaces is 70 cm x 25 cm between row to achieve population at 57,147 plant per ha (12) and 75 cm x 25 cm to gain 53,333 plant per ha (13). The result of Irwan *et al.* (2005) experiments showed that narrow plant row (50 cm x 30 cm) at 133,333 plant ha-1 population provide higher biomass yield compared to wider plant row (75 cm x 25 cm) at 106,666 plant Ha^{-1} population. Furthermore, Enciso et al. (2015) also reported at the water level 336-380 mm per year, high population 292,000 plant Ha^{-1} per ha resulted 58.9 t Ha^{-1} biomass.

The objective of this study was to evaluate optimum population of three sweet sorghum variety to get highest biomass productivity for bioethanol production.

2. Materials and Methods

The experiment was carried out in Indonesian Cereals Research Institute (ICERI) research station Sweet sorghum seed was planted on August 2016 and harvested on December 2016. Split Plot Experimental Design was used with three replications. Sub Plot was three sweet sorghum varieties Super-1,Super-2 and Numbu. Main Plot was six variation population arragement: P1= plant row 60 cm x 10 cm = 166,667 plantHa^{-1}. P2= plant row 60 cm x 15 cm = 111,111 plantHa^{-1}. P3=plant row60 cm x 20 cm = 83,333 plantHa^{-1}. P4=plant row70 cm x 10 cm = 142,857 plantHa^{-1}. P5=plant row70 cm x 15 cm = 95,238 plant per ha. P6=plant row 70 cm x 20 cm = 71,429 plantHa^{-1}.There were18 treatments combination and each treatment combination was planted on 5 m x 9 m plot square, with each plot distance to another was 1 m, so all treatment combination was laid on 3,115 m^2 of land.

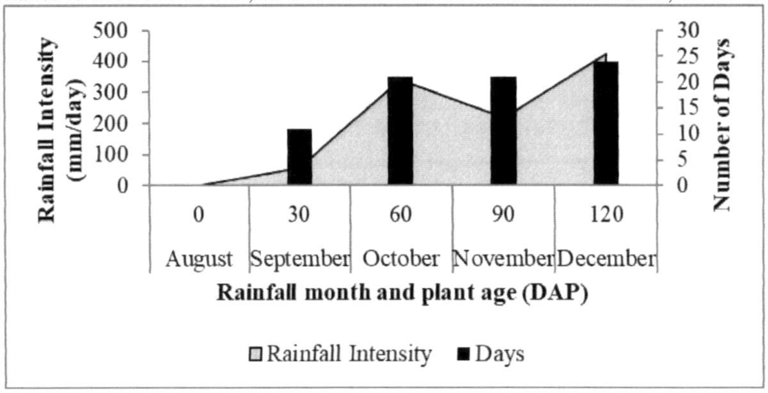

Figure 1. Rainfall intensity and rainfall days data during crops season on ICERI research station Maros, 2016.

The fertilizer dosage of nitrogen, phospat and potassium was 112,5 $kgHa^{-1}$ N, 100 $kgHa^{-1}$ P and 75 $kgHa^{-1}$ potassium (K). Fertilizer applied in urea form (source of N) and compound fertilizer (source of N, P, and K). The application time of fertilizer were splitted at 10 and 35 days after planting (DAP). First application, dosage were half of N and entire P and K dosage was applied. Second fertilize application was to put half remain nitrogen. Weeding were carried out at 7 and 20 DAP. Plant watering conducted 4 times during the season with *furrow irrigation* system using water pump. During the crops season, high intensity of rain occured due to La nina effect when the plant age was at 60 DAP until harvest day (figure 1).

The parameter in this observation were plant height (cm), stalk diameter (mm), node number, yield in stover biomass and stalk biomass measured in weight (tHa^{-1}), leaf biomass weight (tHa^{-1}) sugar percentage (% brix), and ethyl alcohol was measured with alcohol relative meter. Data recorded on the plant age at 105 DAP and 110 DAP.

2.1 Stalk harvesting and Juice Processing

Stalk biomass harvesting was performed by collecting 2 lines of sample plot in 4 m length (about 11-12 percent of the each plot area) at the plant age of 105 DAP and collect 4 rows sample row x 4 m length at 110 DAP sample collection. Number of stalk collected from 110 DAP crops was P1 and P4 = 160; P2 and P5 = 107; P3 and P6 = 80, number of stalk collected from 105 DAP crops was P1 and P4 = 80; P2 and P5 = 54; P3 and P6 = 40). Biomass measured directly after harvesting to record biomass weight. Sorghum plants were then cleaned manually using knife to discard the leaves, sheath and then to be measured again to record clean stalk weight. The next process was to extract cleaned stalk into mechanically for three times, then measured the juice volume. Brix percentage and relatif alcohol percentage observed immediately after fresh juice obtained using digital brix refractometer and ethyl alcohol refractometer.

2.2 Juice Fermentation

Forty litre sweet sorghum juice was collected into drum for the fermentation process. During juice fermentation process, maximum brix value have to be at 14%, if the brix value of juice more than 14%, water was added to dilute and reduce brix. However if brix value were less than 14% adding sugar was required until brix value reached 14%. After yeast from *Saccramomyces cereviciae* strain added, urea and NPK were put into the fermentor, with 40 litre capacity and fermentation minimum for 7 days until the brix value declined to 4.5%-5.5%.

2.3 Destilation

Fermented juice poured into destilator drum, and gas stove was used for heating until temperature reached 80°C (destilation temperature). Ethanol will then drip after required temperature reached. The destilation temperature maintained stable by controlling stove fire until alcohol drips were stopped.

2.4 Statistical Analysis

Sweet sorghum was grow in a split plot experimental design which included 18 combination of six plants densities x three varieties in three replicaties each combination. The data were subjected to two-way ANOVA followed by a Least Significant Different Test of means test was used to determine if means were significantly different (α=0.05) from each other. Statistical analysis was conducted with the software package SAS 9.3. Correlation test were used to determine the relationship between variables.

3. Results and Discussions

3.1 Statistical analysis result

Mean square of agronomic character, yield and yield components of sweet sorghum at 110 DAP and 105 DAP showed on table 1. Interaction within main plot (plant population) and sub plot (sweet sorghum variety) on the all parameter was not significant based on ANOVA analysis. Sweet sorghum variety were highly significant to ethyl alcohol contain at 110 DAP, plant height, node number and panicle weight 105 DAP, highly significant to brix value at 110 DAP, stalk diameter, leaf weight, biomass weight, brix value and ethyl alcohol percentage.

Population highly significant to plant height, stalk diameter, node number and leaf weight, brix value, ethyl alcoholand juice volume at the 110 DAP harvest. Results at the 105 DAP harvest showed that, population had highly significant to stalk weight, bagasse weight, ethyl alcohol and significant on juice volume. Interaction between population and replication result were significant with bagasse weight, ethyl alcohol percentage and juice volume. This result is contrast to (16), which reported that sorghum plant density was not affected significantly to plant height or dry matter yield at either harvest, but did affect dry matter digestibility at early heading. According to (17) experiment suggested that high density combinated with water deficit supply could affected to decrease dry mass of sorghum.

Table 1. Mean square of yield and yield component sweet sorghum at 110 DAP and 105 DAP harvested age.

Source of Variation	Replication (R)		Population (P)		P*R		Variety (V)		P*V	
Harvest at 110 DAP										
Plant height	184.32	ns	7182.62	**	820.21	ns	1313.49	ns	814.58	ns
Stalk diametre	1.52	ns	8.68	**	1.16	ns	2.40	*	0.65	ns
Node number	8.21	**	11.58	**	1.82	ns	1.35	ns	1.19	ns
Biomass weight	1.66	ns	6.66	ns	1.56	ns	2.66	ns	3.11	ns
Stalk weight	1.04	ns	4.34	ns	1.92	ns	2.16	ns	1.11	ns
Leaves weight	0.22	ns	1.08	**	0.29	ns	0.07	ns	0.08	ns
Bagasse weight	2.20	**	0.59	ns	0.12	ns	0.85	ns	0.32	ns
Brix	2.57	ns	15.65	**	1.04	ns	4.37	*	0.52	ns
Ethyl Alcohol	11.05	ns	149.24	**	9.66	ns	28.95	**	5.08	ns
Juice volume	355801.39	tn	6258837.50	**	176697.22	tn	201359.72	tn	307993.06	tn
Harvest at 110 DAP										
Plant height	62.57	ns	106.65	ns	456.95	ns	11245.35	**	284.20	ns
Stalk diametre	21.35	**	9.27	ns	1.86	ns	12.52	*	2.83	ns
Node number	4.02	**	1.23	ns	0.26	ns	14.69	**	0.26	ns
Biomass weight	787.63	*	489.90	ns	388.74	ns	25.80	ns	90.57	ns
Stalk weight	90.02	ns	313.31	**	106.02	ns	188.35	ns	45.62	ns
Panicles weight	1.06	ns	1.11	ns	0.63	ns	12.39	**	1.17	ns
Leaves weight	9.39	*	6.36	ns	4.08	ns	9.39	*	1.94	ns
Bobot Bagasse	13.72	ns	68.74	**	36.03	*	56.17	*	13.61	ns
Brix	1.97	ns	2.90	ns	0.84	ns	6.97	*	0.88	ns
Ethyl Alcohol	13.72	ns	68.74	**	36.03	*	56.17	*	13.61	ns
Juice volume	8282936.70	ns	19536564.40	*	18338705.40	*	167599682.10	**	8945163.20	ns

**=significantly different on P value 1% LSD test ,
*=significantly different on P value 5% LSD test,
ns=not significant.

3.2. Yield in juice volume and biomass form

Data in table 2, showed that juice volume rate from sweet sorghum stalk at 110 DAP harvested age were highest at P1V1 treatment which resulted 19,445 l/ha. The next highest result were obtained on P3V3 treatment non significantly different to P2V3 and P3V1 treatment with the juice volume result 18,796 l/ha 18,315 l/ha and 17,648 l/ha. Different result of juice volume were obtained from 105 DAP harvested age with the value 15,555 l/ha, and P1V1 resulted 14,546 l/ha respectively non significantly different to other treatments. The data suggested that juice volume was affected by population. High population tends to increase juice volume.

Juice volume derived from sweet sorghum stalk at the 110 DAP harvested age were higher than juice volume at the 105 DAP harvested age which average resulted at 14,488.72l/ha compare to 11.739.33 l/ha repectively. The 110 DAP plant age is physiology mature for grain, harvesting sorghum at this age will optimalize the resut of yield both from grain and also from stalk. Brix and ethyl alcoholpercentage were declined for 0.28% and 3.15% respectively. Declined of sugar content in sorghum juice assumed that trigger by stalk plant tissue were aging.

Table 2. Juice volume, brix, ethyl alcohol of sweet sorghum on 105 DAP and 110 DAP harvested age.

Treatment Combination	105 DAP			110 DAP		
	Juice Volume (lHa^{-1})	Brix (%)	Ethyl alcohol(%)	Juice Volume (l/ha)	Brix (%)	Ethyl alcohol(%)
P1V1	14546.00	14.83 ab	40.00 ab	19445.00 a	13.73 c-f	33.37 c-e
P1V2	12222.00	15.33 ab	42.40 a	11759.00 c-e	16.27 ab	44.43 a
P1V3	11667.00	17.17 a	43.93 a	15185.00 a-d	16.47 a	43.20 ab
P2V1	10648.00	14.50 b	37.40 ab	16019.00 a-c	12.97 ef	32.40 c-e
P2V2	11611.00	14.77 b	41.87 a	8796.00 e	15.43 a-d	40.33 a-d
P2V3	14259.00	15.97 ab	46.00 a	18315.00 ab	14.77 a-f	37.03 a-e
P3V1	12778.00	13.90 b	36.77 ab	17648.00 ab	13.47 d-f	31.60 de
P3V2	11213.00	13.80 b	39.50 ab	14445.00 a-d	13.97 b-f	34.67 b-e
P3V3	15555.00	15.57 ab	38.83 ab	18796.00 ab	15.40 a-d	39.10 a-e
P4V1	12619.00	15.07 ab	43.10 a	15183.00 a-d	15.37 a-d	35.27 a-e
P4V2	8444.00	15.70 ab	41.83 a	10397.00 de	15.87 a-c	41.33 a-c
P4V3	10794.00	15.57 ab	39.43 ab	14047.00 b-d	15.07 a-e	38.03 a-e
P5V1	11587.00	14.83 ab	39.70 ab	14397.00 a-d	14.87 a-f	36.23 a-e
P5V2	10397.00	13.70 b	38.67 ab	11587.00 c-e	15.13 a-e	38.30 a-e
P5V3	12619.00	14.80 ab	40.60 ab	14127.00 b-d	13.33 d-f	31.57 de
P6V1	8889.00	14.57 b	38.93 ab	15095.00 a-d	14.73 a-f	36.13 a-e
P6V2	8730.00	14.20 b	29.73 b	8810.00 e	12.60 f	29.77 e
P6V3	12730.00	15.00 ab	37.57 ab	16746.00 a-c	14.80 a-f	36.87 a-e
Mean	11739.33	14.96	39.79	14488.72	14.68	36.65
StDev	1953.49	0.86	3.49	3200.32	1.11	4.08
Min	8444.00	13.70	29.73	8796.00	12.60	29.77
Max	15555.00	17.17	46.00	19445.00	16.47	44.43
CV	31.45	8.09	14.63	18.38	8.02	12.86

same letter means no different between values in the same coloum

3.3. Total biomass weight and stalk weight

Sweet sorghum biomass at 110 DAP harvested age were decreased 0.05 tHa-1 compared to 105 DAP harvested age, however stalk weight increased 3.07 tHa-1 (table 3). This result describe that there was soft tissue stalk were formed into hard tissue stalk. Biomass weight at 105 DAP harvested age were rangedat 40 tHa^{-1} at P6V2 to 68.33 tHa^{-1} P1V3. Highest stalk weight at 105 DAP harvested age was obtained by P1V2 treatment (47 tHa^{-1})and was not significanty different to P1 dan P2 all three variety, also to P3V3, P4V1, P4V3, P5V2 and P5V3 treatment. It was showed that at 105 DAP harvested age, the result at dense population ($\geq$ 95,238 plants/ha) higher than less population ($\leq$ 71,429 plants/ha). This result was contrast to (18) experiment result, which suggested that biomass weight has sharply reduced by increasing of plant densities.

Non significant result were obtained from total biomass weight at 105 DAP harvesting. The P1V2 treatment was highest result of stalk weight from 105 DAP harvesting(47 tHa^{-1}). Highest total biomass weight at 110 DAP were obtained from P1V3 treatment with the result 80.67 tHa^{-1}. Stalk weight at 110 DAP yield were ranged from 26.67 tHa^{-1} – 56.67 tHa^{-1}, and the highest result were P1V2 and P1V3. Hemicelulosa B and cellulose-rich residue (CRF) which is isolated from sorghum biomass are useful as oil-in-water emulsifier and water holding agent respectively. Those product is considered economical as they are isolated from by product of sweet sorghum processing for making biofuel or directly from the sorghum biomass and also from the by products of sorghum grains milling (18).

Table 3. Total biomass weight and stalk weight 105 DAP and 110 DAP of sweet sorghum yield

Treatment Combination	105 DAP		110 DAP	
	Total biomass weight	Stalk weight	Total biomass weight	Stalk weight
P1V1	58.33 a	41.00	68.00 abc	40.67 ab
P1V2	66.00 a	47.00 a	79.33 ab	56.67 a
P1V3	68.33 a	43.00 ab	80.67 a	56.67 a
P2V1	46.00 a	35.00 a-c	43.00 cd	32.33 b
P2V2	57.67 a	37.67 a-c	49.67 cd	40.33 ab
P2V3	51.00 a	38.00 a-c	55.67 b-d	41.67 ab
P3V1	41.33 a	30.00 bc	41.67 cd	31.00 b
P3V2	50.00 a	27.67 bc	49.00 cd	36.67 b
P3V3	51.33 a	37.33 a-c	50.33 cd	34.00 b
P4V1	59.33 a	36.67 a-c	48.33 cd	36.00 b
P4V2	52.00 a	28.33 bc	56.00 b-d	40.67 ab
P4V3	58.00 a	43.33 ab	47.67 cd	37.00 b
P5V1	46.00 a	27.33 bc	48.00 cd	35.67 b
P5V2	46.00 a	31.33 a-c	50.67 cd	36.33 b
P5V3	48.67 a	38.00 a-c	47.33 cd	36.00 b
P6V1	52.67 a	21.67 c	38.00 d	26.67 b
P6V2	40.00 a	28.00 bc	40.00 d	30.00 b
P6V3	40.67 a	29.00 bc	39.00 d	27.33 b
Mean	51.85	34.46	51.80	37.54
StDev	8.17	6.86	12.46	8.22
Min	40.00	21.67	38.00	26.67
Max	68.33	47.00	80.67	56.67
CV	27.36	24.03	25.73	23.51

same letter means no different between values in the same column

3.4. Plant growth and by product

Data in table 4 shows that 105 DAP harvest crops with the highest stalk obtained at P1V2, P4V2, and P6V2 treatment with 317.33 cm, 316.67 cm and 312.33 cm height respectively. At 110 DAP harvested age, the highest plant weight was obtained at P5V2 treatment, which also had large stalk diameter compared to the other treatments and average node number were more than other treatments with each valued (362.27 cm, 19.7 mm and 13.87 node). In the P2V2 treatment, the stalk diameter (20.27 mm) and the average number of sections (14.07 nodes) had the highest value compared to other treatments, as well as at plant height parameter, P2V2 treatment produced a higher stalk than other treatments. Based on the data in table 1, plant height, stalk diameter and number of node at age 105 DAP and stalk diameter at age 110 DAP were more affected by varieties than plant spacing treatment, so for those parameters, it can be concluded that V2 (Super 2) is superior compared to two other varieties.

Varieties were affect to sorghum plant height, V2 (Super 2) considered as higher plant compared to other varieties under various densities. At the other cropping system, the result of (19) experiment suggested that intercrop sorghum x cowpea reduce height than sorghum sole. Plant height supported by node number and length of stem between nodes, rate of Super 2 nodes number was 13.35 wherease Super 1 and Numbu was 11.47 dan 12.76 respectively.

Table 4. Agronomic components of plant height, stalk diameter, and node number of sweet sorghum at 105 DAP and 110 DAP age.

Treatment combination	105 DAP			110 DAP		
	Plant height	Stalk diameter	Node number	Plant height	Stalk diameter	Node number
P1V1	267.67 b-d	19.67 a-c	11.00 d	289.47 d-g	15.77 b	10.73 f
P1V2	317.33 a	22.67 a	13.00 a-c	336.73 a-c	18.97 ab	12.60 a-e
P1V3	252.00 cd	18.67 bc	12.67 a-c	266.77 f-h	21.77 a	12.83 a-d
P2V1	292.33 a-d	20.00 a-c	11.67 cd	301.33 c-f	18.20 ab	11.07 ef
P2V2	293.33 a-c	21.33 a-c	13.33 ab	339.43 ab	20.27 a	14.07 a
P2V3	259.00 b-d	20.00 a-c	12.67 a-c	261.27 gh	19.27 ab	13.43 a-c
P3V1	286.00 a-d	20.67 a-c	11.67 cd	316.67 b-d	19.30 ab	11.67 d-f
P3V2	296.67 ab	22.00 ab	12.67 a-c	323.50 b-d	20.67 a	13.50 ab
P3V3	250.67 d	23.00 a	12.67 a-c	259.43 gh	20.40 a	13.57 ab
P4V1	284.67 a-d	17.67 c	11.00 d	292.83 d-g	19.23 ab	11.60 d-f
P4V2	316.67 a	20.33 a-c	13.00 a-c	348.70 ab	19.47 a	13.03 a-d
P4V3	253.67 cd	19.33 a-c	12.00 b-d	270.17 e-h	18.13 ab	11.70 d-f
P5V1	291.33 a-d	20.33 a-c	12.00 b-d	318.13 b-d	20.93 a	11.97 a-e
P5V2	297.67 ab	21.33 a-c	13.67 a	362.27 a	19.57 a	13.87 a
P5V3	263.00 b-d	21.33 a-c	12.67 a-c	261.50 gh	20.10 a	12.63 a-e
P6V1	289.67 a-d	21.00 a-c	11.67 cd	303.50 c-e	19.80 a	11.80 c-f
P6V2	312.33 a	21.67 ab	14.00 a	350.63 ab	20.87 a	13.00 a-d
P6V3	257.33 bcd	22.33 ab	13.33 ab	247.23 h	20.27 a	12.40 a-e
Mean	282.30	20.74	12.48	302.75	19.61	12.53
StDev	22.52	1.40	0.86	36.24	1.34	0.97
Max	250.67	17.67	11.00	247.23	15.77	10.73
Min	317.33	23.00	14.00	362.27	21.77	14.07
CV	7.54	9.09	5.82	6.26	9.59	6.77

same letter means no different between values in the same column

By product formed as empty panicle, leaf and bagasse from juice squeeze is presented in table 5. Bird pest attacks resulted in the number of panicles of 110 DAP reduced significantly, resulting in poor data processing. Highest leaves weight at 110 DAP crops weight were obtained at P1V3 combination treatment (25.73 tHa^{-1}), meanwhile highest panicle and bagasse weight at 105 DAP crops, and bagasse weight at 110 DAP crops obtained on P1V2 combination treatment.

Table 5. By product of ethanol production from sweet sorghum at the 105 DAP and 110 DAP crops age.

Treatment combination	105 DAP		110 DAP		
	Panicle weight	Bagasse weight	Panicle weight	Leaves weight	Bagasse weight
P1V1	1.67 c	17.33 bc	1.38 b	25.73 a	19.92 b
P1V2	5.33 a	27.33 a	3.84 a	18.62 a-c	46.75 a
P1V3	2.00 c	21.67 ab	1.91 ab	21.82 ab	27.18 b
P2V1	2.00 c	14.33 bc	1.54 b	9.14 bc	14.93 b
P2V2	3.67 a-c	17.00 bc	3.01 ab	7.04 c	24.15 b
P2V3	2.67 bc	20.33 a-c	1.83 ab	11.72 bc	18.60 b
P3V1	3.00 bc	14.33 bc	2.11 ab	8.30 bc	12.08 b

P3V2	4.33 ab	18.33 bc	2.94 ab	9.52 bc	16.65 b
P3V3	3.00 bc	15.67 bc	1.90 ab	14.56 a-c	14.37 b
P4V1	3.00 bc	18.33 bc	1.64 ab	10.80 bc	18.93 b
P4V2	3.67 a-c	20.33 a-c	2.86 ab	12.23 bc	23.77 b
P4V3	3.00 bc	20.67 a-c	2.09 ab	8.99 bc	16.17 b
P5V1	2.67 bc	16.33 bc	1.79 ab	10.43 bc	15.04 b
P5V2	3.67 a-c	17.67 bc	3.29 ab	11.27 bc	16.37 b
P5V3	3.00 bc	16.00 bc	3.46 ab	7.74 c	15.35 b
P6V1	2.00 c	14.33 bc	1.56 b	9.66 bc	12.70 b
P6V2	3.00 bc	15.33 bc	2.86 ab	7.59 c	15.16 b
P6V3	2.33 bc	13.67 c	1.38 b	10.28 bc	10.83 b
Mean	3.00	17.72	2.30	11.97	18.83
StDev	0.91	3.41	0.78	5.13	8.20
Max	1.67	13.67	1.38	7.04	10.83
Min	5.33	27.33	3.84	25.73	46.75
CV	34.40	22.36	48.19	57.43	54.04

same letter means no different between values in the same coloum

3.5 Ethanol production of sweet sorghum

Rate of brix percentage and ethyl alcohol during the fermentation time showed on figure 2. Brix percentage rapidly declined at the 2nd day of fermentation from 13.73 % to 8.98 % indicated that fermentation process were goes well. After 10 days of fermentation, liquid were transfered into destilation drum for the next step.

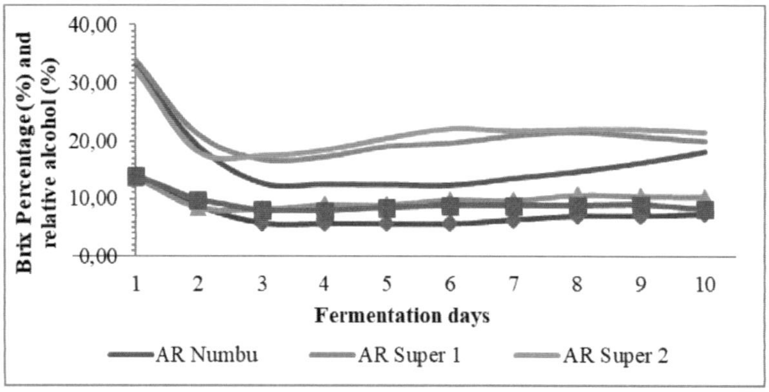

Figure 2. Rate of brix percentage and ethyl alcoholduring fermentation process of bioethanol

The result of ethanol production presented on figure 3. Ethanol production was lower than potential outcome based on previous study (5000 lHa^{-1})(13). On the different process way, (20) study, reported ethanol 91.94 g per kg sorghum was acheived without any detoxification and washing of pre treated biomass after mild acid pretreatment followed by enzimatic hydrolisys and fermentation. The result of this study presented in diagram, higher ethanol production in the Numbu variety, followed by the Super-1 variety, and the lowest Super-2 variety. While based on plant spacing, the Numbu variety and the Super-2 varieties of ethanol production were relatively higher at 60 cm x 20 cm plant spacing, while the Super-1 varieties were higher at 60 cm x 10 cm. On the Super-2 variety showed ethanol yield the lowest despite the high stalk length and stalk weights, but the juice rendementof juice is lower.

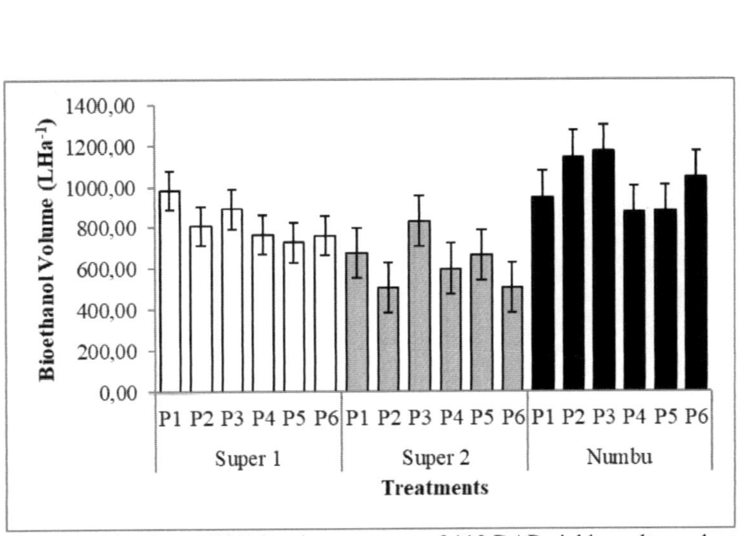

Figure 3. Ethanol production at 70% level percentage of 110 DAP yield result on plant population combination treatment.

Highest significant correlation within sweet sorghum juice volume with all other parameters obtained with total biomass weight r=0.73 (presented on table 6). Positive correlation value indicated that total biomass weight influenced a increasingly of juice volume. The highest significant correlation on all parameters were within brix value and ethyl alcohol content which vauled r=0.91.

Table 6. Correlation of juice volume with other agronomic character

Agronomic Variable	JV	PH	SD	JN	TBW	SW	PW	BBDn	BBgs	JlBt	Brix
PH	0.06ns										
SD	-0.18ns	0.25ns									
JN	0.04ns	0.28*	0.35*								
TBW	0.73**	0.23ns	0.13ns	0.37**							
SW	0.60**	0.13ns	-0.02ns	0.23ns	0.68**						
PW	0.46**	-0.02ns	-0.04ns	0.14ns	0.46**	0.38**					
LW	0.13ns	0.12ns	0.42**	0.50**	0.60**	0.38**	0.13ns				
BBgs	0.27ns	0.44**	0.25ns	0.46**	0.74**	0.46**	0.17ns	0.63**			
JlBt	0.04ns	-0.01ns	-0.40**	-0.17ns	0.14ns	0.23ns	-0.17ns	0.02ns	0.29*		
Brix	0.40**	-0.19ns	-0.44**	-0.32*	0.19ns	0.25ns	0.08ns	-0.24ns	0.03ns	0.49**	
EA	0.51**	-0.10ns	-0.46**	-0.32*	0.30*	0.35*	0.16ns	-0.20ns	0.08ns	0.47**	0.91**

JV=juice volume, PH=plant height, SD=stalk diameter, JN=node number, TBW= total biomass, SW=stalk weight, PW=panicle weight, LW=leaves weight, BgW=bagasse weight, NS=number of stalk, EA=ethyl alcohol

3. 4. Conclusions

Conclusion of the study showed that plant population was high significantly affect to stalk weight, total biomass yield, leaf weight, and also significantly affect bagass weight and juice volume. Varieties were high significantly different in plant height, juice volume, and number of nodes. Super 1 variety on population at 166,667 plants/ha (P1) was obtained the highest juice volume (19,445 lHa^{-1}), meanwhile the highest brix value obtained from Numbu at the same plants population. Furthermore juice volume had significant correlation with biomass weight at the r=0.73. Based on ethanol production, Super 2 and Numbu had the highest volume at 83,333 plants/ha density (P3) and Super 1 at 166,667 plants/ha density with the ethanol volume were 827.68 lHa^{-1}, 1116.50 l/ha and 993.62 lHa^{-1} respectively.

4. Acknowledgement

We deeply thanks, Mr. Haris, Mrs. Murniati and all ICERI TSP team who helped in this experiment and Mr. Sigit Budi Santoso for reading the manuscript

References

1. Subagio H, Aqil M. Pengembangan Produksi Sorgum Di Indonesia. Pros Semin Nas Inov Teknol Pertan. 2013;199–214.
2. Dahlberg J, Berenji J, Sikora V, Latković D. Assessing sorghum [Sorghum bicolor (L) Moench] germplasm for new traits: Food, fuels & unique uses. Maydica. 2011;56(1750):85–92.
3. Yu J, Zhang T, Zhong J, Zhang X, Tan T. Biorefinery of sweet sorghum stem. Biotechnol Adv. 2012;30(4):811–6.
4. Appiah-Nkansah NB, Zhang K, Rooney W, Wang D. Model study on extraction of fermentable sugars and nonstructural carbohydrate from sweet sorghum using diffusion process. Ind Crops Prod [Internet]. Elsevier B.V.; 2016;83:654–62. Available from: http://linkinghub.elsevier.com/retrieve/pii/S0926669015306476
5. Andrzejewski B, Eggleston G, Powell R. Pilot plant clarification of sweet sorghum juice and evaporation of raw and clarified juices. Ind Crops Prod [Internet]. Elsevier B.V.; 2013;49:648–58. Available from: http://dx.doi.org/10.1016/j.indcrop.2013.06.027
6. Agung I, Sardiana I, Diara IW, Nurjaya IGMON. Adaptation, Biomass and Ethanol Yields of Sweet Sorghum (Sorghum bicolor (L.) Moench) Varieties at Dryland Farming Areas of Jimbaran Bali, Indonesia. J Biol Agric Healthc. 2013;3(17):110–5.
7. Ananda N, Vadlani P V., Prasad PVV. Evaluation of drought and heat stressed grain sorghum (Sorghum bicolor) for ethanol production. Ind Crops Prod [Internet]. Elsevier B.V.; 2011;33(3):779–82. Available from: http://dx.doi.org/10.1016/j.indcrop.2011.01.007
8. Moraes EA, Marineli R da S, Lenquiste SA, Queiroz VAV, Camargo RL, Borck PC, et al. Whole sorghum flour improves glucose tolerance, insulin resistance and preserved pancreatic islets function in obesity diet-induced rats. J Funct Foods. 2017;xxx(xxx):1–11.
9. Barbanti L, Grandi S, Vecchi A, Venturi G. Sweet and fibre sorghum (Sorghum bicolor (L.) Moench), energy crops in the frame of environmental protection from excessive nitrogen loads. Eur J Agron. 2006;25(1):30–9.
10. Barcelos CA, Maeda RN, Santa Anna LMM, Pereira N. Sweet sorghum as a whole-crop feedstock for ethanol production. Biomass and Bioenergy [Internet]. Elsevier Ltd; 2016;94:46–56. Available from: http://dx.doi.org/10.1016/j.biombioe.2016.08.012
11. Marta AD, Mancini M, Orlando F, Natali F, Capecchi L, Orlandini S. Sweet sorghum for bioethanol production : Crop responses to different water stress levels. Biomass and Bioenergy [Internet]. Elsevier Ltd; 2014;64:211–9. Available from: http://dx.doi.org/10.1016/j.biombioe.2014.03.033
12. Capriyati R, Tohan, Kastono D. Pengaruh Jarak Tanam dama Tumpangsari Sorgum Manis (Sorghum bicolor L. Moench) dan Dua Habitus Wijen (Sesamum Inicum L.) Terhadap Pertumbuhan dan Hasil. Vegetalika. 2014;3(3):49–62.

13. Pabendon MB, Sarungallo RS, Mas'ud S, Nur A. Penampilan Fenotipik dan Stabilitas Sorgum Manis untuk Bahan Baku Bioetanol. Penelit Pertan Tanam Pangan. 2012;31(1):60–9.

14. Irwan AW, Wahyudin A, Susilawati R, Nurmala T. Interaksi jarak tanam dan jenis pupuk kandang terhadap komponen hasil dan kadar tepung sorghum (Sorghum bicolor [Linn .] Moench) pada Inseptisol di Jatinangor MH 2004 Interaction of spacing and kind of manure on growth and yield of sorghum plant (Sorg. Kultivasi. 2005;4(2):128–36.

15. Enciso J, Jifon J, Ribera L, Zapata SD, Ganjegunte GK. Yield, water use efficiency and economic analysis of energy sorghum in South Texas. Biomass and Bioenergy [Internet]. Elsevier Ltd; 2015;81:339–44. Available from: http://dx.doi.org/10.1016/j.biombioe.2015.07.021

16. Carmi A, Aharoni Y, Edelstein M, Umiel N, Hagiladi A, Yosef E, et al. Effects of irrigation and plant density on yield, composition and in vitro digestibility of a new forage sorghum variety, Tal, at two maturity stages. Anim Feed Sci Technol. 2006;131(1–2):121–33.

17. Habyarimana E, Bonardi P, Laureti D, Di Bari V, Cosentino S, Lorenzoni C. Multilocational evaluation of biomass sorghum hybrids under two stand densities and variable water supply in Italy. Ind Crops Prod. 2004;20(1):3–9.

18. May A, de Souza VF, Gravina GD, Fernandes PG. Plant population and row spacing on biomass sorghum yield performance. Cienc Rural. 2016;46(3):434–9.

19. S. M. Karanja SM, Kibe AM, Karogo PN, Mwangi M. Effects of Intercrop Population Density and Row Orientation on Growth and Yields of Sorghum - Cowpea Cropping Systems in Semi Arid Rongai, Kenya. J Agric Sci [Internet]. 2014;6(5):34–43. Available from: http://ccsenet.org/journal/index.php/jas/article/view/31809

20. Akanksha K, Sukumaran RK, Pandey A, Rao SS, Binod P. Material balance studies for the conversion of sorghum stover to bioethanol. Biomass and Bioenergy [Internet]. Elsevier Ltd; 2016;85:48–52. Available from: http://dx.doi.org/10.1016/j.biombioe.2015.11.027

Consumer's market analysis of products based on cassava

Bina Unteawati[1)], Fitriani[*2)], Cholid Fatih[3)]

[1) 2) 3)]Agribusiness, Economics and Business, Lampung State of Polytechnic, Jl. Soekarno Hatta No. 10 Rajabasa, Bandar Lampung, Lampung, Indonesia
[*2)]fitriani@polinela.ac.id as Corresponding Author; [1)]bina_nurman@yahoo.co.id

Abstract

Cassava product has the important role for enhancing household's income in rural. Cassava as raw material food is plentiful as local food in Lampung. Cassava product is one of strategic value addition activities. Value additional activities are a key to create income source enrichment in rural. The householdwas product cassava as a snack or additional food. Their product cassava was operated in small-scale, traditional, and discontinuous production. They have been lacked in technology, capital, and market access. Measurement the sustainability of their business is important. The market has driven the business globally. This research aims to (1) describe the cassava demand to locally product cassava in rural and (2) analysis the consumer's perception of cassava product. Research take placed in Lampung Province, involved Bandar Lampung and Metro City, Pringsewu, Pesawaran, Central Lampung, and East Lampung district. It is held in February until April 2017. Data were analyzed by descriptive statistic and multidimensional scaling. Based on the analysis conclude that (1) the demand of product cassava from rural was massive in volume and regularity with the enormous transaction. This fact is very important to role business cycles. Consumers demand continuously will lead the production of cassava productsustain. Producers of product cassava will consume fresh cassava for the farmer. Consumption of fresh cassava for home industry regularly in rural will develop balancing in fresh cassava price in the farming gate (2) The consumer's perception on cassava product in the different market showed that they prefer much to consume cassava chips as cassava product products than other. Next are crackers, opak, and tiwul rice. Urban consumers prefer product products as snacks (chips, crumbs, and opak), with consumption frequency of 2-5 times per week and volume of 1-3 kg purchases. Consumers in rural were more frequent with daily consumption frequency. Multidimensional scaling analysis performed that price of cassava product was the centre of product attribute.
Keywords: cassava product, market, consumer, perception

1. Introduction

Development of product cassava product could become an alternative as income improvement for fresh cassava producer. Fresh cassava produced dominantly at North of Lampung, Central of Lampung, and East of Lampung districts. Those districts still faced rural poverty alleviation problem. More than 23% population in North of Lampung lived under the poverty line, at Central of Lampung attaint to 17%, and East of Lampung reached 13% (Lampung Statistical Agency, 2015).

Fresh cassava copes with price fluctuation at the farming gate. There has been failure market as a result of the excess supply. The supply was abundant, as well as being easily damaged and not durable were main causes the price of cassava very low. Furthermore, uncertainty price has decreased the income of farmers.An important step needs to reduce poverty in the center of cassava production. Creating superior local production value addition will be important to improve the farmers' income structure. Strengthening the downstream product line of cassava as agro-industry food at production center area will become one of the solutions.

IOP Conf. Series: Earth and Environmental Science **141** (2018) 012033 doi:10.1088/1755-1315/141/1/012033

Small-scale cassava-based agroindustry involves more domestic actors. Cassava-based industry have been developed include: tapioca, siger/tiwul/oyek rice, mocaf, klanting, opak, crackers, and chips (Hidayat, 2016; Indiako, dkk., 2014; Sagala, dkk., 2013; Sari, 2011; Surfiana, dkk., 2014; Yunus dan Utami, 2012). Small and medium-sized food agroindustry (SME) plays an important role in increasing the income of rural households. The development of SME food agroindustry supports the concept of equity and economic growth (6)(7).

Food and beverage agroindustry at Lampung's GRDP contributes 12.50%, with the growth of 4.2% year-1. Manpower in large agroindustry enterprises reached 48,735 people in 2014 from 222 existing companies (Lampung BPS, 2015). Food agroindustry provides a multiplier effect on the local economy through increased value added, diversity of sources of income, availability of business, increasing income, and ensuring food security for households, as well as reducing rural poverty. Product product based on cassava has an important role in enhancing household's income in rural. Cassava as raw material food is plentiful as local food in Lampung. Product cassava is one of strategic value addition activities. Value additional activities are a key to create income source enrichment in rural.

Most households are product cassava as a snack or additional food. Their product cassava was operated in small-scale, traditional, and discontinuous production. They have been lacked in technology, capital, and market access. Measurement the sustainability of their business is important. The market has driven the business globally. This research aims to (1) describe the cassava demand to locally product cassava in rural and (2) analysis the consumer's perception of cassava product.

2. Methods

2.1 Survey design

Research take placed in Lampung Province, involved Bandar Lampung and Metro City, Pringsewu, Pesawaran, Central Lampung, and East Lampung district. Field survey held in February until April 2017. Snowball sampling methods used to select respondent market trader. Respondents were direct buyers who coming at the market, 5 people for every market. Target market divided three categorize traditional market, cluster, and retail market modern. Traditional market represented by Pasar Tugu, Pasar Koga, and Pasar Tani at Bandar Lampung City and. They also represent the traditional market at Pringsewu, Metro, Punggur, and Way Jepara. Site map location is described at figure1. Target market placed informed as for the table 1 below:

Table 1. Consumer's market

Market	Location	Description
Pasar Tugu, Pasar Koga, and Pasar Tani	Bandar Lampung City	Traditional market
Punggur	Central of Lampung	Traditional market
Metro	Metro City	Traditional market
Way Jepara	East of Lampung	Traditional market
Panjang, PU Chips cluster area	Bandar Lampung City	Cluster market
Sidoarjo	Pringsewu District	Cluster market
Chandra superstore, Fitrinove, dan Surya minimarket	Bandar Lampung City	Modern market

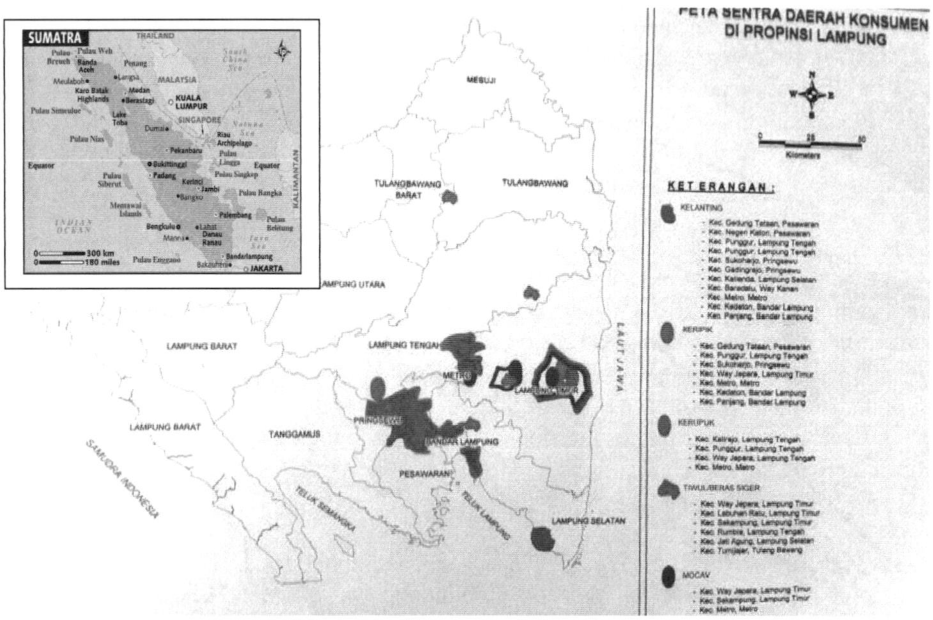

Figure 1. Site research location

2.2 Data analysis

Consumer's perception and demand for cassava product were analyzed by descriptive statistic. The consumer's perception of cassava productwas analyzed by multidimensional scaling (MDS). MDS is an exploratory technique used to identify unrecognized dimensions affecting behavior. MDS is an exploratory data analysis technique that attains this aim by condensing large amounts of data into a relatively simple spatial map that relays important relationships in the most economical manner (8). MDS provides general solutions to many problems in perception, emotion, and cognition, where the stimuli are too complex to be quantified by other means(8),(9), (10). The data for MDS was design using Likert-type scales for rating.Distance model will be represented in geometrical space. Most MDS algorithms were employing Euclidean principles. Distance (d_{ij}) between points i and j is defined as (Equation 1).

$$dij = \sqrt{\Sigma_a (xia - xja)\,2} \qquad (1)$$

Specify coordinates of points i and j on dimension a respectivelynoted as x_i and x_j. Positive monotone transformation is applied to dissimilarity data for scaling into spatial distances while for metric MDS a linear transformation function is applied(11). A stress function represented the fit between input proximities and distances. The Likert-type of MDS model in this research was designed in Table 2. The consumer's perception map on the stimuli response toward purchase attributes as follows:

Table 2. Stimuli response toward purchase attributes:

No	Attribute	VD (1)	D (2)	S (4)	VS (5)
1	Price				
2	Flavor				
3	Savor				
4	Package				
5	Texture				

VD=Very dissatisfied; D=dissatisfied; S=Satisfied; VS=Very satisfied

IOP Conf. Series: Earth and Environmental Science **141** (2018) 012033 doi:10.1088/1755-1315/141/1/012033

3. Results and Discussions

3.1 Cassava demand

Cassava production in Indonesia was broadly cultivated as afood, feed, and bioenergy material. Cassava as food and feed have been supporting the food security, especially in rural livelihood. Cassava also will become an alternative feedstock strategy for several rural biogas providing opportunities for decentralized cheaper fuel and electricity systems in rural areas (12). Cassava biomass potentially offers multiple benefits for producing biofuels (13).The expansion of cassava production in Indonesia 1993-2015 seen in figure 2. Figure 2 described the production center of cassava in Indonesia. The best three regions were Lampung, Central Java, and East of Java. Lampung holds astrategical position in fulfilling domestic and international cassava demand. Lampung contributed 35% of the total national cassava production.Cassava productivity during 2007-2014 was 20.1 ton.ha$^{-1.}$

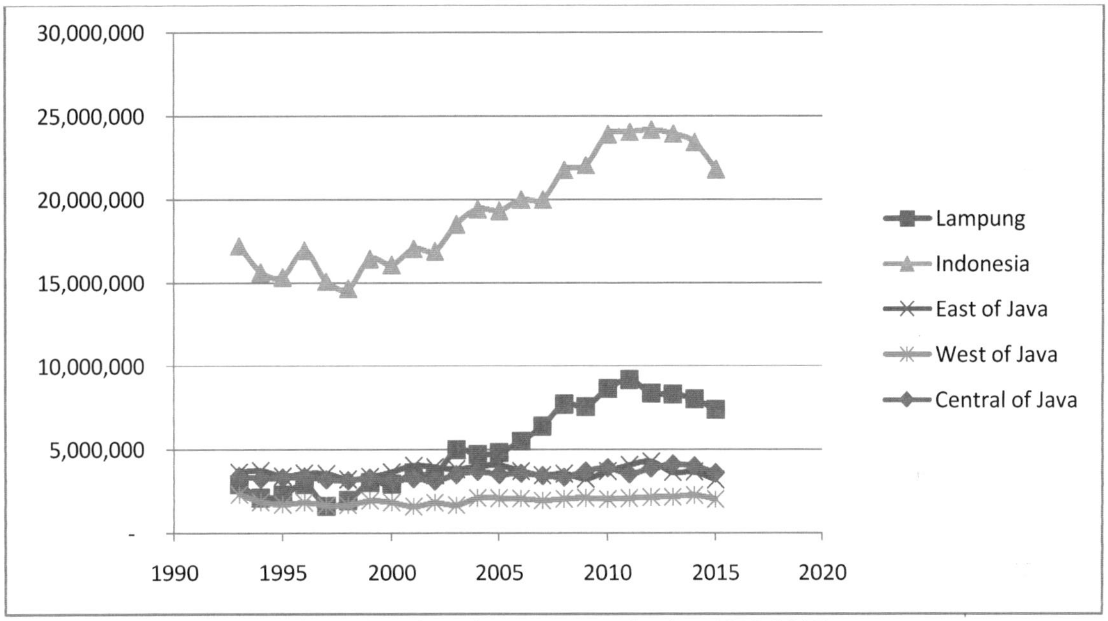

Figure 2. Indonesia cassava production 1993-2015
Sources: Indonesia Statistical Board, 2016

Development the cassava economic is strategic in rural farmer empowerment. Especially on improving the smallholding nature of production, processing, marketing and utilization persist (14). Unfortunately, price volatility as the main problem was always come up. The moving condition of price data forthe main food commodities in Indonesia 2000-2014 seen in figure 3.

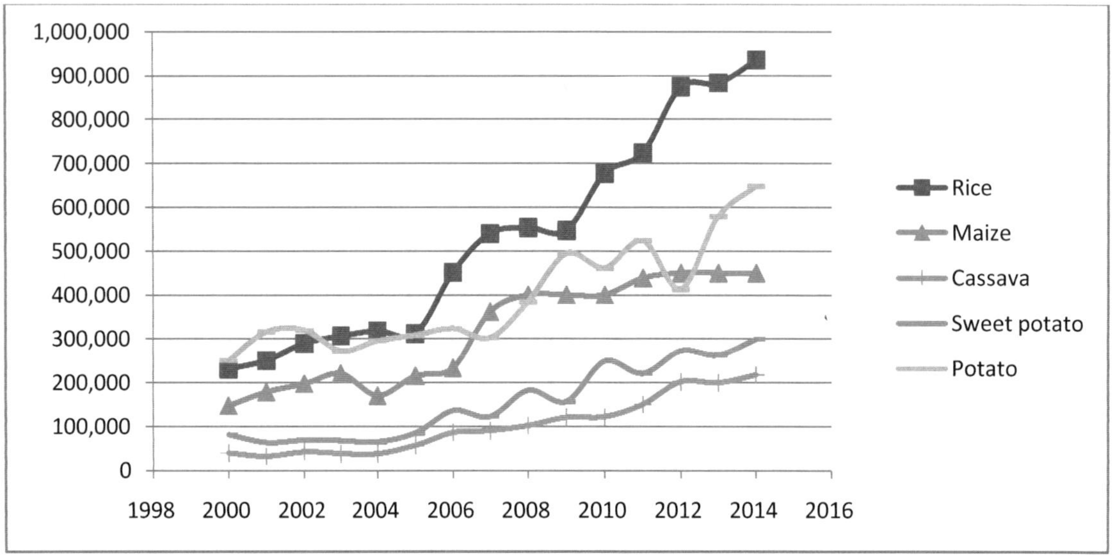

Figure 3. Grocery price of primary agricultural product in Indonesia 2000-2014
Sources: Indonesia Statistical Board, 2016

Based on figure 3 showed that the cassava groceries gate price was the slowest and the lowest compare with others food commodities. This condition affected the farmer income in rural. Cassava processing and marketing in rural have developed bysmallscale business (micro/small medium enteprises = SME). Price volatility could become threat their farmer income and rural food security. Strenghthen the farmers with additional activities on processing cassava is very important. Enhancing the SME medium to large scale enterprises is necessary by taking up the activities using improved semi-mechanized technology (ISMT) and Improved Fully Mechanized Technology (IFMT) (14).

Farming market facilities for farmers are important. The existence of the market of agricultural and agro-industries products requires not only physical facilities, the main is the network and the integrated market linkage between the line of producer farmers with other market institutions, to the final consumer. The market becomes an indicator of guarantee of price incentive and certainty of the sale of farmer producer yield (15). Consumer's purchase based ofproduct cassava product categorize as follows:

Table 3. Purchase volume of product cassava product

Nu	Cassava product product	Purchase Max (kg)	Purchase Min (kg)
1	Cassava rice (Beras siger/tiwul/oyek)	95	60
2	Chips	2452	2079
3	Crackers	100	162
4	Opak	30	56
5	Kelanting	300	190
6	Cassava modified flour/mocav	16	24
	Total	2993	2571

Consumer's purchase frequency is described in Table 4 below:

Table 4. Consumer's purchase frequency

Consumption Time	Frequency	(%)
Daily	13	21.66
2-5 times a week	24	40
6-10 times a week	1	1.66
>10 times a week	22	36.66
Total	60	

The demand for product cassava that appears per month based on the data was ammount 24,716.56 kg until 28,773.5 kg. Urban consumers prefer productcassava as snacks (chips, crumbs, and opak), withconsumption frequency of 2-5 times per week and volume of 1-3 kg purchases. Consumers in rural were more frequent with daily consumption frequency. Based on aNational survey on consumption of fresh cassava product, each person consumed attaint to 3.598 kg.person^{-1}.year^{-1}.The consumption of cassava chips reached 0.209, then tapioca flour 0.039 kg.person^{-1}.year^{-1}. The cassava consumption in Lampung could be predicted as follows (Table 5):

Table 5. The cassava consumption in Lampung, 2015

Consumption of Cassava Product	(kg/person/year)	Ton
Fresh cassava	3.589	27,398.44
Cassava chips	0.209	1,595.51
Tapioca flour	0.034	259.56
Lampung people based on census 2010	7,634,005	
Total cassava consumption		29,253.51
Lampung cassava production in 2015		8,034,016
Rasio input-output cassava (100:30)		26,780,053
(%) Cassava production		0.11%

Based on the value of forwarding linkage indexes of cassava, the value was less than one (0.779). This condition shows that one rupiah output issued by food crop sectors, impacted the multiplier effect generated less than one rupiah. This condition also shows that food production in Lampung is still weaker due to its downstream processing line. Agroindustry with the raw material of food crop not yet optimal developed. Most food production is still consumed as a final product. On the others side, backward index of cassava was greater than 1. It indicates that the presence of food crop sub-sectors provides a growth effect for other sectors of the economy, in this case directly related to the agricultural input industry sector. Cassava gives the highest impact on growth. The index value of 1.2131 means the demand for one rupiah of cassava sector creates a multiplier value of 1,2131 rupiahs for the agricultural input industry sector.

Ratio input-output fresh cassava and product cassava on average was 100:30. It means that minimum fresh cassava for food cassava home industry could reach triple. The fresh cassava not only market as raw material for tapioca industries but could broadly process by household in rural. Furthermore, it will become income enlargement, for minimized poverty alleviation in rural. This condition also reveals at HQCF (high-quality cassava flour) industries in Tanzania (16). Most of cassava productindustries in Lampung work on thesmall-scale business, small manpower, main raw material mostly from local, with small scope marketing spreading (17).

Based on the analysis result it could explain that the demand ofproduct cassava from rural was massive in volume and regularity with the enormous transaction. This fact is very important to role business cycles. Consumers demand continuously will lead the product cassava production sustain. Producers of product cassava will consume fresh cassava for the farmer. Consumption of fresh cassava for home industry regularly in rural will develop balancing in fresh cassava price in farming gate

3.2 Consumer's perception of cassava product

Multidimensional scaling analysis performed that price of cassava product product was the centre of product attribute. The preferred product chosen by consumers was display at Table 6 and Figure 4.

Table 6. Cassava product preferred product

		Frequency	Percent	Valid Percent	Cumulative Percent
Valid	Cassava rice (Beras siger/tiwul/oyek)	4	6.7	6.7	6.7
	Chips	30	50.0	50.0	56.7
	Crackers	10	16.7	16.7	73.3
	Opak	7	11.7	11.7	85.0
	Kelanting	7	11.7	11.7	96.7
	Cassava modified flour/mocav	2	3.3	3.3	100.0
	Total	60	100.0	100.0	

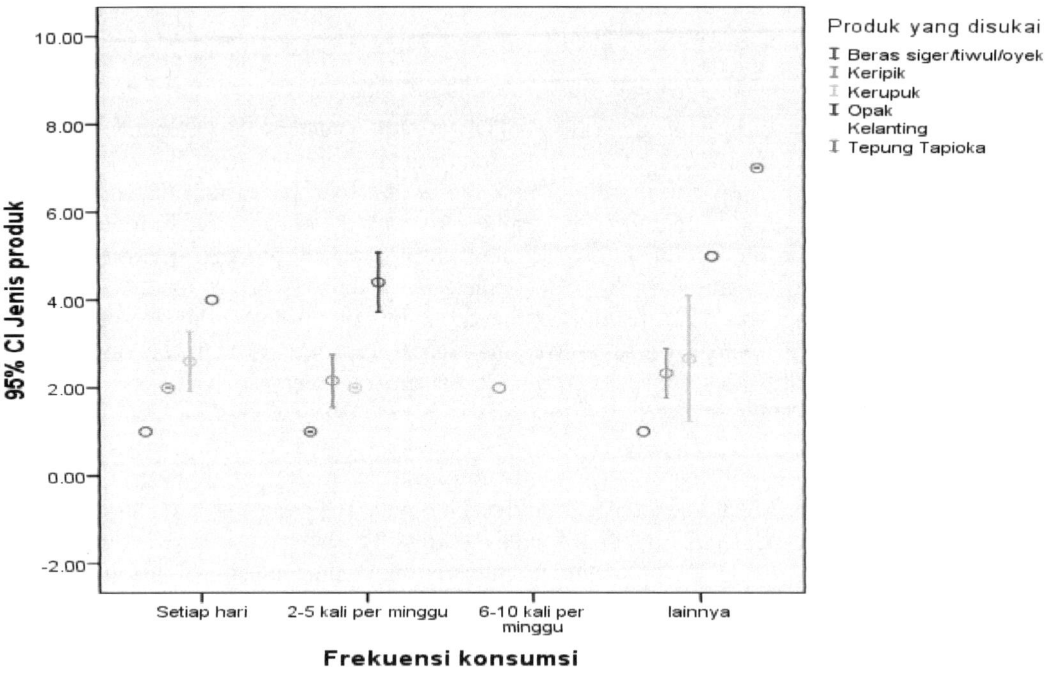

Figure 4. Cross tabulation between consumption frequency and preferred products

The consumer's perception on cassava product in the different market showed that they prefer much to consume cassava chips as cassava product products than other. Next are crackers, opak, and tiwul rice. The goodness of fit MDS analysis performed as follows (Table 7).

Table 7. Stress and Fit Measures

S-Stress	.05443[b]
Dispersion Accounted For (D.A.F.)	.97793
Tucker's Coefficient of Congruence	.98890

PROXSCAL minimizes Normalized Raw Stress.
a. Optimal scaling factor = 1.023.
b. Optimal scaling factor = .968.

The measure of goodness of fit in multidimensional scaling is called S(called)-Stress. Dimensions are similar to factors as they give the number of facets of the relationships between stimuli and may be plotted. S-Stress was standardized to take values between 0 and 1. The rules of thumb for what values of S-Stress make a good fit (8,9). Based on the Table 7, S-Stress value was 0.05443. It means that the MDS fulfil the goodness of fit model. Dispersion Accounted For (DAF) and Tucker's Coefficient of Congruence also have values 'close to 1' (18).

Table 8. Discrimination Measures

	Dimension		
Atribute	1	2	Mean
Price	.337	.338	.338
Flavor	.615	.000	.308
Savor	.822	.116	.469
Package	.776	.657	.717
Texture	.627	.197	.412
Active Total	3.177	1.310	2.243
% of Variance	63.540	26.192	44.866

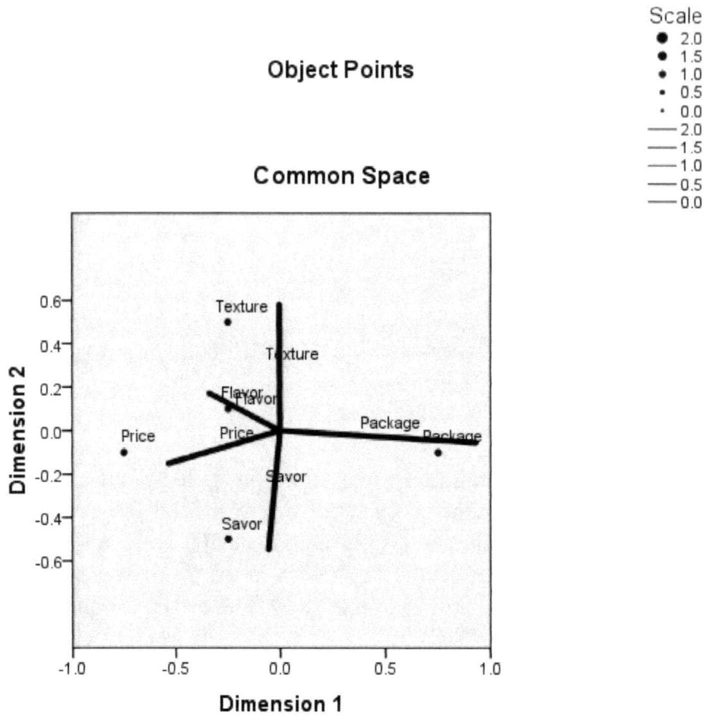

Figure 5. Common space of consumer's perception toward response stimuli

Based on Figure 5, explained that price and flavour were centre stimuli to impulse buying consumers. Convenience and lower relative prices of food items purchased at supermarkets, in comparison to other food stores, are relevant to explain higher share of purchases of ultra-product foods and beverages at supermarkets (19).

On the other hand, design package product was also important. Package design development is a process for creating new products to meet market needs, which focuses on creating effectiveness, efficiency, and idea development. High competitive products will also create the image of Indonesian products as high quality, creative, and environmentally friendly products. With the development of these designs, can produce quality products, unique, and original and well-received by the market (20)

4. Conclusions

Based on the analysis conclude that (1) the demand of product cassava from rural was massive in volume and regularity with the enormous transaction. This fact is very important to role business cycles. Consumers demand continuously will lead the product cassava production sustain. Producers of product cassava will consume fresh cassava for the farmer. Consumption of fresh cassava for home industry regularly in rural will develop balancing in fresh cassava price in the farming gate (2) The consumer's perception on cassava product in the different market showed that they prefer much to consume cassava chips as cassava product products than other. The next products are crackers, opak, and tiwul rice.

References

1. Sagala IC, Affandi MI, Ibnu M. Performance of agroindustry kelanting in Anyar village of Gedongtataan Sub-district, Pesawaran Regency. J Ilmu-Ilmu Agribisnis. 2013;1(Januari):60–5.
2. Yunus HAR, Utami DP. The performance of Cassava Opak Agroindustry in Jolontoro Village Kec. Saporan, Kab. Wonosobo. Surya Agritama. 2012;I(Maret):59–70.
3. Surfiana, Hidayat B, Akmal S. Effectiveness of Transfer of Siger Rice Processing Technology to Increase Productivity of Tiwul Traditional Rice Business. Inov dan Pembang. 2014;2(3):32–45.
4. Indiako MISDV, Ismono RH, Achdiansyah. Comparative study of land allocation pattern, rice expenditure and food consumption pattern among cassava farmers in implementing and non-implementing villages of MP3L Progral in Lampung Selatan District. J Ilmu-Ilmu Agribisnis. 2014;2(4):331–6.
5. Sari RP. Analysis of Added Value and Feasibility of Cassava Chip Agroindustry as Raw Material of Mocaf Maker in Kab. Trenggalek. Brawijaya; 2011.
6. Affandi IMI.Food Cluster Agroindustry Cluster Development Model for Small and Medium Enterprises (Ukm) to Reduce Poverty by 10% in Lampung Province. 2010;7:301–2.
7. Elizabeth R. Development of Foodstuff Agroindustry for Increasing Added Value through Institutional Transformation in Rural Areas. 2010;5(1):102–12.
8. Mugavin ME. Multidimensional Scaling. Nurs Res. 2008;57(1):64–8.
9. Jaworska N, Anastasova AC. A Review of Multidimensional Scaling (MDS) and its Utility in Various Psychological Domains. Tutor Quant Methods Psychol. 2009;5(1):1–10.
10. Ginanjar I. MDS Application fo teachingIn: Prosiding Seminar Matematika dan Pendidikan MIPA UNY. Universitas Negeri Yogyakarta; 2008. p. 1–8.
11. Giguère G. Collecting and analyzing data in multidimensional scaling experiments: A guide for psychologists using SPSS. Tutorials Quatitative Methods Psychol. 2006;2(1):26–37.
12. Okudoh V, Trois C. The Potential of Cassava Biomass as a Feedstock for Sustainable Biogas Production in South Africa. J Energy Power Eng. 2014;8(400):836–43.
13. Okudoh V, Trois C, Workneh T, Schmidt S. The potential of cassava biomass and applicable technologies for sustainable biogas production in South Africa: A review. Renew Sustain Energy Rev [Internet]. 2014;39:1035–52. Available from: http://dx.doi.org/10.1016/j.rser.2014.07.142
14. Adeyemo TA, Amaza P, Okoruw V. The food security effect of a biomass value web concept among smallholder cassava households in Edo State Nigeria . 2016;
15. Fitriani, Sutarni, Ismono H, Haryono D. Performance of Food Crops Sub Sector in Lampung Agriculture Sector. In:Proceedings of the National Seminar on Food Sovereignty and Agriculture. Diterbitkan, editors. Yogyakarta: Jurusan Sosial Ekonomi Pertanian Faperta UGM; 2014. p. 212–20.
16. Amaza P., Abbas A., Bachwenkiz B, Towo E. Adoption of Mechanized Postharvest Cassava Processing Technologies, and the Determinan of High Quality Cassava Flour (HQCF) Processing in Tanzania. Tropicultura. 2016;34(September):411–23.
17. Prianto FW. Pola Pengembangan Agroindustri Yang Berdaya Saing (Studi Kasus Kabupaten Malang) Fajar Wahyu Prianto 1. JEAM. 2011;X(1):48–71.
18. Todd PE, Wolpin KI. Assessing the Impact of a School Su Program in Mexico : Using a Soc Experiment to Validate a Dynam Fertility Behavioral Model of Child Schooling. Am Econ Rev. 2006;96(5):1384–417.
19. Claro RM, Canella D, Levy RB. Price and convenience : The influence of supermarkets on consumption of ultra- processed foods and beverages in Brazil. Appetite [Internet]. 2017;116(May):381–8. Available from: http://dx.doi.org/10.1016/j.appet.2017.05.027
20. Yuhartono N. Export Value Improvement Factors For Product Based Design In Jabotabek. J Cendekia Niaga. 2016;1(December):58–68.

Farmers' decision analysis to select certified palm oil seedlings in Lampung, Indonesia

Bambang Utoyo[1], Clara Yolandika[2]

[1]State Polytechnic of Lampung, Lampung, Indonesia
bambang.polinela@gmail.com

Abstract. This research aimed to analyse (1) decission making process of certified and uncertified palm oil seedlings and (2) factors that influence farmer decision to select certified and uncertified palm oil seedlings. This research was conducted in some districts in Lampung, such as Mesuji, Central Lampung, Tulang Bawang, North Lampung, Way Kanan and South Lampung. The respondents consisted of 30 farmers using certified seeds and 30 farmers using uncertified seeds. The study was conducted from January to May 2017. In addition, factors that influence farmer decision was analysed by logistic regression model. The results showed that decision making on the use of certified or uncertified palm seeds by farmers through the stages: introduction of problems or needs, searching of information, alternative evaluation, purchasing decisions, and post-purchase behaviour. Factors that significantly influence farmer's decision to use certified seeds were land area, seeds price, external influenced, and farmers' perception.

Keywords: certified, decison, palm oil, seedling

1. Introduction

Plantations are all activities of natural resource management, human resources, production facilities, tools and machinery, cultivation, harvesting, processing, marketing, and managing to achieve prosperity for plantation and community (UU RI No. 18, 2004). In maintenance estates play an important role; while in agroforestry and silviculture, plants tend to be allowed to grow according to natural conditions. Due to its intensive nature, plantations almost always apply the way of monoculture cultivation.

Palm oil (*Elaeis*) is one of the plants that produce cooking oil, industrial oil, and fuel (biodiesel). Palm oil grows as a cultivated plant spread in tropical climates even near subtropics in Asia, South America and Africa [8]. Palm oil is the second most successful agricultural product in Indonesia after rice. Palm oil is also the largest agricultural export commodity in Indonesia. The world's palm oil demand has grown rapidly in recent decades with current palm oil production estimated at more than 45 million tonnes. Indonesia is one of the largest producers and exporters of palm oil in the world, with more than 18 million tonnes of palm oil production per year [5].

Palm oil is one of the leading plantation crops of Lampung which has increased in terms of the area. The centre of palm oil plantation of Lampung is located in Mesuji and Central Lampung. According to Statistic Indonesia (2015), Central Lampung has the second largest palm oil plantation area in Lampung with an area of 29. 180 ha. Information on the area of palm oil plantations in each district in Lampung can be seen in Figure 1.

By the vast area of palm oil, the needed for quality palm seeds to produce quality production as well. Palm oil seed quality can be proven by using a certificate of seeds, commonly called certified seeds. Currently the use of certified seeds at the farm level is still very low. Causes of low levels of use of seeds such as the price of quality seeds / certified is still considered more expensive than regular seeds (uncertified), thus causing poor perception of farmers to certified seeds. Meanwhile, the use of certified seeds will be very beneficial for farmers. In addition to rapid production, quality of production is also better because it has been tested previously. Therefore, it is necessary to carry out the seedling of certified seeds. Before doing the logging, it is necessary to see perceptions and factors affecting farmers in the selection of palm oil seedlings.

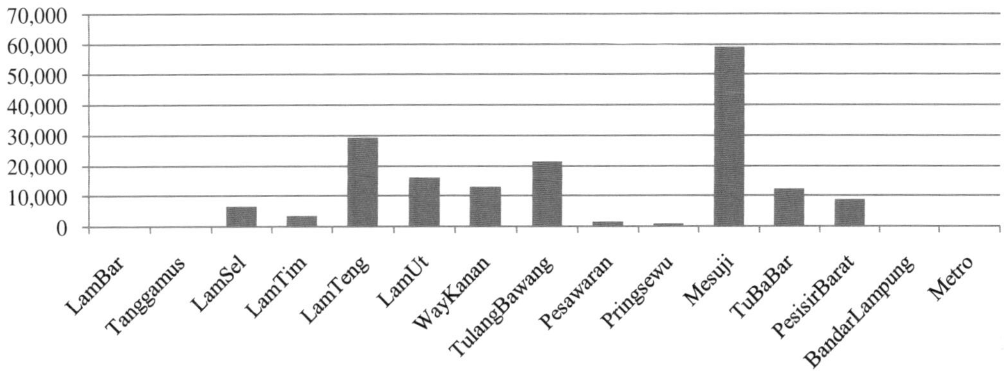

Figure 1 Palm oil plantation area in Lampung by district

2. Materials and Methods

This research was conducted at the centre of palm oil in Lampung, such as Mesuji, Central Lampung, Tulang Bawang, North Lampung, Way Kanan and South Lampung, where in each district were taken 10 farmers as respondents. The respondents consisted of 30 farmers using certified seeds and 30 farmers using uncertified seeds. The study was conducted from January to May 2017. Data analysis method used is quantitative analysis. Quantitative analysis is used to analyse the factors that influence the purchasing decision of farmers in choosing palm oil by using Logistic Regression method.

Farmers Decision Making Analysis

To know the factors that influence the farmers' decision of purchasing in selecting certified and uncertified palm seeds, quantitative analysis was done by using logistic regression model. Respondents in this study were categorized for the group of certified palm oil seedlings farmers and uncertified palm oil seed farmers. The factors thst influence farmers' purchasing decisions in selecting certified and uncertified palm seedlings are suspected to be influenced by educational factors, age, farming experience, land area, seed price, external influence, perception, seed resistance to plant pests and diseases, and seed access. The independent variables of X will affect the dependent variable Y. The dependent variable Y has two possible values of 1 and 0. The value of Y denoted by 1 means the purchase decision of certified palm oil seedlings and 0 means the purchase decision of the uncertified palm oil seedlings. To know the relationship between independent variables with dependent variable, the model used is as follows:

$$Yi = \alpha + \beta 1\ X1 + \beta 2\ X2 + \beta 3\ X3 + \beta 4\ X4 + \beta 5\ X5 + \beta 6\ D1 + \beta 7\ D2 + \beta 8\ D3 + \beta 9\ D4 + e$$

That:

Yi	= The decision to purchase palm oil seedlings Yi = 1 means purchasing decision of certified palm oil seedlings Yi = 0 means purchasing decision of uncertified palm oil seedlings
α	= Intercept
X1	= Education Factor (year)
X2	= Age (year)
X3	= Farming Experience (year)
X4	= Land Area (ha)
X5	= Seed Price (Rp/seed)
D1	= Dummy of external influenced (1=has influence; 0= doesnt's have influence)
D2	= Dummy of farmers' perception (1=certified seed is better than uncertified seed, 0=uncertified seed is better than uncertified seed)
D3	= Dummy of seed resistance to plant pests and diseases (1=resist, 0=not resist)
D4	= Dummy of seed access (1=easy, 0=difficult)
βi	= Parameter variables of Xi
e	= Error

3. Results and Discussions

3.1 Respondent Characteristic

Characteristics of respondents are the specific characteristics of a person such as age, education level, farming experience, land area, and land tenure status (Table 1). Asih (2009) reported that Sulawesi revealed the characteristics of age, education, and farming status affect farmers' skills in managing palm oil farming.

Palm oil farmer was either using certified seeds or uncertified dominated by farmers who are in the range of productive age is between 20 - 55 years. In general, productive people have a high spirit to develop their business, driven by high demand and able to make onion farming better compared with relatively older farmers.

The education level of both respondents who use certified seeds are dominated by undergraduate or diploma education, while respondents who use uncertified seeds are still dominated by junior high school. It means that most of the farmers are uncertified respondents have a standard level of formal education. It will certainly affect the level of courage to take decisions and risks in the management of palm oil farming. This is in accordance with the research of Emiriaet. al. (2014) which states that limited funds result in many farmers who didn't choose to go to school again and continue the work of their parents as farmers.

In palm oil farmers who use both certified and uncertified seedlings, have more experience of farming between 5-16 years. The farming experience shows that the length of farmers engaged in palm oil farming. The longer the farming experience, it can be concluded that the farmers have understood the cultivation techniques in the activities of farming.

Table 1. Characteristics of palm oil farmers in Lampung

No	Characteristic	Farmer who used certified seed		Farmer who used uncertified seed	
		Category	Percentage (%)	Category	Percentage (%)
1	Age (year)	30 – 55	83.33	30 – 45	66.67
2	Education Level	Undergraduate/ Diploma	33.33	JHS	56.67

3	Farming Experience (year)	9 – 16	50.00	5 – 8	70.00
4	Land Area (Ha)	1 – 15	100.00	0.5 – 1	83.33
5	Land Ownership	Own	100.00	Own	100.00

Land is a base in farming activities that acts as one of the capital in agriculture in addition to labor and capital. Respondents of palm oil farmers using certified seedlings in Lampung are generally classified into large-scale farmers with land concessions between 1-15 hectares, while palm oil farmers using uncertified seedlings are mostly classified into medium-sized farmers with land cultivation more than 0.5 - 1 hectare.

The land ownership status of farmers who use certified or uncertified seeds most of their land were their own land. The main reason farmers had their own land because palm oil is an annual plant with long economic life, so it will be more efficient if the land is a land of its own.

3.2 Decision Making Process

The decision-making process of farmers in using certified palm seeds or not through 5 stages, namely: the introduction of problems or needs, searching of information, alternative evaluation, purchasing decisions, and post-purchase behaviour.

Table 2. Indicators of problems or needs introduction

No	Description	Farmer who used certified seed		Farmer who used uncertified seed	
		Category	Percentage (%)	Category	Percentage (%)
1	Motivation of Palm Oil Farming				
	• Get the benefit	25	83.33	18	60.00
	• Heredity	2	6.67	10	33.33
	• Get the own needs	3	1.00	2	6.67
2	Motivation of seed using				
	• Affordable Price	-	-	24	80.00
	• Easy to get	-	-	5	16.67
	• Good Quality	30	100.00	1	3.33
3	Benefit that is found				
	• Production increasing	18	60.00	26	86.67
	• Quality increasing	10	33.33	4	13.33
	• Pests exposure reducing	2	6.67	-	-

Introduction of problems or needs. The process of using certified palm seeds didn't start when farmers begin to feel and recognize the needs for a seed product, and strive to meet those needs (Table 2). For some communities in Lampung Province, palm oil cultivation has become their main occupation. The motivation of farmers in palm oil cultivation is very diverse. In farmers who use certified seeds or not most or about 83.33 percent and 60.00 is to gain profit.

The need for palm oil seedlings was driven by several factors and it was influenced by the motivation of farmers in using seeds. The results showed that in farmers who use certified palm oil seedlings, the incidence of all farmers' motivation in using certified palm oil seeds because of good quality. Unlike farmers who use uncertified seedlings, the incidence of motivation is mostly (80.00 percent) of farmers in using uncertified palm oil seedlings because of their affordable price. The benefits that farmers are looking for in the use of seeds for certified seed farmers and not most (60.00 percent and

86.67 percent) are to increase the amount of production. According to Ancok (1997), the knowledge about the benefits of things will cause a person to be positive about it.

Searching of information. At this stage, the farmers would conduct information about the benefits and weaknesses of seeds that will be used in the farm. The acquisition of information will affect the perception of the farmer even his belief in the seeds, so that ultimately will affect the decision in the use of seeds. In trying to farm the farmers determine the type of seed that will be used. In addition to having their own knowledge, is also strongly influenced by other parties. Sources of interpersonal information commonly used by farmers to obtain information are fellow farmers or parents, extension officers, traders, and distributors.

Table 3. Indicators of searching information

No	Description	Farmer who used certified seed		Farmer who used uncertified seed	
		Category	Percentage (%)	Category	Percentage (%)
1	Own self	4	13.33	15	50.00
2	Farmer Group/ Friends/ Family	25	83.33	5	16.67
3	Seedling Seller	1	3.33	10	33.33

From Table 3 it can be seen that other farmers / farmer groups (83.33 percent) are the most influential sources of information for certified palm oil seedlings farmers. Farmers and farmer groups are the most reliable source of information for farmers, as they see for themselves the productivity of the seeds used. While the largest source of information the farmers users of palm oil seedlings are not certified, comes from self (56.67%). This is because the farmers are already feeling confident with the quality of seeds and affordable prices. This is consistent with Setiadi (2010) statement that information obtained by consumers is largely derived from commercial sources, but the most effective information comes from personal sources.

Table 4. Indicators of alternative evaluation

No	Description	Farmer who used certified seed		Farmer who used uncertified seed	
		Category	Percentage (%)	Category	Percentage (%)
1	Important Information				
	• Price	-		28	93.33
	• Productivity	13		-	-
	• Quality	17		2	6.67
2	Choosing Consideration				
	• Habbits	2	6.67	21	70.00
	• Following the other farmer	28	93.33	5	16.67
	• Have good market	-	-	4	13.33

Alternative evaluation.At this stage, farmers make the best consideration to be taken in meeting their needs by choosing certain criteria relevant to the wishes and needs to make a decision on the use of seeds. Based on Table 4, the seeds of palm oil that circulating in Lampung Province quite a lot of species and can be differentiated into certified seeds and not certified. Important information sought by farmers in the use of certified seeds is mostly due to quality (56.67 percent), while for seeds are not certified, mostly because of the price (93.33 percent). Factors considered by farmers to choose to use certified palm oil seedlings are mostly due to other farmers (93.33 percent), while the consideration for farmers choosing to use uncertified palm oil seedlings is largely due to the habit that has been done (70.00 percent).

Purchasing decisions. At this stage, the farmer decides what seeds to use in his farm. According to Engel et al. (1994), purchase intentions are classified into two categories: planned purchases and unplanned purchases. The results showed that the farmers are certified seed users or do not purchase the seedlings mostly in a planned manner. Farmers make purchases of seedlings in a planned because it already has a schedule of palm oil planting that will be done at a certain time, so the need for seed stocks before planting time begins.

The use of seeds by farmers in Lampung, in addition to being determined by yourself, is also influenced by outsiders. The result of the research shows that for the farmers, the certified seedlings that are influential in the process of purchasing the seedlings are mostly other parties, whereas for the uncertified seed growers who are influential in the process of purchasing the seedlings are mostly self.

Table 5. Indicators of purchasing decisions

No	Description	Farmer who used certified seed		Farmer who used uncertified seed	
		Category	Percentage (%)	Category	Percentage (%)
1	Satisfaction of seed purchase				
	• Satisfied	30	100.00	18	60.00
	• Not satisfied	-	-	12	40.00
2	If the seed price rise				
	• Keep buying	30	100.00	10	33.33
	• Not buying	-	-	20	66.67
3	If the seed unavailable				
	• Find in another place	30	100.00	12	40.00
	• Buy another seed	-	-	18	60.00

Distance where the purchase of seedlings with farmers' residence in Lampung is relatively far away. In certified seed farmers, the spacing of seed purchases is more than 5 km (100.00 percent). This is because most of the farmers using certified palm seeds buy seedlings at the Palm Oil Breeding Centre located in Bandar Lampung. Similarly, in uncertified seed farmers, the proximity of most purchases is more than 5 km (60 percent).

Post-purchase behaviour. After the farmers of the respondents have used the seeds, there would be a change in assessing the results of their farming based on the appropriate facts in the field (Table 6). All farmers of respondents who use certified palm seeds are satisfied with the seeds they use. They claim that the seeds they use have good quality and also the products that are in accordance with the wishes of farmers. But unlike farmers who use uncertified seeds of palm oil seedlings, only 60.00 percent of farmers are satisfied with the seeds that are not certified because it is good quality. They have complaints because of poor seed quality and uneven growth. The higher the level of farmers' satisfaction in using a type of seed, the higher the farmer's desire to use the seeds. Positive responses of farmers to seedlings will affect the use of the next seed.

The state of consumers in buying a product is strongly influenced by the price of the product itself. The results showed that if the price of certified palm oil seeds increased, then all farmers will still buy the seeds. This is because the farmers have been very confident with the quality is guaranteed. Unlike farmers who use seeds are not certified. If the price of palm oil seedlings is not certified increases, then most farmers (66.67 percent) will not buy the seeds.

The decision on the use of a seed type is also strongly influenced by the availability of seeds. The results show that if the certified seeds to be used are not available in the field, then all certified seed farmers will find the seeds elsewhere. It means that both local and imported seed farmers have a high degree of loyalty to the seeds they normally use, so that if the seeds are unavailable in the field they will look elsewhere. Unlike farmers who use seeds are uncertified. If uncertified palm oil seedlings are not available in the field, most (60.00 percent) of farmers will look for another varieties.

3.3 Factors of Affecting Decision Making

Logistic regression analysis was used to determine the factors that influence the decision of farmers in using certified palm seeds and not. The response variable (Y) in this analysis is categorical, where the palm oil farmers using certified seeds are rated 1 and the palm oil farmers using the uncertified palm oil seeds are rated 0. Based on the previous literature and on the considerations at the study sites, there are nine variables free ones. The nine independent variables are education, age, farming experience, land area, seed price, external influenced, farmer perception, seed resistance to pests and diseases, and seed access.

To see the accuracy of the model used Nagelkerke R square value of 0.943. This value indicates that the ability of independent variables in explaining the dependent variable is 0.943 or 94.3 percent while the rest of 5.7 percent is explained by other variables outside the model. To test the significance of the model in the model, a model goodness-of-fit test was performed using the Hosmer-Lemeshow (H-L) method. The result from the model with the H-L test is 1,229 and the p-value or significance is 0.998. The p-value value is greater than the 5 percent, so it can be concluded that the logistic model is feasible to use because it is able to explain or predict the farmer's decision in using the palm oil seeds.

Table 6. The result of logistic regression analysis

No	Variable	B	S.E	Wald	df	Sig.	Exp (B)
1	Education	0.114*	2.029	1.484	1	0.072	1.104
2	Age	-0.804	8.243	6.058	1	0.213	0.670
3	Farming Experience	0.052	1.721	0.472	1	0.172	0.005
4	Land Area	0.364**	9.779	1.601	1	0.001	20.137
5	Seeds Price	-0.513**	0.758	2.733	1	0.027	1.001
6	External Influenced	0.276**	1.610	4.508	1	0.001	29.944
7	Farmers' Perception	0.441**	1.864	4.224	1	0.001	1.859
8	Seed resistance to pests and diseases	0.057**	1.817	4.139	1	0.017	2.398
9	Seeds Access	-0.545	1.022	1.134	1	0.065	0.010

Note: ** = α = 5 % ; * = α = 10%

The result of logistic regression model estimation at 95% confidence level (α= 5%) there are five variables that give real effect to farmer decision in using certified palm seeds. These variables are the land area, seeds price, external influenced, and farmers' perception, than the endurance of seeds against pests and diseases. Four other variables: age, farming experience, education level, and seeds access have no real effect. The results of data processing to see the independent variables that significantly affect the dependent variable by using logistic regression are presented in Table 6.

Land Area. Land area has a significance value of 0.001, so this variable has a significant effect on the decision of farmers using certified palm seeds. Variable land area has a positive coefficient value, this means that the opportunity of farmers using certified palm seeds is positively related to land area, so the more land owned by farmers, the possibility of farmers using certified palm seeds will be greater. The value of odds ratio on variable land area is 20,137 meaning that every existence of increase of land area equal to 1 hectare hence opportunity of farmer to use local onion seeds equal to 20,137 times (ceteris paribus). The average certified palm oil farmer has a land area of 3.16 hectares, while farmers who use uncertified palm oil seeds have an average land area of 0.80 hectares.

This indicates that certified palm oil seedlings have larger land than farmers who use uncertified seedlings. Farmers who have narrower land have a tendency to maintain a pre-existing cropping pattern by using uncertified palm oil seedlings for reasons of the potential risks and uncertainties of production and marketing that may occur in certified palm oil seedlings. In general, smallholder farmers have a small farming capital, so their chances of using palm oil seedlings are uncertified because the purchase price of uncertified seeds are cheaper.

Seeds Price. Variable of seed price has significance value equal to 0,027, so this variable has significant effect to farmer decision using certified palm oil seedlings. The seed price variable has a negative coefficient value, it means that the chance of farmers using certified palm seedlings is negatively related to the price of seedlings; The higher the price of seeds, the possibility of farmers using certified palm seedlings will be smaller. The value of odds ratio on the variable of seed price is 1,001 which means that every increase of seed price is Rp 1, the farmer's chance to use certified palm seed oil decreases 1,001 times (ceteris paribus).

According to Schiffman and Kanuk (2008), Joni et al. (2001), price is one of several factors that influence consumers in the decision to buy seeds. Likewise with Theresia (2016) which states that one of the factors that influence the decision of farmers using rice seed is the price of seed. Based on the results, if the price of oil palm seeds is uncertified to increase, the farmers will tend to replace the seeds that will be used.

External Influence. The external influence variable has positive effect to farmer decision using certified palm oil seedlings. The external influence variable has positive coefficient value, it means that the chance of farmers using certified palm seedlings is positively related to the external influence, so the more external influence hence the chance of farmer use the certified seed will be bigger. The odds ratio value on external influence variables is 29.944 which means that if the external influence is large, then the chance of farmers using certified palm seeds is 29.944 times bigger than that of external influence (ceteris paribus).

Farmers' Perception. Variable of farmers' perceptions have positively affect the decision of farmers using certified palm seeds. Farmers' perceptions have positive coefficient value, it means that the opportunity of farmers using certified palm seedlings is positively related to the farmers' perception, so the higher farmers' perception to certified palm seeds, then the chance of farmers using certified seeds will be greater. The value of odds ratio on farmers' perception variable of 1,859 which means that if the farmer perception increase, then the opportunity of farmer use certified palm seeds is 1,859 times bigger than that small farmers' perception (ceteris paribus).

Seed resistance to pests and diseases. The seed resistance variable has a positive effect on the farmers' decision using certified palm oil seedlings. The seed resistance variable has a positive coefficient value, it means that the chance of farmers using certified palm seedlings is positively related to seed resistance, so the higher the seed resistance, the more chance the farmers use the certified seeds will be greater. The odds ratio value of the seed resistance variable is 2,398, which means that if the seed resistance increases, the chance of farmers using certified palm seeds is 2,398 times bigger than those with low seed resistance (ceteris paribus).

Other Variables. The result of logistic regression test shows that the age, farming experience, education level, and seed access have significance value greater than 0,05. It means that the variables have no significant effect to participate in determining farmers' opportunities in using certified palm oil seedlings.

Given the facts on the ground, the process of changing farmers' attitudes toward the use of local onion seeds is not influenced by the experience of farming but is influenced by evidence that has been seen in the surrounding environment. Farmers tend to observe and assess the results of other farmers who have previously used certified palm oil seedlings. After knowing the superiority of certified oil palm seeds compared to uncertified oil palm seedlings, they are only interested to use certified seeds.

4. Conclusions

Decision making on the use of certified or uncertified palm seeds by farmers through the stages: introduction of problems or needs, searching of information, alternative evaluation, purchasing decisions, and post-purchase behaviour. Differences in decision making between certified and certified palm seed farmers are not found in the motivation to use seeds, namely the motivation of certified seed farmers because of good quality, while the farmers are not certified because the price is affordable.

Factors that significantly influence farmer's decision to use certified seeds are land area, seeds price, external influenced, and farmers' perception. The seeds price has a negative and significant effect, while age, farming experience, education level, and seeds access have no real effect have a positive and significant effect on the farmer's decision to use certified palm oil seedlings.

References

[1] Ancok D 1997 *Teknik Penyusunan Skala Pengukur* (Yogyakarta: Pusat Penelitian Kependudukan UGM)

[2] Asih D N 2009 Analisis karakteristik dan tingkat pendapatan usahatani bawang merah di Sulawesi Tengah *Jurnal Agroland* 16 (1): 53-59.

[3] Emiria F, H Purwandari 2014 Pengembangan pertanian organik di Kelompok Tani Madya, Desa Kebonagung, Kabupaten Bantul, Daerah Istimewa Yogyakarta *Jurnal Penyuluhan* 10 (2)

[4] Engel J F, R D Blackwell, P W Miniard 1995 *Perilaku Konsumen Jilid 1* (6th ed). (Jakarta: Binarupa Aksara)

[5] Hasan M 2015 Peranan Kelapa sawit bagi perekonomian bangsa http://www.investasikelapasawit.com/peran-industri-dan-perkebunan-sawit-bagi-perekonomian-bangsa/ access on Mei 2nd 2017

[6] Joni M M A, M F R Rubzen, P J Batt 2001 Factors influencing a farmer's decision to purchase seed potatoes in East Java. *Paper presented at the 45th Annual Conference of the Australian Agricultural and Resource Economics Society* Adelaide South Australia

[7] Schiffman, Kanuk 2008 *Perilaku Konsumen Ed 7* (Jakarta: PT. Indeks)

[8] Setiadi N J 2003 *Perilaku Konsumen Konsep dan implikasi untuk Strategi dan Penelitian Pemasaran* (Jakarta: Prenada Media)

[8] Setyamidjaja D 2006 *Teknik Budi Daya dan Pengolahan Kelapa Sawit* (Yogyakarta: Kanisius)

[9] Statistic Indonesia 2015 *Statistik Lampung Tengah dalam angka* (Bandar Lampung: BPS)

[10] Theresia V, Fariyanti A, Tinaprilla N 2016 Pengambilan Keputusan Petani Terhadap Penggunaan Benih Bawang Merah Lokal dan Impor di Cirebon Jawa Barat. *Jurnal Agraris* 2(1):50-60

[11] Undang- Undang Republik Indonesia No 18 2004. *Undang - Undang tentang Perkebunan* Jakarta

IOP Conf. Series: Earth and Environmental Science **141** (2018) 012035 doi:10.1088/1755-1315/141/1/012035

Application of small digester biogas for energy supply in rural areas

S Wahyuni[1], S H Sutjahjo[2], Y A Purwanto[3], A M Fuah[4], R Kurniawan[5]

[1] Graduate School, Bogor Agricultural University, Kampus IPB Darmaga, Bogor 16680
[2] Department of Agronomy and Horticulture, Bogor Agricultural University, Kampus IPB Darmaga, Bogor 16680
[3] Departemen mechanical and Biosystem Engineering Bogor Agricultural University Kampus IPB Darmaga, Bogor 16680
[4] Department of Animal Nutrition and Feed Science, Bogor Agricultural University, Kampus IPB Darmaga, Bogor 16680
[5] Ministry of National Development Planning the Republic Indonesia, Jakarta
Email: swenitrans@yahoo.com

Abstract. Village's self-fulfillment of energy supply is a new concept being developed in Indonesia. Villages are expected to meet their own energy needs without relying on fossil-fuel and other non-renewable energy materials. The concept of village's self-fulfillment of energy here includes two types of provision: the provision of electrical energy and the provision of bio-fuel by utilizing environmentally friendly renewable energy. Biogas is an alternative energy sources produced by anaerobic activities of organic materials such as human and animal feces, or domestic waste in rural communities. In general, rural farmers have 2-4 cows. It is assumed that cows/cattle produces ± 15 kg feces/head/day. A digester with $4m^3$ capacity potentially produces bio-gas per day that reaches 1.34 m^3 equivalent to 6.30 kWh/day. This volume of bio-gas can be used for lamp lighting of 60-100 watts for 8 hours. Other applications that can be utilized are biogas stove. A family of farmer can use 2 stoves to cook 3 food recipes for 8 people and a generator for 1 hour. There is also potential solid mud of 40 kg that produces 8 kg of compost/day and potential sludge fluid as much as 160 liters including bio-urine of 32 liters/day. This paper discusses about the potency of small biogas digester application for village's self-fulfillment of energy supply.

1. Introduction

Energy demand is obtained through global energy supply where its availability is limited or irreversible. Energy consumption during the year 2000 - 2014 is still dominated by fossil fuel (gasoline, diesel fuel, kerosene). The consumption of electricity during 2000 - 2014 in average grew 6.8% per year. This excessive utilization of energy drives energy crisis that systemically give an impact to all aspects.

To date, villages in remote areas have not been reached by the power grid from State Electricity Company or PLN/Perusahaan Listrik Negara or State Electricity Company. According to Bureau Central of Statistics/ BPS (2015), there are about 15.4% or more than 12,000 villages have not been reached by

IOP Conf. Series: Earth and Environmental Science **141** (2018) 012035 doi:10.1088/1755-1315/141/1/012035

electricity till end of 2014. A total of 31,387 villages are having only minimum lighting along their main roads.

Several limitations faced to provide electricity for a village are : 1) State Electricity Company requires road access to the village destination for developing its electric network, while in Indonesia, most of villages have no road access; 2) The use of fossil fuel for a generator makes it expensive though most of the villages are poor; 3) Potential renewable energy sources in all rural areas also require an expensive preparation cost plus the lack of knowledge of the technology; 4) Private investors are not interested in funding electricity in rural areas because it is not profitable; 5) Technical issues of extending the distribution including land acquisition and legal certainty; 6) The relatively small allocation of government budget comparing to the total area coverage.

The complexity of energy problems in Indonesia requires a comprehensive national energy management through a clear and measurable National Energy Policy. On that basis, Act no. 30 of 2007 on Energy mandates the drafting of National Energy Policy (KEN) as a guide in national energy management. The basic principles that become the reference in the drafting process of KEN as mentioned in Act no. 30 year 2007 article 1 point 25 are the principle of justice, sustain-ability and environmentally friendly in order to achieve the self-fulfillment and national energy resilience with policy direction of realizing energy security in order to support sustainable development.

Village's self-fulfillment of energy supply is a new concept developed in Indonesia. It means as a village where its people have the capability to meet more than 60% of their energy needs (electricity and fuel) of renewable energy generated through the utilization of local resource potencies.

One of the alternative energy is the utilization of biogas energy. Biogas can be categorized as bioenergy, because the energy comes from biomass. Biomass is a relatively young organic material derived from living things or cultivated products and industrial wastes (agriculture, plantations, forestry, animal husbandry, and fisheries).

The potency of livestock waste as one of raw materials required for biogas production can be found in rural areas where most people have livestock. Waste farms such as feces, urine and cattle feed residue is one source of materials that can be utilized to produce biogas. Biogas is a renewable energy that can be used as an alternative fuel to replace fossil fuels such as kerosene and natural gas. The issues to be discussed are:

1. What is a biogas
2. How is the process of cow feces to become biogas energy as an alternative energy supply in rural areas
3. How to utilize biogas from cow feces as an alternative energy

2. Literature Review

2.1. Biogas

Biogas is a gas fuel (biofuel) and renewable fuel produced by anaerobic digestion or anaerobic fermentation of organic matter with the help of methane bacteria such as *Methanobacterium* sp. The materials that can be used as raw material for biogas production are biodegradable materials such as biomass (non-fossil organic material), dirt, solid waste from rural activities and others. However, biogas is usually made from livestock feces such as cattle, buffalo, goats, horses, and others. The main content of biogas is methane gas (CH_4) with a concentration of 50-80 vol. Gases in biogas that can act as fuel are methane gas (CH_4), Hydrogen gas (H_2), and Carbon monoxide (CO) gas.

IOP Conf. Series: Earth and Environmental Science **141** (2018) 012035 doi:10.1088/1755-1315/141/1/012035

Methane (CH_4) is an important and major component of biogas because of its usefulness and has a high calorific value (Table 1). It is an odorless and colorless properties. If the gas produced from this anaerobic fermentation process can burn, it contains at least 45% of methane gas. For pure gas (100%), it has a calorific value of 8,900 kcal/m^3. With high heating value, biogas can be used for cooking purposes, lighting, and the source of prime movers.

Table 1. Main compotition of biogas

No.	Gas	Hadi (1981)	Price (1981)
1	Methan (CH_4)	54-70 %	65-75%
2	Carbon dioksida (CO_2)	27-35 %	25-30%
3	Nitrogen (N_2)	0.5-2.0 %	<0.1%
4	Hidrogen (H_2)	-	<0.1%
5	Carbon monoksida (CO)	0.1 %	-
6	Hidrogen sulfida (H_2S)	Low	<0.1%

Source: Hambali (2007)

2.2. Materials Producing Biogas

A small-scale farms or rural household can keep 2-4 cows. Biogas from livestock manure has a nutritional balance, easy to be diluted and biologically processable. The biological processing range between 28-70% of organic matter depends on its feeding condition. The production of fresh livestock feces shown in Table 2.

Table 2. The production of fresh livestock feces per day

No.	Categories	Weight (kg/unit)	Production (kg/day)
1	Beef Cows	400-500	20-29
2	Milk Cows	500-600	30-50
3	Eggs Hen	1.5-2.0	0.10
4	Meat Hen	1.0-1.5	0.06
5	Pig (mature)	80-90	7.00
6	Sheep	30-40	2.00

2.3. Biogas Process

Biogas physically is a gas. The process of its formation requires a tight or closed condition of space in order to be stable. Biogas is formed through several processes that take place in an anaerobic space or without oxygen. This ongoing process also gives ecology advantage because it does not produce odor.

$$\text{Organic materials} \xrightarrow{\text{Anaerobic microorganisms}} CH_4 + CO_2 + H_2 + NH_3$$

2.4. Biogas Digester

Figure 1. Types of biogas digester

To produce biogas, a biogas generator called a digester is needed. There are several types of digester based on the raw material of manufacture that is dome type fixed digester (fixed dome), cylinder digester, balloon digester, and glass fiber digester (Figure 1).

2.5. The main components of bio-digester
The main components of biogas include digester equipped with an inlet and outlet, gas shelter, and sludge shelter (waste in solid and liquid form) (Wahyuni 2017).

Figure 2. Biogas components

2.6. Bio-digester supporting components
Supporting components of bio-digester are added to produce biogas in large quantities and safe to be used such as installation of biogas pipes, biogas pumps, biogas container bags (Figure 3), manometers, and valves or gas taps (Wahyuni 2017).

Figure 3. Biogas container bags

2.7. Biogas Process
According to Wahyuni (2017), measures of making biogas from waste material of livestock manure.
1. Prepare the raw material of fresh manure (2-3 days)
2. Add water with a ratio of 1: 2 (one part of dirt: two parts water). Mix them well, then in-sert or wash them into biogas digester

3. Perform filling through the inlet channel continuously until the biogas digester is filled with 60% of its volume capacity. Full digester condition is detected by the presence of a mixture of material out through the outlet discharge at the time of filling
4. After the digester is full, let it stand for 13-20 days. Be sure that the position of the con-trol gas faucet and the spraying gas faucet extending into the stove or application tool are closed. It aims to produce a room in anaerobic conditions to support the fermentation process of microorganisms.
5. On day 14-21, the result of the fermentation process will be visible and usually biogas in the form of methane gas (CH_4) will accumulate at the top of the biogas digester dome and ready for use for cooking or lighting.
6. Filling of biogas materials can then be done daily, about 10% of the digester volume. The remaining processing of biogas material in the form of sludge will automatically come out from the outlet whenever the biogas material is filled. The residues of biogas processing materials are manure/organic fertilizer, both in wet and dry conditions.

3. Results and Discussions

3.1. Biogas energy conversion and electricity

A cow with a weight of 450 kg produces a waste of 25 kg/day of feces. The number of cows owned by the people in rural areas on average are 2-4 cows. So the production of cow feces per day is 2 x 25 = 50 kg/day. The content of dry matter (BK) for cow feces is 20%, then the total dry matter content is: 50 x 0.20 = 10 kg. BK. So the potency of biogas of 2 cows is 10 x 0.04 = 0.4 m^3/day. The potency of electrical energy generated from feces of 2 cows owned by villagers on average is 0.4 m^3 x 4.7 Kwh = 1.88 Kwh/day. The potency of gases generated from different types of feces and conversion of biogas into the energy are presented in Table 3 and Table 4.

Table 3. The potency of gases generated from different types of feces

No.	Types of feces	Dry materials (BK)	Gas produced (m^3)
1	Cows or buffalo	20	0.023-0.040
2	Pig	9	0.040-0.059
3	Hen	28	0.065-0.116
4	Human	11	0.020-0.028

Table 4. Conversion of biogas into other energy

No.	Utilization	Energy 1m^3 biogas
1	Lighting	Lamp of 60-100 watt for 6 hours
2	Cooking	Cooking 3 recipes of food for 5-6 people
3	Horse Power	Running a motorcycle 1 *hp* for 2 hours
4	Electricity	4.7 Kwh electric energy

3.2. Concersion of biogas to other substitution of energy

The biogas produced is used as an alternative energy substitution for personal needs. In this activity, income estimation is derived from liquid petroleum gas (LPG) expenditure which can be saved through the use of biogas generated and based on biogas energy equivalence on LPG. A cow produces 25 kg of feces/day. 1 kg of feces is equivalent to 0.06 m^3 of gas. If a household in rural areas has 2 cows, then 0.06 m^3 x 50 kg = 3 m^3 gas. In 30 days (one month), 2 cows produce gas equivalent to 90 m^3. For conversion, 1 m^3 biogas is equivalent to 0.46 kg LPG gas. Then 90 m^3 biogas is equivalent to 41.4 kg LPG. The highest retail price of LPG at this time is IDR150,000.00/12 kg (12 kg LPG contains 26.2 liter) or equal to IDR5,700.00/liter. The LPG energy content is 49.51 MJ/kg, while the biogas (70% methane) is 35 MJ / kg, so the biogas energy equivalence is 70% to LPG energy. Biogas production of

2.4 Nm^3/day or 72 Nm^3/month, the usage of LPG which can be saved in one year is IDR2,297,872.00 or about IDR191,490.00/month. Conversion with kerosene for 1 m^3 biogas is equivalent to 0.62 liter. Biogas production for a month from 2 cows by 90 m^3 is also equivalent to about 55.8 liters of kerosene.

3.3. Side results of biogas
The material from livestock digestion is used as raw material for composting with the use of effective microorganism (EM). This compost is sold IDR400.00/kg or IDR20,000.00/sack. It is assumed that the compost from digestive waste is 40% of the weight of the input feces per day. If 2 cows produce 40 kg/day of feces, the compost produced per year is equivalent to 3650 kg. With the sale price mentioned above, then the additional income per year from the sale of compost in average is IDR1,460,000.00/year or about IDR121,667.00/month. Summary of income estimated from bio gas development activities derived from the feces of 2 cows is in Table 5.

The potency of urine. A cow with a weight of ± 300 kg produces urine 8-12 liters per day, while ± 250 kg cows produce urine 7.5-9 liters per day (Adijaya *et al.* 2008). Cow's urine can be used as organic fertilizer (Maurer *et al.* 2006). Cow's urine is converted into liquid fertilizer known as biourine through a fermentation process involving the role of microorganisms. The amount of biourine produced reaches 40 liters x IDR700.00 (price) x 30 days then the income from biourine reaches IDR840,000.00/month.

Table 5. Revenue from the development of biogas with the capacity of digester of 2 cows

Revenue	Total	Price per unit	Total revenue per year
Saving LPG	876 Nm^3/tahun	IDR3,220.00/Nm^3	IDR 2,820,720.00
Selling compost	3650 kg/year	IDR400.00/kg	IDR 1,460,000.00
Biourine	14,400liter/tahun	IDR700.00/liter	IDR10,080,000.00
Total revenue per year			IDR14,360,720.00

Figure 4. Biogas application

4. Conclusions
Biogas is gas fuel and renewable fuel from organic matter and resulted by anaerobic fermentation. Manures which produce by two cows can result on average is 0.4 m^3 biogas/ day and equivalent to elictrical energy equal to 1.88 Kwh/ hari. Biogas can used as alternative energy to substitute the use of LPG. In addition, biogas produced side result such as compost and biourine.

5. References

[1] [BPS] Badan Pusat Statistik 2015 *Statistik Listrik* (Jakarta: Badan Pusat Statistik)
[2] Adijaya IN *et al.* 2008 Laporan Prima Tani LKDRIK Desa Sanggalangit, Kec. Gerogak, Kab.

Buleleng, Bali (Denpasar: Balai Pengkajian Teknologi Pertanian)
[3] Hadi N 1981 Gas bio sebagai bahan bakar *Proyek Laboratorium Protein Sel Tunggal Pusat Penelitian dan Pengembangan Teknologi Minyak dan Gas Bumi* (Cepu: Lembaga Minyak dan Gas Bumi (LEMIGAS))
[4] Hambali E 2007 *Teknologi Bioenergi* (Jakarta: PT Agromedia Pustaka)
[5] Maurer M *et al.* Treatment processes for source-separated urine *Water Res* 40 3151-3166
[6] Price EC and Chemisinoff PN 1981 *Biogas Production and Utilization* (United States of America: Arbor Science Publishers, Inc.)
[7] Wahyuni S 2017 *Biogas* (Jakarta: PT Agromedia Pustaka)

IOP Conf. Series: Earth and Environmental Science **141** (2018) 012036 doi:10.1088/1755-1315/141/1/012036

Application of chitosan as biomaterial for active packaging of ethylene absorber

E Warsiki[a*]

[a]Department of Agroindustrial Technology, Faculty of Agricultural Engineering and Technology, Bogor Agricultural University, Bogor
2nd Floor FATETA Building, IPB Darmaga Campus, Bogor, Indonesia 16680

*Email : endangwarsiki@apps.ipb.ac.id

Abstract. Chitosan was used for active packaging of ethylene absorber that can change the head space of food packaging to extend the shelf life. The purpose of this study was to develop active packaging from chitosan and KMnO4 and apply the active filmto package tomatoes. Active film was prepared by mixing chitosan 6 g, 140 mL of acetic acid 1%, 60mL of aquadest, 2mL sorbitol, and $KMnO_4$ with concentration of 3 g, 5 g, and 7 g. After 5 days of storage, the film used to wrap the tomatoes was sweated. The best formulation to absorb the ethylene was made from 7 g of KMnO4 since it could inhibit tomatoes from shortly ripening compared to other formulations. The fruits were packed at room temperature had a high hardness as much as 17,79 mm/50mg/5s compared to the control of 3,47 mm/50mg/5s, while at the refrigerator, the tomato had lower hardness value of the 2,72 mm/50mg/5s compared to the control of 4,29 mm/50mg/5s. Addition of KMnO4 on the film could maintain the value °Hue either for all treated sample and control in the range of the yellow to red.

Key words : Biomaterial; chitosan; ethylene absorber; tomatoes

1. Introduction

Packaging has important role in food processing, mainly to protect and distribute of the product. Packaging should protect the food from unpleasant environment such as light, gas, moisture, microbe, mechanical stress and dirt [1]. So far, modern packaging have been developed to be more effectively integrated with the needs of the distribution supply chain [1] include modified atmospheric packaging (MAP), active packaging and smart packaging. This packaging technology was aimed to maximally improve the safety and the quality of the food product. Regarding to the active packaging, Warsiki *et al.* [12] has been researched the development of chitosan-based active anti-oxidant package with the addition of anti-microbial agents such as garlic and betel leaves extract. Terry *et al.* [10] developed a new material to absorb ethylene by using Palladium (Pd)-promote powder. Warsiki *et al*[12] has also examined the evaluation of physical-mechanical properties and the permeability of chitosan films.

Chitosan has been commonly selected as the base for filmmakers because it can be formed into films easily in low temperature with a good permeability [6]. Chitosan film has strong properties i.e. elastic, flexible and difficult to be torn [4]. In addition, this chitosan film has a high permeability, sealableand easy to be added some additives [12] thus it is possible to use this chitosan film as a matrix to carry such potassium permanganate as active packaging of ethylene absorber. According to Day [5] potassium permanganate ($KMnO_4$) is one of the additional ingredients that serve as an ethylene absorbent material. This compound has been applied commercially to prolong the shelf life of climacteric fruits. Widayani *et al.*[13] and Syamsu *et al.* [9] has demonstrated that potassium permanganate could effectivelyprolong the shelf life of banana as 27 days. Commonly, ethylene absorber was formed in the sachet, however in this study, the absorbent compounds of potassium permanganate and chitosan are mixed to be made into an ethylene absorbent wrap. The film was used to package fruits to slow the ripening process and keep its freshness longer.

2. Materials and Methods

2.1 Materials and apparatus

The materials are used to make the film of active packing were chitosan, glacial acetic acid 1%, distilled water, and $KMnO_4$. While the apparatus were magnetic stirer, thermometer, analytical balance, aluminum and glass plates as well as oven.

2.2 Ethylene Absorbent Film

Chitosan of 6 g was dissolved in 1% acetic acid as much as 140 mL and stirred it until homogenous using magnetic stirrer for 1 hour at 40°C. Then as much as 60 mL of aquadest was introduced into chitosan solution and added with sorbitol of 2 mL. After it was homogen, the film solution was well mixed with KMnO4 powder with a concentrations of 3 g, 5 g, and 7 g. The solution was continue stirred until homogen, then it was cooled at room temperature (32°C). The film was casted on a glass plate with size of 20 cm × 30 cm, then stored in a 50°C oven for 24 hours. The film was characterized visually. Table 1 showed the formula code based on the treatments.

Table1.Code of sample

Sample Code	Formula			Drying temperature (°C)
	Chitosan (g)	$KMnO_4$ (g)	Sorbitol (mL)	
F1	6	3	2	50 °C
F2	6	5	2	50 °C
F3	6	7	2	50 °C

2.3 Application of ethylene absorbent packaging

Active film was used to wrap tomatoes. Every single tomato was packagedinto ethylene absorber film and then store into room (32°C) and refrigerator temperature (9°C) for 7 days. The stored tomatoes and then analyzed its quality including the hardnessand the degree of color.

2.4 Data Analysis

The experimental was designed in a complete factorial randomized design with two factors and two replications. The factor used was the concentration of $KMnO_4$ added to film solution storage temperature. In this experiment it would see the effect of these factors on the performance of the ethylene adsorbent packaging. The measured variable data was then analyzed using the SPSS version 16.0 statistical program

3. Results and Discussions

3.1 Visually film characteristics

The three formulas were prepared with different concentrations of KMnO4 i.e. F1 (3 g), F2 (5 g), and F3 (7 g). KMnO4 is an aggressive organic material thus when it was added into film solution, a clumping and rising temperature has occurred. This is due to the reactive nature of this compound (Anonymous 2013). Based on the visual characterization of the film, the results obtained that Formula 1 (F1), Formula 2 (F2), and Formula 3 (F3) have different characteristics. The thickness and fragility of the resulting ethylene adsorbent film was affected by the difference in the amount of KMnO4 added to each of the formulas. Table 2 shows that F3 has characteristics of less thickness and breakable. The more $KMnO_4$ added, the film has an increasingly fragile. It seems that $KMnO_4$powder was not really compatible with the chitosan powder thus the bond between $KMnO_4$and chitosan was not strong enough. The characterization results in Table 2 also show that the three formulas have elastic properties. This elastic was caused naturally by sorbitol which was used as plasticizer material.

Plasticizer can reduce the intermolecular forces along the polymer chains thus it accentuates the increasing of film flexibility [3].

Table 2.The characteristic of ethylene absorber film

Code	Results
F1	Thick, elastic
F2	Thick, fragile, elastic
F3	Thin, fragile, elastic

3.2 Application of ethylene absorber film

Application of ethylene absorbent packaging was performed to see the performance the film in absorbing ethylene gas produced from fruit. The experiment was carried out by wrap tomatoes using an ethylene absorbent film. The packaged was sealed then stored at room temperature (32°C) and refrigerator temperature (9°C) for 5 days. The tomatoes then analyzed for their physical and chemical quality changes. Based on visual observations, the film used to pack the tomatoes becomes sweaty. Sweat was a water vapor produced from tomatoes due to respiration process. Both of these storage treatments show a difference results.

At room temperature, fruits tended to has faster maturation when compared with tomatoes stored at refrigerator storage temperature (9°C). The tomato stored at room temperature was shown in Figure 1 (a1), while the tomatoes stored at the refrigerator temperature are shown in Figure1 (a2). Additionally, the plastic film storage at 32°C has more moisture than the film kept at refrigerator temperature (9°C). This occurs because of the difference in respiration rate of the fruit at each of the temperature treatments. According to Pantastico *et al.*[7] the temperature was one of the external factors that affect the rate of respiration. Turiska [11] stated that the higher the storage temperature, the faster the rate of respiration and discoloration of the fruit. A high rate of respiration would result on faster ripening of the fruits which it was indicated by changing color of the fruit skin.

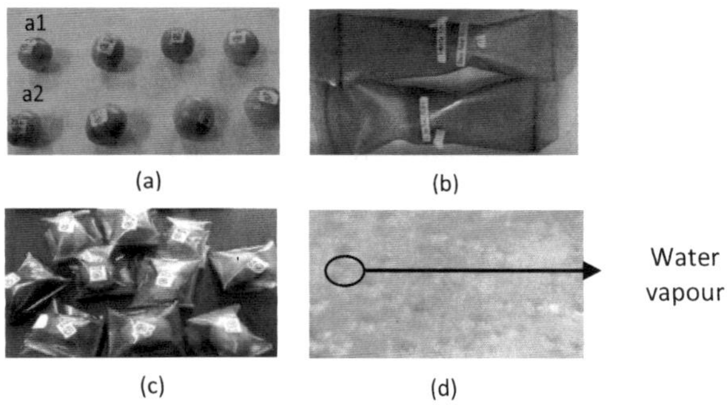

Figure1. (a) Tomatoesstore at : (a1) room temperature (32°C); (a2) refrigerator (9°C); (b) plastic film of ethylene absorber; (c) tomatoes wrapped into film; (d) visually changing of ethylene absorber film

3.3 Changing fruits quality during storage

3.3.1 Hardness

The hardness of fruits is affected by turgor from living cells that are constantly changing in the process of maturation. This is due to the changing cell wall components, these changes will affect the fruit hardness that is often physically observed by fruit firmed [14].

Hardness is one of the criteria used by consumers to determine the fruit maturity level which is observed by the depth of needle into the fruit thus the greater the value of fruit hardness the lower the quality of fruit [8]. The higher the hardness means the more soft fruit that indicates that the fruit is already mature. Hardness test at room temperature for 4day of storage showed that the control sample had lower hardness value than other sample, that was equal to 3,47mm/50mg/5s, while formula F1, F2, and F3 on the same day respectively had higher value of 15.57; 7.30 and 17.79mm/50mg/5s. The relationship between hardness value and storage time (day) at room (32°C) and refrigerator (9°C) can be seen in Figure 2.

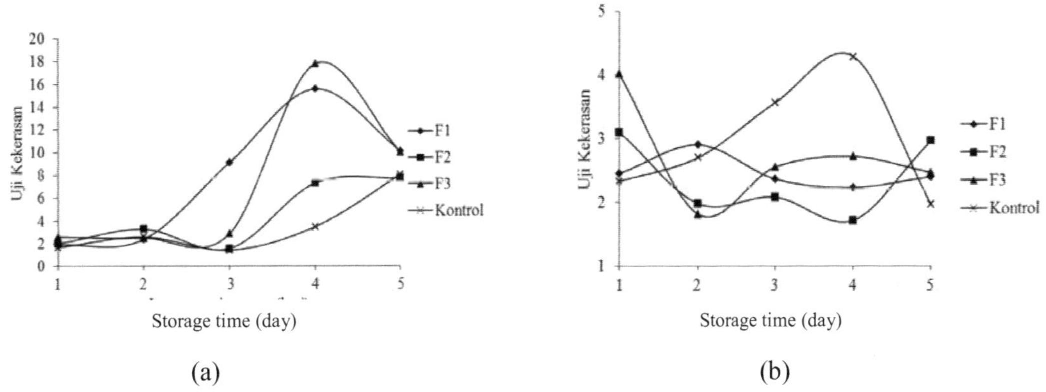

Figure 2. Hardness value (mm/50mg/5s) along the storage at 32°C (a) and 9°C (b)

The fruit will be softer with increasing storage time. The softening of this fruit was indicate the beginning of the maturation process. Pantastico [7] suggested that fruit hardness was reduced due to the destruction of insoluble protopectin to a soluble acid and pectin. The observation of the 4th day of the control sample has a higher hardness value of 4.29mm/50mg/5s when compared to the formula F1, F2, and F3 with the value of 2.23, 1.72, and 2.72mm/50mg/5s respectively. The more KMnO4 concentration would resulted on decreasingthe hardness of the fruit. This showed that the ethylene absorber film in F1, F2, and F3 may inhibit the rate of tomato respiration. Based on statistical analysis on the confident level of 0.05 indicated that the temperature storage influences the hardness of the tomato fruit, but no significant effect for the addition of KMnO4 or the interaction between KMnO4 concentration and storage temperature.

3.3.2 Value of °Hue

The relation between °hue and storage time (days) at refrigerator was shown in Figure 3. Based on this observation, the control has °hue lower (67.86) compared to sample of F1 (69,68), F2 (69,37), and F3 (68,47). However, the difference between these values between F1, F2, F3 and control was not significant difference. It mean that the adding of $KMnO_4$ was inadequately inhibited the fruit ripening. It looks that KMnO4 used in the formulation was too small thus it was not enough to slow the maturing process of the fruit.

The observations showed that treated sample and control has °hue values in the range of 54°-90°, which mean the tomato skin color changedfrom yellow to red. The more mature the tomato, then the value of °hue will be bigger and closer to the value of 90° or the skin fruit visually change ro red. Based on this observations, the control tomatoes store in room temperature has a °hue of 67.87°, while the sample of F1 and F3 have value of 67.77° and 67.65°. This indicated that the control sample was facing faster maturation. With the presence of ethylene absorbent packaging for wraping the tomato, it can be conclude that the film could slightly inhibit fruit ripening. The relationship between °hue value during storage ws shown in Figure 3.

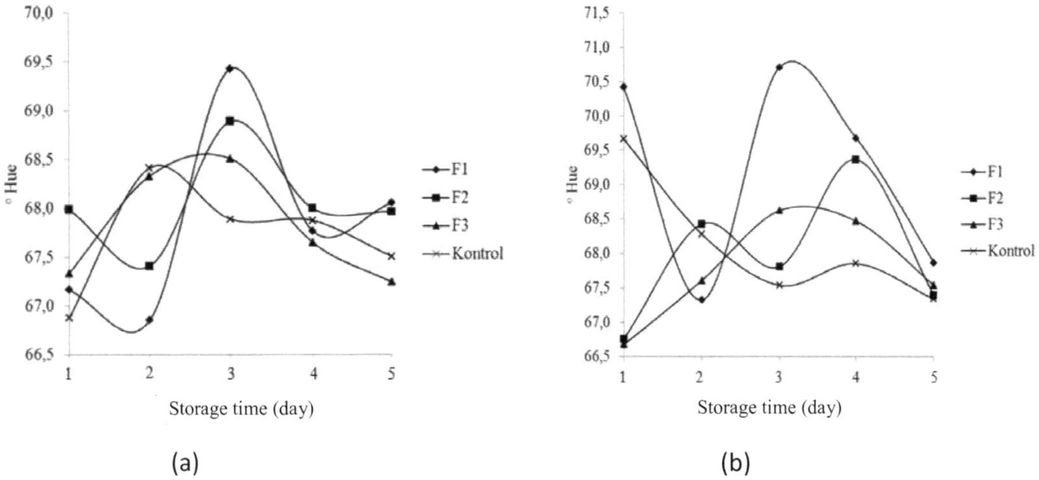

Figure 3. Value °hue during storage at 32°C (a) and 9°C (b)

4. Conclusions

Chitosan and KMnO4 can be used as a base material for the preparation of ethylene absorber film. After storage for 5 days it was obtained that the film was sweated. The ethylene absorber packaging has less thickness and elastic. The best formulation was the addition KMnO4 of 7 g. The packed fruit at room storage temperature has a higher hardness value than the control sample, whereas at the storage temperature, the packed sample has a hardness value lower than the control. Both sample and control has a °hue values were in yellow-red color range. Further study is required in order to produce better ethylene absorber film. It is recommended to conduct further research on KMnO4 concentrations greater than 7 g and observation time longer than 2 weeks.

References

1. Ahvenainen R. 2003. Active and Intelligent Packaging. Ahvenainen R (ed). *Novel Food Packaging Techniques*. Abington (USA) : Woodhead Publishing, p 5-21.
2. Anonim. 2013. Material Safety Data Sheet. http://www.sciencelab.com/msds.php?msdsId=9927406Accessed 4 Juli 2013.
3. Banker GS. 1996. Film Coating, Theory and Practises. *J Pharm Sci*. 55:81.
4. Butler BL, Vergano RF, Testin JMB, Wiles JL. 1996. Mechanical and barrier properties of edible chitosan films as affected by composition and storage. *J Food Sci*. Vol 61(5):953-955.
5. Day BPF. 2008. Active Packaging of Food. Willey John (ed). *Smart Packaging Technologies for Fast Moving Consumer Goods*. P 75-96. John Willey & Sons Limited, England (UK) :
6. Hoagland PD, Paris N. 1996. Chitosan/ Pectin Laminated Films. *J Agric Food Chem*. Vol 44:1915-1919.
7. Pantastico EB, Matto AK, Murata T dan Ogata K. 1986. Postharvest Physiology Handling and Utilization of Tropical and Subtropical Fruits and Vegetables.Avi Pub. Co., Madison.
8. Sabrina B. 2012. Effectiveness of the Ethylene Oxidator Packaging Material To Extend Banana Shelf Life. [Skripsi]. Bogor (ID) : Bogor Agricultural University.
9. Syamsu K, Warsiki E, Yuliani S, Widayanti SM. 2016. Nano Zeolite-kmno4 as ethylene adsorber in active packaging of banan (Musa Paradisiaca). International Journal of Sciences: Basic and Applied Research (IJSBAR).Vol 30(1): 93-103.

10. Terry LA, Ilkenhans T, Poulston S, Rowsell L, Smith AWJ. 2007. Development of new palladium-promoted ethylene scavenger. *J. Postharvest Biology and Technology*. Vol 45:P214 – 220.
11. Turiska S. 2007. Effect of Temperature and Time of Storageto the Quality of Banana [Skripsi] Department of Agriculture Engineering, Faculty of Agricultural Technology and Engineering. Bogor (ID). Bogor Agricultural University.
12. Warsiki E, Sianturi J dan Sunarti TC. 2011. Evaluation of physical-mechanical properties and permeability of chitosan films. Journal of Agroindustrial Technology. Vol. 21 (3) : 139 – 145.
13. Widayanti SM, Syamsu K, Warsiki E, Yuliani S. 2016. Effect of natural Bayah zeolite particle size reduction to physico-chemical properties and absorption against potassium permanganate (KMnO4). The 6th Nano science and Nanotechnology Symposium (NNS2015). 1710, 030029 (2016); doi: 10.1063/1.4941495.
14. Winarno FG,Wirakartausumah MA. 1979. Postharvest Technology.Sastra Hudaya, Jakarta.

The production of paper soaps from coconut oil and Virgin Coconut Oil (VCO) with the addition of glycerine as plasticizer

Asri Widyasanti[1], Anastasia Miracle Lenyta Ginting[2], Elgina Asyifani[3],Sarifah Nurjanah[4]

[1,2,3,4]*Departement of Agricultural and Biosystem Engineering , Faculty of Agro-Industrial Technology, Padjadjaran University, Bandung Sumedang km 21 St., Jatinangor, Bandung 40600*
Corresponding e-mail: asri.widyasanti@unpad.ac.id

Abstract. Hand washing with soap is important because it is proven to clean hands from germs and bacteria. The paper soapswere made from coconut oil and virgin coconut oil (VCO)with the addition of glycerin as a plasticizer. The aims of this research were to determine both formulation of paper soap using coconut oil and VCO based with addition of glycerin, and to determine the quality of the paper soapswhich is a disposable hand soap.This research used laboratory experimental method using descriptive analysis. The treatments of this research were treatment A (paper soap without the addition of glycerin), treatment B (paper soap with the addition of glycerin 10% (w/w)), treatment C (paper soap with the addition of glycerin 15% (w/w)), treatment D (paper soap with the addition of glycerin of 20% (w/w)). Parameters tested were moisture content, stability of foam, pH value, insoluble material in ethanol, free alkali content, unsaponified fat, antibacterial activity test, and organoleptic test. The result of physicochemical characteristics for bothcoconut oil-paper soap and VCO-paper soap revealed that treatment C (the addition of glycerin 15% (w/w) was the best soap formulation. Coconut Oil papersoap 15% w/w glicerin had water content 13.72%, the content of insoluble material in ethanol 3.93%, the content of free alkali 0.21%, and the content of unsaponified fat 4.06%, pH value 10.78, stability of foam 97.77%, and antibacterial activity against S. aureus 11.66 mm. Meanwhile, VCO paper soap 15% w/w glicerin had the value of water content of 18.47%, the value stability of foam of 96.7%, the pH value of 10.03, the value of insoluble material in ethanol of 3.49%, the value of free alkali content 0.17%, the value of unsaponified fat 4.91%, and the value of inhibition diameter on the antibacterial activity test 15.28 mm. Based on Mandatory Indonesian National Standard of solid soap SNI 3532:2016 showed that both of paper soap had not been accorded with SNI 3532:2016, unless the value of the insoluble material in ethanol. Moreover, organoleptic tests performed thatboth paper soap treatment D (20% w/w glicerine) were preferred by the most panelists.

Keywords: Paper Soap, Coconut Oil, Virgin Coconut Oil, Glycerin, Plasticizer

1. Introduction

Cleanliness is a very important thing due to the increasing number of diseases caused by bacterias and germs[1]. Soap is a substance used with water for washing and cleaning, made of a compound of natural oils or fats with sodium hydroxide or another strong alkali, and typically having perfume and coloring added. Even today, soap is not just used for cleaning to maintain the health of the skins, there are also some soap that also serves as softening soap and whiteningsoap. In making soap often used various kinds of fats or oils as raw material. For being used in soap manufacture the type of oil needs to be selectedin accordance with the use of soap itself.

Based on the making process, soap is made in two ways, namely saponification process and oil neutralization process. Saponification process occurs because the reaction between triglycerides with alkali, while the neutralization process occurs due to free fatty acid reaction with alkali[2]. Oil saponification process will obtained by-product namely glycerol, while neutralization process could not get any glycerol.

Along with the advancement of the era, then also developed a kind of soap that circulates in the market. Bath soap that circulates in the market based on its shape is divided into two forms, namely solid soap and liquid soap[3]. Solid soap itself is divided into several types based on their appearance, ie, opaque soap (non-transparent solid soap), and translucent soap (solid soap with almost transparent color), and transparent soap (solid soap with transparent color).

The two main components of soap constituents are fatty acids and alkali. Selection of the type of fatty acids determines the characteristics of the resulting soap, because each type of fatty acids will give different properties to the soap [4]. Fatty acids are the main components of fat and oil composites, so the selection of the type of oil to be used as a raw material for making soap is very important. To produce soap with good quality, it must use raw materials with good quality too.

Today, people want hand-washing soap that is practical to carry anywhere. Paper soap itself is one solid soap product innovation that printed or molded as thin as paper, paper soap will dissolve when exposed to water, it turns out to be foam. Soap paper is generally used as a disposable hand-washing soap because of its small size and thin so it easy to carry anywhere and suitable for use during at any outdoor activities. In Indonesia there is still difficulty to find a factory that produces paper soap. Paper soap majority is produced in China and the paper soap being produced in a big industrial scale. The paper soap made from materials, such as *Akyl Sodium Sulfate, Cocoamido Propyl Betaine, Cocoamide DEA, Paraffnum Liquidum, Perfume, Glyserin, Metyl Paraben, Propyl Paraben, Cabomer, Dimethicone, Sodium Starch Polucrylate, Aqua,* as well as *Alternatifolia Melaleuca Oil*. Until now there is no paper soap that made from natural ingredients. Therefore, in this research try to develop paper soap that made from natural ingredients such as coconut oil and chemical wich are safe and healthy for the skin.

Coconut oil rather than used for cooking, it alsocan be used for raw materials in cosmetics and some health products such as soaps, ointments, and others. Based on research wasdone by [5], states that the excess of soap made from coconut oil compared to other oils is that coconut oil soap has a good cleaning power due to the presence of lauric acid as the dominant fatty acids in coconut oil. The addition of other ingredients as a mixture in paper soap making can also maximize the benefits of the paper soap used. One of the ingredients of the mixture used as a reference treatment in the process of making paper soap is glycerine. Based on experiments that have been done glycerin as a plasticizer can make soap become elastic in texture such as industrial paper soap product.

To know the best glycerine concentration, proper formulation, and quality of resulting from paper soap, it is necessary to do this research. Until now there has been no research on making coconut oil-based paper soap and using the concentration of glycerine as a treatment. So there is no provision regarding soap formulation of palm oil based paper and how the printing soap process. The purpose of making paper soap is expected to produce soap paper that can be used everyday whenever and wherever, and able to cause a sense of comfort on the skin and able to preventthe skin from infections caused by bacteria.

2. Materials and Methods

The materials used were coconut oil and virgin coconut oil (VCO) obtained from Balai Besar Industri Agro, Bogor. The chemical materials equipped were30% NaOH solutions, stearic acid, alcohol, perfume soap (fragrance oil) citrus scented, glycerine and aquadest. While the supporting ingredient for the quality test were 96% neutral ethanol, 0.9 N and 2 N, 0.1 N, 0.1 N hydrochloric HCl, 1% phenolflatein indicator, sodium hydrogen carbonate, and n-hexane (technical grade). The test bacteria used in soap bacterial activity test was *Staphylococcus aureus* bacteria. Bacteria media used was Nutrient Agar (NA). The supporting materials consist of pH indicator paper, tissue, plastic wrap, aluminum foil, bread paper, disc paper, and transparent plastic.

The equipment used were beaker glass, hot plate stirrer, stirrer rod, thermometer, digital scales, dropper drops, spiral cake paste, spatula, petri dish, desiccator, oven, analytical scales, threaded tube, vortex, Erlenmeyer flask, acid chamber, separator funnel, microburet, upright cooler, tweezers, micro pipettes, test tubes, ose needles, bunsen burners, autoclaves, incubators, and slicers.

This research used laboratory experimental method using descriptive analysis. The treatments of this research were treatment A (paper soap without the addition of glycerin), treatment B (paper soap with the addition of glycerin 10% (w/w)), treatment C (paper soap with the addition of glycerin 15% (w/w)), treatment D (paper soap with the addition of glycerin of 20% (w/w)). This research consists of several stages, the first stage was the preparation of the raw materials for making transparent soap, the second stage was the making of paper soap, the third stage was the quality analysis of the coconut oil and VCO paper soap.

2.1 Raw Materials Preparation

The first step in this research wascreate 30% NaOH solutions, then prepared the raw material needed for making paper soap which includes pure coconut oil (or Virgin Coconut Oil) as fatty acids for about 23 gram, 30% NaOH as alkali 25 gram, stearic acid as additional material 11 grams, aquades as a solvent 22.5 grams; 12.5 grams; 7.5 grams; and 2.5 grams depends on the soap treatments, fragrance oil as a perfume of 0.5 grams, 96% ethanol as a solvent 18 grams, and glycerin as plasticizer ingredient of 0 grams, 10 grams; 15 grams; 20 grams along with the variation in glycerine concentration used 0% as control, 10%, 15%, and 20%.

2.2 Paper Soap Making

In the process of making paper soap there were several stages, first heating pure coconut oil or VCO upto 60°C, then mixed with 30% NaOH solution. Then it is stirred until the solution become more thicken to produce the mixture solution between the oil and 30% NaOH solution. The next step was heating and melting the stearic acid, after thatglycerine was added, followed by 96% ethanol and aquades, then stirred until homogeneously. The mixed solution in the previous step were combine with the stearic acid solution which is then heated for approximately one hour. The soap solution can be poured into the silicone mold, wait for approximately 24 hours for the soap to be removed from the mold. After the soap hardens, the soap is cut using a wooden slicer. The size of paper soap produced is 2x2 cm with a thickness of 1 mm. Next curing process is done to reduce the pH value, this process is done by storing paper soap at room temperature for about two weeks. Flowchart of process of making pure coconut oil or VCO based paper soap can be seen in Figure 1.

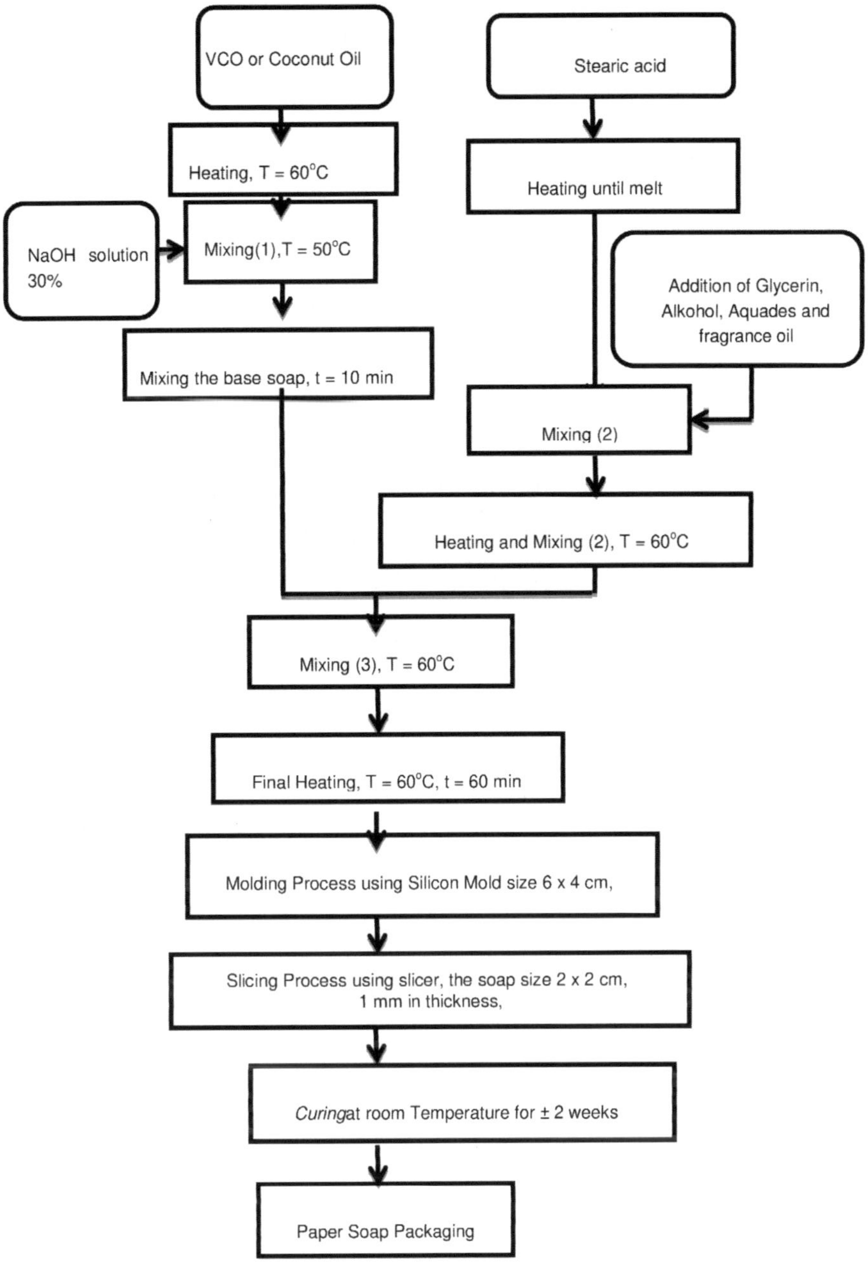

Figure 1. Flow diagram of the making paper soap process.

2.3 Quality Test

Observations to be made on paper soap produced were organoleptic test, physicochemical properties of soap, and antibacterial test. For organoleptic test conducted that is covering color, aroma, shape, texture, and many foam. For the physical properties of the soap, the foam stability test was performed, while for soap chemical properties, the water content, the insoluble material in ethanol, the free alkali, the unsaponified fat, and the pH contained in the resulting soap. For test of chemical properties based on SNI 3532: 2016 regarding the standard of solid soap bath quality. While organoleptic testing determined by the subjective value from 1 to 4 based on an assessment of the importance of texture, aroma, amount of foam, power of detergency, and after usage impression.

3. Results and Discusions

3.1 The Characteristic of Coconut Oil based Paper Soap and VCO based Paper Soap

The resulting paper soap product was the result of a transparent soap formulation with a combination of variation in glycerine concentration used 0% as control, 10%, 15%, and 20%. The transparent soap produced after cutting into paper soap can be seen in Figure 2.

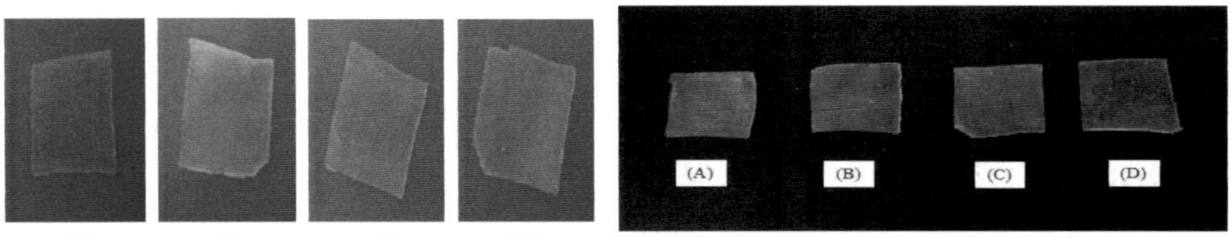

(A) (B) (C) (D)

Figure 2. Coconut Oil Paper Soap (right) and Virgin Coconut Oil Paper Soap (left)

3.2 The Quality of Paper Soaps

3.2.1 Water Content

Moisture content of solid soap according to SNI 3532: 2016 is 15% (wb) maximum. From the results of the measurement of water content on the paper soap, soap with no glycerine (treatment A) revealed the highest water content. From Figure 3 it can be seen that the more concentration of glycerin added to paper soap making, the moisture content contained in the soap is lower.Glycerine has the ability to bind water. From these statements it is suspected that the more glycerin added to the paper soap formulation the more water the glycerin bound together to cause the water content to decrease[6].

The water content of VCO paper soap (17.19-23.13%) were slightly higher than the water content of coconut oil paper soap (13.38-22.01%). Only treatment C dan D of coconut oil paper soaps were met the requirement of SNI. In paper soap making, several things are thought to affect the water content of paper soap, one of which is some type of fatty acid contained in coconut oil and volatile substances, such as alcohol. Several types of fatty acids, for example lauric acid, are water soluble and volatile when using water or steam[7].

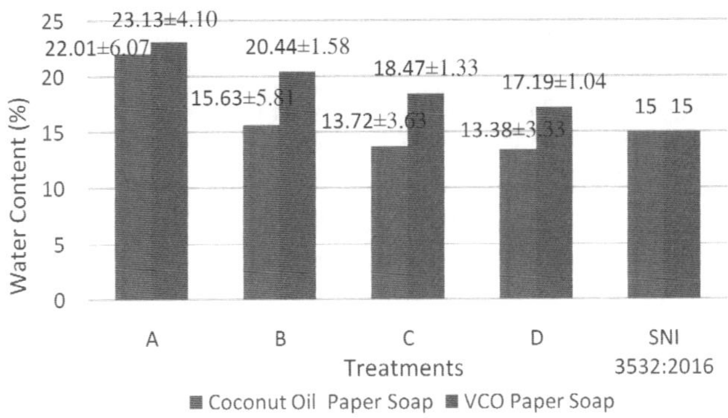

A (control): addition of 0% of glycerine
B : addition of 10% of glycerine
C: addition of 15% of glycerine
D: addition of 20% of glycerine

Figure 3. The relations between glycerine addition to the water content of paper soaps.

3.2.2 Insoluble Materialsin Ethanol

Based on SNI 3532: 2016 on the standard of solid soap baths states that the value of insoluble material in ethanol is a maximum of 5.0%. The results of insoluble material testing in ethanol on paper soap can be seen in Figure 4. The range value of insoluble materials in ethanol presence in the coconut oil paper soap were from 2.57-3.93%, where as in the VCO paper soap were around 1.26-3.49%.

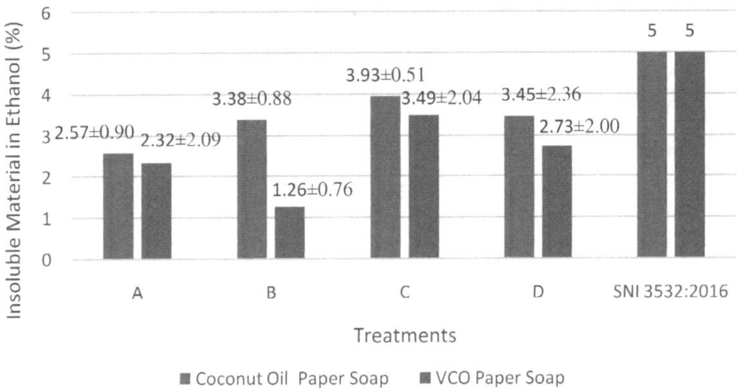

A (control): addition of 0% of glycerine
B : addition of 10% of glycerine
C: addition of 15% of glycerine
D: addition of 20% of glycerine

Figure 4. The relation between the glycerine addition to the insoluble materials in ethanol.

The amount of insoluble material in ethanol contained in the resulting of both paper soap was in accordance with SNI 3532: 2016 which is a maximum of 5.0%. From the results of the measurements showed that the addition of glycerin to paper soap formulations was not expected to affect the value of insoluble content in ethanol, glycerin is a soluble ingredient in alcohol[8]. The presence of an insoluble material in ethanol is suspected by the use of ingredients in the soap-making process. These substances are thought to contain insoluble molecules in ethanol. The value of insoluble material in ethanol in this paper soap is thought to be influenced by the presence of oil which is not well saponified and the oil is not completely soluble in water and ethanol.

3.2.3 Free Alkali Content

Free alkali is an unattached alkaline compound as a compound at the time of soap making, it is due to the excessive addition of alkali during the saponification process [5].The percentage of free alkali content (calculated as NaOH) on paper soaps can be seen in Figure 5.

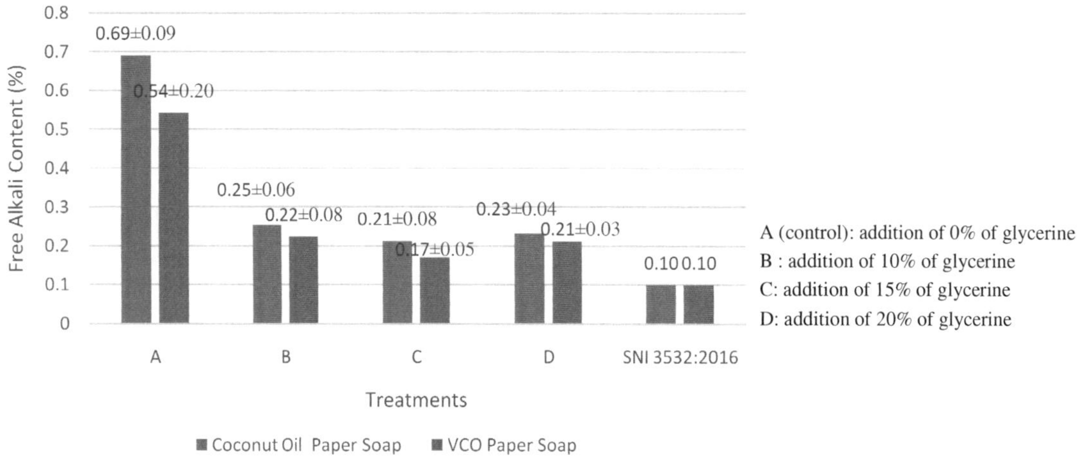

Figure 5. The relation between the glycerine addition to the free alkali content.

In Figure 5 it can be seen that treatment A (soap without addition of glycerin) has the highest free alkali content, that are equal to 0.69% (coconut oil papersoap) and 0.54% (VCO paper soap). The free alkali content contained in the resulting paper soap is quite high, presumably because the added concentration of the alkali is too thick or due to the excessive addition of alkali. In addition, the difference in the amount of NaOH usage in paper soap formulations with solid soap bath formulations generally makes the alkali content of paper soap has a higher mean value compared to the standard of SNI 3532: 2016.

Free alkali test on the paper soap it showed the effect of adding glycerin to paper soap formulations to free alkali values. The addition of glycerin to a paper soap formulation is thought to bind OH^-ions so as to decrease the free alkali value of the paper soap. Treatment C paper soap with 15% (w / w) addition of glycerin has a lower free alkali value compared with soaps of B and D treatments with 10% (w / w) and 20% (w / w) glycerine additions. This shows that paper soap with the addition of 15% (w/w) of glycerin addition is the optimum treatment to obtain free alkali values closed to SNI.

3.2.4 Unsaponified Fat Content

The presence of unabsorbed fraction or fat can decrease the ability to clean (detergency) on soap (Fachmi, 2008). The percentage of unabsorbed fat content in the resulting paper soap can be seen in Figure 6. From Figure 6, it can be seen that treatment A (coconut paper soap) has the lowest unsaponified fat content of 1.15%. The amount of fat not covered with soap paper has not fulfilled the quality criteria of unsaponified fat content according to SNI 3532: 2016, which is maximal 0.1%. Based on these observations, an increase in the added concentration of glycerin used may increase the unsaturated fat content in the resulting paper soap. The higher the addition of glycerin concentration, the higher the unabsorbed fat content contained in the paper soap. This is presumably because glycerine is a by-product of breaking oils or fats to produce fatty acids as well. The more glycerin means fatty acids that react with coconut oil are also more and more so that the amount of unsaponified fat contained in paper soap is higher. The amount of OH (*hydroxyl compound*) contained in glycerin producea lot of unsaponified fat in the saponification process.

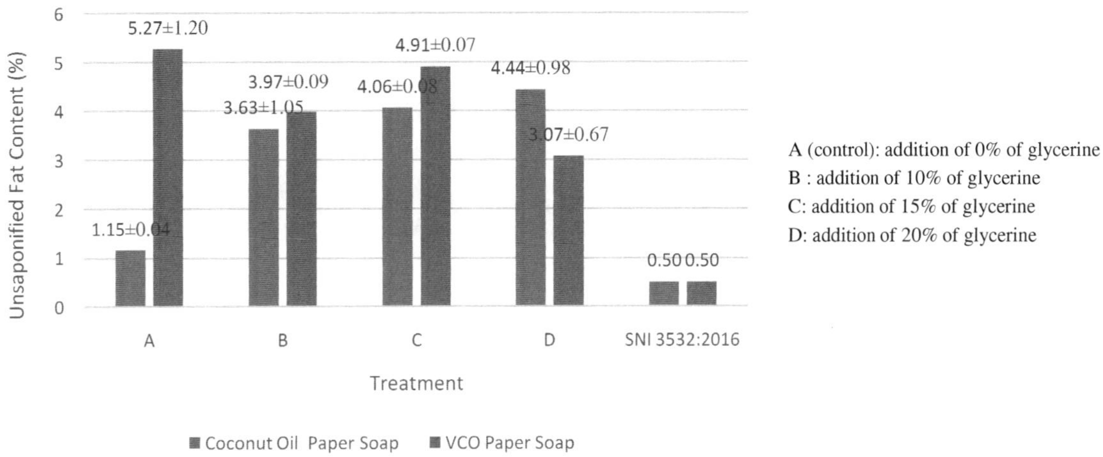

Figure 6. The relation between the glycerine addition to the unsaponified fat content.

Meanwhile, VCO paper soap treatment A (0% glycerine) had the highest unsaponified fat content 5.27%. The more glycerine added the VCO paper soap tend to decrease the unsaponifed fat content value. The addition of glycerine to paper soap formulations has an effect on the value of unabsorbed fat. The addition of glycerine to paper soap formulations is thought to assist the saponification process so as to decrease the value of unabsorbed fat on paper soap.This may be due to an inadequate saponification process between oil and NaOH, so that there is an oil that is not well-stuffed / unsaponified. The stirring process between oil and NaOH was carried out for about 10 minutes, presumably the stirring time was shortened so that it still left unreacted fat forming soap with alkali.

3.2.5 pH

The degree of acidity or pH is a chemical parameter to determine whether the resulting paper soap is acidic or alkaline. The soap that is too alkaline can cause skin irritation, so based on ASTM D 1172-95 the standard pH value for bath soap is 9 – 11.It shows that paper soap in this study already meet the pH standard of bath soap[9]. The pH value measurements of paper soap are done after the paper soap has been cured for about two weeks. From the results of pH measurements that have been done show that the addition of glycerin in paper soap formulations does not affect the pH value, glycerine has a neutral pH value[8]. The pH value of paper soap is influenced by the curing process, the longer the curing process will decrease the pH value of the paper soap. In addition, the pH value of the paper soap is also influenced by the value of free alkali, the higher the free alkali value the higher the pH value and the lower the free alkali value the lower the pH value of paper soap.The pH value contained in the resulting paper soap can be seen in Figure 7.

IOP Conf. Series: Earth and Environmental Science **141** (2018) 012037 doi:10.1088/1755-1315/141/1/012037

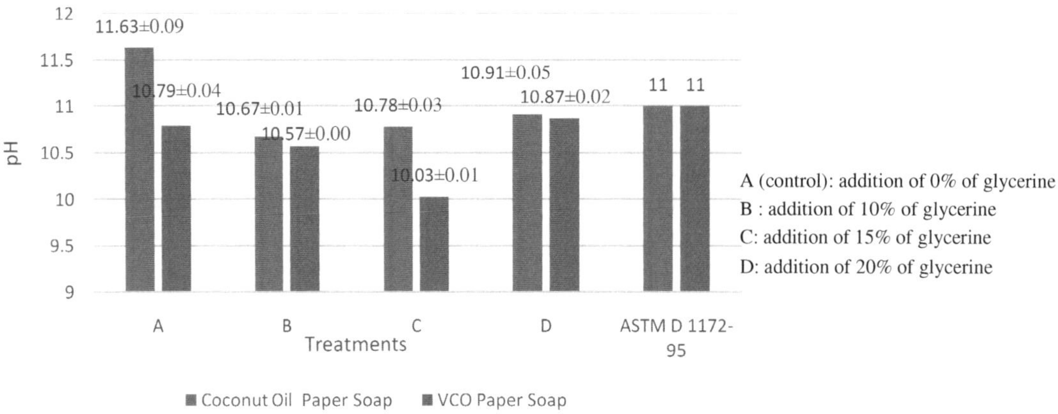

Figure 7. The relation between the glycerine addition to the pH value.

From Figure 7 it can be seen that the treatment A (coconut paper soap) has the degree of acidity or the highest pH value of 11.63. Overall the degree of acidity or pH value resulting from this paper soap is in accordance with the criteria of soap bathing quality, which is 9 - 11 based on ASTM D 1172-95.Results of pH measurements that have been done show that the addition of glycerin in paper soap formulations does not affect the pH value, because according to experiments that have been done using pH indicator states that glycerine has a neutral pH value.

The pH value measurements of paper soap are done after the paper soap has been cured for about two weeks. The amount of alkali present in the paper soap may affect the pH value. This paper soap manufacture involves the use of large amounts of NaOH, which accounts for 25% of all paper soap making materials. The pH value of the paper soap is affected by the curing process, the longer the curing process, the more pH value will decrease on the paper soap. In addition, the pH value of paper soap is also influenced by the value of free alkali, the higher the free alkali value the higher the pH value because alkali is a strong base.

3.2.6 *Antibacterial Activity against Staphylococcus aureus*

The bacteria used for this paper soap test is the gram-positive bacteria *Staphylococcus aureus* that can attack the skin. Antibacterial activity test was done by paper disc method, while the used disc paper had a diameter of 6 mm. The results of the analysis of paper soap clearance testing against *Staphylococcus aureus* bacteria can be seen in Figure 8.

In this antibacterial activity test used control with the test sample using aquadest. From Figure 8 it can be seen that the control treatment has the clearest zone or the lowest inhibitory diameter (DDH) against the *Staphylococcus aureus* bacteria of 0.32 mm. The control treatment was carried out to find out the comparison when washing hands using water alone using paper soap. Clear zones or inhibitory forces against the growth of *Staphylococcus aureus* bacteria on paper soaps may be derived from antiseptic and antimicrobial ingredients such as alcohol, glycerine, and coconut oil used in such paper soap formulations. The largest clear zone is owned by the soap of the third treatment paper (paper soap with 15% (w / w) glycerine addition), which is 11.66 mm. This proved that the ability of soap treatment C to inhibit *Staphylococcus aureus* bacteria is greater than other paper soap.

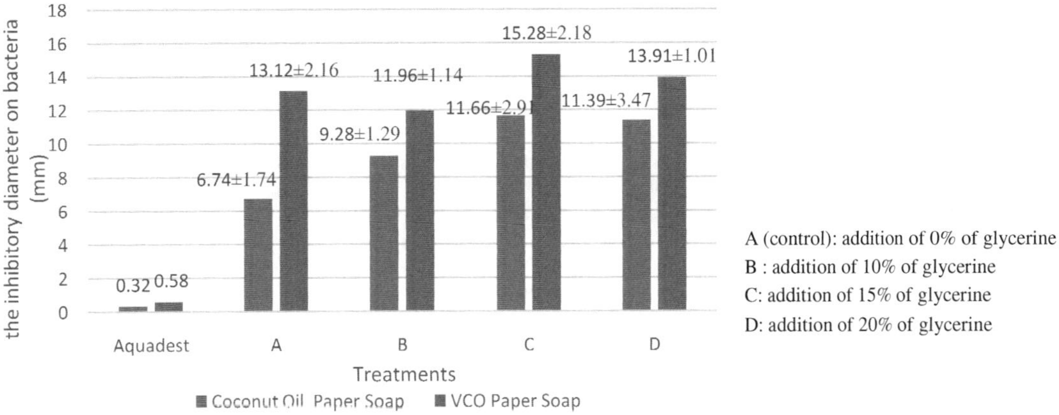

Figure 8. The relation between the glycerine addition to the inhibitory diameter on bacteria

Based on Fig. 8 also shows that the higher concentration of glycerin added to paper soap, clear zone or bacterial inhibition zone is also greater. Glycerine can be used as a solvent, sweetener, lubricant, plasticizer, and as an antimicrobial agent as well[10]. In this study, in addition to soap treatment C (15.28 mm) which has the largest clear zone for coconut oil paper soap, while treatment B (11.96 mm) gained the largest inhibitory diameter on VCO paper soap.It is assumed that themore glycerineadded, it will increase the antibacterial activity up to certain concentration then it will decrease because the bacteria will difficult to diffuse.

3.2.7 Foam Stability

Foam stability is the consistency of the amount of foam produced by paper soap. The resulting foam on paper soap is smoother compared to regular soap. The foam can be stable in the presence of a foaming agent. Solutions containing surface-active agents will produce stable foams when mixed with water. Glycerin generally does not contain surface active ingredients so it does not have a significant effect on foam stability. The percentage of foam stability for treatment A, treatment B, treatment C, and D treatment was presented in Figure 9.

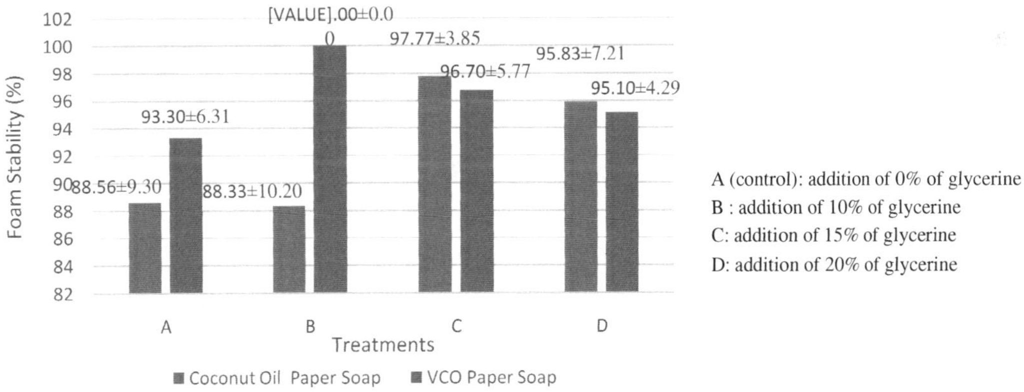

Figure 9. The relation between the glycerine addition to the foam stability.

The highest percentage of foam stability of paper soap produced is in treatment Cthat is 97.77%. This is assumed because in the second treatment, the value of each loop of high measurement of paper soap foam is generally unchanged after being silenced for approximately 1 hour. A stable foam is also affected by the type of fatty acids used. In coconut oil there are lauric and myristic acids that can produce soft foams, while palmitic and stearate acids have a stabilizing properties of foams. While oleic acid and risinoleat can produce a stable and soft foam [11].

3.2.8 Overall Results Paper Soap in term of Physicochemical Properties and Organoleptic Quality

The paper soap with glycerine addition was tabulated in each treatment and compared with solid soap's standard including Indonesia National Standard SNI3532[12]and ASTM D 1172-95. Here are the recapitulated results which were presented in Table 1 and Table 2. The result of physicochemical characteristics for bothcoconut oil-paper soap and VCO-paper soap revealed that treatment C (the addition of glycerin 15% (w/w) was the best soap formulation. Table 3 and Table 4 revealed the organoleptic results. For organoleptic tests performed thatboth paper soap treatment D (20% w/w glicerine) were preferred by the most panelists.

Table 1. Overall Results of Physicochemical Properties and Quality of VCO based Paper Soap

Parameters	Analysis Result The addition of glycerine (w/w)				Standard
	A 0%	B 10%	C 15%	D 20%	
Water content (%)	23.14	20.44	18.47	17.32	Max. 15*
Free Alkali content (%)	0.54	0.22	0.17	0.21	-
Insoluble Material in Ethanol (%)	2.32	1.26	3.49	2.73	-
Unsaponifiable Fatty matter (%)	5.27	3.97	4.91	3.07	-
pH	10.79	10.57	10.03	10.87	9 – 11**
Foam stability (%)	93.30	100.00	96.70	95.10	-
Antibacterial activity (mm)	13.12	11.96	15.28	13.91	-

*: (SNI 3532:2016) , **: ASTM D 1172-95 (2001)

Table 2. Overall Results of Physicochemical Properties and Quality of Coconut Oil Based Paper Soap

Parameters	Analysis Result The addition of glycerine (w/w)				Standard
	A 0%	B 10%	C 15%	D 20%	
Water content (%)	22.01	15.63	13.72	13.38	Max. 15*
Free Alkali content (%)	0.69	0.25	0.21	0.23	-
Insoluble Material in Ethanol (%)	2.57	3.38	3.93	3.45	-
Unsaponifiable Fatty matter (%)	1.15	3.63	4.06	4.44	-
pH	11.63	10.67	10.78	10.91	9 – 11**
Foam stability (%)	88.56	88.33	97.77	95.83	-
Antibacterial activity (mm)	6.74	9.28	11.66	11.39	-

*: (SNI 3532:2016) , **: ASTM D 1172-95 (2001)

Table 3. Overall Results of Organoleptic Quality of VCO based Paper Soap

Parameters	Analysis Result: The addition of glycerine (w/w)			
	A 0%	B 10%	C 15%	D 20%
Texture (Elasticity)	2.57	2.87	2.97	2.93
Aroma	2.87	3.07	3.17	3.17
The Amount of Foam	2.27	2.70	3.17	3.43
Power of Detergency	2.53	2.77	3.13	3.23
After Usage Impression	2.47	2.73	3.10	3.13

Table 4. Overall Results of Organoleptic Quality of Coconut Oil based Paper Soap

Parameters	Analysis Result: The addition of glycerine (w/w)			
	A 0%	B 10%	C 15%	D 20%
Texture (Elasticity)	2.63	2.66	2.93	3.10
Aroma	2.60	2.86	3.03	3.16
The Amount of Foam	1.93	2.50	2.90	3.20
Power of Detergency	2.40	2.63	3.03	3.16
After Usage Impression	2.43	2.73	2.93	3.06

4. Conclusions

The process of making paper soap is performed by hot process soap making by using hot plate stirrer as heating medium. The process of molding paper soap, paper soap is first molded as a solid soap by using silicon mold and then the size is reduced using slicer to form paper soap with size 2 x 2 cm and thickness 1 mm.Paper soap formulations in this study consisted of coconut oil (or VCO), 30% NaOH, stearic acid, aquades, 96% alcohol, fragrance oil, and glycerine with concentrations of 0% (w/w), 10% (w/w), 15 % (W/ w), and 20% (w/w). Based on this, the bestpaper soap from organoleptic test was treatment D (paper soap with 20% glycerin (w/w) concentration added. Meanwhile, based on physicochemical properties test result, the treatment C (paper soap with addition of 15% glycerine (w/w) concentration).

Acknowledgements

The author would like to appreciate to Padjadjaran University for funding.

References

[1] Gusviputri, A., Meliana, N., Aylianawati., Indraswati, N. 2012. *Pembuatan Sabun Dengan Lidah Buaya (Aloe Vera) sebagai Antiseptik Alami*. Jurnal Widya Teknik Vol 12 (1):11-21.
[2] Qisti, R. 2009. *Sifat Kimia Sabun Transparan dengan Penambahan Madu pada Konsentrasi yang Berbeda*. Skripsi. Program Studi Teknologi Hasil Ternak Fakultas Peternakan, IPB:Bogor. (Diakses pada :http://repository.ipb.ac.id/handle/123456789/59966)
[3] Hambali, E., Bunasor, T.K., Suryani, A., Kusumah, G.A. 2005. *Aplikasi Dietanolamida dari Asam Laurat Minyak Inti Sawit pada Pembuatan Sabun Transparan*. Jurnal Teknik Industri Pertanian Vol. 15 (2): 46-53

[4] Widiyanti, Y. 2009. *Kajian Pengaruh Jenis Minyak terhadap Mutu Sabun Transparan.* Skripsi. Fakultas Teknologi Pertanian, IPB:Bogor. (Diakses pada :http://repository.ipb.ac.id/handle/123456789/60649)

[5] Purnamawati, D. 2006. Kajian Pengaruh Konsentrasi Sukrosa dan Asam Sitrat Terhadap Mutu Sabun Transparan. Skripsi. Fakultas Teknologi Pertanian, Institut Pertanian Bogor: Bogor. (Diakses pada: http://repository.ipb.ac.id/bitstream/handle/123456789/3491/F06dpu_abstract.pdf;jsessionid=B410118160D9B8615A895854BBC2BCB2?sequence=1).

[6] Laeha, N. A. 2015. Pengaruh Penggunaan Gliserin Sebagai Humektan Terhadap Sifat Fisik dan Stabilitas Vitamin C dalam Sabun Padat. Skripsi. Fakultas Farmasi, Universitas Muhammadiyah: Surakarta. (Diakes pada: http://eprints.ums.ac.id/39949/1/NASKAH%20PUBLIKASI.pdf).

[7] Fachmi, C. 2008. Pengaruh Penambahan Gliserin dan Sukrosa Terhadap Mutu Sabun Transparan. Skripsi. Fakultas Teknologi Pertanian, Institut Pertanian Bogor: Bogor. (Diakses pada: http://repository.ipb.ac.id/handle/123456789/12111).

[8] Badan Pengawasan Obat dan Makanan Republik Indonesia. 2011. Gliserin. BPOM RI: Jakarta. (Diakses pada: http://ik.pom.go.id/v2014/katalog/Gliserin_upload.pdf).

[9] ASTM Standards. 2001. *Designation: D 1172-95, Standard Guide for pH of Aqueous Solutions of Soap and Detergents*. United States: West Conshohocken, PA.

[10] Lodén, M. 2014. *Hand Book of Cosmetic Science and Technology 4th Edition*. Ebook. ISBN: 978-1-84214-564-7.

[11] Cavitch, S. 1995. *The Natural Soap Book, Making Herbal and Vegetable-Based Soaps*. Ebook. ISBN: 978-0-88266-888-8.

[12] Badan Standarisasi Nasional. 2016. *Standar Mutu Sabun Mandi*. SNI 3532:2016. Dewan Standardisasi Nasional: Jakarta.

The characteristics of palm oil plantation solid biomass wastes as raw material for bio oil

RN Yanti[1,2], E Hambali[3,4], G Pari[5], and A Suryani[3,4]

[1] Natural Resource and Environmetal Management Study Program,Bogor Agricultural University, Indonesia.
[2] Forestry of Faculty, Lancang Kuning of University, Riau, Indonesia.
[3] Department of Agroindustrial Technology, Faculty of Agricultural Technology, Bogor Agricultural of University, Bogor, Indonesia.
[4] Surfactant and Bioenergy Research Center, Bogor Agricultural University, Bogor Indonesia.
[5] Center for Research and Development on Forestry Enggineering and Forest Product Processing, Bogor Indonesia.

Author to whom any correspondence should be addressed: rina.fahutan@gmail.com

Abstract. Indonesia is the largest palm oil plantations estate in the world. It reached 11,30 million hectares in 2015 and increased up to 11,67 million hectares in 2016. The advancement of technology recent, the solid waste of palm oil plantation can be re-produced become bio oil through pyrolysis hydrothermal process and utilized for biofuel. The purpose of this research was to analyze the characteristics of feedstock of bio oil of solid waste of palm oil plantations estate. The feedstock used was derived from solid waste of palm oil plantations in Riau Province. Characteristic analysis of waste oil included chemical compound content (cellulose, hemicellulose, lignin), ultimate analysis (C, H, N, O, S) to know height heating value (HHV). The result of analysis of chemical content showed that solid waste of palm cellulose 31,33 – 66,36 %, hemicellulose 7,54 – 17,94 %, lignin 21,43 - 43,1. The HHV of hydrothermal pyrolysis feedstock was 15,18 kJ/gram - 19,57 kJ/gram. Generally, the solid waste of palm oil plantations estate containing lignocellulose can be utilized as bio oil through hydrothermal pyrolysis. The CG-MS analysis of bio oil indicated hydrocarbon contents such as pentadecane, octadecane, hexadecane and benzene.

1. Introduction

Indonesia has the largest palm oil plantation field in the world with 11.67 million ha, that produced 33.50 million tons of Crude Palm Oil (CPO) in 2016 [1]. CPO produces biomass solid wastes not only on the plantation but also in the palm oil mill. The palm oil plantation wastes consist of trunks and fronds, while mill wastes consist of processed fresh fruit bunch that, in turn, produces empty fruit bunches (EFB), palm kernel shells (PKS) and mesocarp fibers (MF) . The biomass solid wastes have already been utilized to fulfill some needs such as trunks and fronds to substitute wood, empty fruit bunches to substitute compost material, kernel shells, and mesocarp fibers as boiler fuel in the palm oil mill.

Fossil fuel used as driving force for the machine has already reported causing environmental pollution through the emission of gases. The limitation in the use of fossil fuel has been implemented by mean of utilizing renewable energy such bio oil. Bio oil can be created by taking advantage of biomass through a process called pyrolysis. The advantage of using bio oil as an energy source is that it reduces gas emission, decreases CO_2 emission level in the atmosphere that leads to a decrease in green house effects.

Biomass is considered as clean energy,because it contains negligible amount of nitrogen, sulphur and ash compared to conventional fossil fuels, which results in lower emission of SO_2, NO_x, and soot than do conventional fossil fuels [2,3]. In addition,CO_2 released from biomass is resolved into plants by photosynthesis.

The chemical composition of palm oil plantation biomass solid waste needs to be investigated in order to obtain characteristics that fit the need of bio oil raw materials. One of the raw material requirements in the burning process is a biomass waste that contains lignocellulose. The aim of the present study was to analyze the characteristics of bio oil raw material from palm oil plantation biomass solid waste and determine the characteristics of bio oil from selected waste materials. The results of the present study are expected to determine raw material criteria and production process of bio oil in a specific and sustained way (zero-waste concept).

2. Materials and Methods

2.1. Biomass Powder Preparation

Palm oil plantation solid waste consisting of fronds, trunks, empty fruit bunch, palm kernel shell and palm mesocarp fiber were brought from a palm oil plantation in Riau Province. Biomass was processed into powder at a size that passed through a 40-mesh size filter and restrained to a 60-mesh size filter (using a hammer mill).

2.2.The Analysis of Pyrolysis Raw Material Chemical Compounds

Prior to chemical components analysis, the generated waste powder was dried to reach a water content of 15 %. Chemical components such as cellulose, hemicellulose, and lignin levels were analyzed by mean of TAPPI (Tappi 1989) [4] procedures. The ultimate analysis was using CHNO/S analyzer to determinate the Carbon, Hydrogen, Nytrogen contents. The oxygen contents was determined by difference as following : O = 100 – (C + H + N) [5]. The evaluation of chemical element components such as carbon, hydrogen, nitrogen, oxygen, and sulfur in palm oil biomass waste was performed in reference to ASTM D 3172 – D 3175 [6].

2.3. Pyrolysis Of Biomass Solid Waste

Biomass solid waste pyrolysis was carried out by mean of a hydrothermal reactor that can operate up to a maximal temperature of 350 °C. BSW powder was processed into a size of 40 mesh which was mixed with EFB liquid smock at a ratio of 1:1 and Hydrogen gas flowed across the reactor at a pressure of 10 bar. The reactor was turned on and left on to reach a temperature of 350 °C and a constant heat was maintained for an hour. Afterward, the mentioned reactor was switched off and left for 24 hours. Subsequently, fractionation was performed to obtain liquid smock and bio oil. During the fractionation process, the reactor was turned on to reach a maximal temperature of 400 °C (until the total absence of condensed smock). The high heating value (HHV) determination was carried out by analyzing the chemical element components of bio oil and the chemical compound content using a GC-MS device.

3. Results and Discussions

3.1 The characteristics of palm oil plantation solid biomass wastes

Palm oil plantation biomass solid wastes such as fronds, trunks, empty fruit bunch (EFB), palm kernel shell, and palm mesocarp fiber can be used as fuel energy source. Energy conversion can be carried out by direct burning, pyrolysis, gasification, and *torefaktion*. The characteristics of a biomass will affect its role as fuel. Palm oil biomass waste contains lignocellulose (cellulose, hemicellulose) and lignin that play roles in energy conversion processes such as pyrolysis. Figure 1 determined the characteristics of palm oil biomass waste, from which the most suitable waste will be determined as raw material for pyrolysis that produces bio oil.

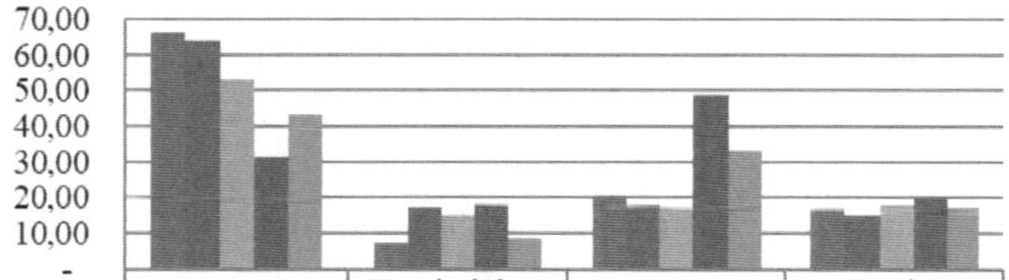

	Selulosa (%)	Hemiselulosa (%)	Lignin (%)	Heating Value (Kj/gr)
Trunks	66,36	7,54	20,49	16,84
Frond	64,12	17,54	18,09	15,18
EFB	52,93	15,27	17,43	18,14
PKS	31,33	17,94	48,83	19,57
MCF	43,14	8,54	33,14	17,61

Figure 1. The characteristics of palm oil plantation solid biomass wastes

The dominant chemical compounds in palm oil biomass waste are lignocellulose (cellulose, hemicellulose) and lignin. The mentioned chemical compounds are the main materials for the pyrolysis process to produce a liquid product, charcoal, and gas. Pyrolysis is a thermal decomposition occuring in the absence of oxygen at high temperatures (300 °C – 500 °C) [7,8]. According to [9], there are fundamental differences in characters when the process of pyrolysis is carried out with cellulose, hemicellulose, and lignin, and during the process, the first to unravel is hemicellulose at 200 – 315 °C, while cellulose starts to unravel at 300 -355 °C with a thick residue. Lignin ravels at a temperature of 160 – 900 °C. Pyrolysis results using cellulose and hemicellulose produced organic compounds (C=C, C-O-C), while lignin was observed to produce more methane gas compound (CH_4) due to the fact that it perfectly unraveled at high temperature (900 °C) [10,11]. The present research was aimed at producing liquid product i.e. bio oil at a temperature of 300 – 350 °C.

The characteristics analysis results of palm oil biomass (Figure 1) through chemical component and proximate analyses revealed that lignocellulose compound was higher in trunks, fronds, and EFB, while the highest heat value was observed in Palm kernel shells and EFB. Lignin percentage was higher in palm kernel shells compared to other wastes, and when used as pyrolysis raw material, it will produce the acid compound. However, if a lignocellulose content material is used as raw material in pyrolysis, it will result in the production of flammable compounds such phenol.

3.2 Pyrolysis of Empty Fruit Bunches (EFB)
Empty palm oil bunches used in the present study were processed into powder with a size of 40 mesh and a water content of 12 %. Pyrolysis uses hydrothermal process (slow pyrolysis) by adding liquid smock from EFB. The ratio between the liquid smock and EFB powder was 1:1, and the maximal temperature was 350 °C with a hydrogen bar of 10. The pressure during pyrolysis process ranged between 60- 85 bar. The purpose of adding Hydrogen gas to the process was to accelerate the termination of carbon chains that influenced the obtained bio fuel. The pyrolysis of EFB using

hydrothermal reactor resulted in 2 types of the liquid i.e. liquid smock and a thick black liquid called bio oill as shown in Figure 2.

Figure 2. Liquid yield from hydrothermal pyrolysis

Hydrothermal pyrolysis of the remaining lignocellulosa solids took please at higher temperature, producing intermediates that could undergo an extraordinary variety of reactions such as isomerization,dehydration, fargmentation and condensation reactions and ultimately form liquid oils, gas and char [12]. The present research focused on the bio oil and the results of the characterization analysis revealed that the liquid had a pH of 5 which was close to the normal pH (in general, bio oil pH is relatively low). In Table 1, a significant difference, in terms of water content, was observed in a research conducted by [13] that resulted in high water content. Indeed, the high the water content is, the low the burning value. Besides water, a high oxygen content is not needed in fuel burning process. Indeed, the HHV value in the present study was close to that of fossil fuel used in motorized vehicles.

Table 1. Properties of bio oil EFB

Component		Bio oil	Bio oil [a]	Light Fuel[b]	Heavy Fuel[b]
Moisture	%	15 – 30	65	-	-
pH		5	-	-	-
C	%	76	86,0	86,0	85,6
O	%	10,69	78,83	0	0,6
H	%	9,51	9,64	13,6	10,3
S	%	1,91	0,02	< 0,18	2,5
N	%	1,90	0,2	0,2	0,6
Higher Heating Value (HHV)	MJ/kg	39,19	30,58	-	
Low Heating Value (LHV)	MJ/kg	37,12	28,48	40,3	40,7

[a] [14], [b] [15]

From Table 1, the chemical composition of bio oil was complex mixture of degradation product of EFB constituens namely cellusa, hecellulosa and lignin. The bio oil composition depends on type of raw material (biomass) and the pyrolysis condition [16]. The GC-MS analysis was carried out to observe the chemical compound content that was close to that of an environmentally friendly fuel. The presence of pioneer compounds that could be developed into fuel was the hydrocarbon group or easily flammable compound. The GC-MS analysis result is presented in Table 2 and phenol compound was observed to be the most dominant compound (56.6%), followed by hydrocarbon (19.5%) and acids (3.43%). The presence of hydrocarbon compounds that are easily flammable such as pentacene, Hexadecene, Heptadecene, Octadecene and other compounds are potential compounds to be developed into alternative fuel with upgrading process. Side product of bio oil are charcoal and gases can use more aplication such as activated carbon using various activation methods and use as absorbents to elimate polllutan in wastewater treatment [17].

Table 2. The GC-MS analysis dominant result of bio oil chemical compound

Chemical compound	Value (%)
Phenol (3,5-Dioctoxyphenol)	56,6
Acetic acid	3,43
Tetradecane	2,71
Pentadecane	8,17
Hexadecane	1,57
Heptadecane	2,19
Tridecane	2,07
Heptadecanone	1,25
Cyclopentanone	1,01
Benzene	3,01
Decane	1,41

4. Conclusions

Research on bio oil raw material characteristics from palm oil plantation solid waste such as fronds, trunks, empty fruit bunch (EFB), palm kernel shell and mesocarp fibers showed that each material contained different lignocellulose and lignin levels. The results of the present study revealed that the raw material required for pyrolysis (to produce bio oil) is the one that contains high cellulose level with a high HHV value and a low lignin. Pyrolysis results of EFB in producing bio oil revealed the presence of hydrocarbon that can be developed into fuel to substitute fossil fuel.

Acknowledgements

The authors would like to gratefully acknowledge support : (1) Ministry of research and Technology Directorate General of Higher Degree Education, Republic of Indonesia. (2) Surfactant Bioenergy Research Center IPB Bogor Indonesia. (3) Center for Research and Development on Forestry Enggineering and Forest Product Processing, Bogor Indonesia

References

[1] [Director General of Plantation Republic of Indonesia]. 2014. Plantation statistik Indonesia palm oil 2013- 2015 [ID]

[2] Zhang Q,Chang J, Wang T. 2006. Upgrading bio oil over differentsolid catalysts. Energy & Fuel. **20** : 2717 2720.

[3] Borjesson PI. 1196. Emission of Co_2 from biomass production and transportation in agriculture and forestry. Energi Conversion and Management. 37: 1235 – 1240.

[4] Bentoro M, Puente G, Sendra U. 2012. Fuel from bio oils : bio oil production from different residual sources, characterization and thermal conditioning. Fuel. **95** : 263 – 271.

[5] [TAPPI] The technical association of the pulp and paper industry. 1996. *TAPPI test methods*. Atlanta : TAPPI Press.

[6] [ASTM] American Society for Testing and Material, D3172. 2007. Standart practice for proximate analysis of coal and coke

[7] Basu B. 2008. Biomass gasification, pyrolysis and torrefaction. Academic Press is an imprint of Elsevier.

[8] Bridgwater. 2012. Review of fast pyrolysis of biomass and product upgrading. biomass and bioenergy. **38** : 68-94.
[9] Yang H, Yan R, Chen H, Lee DH, Zheng C. 2007. Characteristics of hemicellulose, cellulose and lignin pyrolysis Fuel. **86** : 1781 – 1788.
[10] Pasangulapati P, Ramachandriya DK, Kumar A, Wilkins MR, Jones CL, Huhnke RL. 2012. Effects of cellulose, hemicellulose and lignin on thermochemical conversion characteristics of the selected biomass. Bioresource Technology. **114** : 663 – 669.
[11] Yang H, Yan R, Chen H, Zheng C, Lee DH, Liang DT. 2006. In – depth investigation of biomass based the three major components : Hemicelluce, celluloce and lignin. Energy& Fuel. **20** : 388 – 393.
[12] Xiu S, Shahbazi A, Shirley and Cheng D. 2010. Hydrothermal pyrolisis of swine manure to bio oil : effects of operating parameters on products yield and characterization of bio oil. Analytical and applied pyrolysis **88** : 73-79
[13] Sembiring CK, Rinaldi N, Simanungkalit SP. 2015. Bio oil from fast pyrolysis of empty fruit bunch. Conference and Exhibition Indonesia - New, Renewable Energy and Energy Conservation (The 3rd Indo-EBTKE ConEx 2014). Energy Procedia **65** : 162 – 169.
[14] Chiaramonti D, Oasmaa A, Solantausta Y. 2007. Power generation using fast pyrolysis liquids from biomass. Renewable and Sustainable Energy Reviews **11**(6):1056 - 1086.
[15] Bycova MV, Ermakov DY, Kaichev VV, Bulavchenko OA, Saraev AA, Lubedev MY and Yakovlev VA. 2012. Ni – base sogel catalystsas promising systems for crude bio oil upgrading : guaiacol hydrodeoxygenation study. Applied catalysisi B: enviromental (113-114) : 296 – 307.
[16] Choi GG, Oh SJ, Lee SJ and Kim SJ. 2015. Production of bio based pHenolic resin and activated carbon from bio-oil and biochar derived from fast pyrolysis of palm kernel shells. Biorecource Technology **179** : 99-107.
[17] Pimenidou P, Dupont V. 2012. Characterisation of palm empty fruit bunch (PEFB) and pinewood bio-oils and kinetics of their thermal degradation. Bioresource **109** : 198 – 205.

IOP Conf. Series: Earth and Environmental Science **141** (2018) 012039 doi:10.1088/1755-1315/141/1/012039

Review on Bamboo Utilization as Biocomposites, Pulp and Bioenergy

Sulaeman Yusuf[1,2], F. A. Syamani[1], W. Fatriasari[1], Subyakto[1]

[1]Research Center for Biomaterials LIPI, Jl Raya Bogor KM 46 Cibinong Bogor 16911, Indonesia
[2]Author to whom any correspondence should be addressed
sulaeman@biomaterial.lipi.go.id

Abstract. One of potential non wood bioresources utilized in industrial application is bamboos. Bamboos are include in graminae family which have high biomass productivity, easy and rapid production, wide avability and high holocellulose content. Indonesia has a huge potential of bamboos, more than 162 bamboo species are found however only some of them are planted that have a high economic value. Bamboos have some advantages such as can be harvested at 3 years, straight culm, high strength, easy to be processed, and relatively cheap. Research Center for Biomaterials has developed utilization of bamboo culm for ply bamboo product as alternative of plywood since 1995, using gombong bamboo, tali bamboo, sembilang bamboo, andong bamboo with PF resin as adhesive. Other biocomposite products from bamboos include particle board, cement board and polymer-bamboo fiber composites. In term of processing technique and final product quality, bamboo composites from ply bamboo are the most prospectable material to be utilized in industrial application. Yellow bamboo and betung bamboo have also been developed as pulp and paper. Biopulping using soda and kraft pulping after biological pretreatment using white rot fungi to remove lignin was used as pulping method in this conversion. Biokraft pulping with *Trametes versicolor* for 45 days with inoculum loading of 10% resulted better pulp quality compared to the other fungi. Betung bamboo had good morphological characteristics and chemical component content to be converted into bioenergy such as bioethanol. Several pretreatment methods have been developed in order to result high sugar yield. Microwave assisted acid hydrolysis was preferedin producing higher yield from the pretreated bamboo compared to enzymatic hydrolysis. By using this method, the bamboo pretreated by biological-microwave pretreatment results higher improvement to increase sugar yield.

1. Introduction

Among various non-wood lignocellulosic bioresources, bamboo has attracted many researchers attention. Bamboos has high biomass productivity, easy and readly to grow in many various land types, wide availability and high holocellulose content. Indonesia has a huge potential of bamboos, more than 162 bamboo species were found [1]. However only some are planted that have a high economic value. Bamboos have some advantages such as can be harvested at 3 years, straight culms, high strength, easy to be processed, and relatively cheap. Bamboos can be found good growing in Asia including Indonesia, in which Bamboo plantations in Indonesia, India and China include as first, second and third rank [2].

Bamboos can be utilized into many purposes such as reinforced fiber, composites, pulp and paper, lactic acid, methane, construction, textile, food, bioenergy [3]. Bamboos for traditional utilization such as construction and composites, all bamboo culms can be utilizedwithout discarding certain bamboo parts. In advanced utilization of bamboos such bamboo for bioenergy, pulp and paper, and wood plastic composites, chemical component (cellulose, hemicellulose and lignin) of biomass must be removed by pretreatment process. In this case, lignin can become major inhibitor in the conversion process. Even though, all bamboo species can be converted into derived materials, there are variation in their basic properties. Evaluation on its basic properties can be done before further utilization. In advance utilization, the effectivity of pretreatment was affected by bamboo's properties and also the chosen pretreatment method. Many development processes have been reported in bamboo utilization for composites, pulp and bioenergy, and Research Center for Biomaterials LIPI also contributed in this area by conducting some research improvement continuously. This development give possibility in utilization of derived materials from bamboos to be applied in industrial scale. It is in line with growing researcher interest in the world to utilize bamboo optimally. This paper was intended to review some research results on bamboos that have been carried out in RC Biomaterials LIPI especially for biocomposites, pulp and sugar production as material for bioethanol.By this study, it might become an effective media for technology transferring or open mutual research collaboration with other institutions.

2. Utilization of Bamboo for Production of Biocomposites, Pulp and Bioenergy

2.1 Bamboo as Raw Material for Composite Products

Bamboo culms have been processed to produce plies, particles or bamboo pulp then used in composite products. Those treatments were intended to obtain the optimal utilization of bamboos.

2.1.1Ply Bamboo Board

Utilization of bamboo culms for ply bamboo product can be used as an alternative of plywood [4]. Research on ply bamboo made from Talibamboo and Gombong bamboo using phenol formaldehyde (PF) resin has been conducted [5-6]. In 2016, Subyakto et al [7] investigated characteristics of ply bamboo board from Sembilang bamboo and Andong bamboo using PF resin as adhesive.

Sudijono et al. [5] has produced ply bamboo board using Tali bamboo (*Gigantochloa apus*) and PF resin at glue spread variation of 400, 450 and 500 g/m^2. Bamboos were cut to 600 mm (face/back) and 400 mm (core) and soaked in water for 30 days to remove starch content. Then, they were split into 2-3 parts and crushed using bamboo crusher machine to produce zephyr bamboo. Outer and inner parts of zephyr bamboo were sanded to remove bamboo bark about 2-3 mm, resulted in thickness of 4 mm (for face and back) and 7 mm (for core). Zephyr bamboos were dried in dry kiln to reach moisture content of about 7-8 %. Dry zephyr bamboos were mixed with PF resin at the amount of 400, 450 and 500 g/m^2 and hand formed to make 3 layer mat with the core layer oriented perpendicularly. The mat was cold press for 10 min at 10 kg/cm^2 then hot press at 160°C for 15 min at 10 kg/cm^2 to produce 15 x 400 x 600 mm board. After conditioning for2 weeks, the boards were tested for bending, shear strength and thickness swelling in accordance with Japanese Agricultural Standard (JAS) No 516-1992: Structural plywood.

Tali bamboo zephyr (ply) board with the amount of PF resin of 450 g/m^2 gave the best properties [5]. The mechanical properties of bamboo zephyr (ply) board were as follow: MOE = 127.000 kg/cm^2 or 12.45 GPa, MOR = 791 kg/cm^2 or 77.57 MPa. In addition, delamination testing showed that at dry test, shear strength was 23.84 kg/cm^2, while at steam treatment test, the shear strength was 15.86 kg/cm^2 and at cycle boiling test, shear strength was 12.66 kg/cm^2. The thickness swelling of bamboo zephyr (ply) board was 12.3% demonstrated that Tali bamboo zephyr (ply) board was good enough for outdoor application.

Gopar and Subyakto [6] produced zephyr board made from Gombong bamboo bonded with urea formaldehyde (UF) or PF. Gombong bamboo has relatively big diameter, tall and thick wall. Bamboos were split, crushed and planned into ply bamboo with 7 mm thickness. Those bamboos were planned to remove inner barks, while the outer barks were remained. The resin content was applied to about 17% based on oven dry weight of the ply bamboos. Mat of ply bamboo and resin was hot pressed at temperature of 160°C for PF and 140°C for UF. Pressing time was 20 minutes with three steps of pressing schedule; 5 min at pressure of 30 kgf/cm^2, 5 min at pressure of 10 kgf/cm^2 and then 10 min at pressure of 7.5 kgf/cm^2. The target of board density was 0.75 g/cm^3. The boards were tested in accordance with the Japanese Agricultural Standard (JAS) No 516, 1992: Structural Plywood.

Gopar and Subyakto [6] reported that the thickness swelling of the ply bamboo boards which inner and outer barks were removed, and bonded with UF or PF were 6.55% and 11.818% respectively. While the water absorption values were 39.48% and 34.61%. The shear strength, MOR and screw withdrawal of ply bamboo board bonded with PF were higher than that of ply bamboo bonded with UF. The shear strength, MOR and screw withdrawal of ply bamboo bonded with PF were 13.24 kgf/cm^2, 686.82 kgf/cm^2 and 799.47 kgf, respectively. While the shear strength, MOR and screw withdrawal of ply bamboo bonded with UF were 11.18 kgf/cm^2, 436.13 kgf/cm^2 and 184.00 kgf, respectively. The shear strength and MOR of ply bamboo board were higher than minimal values in JAS standard, which were 10 kgf/cm^2 for shear strength and 240 kgf/cm^2 for MOR. It proved that ply bamboo board from Gombong bamboo bonded with UF or PF had fulfilled the requirement for structural application.

Considering the results of Sudijono et al [5] and Gopar-Subyakto [6], we can conclude that the physical and mechanical properties of plybamboo boards were affected by bamboo species and the type of resin used in plybamboo board production.

2.1.2 Bamboo Particle Board

Kusumaningrum et al. [8] employed Betung bamboo (*Dendrocalamus asper*) to produce bamboo particle board. Bamboo culm was cut, splitted and crushed, then processed in hammer mill to produce bamboo particles. Screening was conducted to obtain bamboo particles which were passed a 10-mesh screen and remained on 10-mesh screen. Bamboo particles were air-dried until the moisture content reached about 8%. Two types of adhesives were used i.e UF (produced by PT Palmolite Adhesive Industry, West Java, Indonesia) and isocyanate (produced by PT Polychemia Asia Pacific Permai, Jakarta, Indonesia). Board dimension was 25 x 25 x 1 cm with target density of 0.6 g/cm3. The adhesive content used was 10% based on particle dry weight. Bamboo particles were sprayed with UF or isocyanate in drum mixer using spray gun. Then they were put in forming box, hand pressed and hot pressed for 10 minutes at pressure at 20 kgf/cm^2 and at degree at 140°C for UF and 150°C for isocyanate. The particle boards were conditioned at room temperature for 14 days prior to physical and mechanical testing in accordance with JIS A 5908-2003.

Kusumaningrum [8] also reported that the moisture content of bamboo particle board bonded with UF or isocyanate was 8% and it fulfilled the requirment of JIS. The physical properties of bamboo particle board indicated that thickness swelling after 24 hours water soaking was 25.96% for bamboo particle board bonded with UF and 19.58% for bamboo particle board bonded with isocyanate. The MOE value of bamboo particle board bonded with UF or isocyanate was 1119.17 N/mm^2 and 2173.79 N/mm^2, respectively. The MOR values of bamboo particle board bonded with UF or isocyanate were 12.24 N/mm^2 and 13.44 N/mm^2, respectively. The internal bond values of bamboo particle board bonded with urea formaldehyde or isocyanate were 0.57 N/mm^2 and 0.65 N/mm^2. The screw withdrawal values of bamboo particleboard bonded with urea formaldehyde or isocyanate were 437.99 N and 364.08 N.

Besides using formaldehyde based adhesive, bamboo particles had been used to produce particleboard using natural adhesive, such as citric acid [9]. Three bamboo species particles, i.e. Petung (*D. asper*) bamboo, Wulung bamboo (*G. atroviolacea*), and Apus bamboo (*G. apus*) were used in Widyorini et al [9] study. Bamboo particles were screened passing through a ten mesh (fine materials). All of the particles were then air-dried to a moisture content of around 12 %. Citric acid (anhydrous) was used without further purification. Citric acid was dissolved in water to a solution concentration of 59–60 wt.%. The solution was used as adhesive and sprayed onto each bamboo species particle at 0%, 15% and 30 wt.% resin content based on the weight of the dried particles The sprayed particles were then oven dried over night at 80°C to reduce moisture content. The moisture content of the mat was 4–6 %. The particles were hand-formed into a mat by using forming box, followed by hot pressing into particleboard. The boards were pressed at 180°C for 10 min at a pressure of 3 MPa.

Apus bamboo particles bonded with 30% citric acid produced particleboard with the best thickness swelling value (2%) and water absorption (12%), compared to Petung bamboo and Wulung bamboo(Widyorini et al [9]). The physical properties of bamboo particleboards is presented in Table 1. It indicates that Apus bamboo produced particleboard which is more durable when exposed outdoor, compared to Petung and Wulung bamboo.

Table 1. Physical properties of bamboo particleboard with variation of citric acid content

Raw Material	Water Absorption (%)			Thickness Swelling (%)		
	0 wt%	15 wt%	30 wt%	0 wt%	15 wt%	30 wt%
Petung	98	35	15	35	9	4
Wulung	70	17	13	28	7	3
Apus	85	16	12	45	7	2

Sources: Widyorini et al. [9]

Widyorini et al. [9] reported that modulus of rupture of Apus bamboo particle board was the the highest compared to Petung bamboo or Wulung bamboo particle boards, when bonded with 30 wt% citric acid, which reached 15 MPa. On the other hand, modulus of elasticity of Petung bamboo particle board was the highest, when bonded with 30 wt% citric acid, which was 4.1 GPa. Furthermore, Petung bamboo with 30 wt% citric acid was also produced particle board with highest internal bond (0.53 MPa). Although Widyorini declared that different holocellulose content of three bamboo species had almost no effect on mechanical properties of the particle boards, we assume that mechanical properties of particleboard areaffected by hemicellulose content of bamboos. The hemicellulose content in Petung bamboo and Apus bamboo was 29.58% and 30.27%, respectively. While hemicellulose content in Wulung bamboo was only 24.08%. On the other hand, cellulose and lignin content of those three bamboo species were almost not different [9].

2.1.3.Bamboo Cement Board

Bamboo particles has been also used as cement board. Bakri [10] manufactured cement board from three different species which were Parring bamboo (*Gigantochloa atter*), Betung bamboo (*Dendrocalamus asper*), and Tallang bamboo (*Schizostachyium brachycladum*). The ratio of bamboo particles, Portland cement and water was 1 : 2.5 :1.25. Density of the cement board was targeted at 1 g/cm^3 and thickness was 1 cm. Finer bamboo particles which passed a 10-mesh screen and remained on 20-mesh screen were used for face and back layers. In other hand, coarser bamboo particles which passed a 10-mesh screen and remained on 20-mesh screen were used for core layer in cement board. Particle size ratio of face : core : back was 15% :70 % : 15 %. Bamboo particles were immersed in water for 48 hours to remove the extractives and dried in room temperature until moisture content of 30%. Bamboo particles, Portland cement and water were stirred until homogenous mixture obtained. Mixture poured into iron plate of 25 cm x 25 cm x 1 cm was covered by plastic sheet and pressed for 24 hours to form cement board. It was converted to solid cement board after setting time completed.Injection of CO_2 in liquid phase into the cement board carried out for curing for 30 minutes. Sample of cement board was put into injection tube and CO_2 was flowed into the injection tube. Liquid phase of CO_2can reached by setting the tube temperature at 15° C and tube pressure at 50 kg/cm^2 for 30 minutes. Cement board was removed from the tube and put into desiccator for 15 minutes. Cement board was weighed and put into oven for the next curing at 80° C for 10 hours. Samples of cement board were tested for physical and mechanical properties according to Japanese Industrial Standard (JIS) A JIS A 5417:1992. These properties of bamboo cement board were summarized in Table 2.

Table 2. Physical and mechanical properties of cement board

Raw Material	Physical Properties		Mechanical Properties		
	TS (%)	WA (%)	MOE (kg/cm2)	MOR (kg/cm2)	IB (kg/cm2)
Schizostachyium brachycladum	0.87	49.20	5030	79.59	0.74
Gigantochloa atter	0.65	45.11	2339	49.75	0.41
Dendrocalamus asper	1.35	57.60	1339	40.12	0.18

Sources: Bakri [10]

2.1.4 Composite of polypropylene or polylactic acid and bamboo micro fibers (BMF)

Besides using resin as adhesive to produce bamboo composite, polymer has also been used as matrix in polymer and bamboo composite. In polymer composites, bamboo should be processed mechanically into powder or chemically to obtain pulp or cellulose fibers before blended with polymer. Some polymers such as polypropylene and polylactic acid have been employed as matrix in Betung bamboo (*Dendrocalamus asper*) pulp.

Production of bamboo micro fibers

Fibers of bamboo were processed into pulp as follow. Fresh Betung bamboos were cut into 1 m length, debarked and split into two parts. Bamboo parts were processed with bamboo crusher for 10 times, and then cut into about 5 cm length. Fibers of bamboo at amount of 250 g (oven dried/DW) were cooked in NaOH 2.5% (1 part of fiber soaked in 10 part of solution) for 2 hours. Pulp were separated from the cooking liquid and washed until free alkaline. Pulps then were processed using beater Hollander for 90 minutes. The pulp grinding was continued using stone grinder for 5 times. The pulp was stored in a freezer before use. The morphological characteristics by Scanning Electron Microscope (SEM) was observed to the pulp before and after stone grinder treatment.

Production of PP/bamboo micro fibers

Ratios of bamboo fibers / PP (polypropilene) were 30/70, 40/60, and 50/50. Composites of fiber/PP were prepared by mixing of dry pulp with polymer. Dry pulp was prepared by filtering and drying wet pulp in an oven at temperature of 60°C for 24 hours. Dry pulp was tear into small pieces about 1 cm by 1 cm and mixed with PP in a Laboplastomill at temperature of 175°C, 60 rpm for 20 minutes. Maleic anhydride modified PP (MAPP) at the amount of 5% was added when fiber/PP was mixed. Compound then was hot pressed at temperature of 180°C, pressure of 1 MPa for 30 seconds. Immediately compound was taken and cold pressed at pressure of 1 MPa for 5 minutes. And finished sample was taken.

Production of PLA/bamboo micro fibers

Ratios of bamboo fibers/PLA (poly lactic acid) were 30/70, 40/60, and 50/50. Composite of fiber/PLA was prepared by soaking PLA in dichloromethane, and then wet pulp were added and mixed until homogenous. Triacetin as plasticizer at the amount of 7% was added when mixing fiber/PLA. This compound was oven dried at temperature of 60°C for 6 hours. Dry compound was processed in a Laboplastomill at temperature of 170°C, 60 rpm for 20 minutes. Further process was similar to that of fiber/PP composite (hot pressing and cold pressing).

Size of bamboo fibers/PP or bamboo fiber/PLA composites was 150 mm by 50 mm by 3 mm, with target density of 1.0 g/cm^3 for fiber/PP composite and 1.3 g/cm^3 for fiber/ PLA composite. Composites were conditioned at room temperature for 3 days and tested for mechanical properties using Universal Testing Machine (UTM). The span was 100 mm and crosshead speed of test was 50 mm/min.

Mechanical properties of bamboo fibers/PP or bamboo fibers/PLA

The research of bamboo fibers and polypropylene composite has been conducted by Subyakto et al [11]. SEM observation was done for bamboo fibers prior and after stone grinding process. The average diameter size of fibers is less than 10 μm, and classified as micro fiber. The mechanical properties of bamboo fiber/PP composite and bamboo fiber/PLA composite were presented in Table 3. Micro fibers of bamboo as reinforcement of PP or PLA composites yielded in higher mechanical properties compared to those of pure PP or PLA.

Table 3. Mechanical properties of bamboo fiber /PP or PLA composites

Ratio Bamboo fibers / PP or PLA	Bamboo fiber / PP composite		Bamboo fiber / PLA composite	
	MOR (MPa)	MOE (GPa)	MOR (MPa)	MOE (GPa)
PP	27.52	1.18	-	-
PLA	-	-	70.36	3.17
30/70	52.36	2.80	100.83	6.15
40/60	47.52	2.78	105.23	7.06
50/50	47.71	3.25	73.31	6.11

Souces: Subyakto et al. [11]

Subyakto et al. [11] reported that bamboo fiber/PP composites with ratio of 30/70 and 5% MAPP showed the highest MOR (52.36 MPa). While the highest MOE values (3.25 GPa) was resulted from bamboo fibers/PP composite with ratio of 50/50 and 5% MAPP. This result is slightly different from the result that reported by Okubo et al. [12]

who found that the optimum ratio of bamboo fiber/PP was 50/50 with 5% MAPP. While bamboo fiber/PP composites with ratio of 40/60, Okubo [12] obtained tensile strength of 30.3 MPa and Young's modulus of 3.66 GPa.

Composites of bamboo fibers/PLA have higher mechanical properties compared to those of fibers/PP. This might caused by the difference in density. Composites of fibers/PLA have higher density (1.3 g/cm^3) compared to those of fibers/PP (1.0 g/cm^3). Composites of bamboo fiber/PLA have higher flexural strength compared to those of bamboo fiber/PP composites. MOR ranged 73.31 – 105.23 MPa and MOE was at range of 6.11 – 7.05 GPa. Lee et al. [13] found that bamboo fiber/PLA composites with ratio of 30/70 and added 5% MAPP have tensile strength of 35 MPa and Young's modulus of 2.72 GPa. They observed that increase of fiber ratio decreased tensile strength but Young's modulus was since in versa.

2.2 Bamboo for Pulp Production

2.2.1Determination of the most suitable bamboo species

As reported by Fatriasari and Hermiati [14], 2 year old of six commercial bamboo species from Gunung Sindur Bogor, Indonesia were observed the morphological, physical-chemical properties for determining their potency as pulp and paper raw material. They were Andong bamboo (*Gigantochloa verticillata* (Wild) Munro, Tali/apus bamboo (*Gigantochloa apus* (Bl.ex Schult.f.) Kurz, Hitam bamboo (*Gigantochloa nigrocillata* Kurz.), Ampel bamboo (*Bambusa vulgaris*), Betung bamboo (*Dendrocalamus asper* (Schult.f.) Backer ex Heyne) and Kuning bamboo (*Bambusa vulgaris* Schard var. vitata). To determine the fiber dimension including fiber length, fiber diameter, lumen diameter, and cell wall thickness, the bamboo chips were mercerized using schultze method. The fiber derived values (runkell ratio, felting power, muhlsteph ratio, coefficient of rigidity, flexibility ratio) were also calculated from its fiber dimension. Physical properties including spesific gravity and colour of bamboo were observed. Chemical properties of bamboo including ethanol-benzene solubility (ASTM D 1107-56, Reapproved 1995), hot and cold water solubility (ASTM D–1110-84,Reapproved 1995), 1 % NaOH solubility (ASTM D1190-84, Reapproved 1995), lignin content (klason method), holocellulose (wise method), ash content (TAPPI T 211 cm-86) andsilica content (TAPPI T 245 om-94) were also measured. Rank system in all parameter (the highest and lowest value of 6 and 1, respectively) was used to determine the highest total value identified as the most potential bamboo species for raw material pulp and paper.

Based on this analysis, six bamboo species can be categorized as long fiber (up to 2 mm) including in quality class I based on criteria of LPHH requirement for pulp and paper with high holocellulose content and medium lignin content [14]. Long fibers can serve to create a good interfiber bonding, facilitate the pulp washing easily. They also affected on the smoothness of pulp sheet [15]. A good fiber bonding in bamboo especially Andong, betung and Hitam bamboo due to their long fibers between 4-5 mm played to strengthen the paper sheet from tear strength. Therefore, the sheets cannot penetrated by light easily. Fiber diameter, consisted of cell wall and lumen cell, was the other important characteristics in paper properties, in which the widest lumen cell and the thinnest cell wall belongs to Kuning bamboo. This characteristic facilitates to provide wide suface for interfiber bonding thus the tensile strength and folding power of paper sheet can be obtained [16].

The fiber dimension derivatives as resulted from interaction of the fiber dimension give more complete prediction on pulp and paper sheet properties. These caracteristics showed that Hitam bamboo was superior in runkell ratio, felting power and flexibility ratio, while Ampel bamboo has the best felting power. Besides that observation on the chemical and physical properties of bamboos that summarized in rank system indicated that Betung and Hitam bamboo were predicted producing a good pulp quality considering their lignin content (25-30 %). The lowest lignin, the highest pulp quality. Lignin having rigid properties can become inhibitor during interfiber bonding to create high paper sheet properties. It also causes brown or dark color in the paper sheet surface. Furthermore, all bamboo species have good holocellulose content with the highest content was in Ampel bamboo. Based on the physical properties, Betung bamboo has the lowest density indicating the most suitable for pulp and paper raw material. Summary of all bamboo properties by rank system also showed that Kuning bamboo and Betung bamboo had been selected as the first and second best candidate for pulp and paper raw material. Therefore, these bamboos then were subjected in pulping process [14].

2.2.2 Biopulping of Kuning and Betung bamboo

We have choosed two pulping process to produce pulp including semichemical (soda open hot soda) and chemical pulping (kraft and soda). Prior to pulping process, barkless bamboo chips were pretreated by biological pretreatment using white rot fungi such as *Trametes versicolor* (TV), *Pleurotus ostreatus* (PO), *Phanerochaete crysosporium* (PC) and *Schizophyllum commune* [17], [18], [19], [20], [21]. Application of fungi culture into bamboos was used two methods, i.e single and mixed culture and then incubated for 30 and 45 days. After incubation finished, the treated bamboos were evaluated the anatomical structure changes after single culture fungi pretreatment [22] and mixed culture fungi pretreatment [23].

Except for Muhlstep ratio, an improvement of morphology and fiber derived value characteristics was found after mixed culture pretreatment of white rot fungi. The treated bamboos were predicted resulting a good quality pulp (grade I) based on FAO and LPHH requirements. The best fiber dimension and derived value were resulted after co culture of PC dan TV for 45 days. Delignification of pretreated bamboo was initiated by the fungi hypae colonization on the surface area of bamboo, afterwards the mycelium penetrated into inner bamboo part [23].

Treatment using single culture of white rot fungi on Betung bamboo showed similar finding with mixed culture of white rot fungi. Based on macroscopic image observation, no significant difference degradation pattern of pretreated bamboo was observed. Considering these properties and its fiber derived value, bamboo pretreated TV for 30 days was identified able to produce the best pulp properties including in grade I [22].

In soda pulping, pretreated chips were cooked for 3 hours at 170 °C by 20% of NaOH to dry weight of bamboos and liquor to wood ratio of 10:1. After soda pulping, cooked bamboo chips and black liquor were then separated, washed by trap water until alkaline free, and defiberized using disk refiner 3 times [20], [21] before further analysis. The pretreated chips were also kraft pulped for 3 hour of maximum temperature (170 °C) using 20 and 15% of active alkali and sulfidity to oven dried bamboo with the liquor to wood ratio of 5: 1. To optimize penetration of cooking solution residue, the pulp was soaked for 24 h. Characterization of pulp included total yield (TAPPI T210 cm-93), Kappa number (TAPPI 236 cm – 85), freeness (TAPPI 227 cm-85), selectivity delignification (rasio of carbohydrate and residual lignin in pulp) and brightness level (SNI 14-4733-1998) [20], [21]. The other pulping method of bamboo was open-hot soda in which these chips were cooked for 2 h at 100 °C using NaOH of 20 % to oven dried chips with liquor to wood ratio of 10:1. The cooked chips were then also washed until alkali free and fibrillated by hollander beater for 90 min and stone refiner for 1 times.

Fibrillated pulp was measured the pulp yield, Kappa number and degree of freeness in triplicate.

Biological pretreatment using single culture of PO and TV fungi followed by open-hot soda also affected in improvement of pulp characteristics (increase of pulp yield and reduction of Kappa number and degree of freeness). Prolonging incubation time had correlation in reducing of kappa number. And based on observation of pulp properties, pretreatment of TV on Betung bamboo for 30 days had choosen as the best pretreatment condition [18]. However, compared with chemical pulping, this open-hot soda pulping has less improvement in pulp qualities. *S. commune* pretreatment on Betung and Kuning bamboo prior to open-hot soda pulping was also conducted. Insignificant effect on pulp properties (Kappa number, delignification selectivity and pulp yield) was observed after treatment. In this pulping process, pulp yield of Betung bamboo was lower than that of Kuning bamboo.

Kraft pulping of pretreated bamboo using single culture of TV and PO fungi gave more influence than that of soda pulping. Based on the biokraft pulp properties, TV for 45 days showed the best pretreatment condition indicated by the highest pulp yield and delignification selectivity and the lowest degree of freeness and Kappa number. With the same observation, this pretreatment condition was also demonstrated the best pulp qualities of soda pulping. Therefore, incubation using TV for 45 days was proposed as the best pretreatment condition prior to chemical pulp [20]. As mention before, according the anomical structure analysis, bamboo pretreated TV for 30 days can be predicted as the best pulp properties [22], however the pulping method played important contribution on pulp properties. Thus, the different best pretreatment condition was obtained.

Fatriasari et al. [19] reported that decrease of pulp yield occurred after treatment of bamboo with mixing between two or three fungi culture especially on PO and PC and PC for 45 days. It was caused competition of space and consumption of limited nutrients. These treatment combinations might not degrade selectively on lignin following degradation of some holocellulose mainly on aldehide end grup. From three categories of fungi interaction types that identified by Body [24], POPC and POPCTV incubated for 45 days can be included in antagonist interaction. However, compared to control, biological pretreatment of bamboo proved succesfully to improve the pulp yield. After mixed culture fungi pretreatment, as much as 77-91% of Kappa number of pulp decreased, the highest and the lowest decrease was found in TVPC and TVPCPO for 30 days. It is indicated that these treatments were effective to remove lignin. Evaluation of biological pretreatment effectivity can be accessed by delignification selectivity in which mixed culture treatments. They were able to increase delignification selectivity and the highest value was POPC for 45 days. Prior to kraft pulping, lignin degradation occurred. Thus the pulping become more effective. In general, prolonging incubation time tended to decrease Kappa number and increase delignification selectivity and pulp yield. Regarding the pulp properties, the best mixed fungi culture treatment was co culture of TV and PC for 30 days. It did not in line with evaluation on anatomical characteristics and fiber derived value that reported in Fatriasari et al. [23]. The pulping method might give effect on the final properties of pulp.

During biological pretreatment, addition of corn steep liquor (CSL) in the inoculum of PC was also observed. A better pulp yield was obtained by increasing CSL amount. The longer incubation time tended to increase pulp yield, while the Kappa number was since versa. The treatment caused degradation in cell walls according to SEM analysis. The kraft pulping of bamboo treated by 10% CSL incubated for 30 days showed the best pulp properties [21]. Based on evaluation on pulp properties, kraft pulping was preferred as pulping method compared to the other methods.

2.2.3 Bamboo for production of sugar for bioenergy

Betung bamboo was reported suitable for pulp raw material. Therefore, in bamboo conversion for sugar production, this bamboo species was used. Biological pretreatment with TV on Betung bamboo using single or combination with microwave pretreatment has been reported before [25]. These treatments facilitated to decrease of lignin content and delignification selectivity. This result was inline with biological pretreatment before chemical pulping [20]. Incubation for 30 days showed the higher delignification selectivity compared to incubation for 15 and 45 days. Pretreatment caused more opened biomass structure, soften substrate, fiber fibrillation and more exposed cellulose. Some lignin losses in pretreated bamboo mainly were caused by decrease in lignin aromatic rings. The structure changes in pretreated bamboo facilitated in increasing the enzyme accessibility in breaking down the cellulose into glucose.

The pretreated bamboo was then hydrolyzed using enzymatic and sulfuric acid hydrolysis. The highest reducing sugar yield per biomass (17.06 %) was found after microwave assisted acid hydrolysis of pretreated bamboo. This yield was higher than that of enzymatic hydrolysis (2.53%). Until 12.5 min irradiation time, the longer irradiation time, the higher reducing sugar yield. Inoculum loading of 5% resulted higher reducing sugar yield compared to 10%. Biological pretreatment with 5% inoculum loading irradiated for 12.5 min using 1% sulfuric acid showed the highest reducing sugar yield. In this condition, about 22.75% holocellulose can be converted into reducing sugar.

Besides biological pretreatment, bamboo was also pretreated by microwave pretreatment in water medium. And irradiation time for 5, 10, 12.5 min with power level of 330 watt (W) and irradiation for 5 min with power level of 770 W was selected for combining with bamboo pretreated by biological pretreatment. The combination treatment increased reducing sugar yield as much as 8.4 fold compared to the highest reducing sugar resulted from enzymatic hydrolysis, and it improved 13.7 fold to control. The highest reducing sugar yield about 16.65% was achieved after hydrolysis for 12.5 min of bamboo pretreated 5% inoculum loading followed by microwave pretreatment for 5 min (330 W). About 27.21% holocellulose of bamboo was converted into reducing sugar [26].

The other biological pretreatment method on bamboo that was also developed is two-stage fungal pretreatment method by using white rot fungi pretreatment for 30 days followed by brown rot pretreatment for 30 days or since in versa [27]. The single pretreatment was used to compare the microwave assisted acid pretreatment. In this method, white rot fungi, PC was to degrade lignin, while brown rot (*Fomitopsis polustris*) pretreatment was for depolymerization of cellulose. This pretreatment combination resulted high delignification selectivity. The highest reducing sugar yield was obtained in pretreatment of brown rot (12.19%). The lowest reducing sugar yield (7.66%) was resulted after microwave assisted acid hydrolisis of bamboo pretreated *F.polustris*-PC. Compared to control, two-stage fungal pretreatment can increase the sugar yield. Lignin content was not only one factor affecting in sugar yield.

3. Conclusion

The physical and mechanical properties of ply bamboo boards are affected by bamboo species and the type of resin. Three layers of bamboo zephyr board using PF resin is potential as a substitute of ply wood for pallet in concrete block production. In bamboo particle board production, Betung bamboo bonded with isocyanate produced particle board with higher modulus of rupture, modulus of elasticity and internal bond compare to Betung bamboo bonded with UF. Furthermore, particle board produced with Apus bamboo and citric acid as environmental friendly adhesive, showed higher modulus of rupture than of Petung bamboo or Wulung bamboo. On the other hand, modulus of elasticity and internal bond of Petung Bamboo and citric acid particleboard is the highest. The mechanical properties of Tallang bamboo cement board were higher than that of Atter bamboo and Betung bamboo cement boards. Composites of bamboo fibers/PLA have higher mechanical properties compared to those of bamboo fibers/PP composites. The optimum ratio of bamboo fiber/PLA compositesis 40/60. Kuning bamboo and betung bamboo are proposed to be used aspulp and paper raw material considering the fiber properties and physical and chemical properties, fiber derived value as well. Biopulping of Kuning bamboo and Betung bamboo was conducted succesfully to produce bamboo pulp. Biological pretreatment of bamboo using white rot fungi (TV, PO and PC) in single and mixed culture prior to pulping process facilitates in lignin removal. Biopulping pretreatment improved the pulp qualities mainly increase the pulp yield and delignification selectivity, also reduce the kappa number. In conversion of bamboo for sugar production, biological, biological and microwave method, two-stage fungal pretreatment, have been developed. By biological pretreatment, 17.06% reducing sugar yield can be resulted after microwave assisted hydrolysis for 12.5 min.

Acknowledgements
The authors would like to thank the Surfactant and Bioenergy Research Center (SBRC) for its opportunity in inviting us to share our work in 2nd International Conference on Biomass.

References
[1] Dransfield S, Widjaja E A 1995 *Plant Resources of South-East Asia No. 7 Bamboos* (Backhuys Publisher Leiden) p.189
[2] Lobovikov M, Paudel S, Piazza M, Ren H, Wu J 2007 World bamboo resources. a thematic study prepared in the framework of the global forest resources assessment 2005 Non Wood Forest Products 18 Food and Agricultural Organization of the United Nations Rome p. 73
[3] Fatriasari W 2014 Reducing sugar production through pretreatment process engineering of betung bamboo (*Dendrocalamus asper* (Schult. f)) Doctoral Thesis Graduate School Bogor Agricultural University Indonesia
[4] Subiyanto B, Subyakto 1995 Development of processing technology of semi fiber bamboo board I. shortening the press cycle *Proceedings of The IVth International Bamboo Congress* **3** Bali June 1995 pp. 155-164
[5] Sudijono, Subyakto, Subiyanto B 2002Effects of resin content of phenol formaldehyde on internal bond strength of zephyr bamboo board made from tali (*Gigantochloa apus* Kurs) *Proceedings of National Seminar of MAPEKI IV* 6-9 Agustus 2001 Samarinda pp. IV.220-IV.224 (in Indonesian)
[6] Gopar M, Subyakto 2002 Physical and mechanical properties of zephyr board made from gombong bamboo *Proceedings of The Fourth International Wood Science Symposium* Serpong Indonesia 2-3 September 2002 pp. 257-261
[7] Subyakto, Romansyah E, Zakaria A, Bachtiar G, Nasution N, Prihantono. 2016. Physical and mechanical properties of ply-bamboo made from sembilang (*Dendrocalamus giganteous*) and Andong (*Gigantochloa pseudoarundinacea*) bamboos. *Proceeding of The 7th International Symposium of Indonesian Wood Research Society* 5-6 November 2010 Bandung West Java Indonesia
[8] Kusumaningrum WB, Syamani FA, Subyakto 2010 Papan komposit sisal (*Agave sisalana* Perr.) dan bambu betung (*Dendrocalamus asper* (Schultes f) Backer ex. Heyne). *Proceeding of Seminar Nasional MAPEKI XII*, 23-25 Juli 2009 Bandung West Java Indonesia.

[9] Widyorini R, Umemura K, Isnan R, Putra D R, Awaludin A, Prayitno T A 2016. Manufacture and properties of citric acid-bonded particleboard made from bamboo materials *Eur. J. Wood Prod.***74** 57–65
[10] Bakri 2015 Application of carbon dioxide injection technology in bamboo cement board production *Proceeding of The fifth International Symposium of Indonesian Wood Research* Society 7-9 November 2013 Balikpapan East Kalimantan Indonesia
[11] Subyakto, Hermiati E, Yanto DHY, Masruchin M, Fitria, Prasetiyo KW, Ismadi 2010 Biocomposites of polypropylene or polylactic acid reinforced with sisal or bamboo micro fibers *Proceedings of The First International Symposium of Indonesian Wood Research Society*
[12] Okubo K, Fujii T, Yamamoto Y 2004 Development of bamboo-based composites and their mechanical properties *Composites Part A***35**377-383
[13] Lee SH, Ohkita T, Kitagawa K 2004 Eco-composite from poly(lactic acid) and bamboo fiber. *Holzforschung* **58** 529-536
[14] Fatriasari W, Hermiati E 2008 Analisis morfologi serat dan sifat fisis-kimia pada enam jenis bambu sebagai bahan baku pulp dan kertas *Jurnal Ilmu dan Teknologi Hasil Hutan***1**(2) 67-72
[15] Haygreen J G, Bowyer J L 1996 *Hasil Hutan dan Ilmu Kayu: Suatu Pengantar Terjemahan* (Gajah Mada University Press Yogyakarta)
[16] Casey J P 1980 *Pulping Chemistry and Chemical Technology Volume I Pulping and Paper Making* (Interscience Publisher Inc.New York)
[17] Fitria, Ermawar R A, FatriasariW, FajriutamiT, Yanto DHY, Falah F, Hermiati E 2013 Biopulping of bamboo using white-rot fungi *Schizophyllum commune Proceedings The 2nd International Symposium for Sustainable Humanosphere* Bandung 29th August, 2012pp.8-13
[18] Fatriasari W, Ermawar R A, Falah F, Yanto D H Y, Hermiati E 2009Pulping soda panas terbuka bambu betung dengan praperlakuan fungi pelapuk putih (*Pleurotus ostreatus* dan *Trametes versicolor) J Ilmu dan Teknologi Hasil Hutan* **2**(2) 45-50
[19] Fatriasari W, Anita S H, Falah F, Adi D T N, Hermiati E2010 Biopulping of betung bamboo using mixed culture of white rot fungi (*Trametes versicolor, Pleurotus ostreatus dan Phanerochaete crysosporium*) *Berita selulosa* **45**(2) 44-56
[20] Fatriasari W, Ermawar RA, Falah F, Yanto DHY, Adi DTN, Anita S H, Hermiati E 2011 Kraft and soda pulping of white rot pretreated betung bamboo *Jurnal Ilmu dan Teknologi Kayu Tropis* **9**(1) 42-55
[21] Falah F, Fatriasari W, Ermawar A R, Adi D T N, Hermiati E 2011 Effect of corn steep liquor on bamboo biochemical pulping using *Phanerochaete chrysosporium. Jurnal Ilmu dan Teknologi Kayu Tropis* **9**(2) 111-125
[22] Fatriasari W, Damayanti R, Anita S H 2012aThe pretreatment effect of single culture white rot fungi on the anatomical structure of betung bamboo *Jurnal Ilmu dan Teknologi Kayu Tropis* **10**(1) 66-75
[23] Fatriasari W, Damayanti R, Anita S H 2013 Fiber qualities of pretreated betung bamboo (*Dendrocalamus asper*) by mixed culture of white-rot fungi with respect to its use for pulp and paper *Journal of Forestry Research* **10** (2) 95-106
[24] Boddy L 2000 Interspesific combative interactionsbetween wood-decaying *FEMS Microbiology Ecology* **31** 185-194
[25] Fatriasari W, Syafii W, WistaraN J, Syamsu K, Prasetya B 2014a Digestibility of betung bamboo fiber following fungal pretreatment *Makara Journal Technology* **18**(2):51-58

[26] FatriasariW,Syafii W, WistaraN J, Syamsu K, Prasetya B 2014b Performance of microwave pretreatment on enzymatic and microwave hydrolysis of betung bamboo (*Dendrocalamus asper*) *Teknologi Indonesia* **37**(3) 162-167
[27] Fatriasari W, A S Heris 2012b Evaluation of two-stage fungal pretreatment for the microwave hydrolysis of betung bamboo *The 2nd Korea – Indonesia Workshop & International Symposium on Bioenergy from Biomass* P 83-86